Edited by
Daniel Abou-Ras,
Thomas Kirchartz,
and Uwe Rau

Advanced Characterization Techniques for Thin Film Solar Cells

Related Titles

Würfel, P.

Physics of Solar Cells
From Basic Principles to Advanced Concepts

2009
ISBN: 978-3-527-40857-3

Poortmans, J., Arkhipov, V. (eds.)

Thin Film Solar Cells
Fabrication, Characterization and Applications

2006
ISBN: 978-0-470-09126-5

Luque, A., Hegedus, S. (eds.)

Handbook of Photovoltaic Science and Engineering
Second Edition

2010
ISBN: 978-0-470-72169-8

Edited by
Daniel Abou-Ras, Thomas Kirchartz, and Uwe Rau

Advanced Characterization Techniques for Thin Film Solar Cells

WILEY-VCH Verlag GmbH & Co. KGaA

The Editors

Dr. Daniel Abou-Ras
Helmholtz-Zentrum Berlin
für Materialien und Energie
Berlin, Germany
daniel.abou-ras@helmholtz-berlin.de

Dr. Thomas Kirchartz
Imperial College London
London, United Kingdom
t.kirchartz@imperial.ac.uk

Prof. Dr. Uwe Rau
Forschungszentrum Jülich
Jülich, Germany
u.rau@fz-juelich.de

All books published by **Wiley-VCH** are carefully produced. Nevertheless, authors, editors, and publisher do not warrant the information contained in these books, including this book, to be free of errors. Readers are advised to keep in mind that statements, data, illustrations, procedural details or other items may inadvertently be inaccurate.

Library of Congress Card No.: applied for

British Library Cataloguing-in-Publication Data
A catalogue record for this book is available from the British Library.

Bibliographic information published by the Deutsche Nationalbibliothek
The Deutsche Nationalbibliothek lists this publication in the Deutsche Nationalbibliografie; detailed bibliographic data are available on the Internet at http://dnb.d-nb.de.

© 2011 WILEY-VCH Verlag GmbH & Co. KGaA, Boschstr. 12, 69469 Weinheim, Germany

All rights reserved (including those of translation into other languages). No part of this book may be reproduced in any form – by photoprinting, microfilm, or any other means – nor transmitted or translated into a machine language without written permission from the publishers. Registered names, trademarks, etc. used in this book, even when not specifically marked as such, are not to be considered unprotected by law.

Typesetting   Thomson Digital, Noida, India
Printing and Binding   Strauss GmbH, Mörlenbach
Cover Design   Adam-Design, Weinheim

Printed in the Federal Republic of Germany
Printed on acid-free paper

ISBN: 978-3-527-41003-3

Contents

Preface *XVII*
List of Contributors *XXI*
Acknowledgments *XXVII*
Abbreviations *XXXI*

Part one   Introduction *1*

1   **Introduction to Thin-Film Photovoltaics** *3*
          Thomas Kirchartz and Uwe Rau
1.1   Introduction *3*
1.2   The Photovoltaic Principle *5*
1.2.1   The Shockley–Queisser Theory *5*
1.2.2   From the Ideal Solar Cell to Real Solar Cells *9*
1.2.3   Light Absorption and Light Trapping *10*
1.2.4   Charge Extraction *12*
1.2.5   Nonradiative Recombination *16*
1.3   Functional Layers in Thin-Film Solar Cells *18*
1.4   Comparison of Various Thin-Film Solar-Cell Types *20*
1.4.1   Cu(In,Ga)Se$_2$ *20*
1.4.1.1   Basic Properties and Technology *20*
1.4.1.2   Layer-Stacking Sequence and Band Diagram of the Heterostructure *22*
1.4.2   CdTe *23*
1.4.2.1   Basic Properties and Technology *23*
1.4.2.2   Layer-Stacking Sequence and Band Diagram of the Heterostructure *24*
1.4.3   Thin-Film Silicon Solar Cells *25*
1.4.3.1   Hydrogenated Amorphous Si (a-Si:H) *25*
1.4.3.2   Metastability in a-Si:H: The Staebler–Wronski Effect *27*
1.4.3.3   Hydrogenated Microcrystalline Silicon (μc-Si:H) *27*

1.4.3.4	Micromorph Tandem Solar Cells 27
1.5	Conclusions 28
	References 28

Part Two	Device Characterization 33

2	**Fundamental Electrical Characterization of Thin-Film Solar Cells** 35
	Thomas Kirchartz, Kaining Ding, and Uwe Rau
2.1	Introduction 35
2.2	Current/Voltage Curves 36
2.2.1	Shape of Current/Voltage Curves and their Description with Equivalent Circuit Models 36
2.2.2	Measurement of Current/Voltage Curves 41
2.2.3	Determination of Ideality Factors and Series Resistances 42
2.2.4	Temperature-Dependent Current/Voltage Measurements 44
2.3	Quantum Efficiency Measurements 47
2.3.1	Definition 47
2.3.2	Measurement Principle and Calibration 49
2.3.3	Quantum Efficiency Measurements of Tandem Solar Cells 51
2.3.4	Differential Spectral Response (DSR) Measurements 52
2.3.5	Interpretation of Quantum Efficiency Measurements in Thin-Film Silicon Solar Cells 53
	References 58

3	**Electroluminescence Analysis of Solar Cells and Solar Modules** 61
	Thomas Kirchartz, Anke Helbig, Bart E. Pieters, and Uwe Rau
3.1	Introduction 61
3.2	Basics 62
3.3	Spectrally Resolved Electroluminescence 65
3.4	Spatially Resolved Electroluminescence of c-Si Solar Cells 68
3.5	Electroluminescence Imaging of Cu(In,Ga)Se$_2$ Thin-Film Modules 71
3.6	Modeling of Spatially Resolved Electroluminescence 75
	References 77

4	**Capacitance Spectroscopy of Thin-Film Solar Cells** 81
	Jennifer Heath and Pawel Zabierowski
4.1	Introduction 81
4.2	Admittance Basics 82
4.3	Sample Requirements 83
4.4	Instrumentation 84
4.5	Capacitance–Voltage Profiling and the Depletion Approximation 85
4.6	Admittance Response of Deep States 86
4.7	The Influence of Deep States on CV Profiles 90
4.8	DLTS 91

4.8.1	DLTS of Thin-Film PV Devices	94
4.9	Admittance Spectroscopy	95
4.10	Drive Level Capacitance Profiling	97
4.11	Photocapacitance	98
4.12	The Meyer–Neldel Rule	99
4.13	Spatial Inhomogeneities and Interface States	100
4.14	Metastability	102
	References	102

Part Three Materials Characterization *107*

5 Characterizing the Light-Trapping Properties of Textured Surfaces with Scanning Near-Field Optical Microscopy *109*
Karsten Bittkau

5.1 Introduction *109*
5.2 How Does a Scanning Near-Field Optical Microscope Work? *110*
5.3 Light Scattering in the Wave Picture *112*
5.4 The Role of Evanescent Modes for Light Trapping *113*
5.5 Analysis of Scanning Near-Field Optical Microscopy Images by Fast Fourier Transformation *116*
5.6 How to Extract Far-Field Scattering Properties by Scanning Near-Field Optical Microscopy? *120*
5.7 Conclusion *122*
References *122*

6 Spectroscopic Ellipsometry *125*
Sylvain Marsillac, Michelle N. Sestak, Jian Li, and Robert W. Collins

6.1 Introduction *125*
6.2 Theory *127*
6.2.1 Polarized Light *127*
6.2.2 Reflection from a Single Interface *128*
6.3 Ellipsometry Instrumentation *129*
6.3.1 Rotating Analyzer SE for *Ex-Situ* Applications *130*
6.3.2 Rotating Compensator SE for Real-Time Applications *131*
6.4 Data Analysis *133*
6.4.1 Exact Numerical Inversion *133*
6.4.2 Least-Squares Regression *134*
6.4.3 Virtual Interface Analysis *134*
6.5 RTSE of Thin Film Photovoltaics *134*
6.5.1 Thin Si:H *135*
6.5.2 CdTe *139*
6.5.3 $CuInSe_2$ *141*
6.6 Summary and Future *145*
6.7 Definition of Variables *145*
References *146*

7	**Photoluminescence Analysis of Thin-Film Solar Cells** *151*
	Thomas Unold and Levent Gütay
7.1	Introduction *151*
7.2	Experimental Issues *154*
7.2.1	Design of the Optical System *154*
7.2.2	Calibration *156*
7.2.3	Cryostat *156*
7.3	Basic Transitions *157*
7.3.1	Excitons *158*
7.3.2	Free-Bound Transitions *159*
7.3.3	Donor–Acceptor Pair Recombination *160*
7.3.4	Potential Fluctuations *162*
7.3.5	Band–Band Transitions *163*
7.4	Case Studies *164*
7.4.1	Low-Temperature Photoluminescence Analysis *164*
7.4.2	Room-Temperature Measurements: Estimation of V_{oc} from PL Yield *168*
7.4.3	Spatially Resolved Photoluminescence: Absorber Inhomogeneities *170*
	References *173*
8	**Steady-State Photocarrier Grating Method** *177*
	Rudolf Brüggemann
8.1	Introduction *177*
8.2	Basic Analysis of SSPG and Photocurrent Response *178*
8.2.1	Optical Model *178*
8.2.2	Semiconductor Equations *180*
8.2.3	Diffusion Length: Ritter–Zeldov–Weiser Analysis *181*
8.2.3.1	Evaluation Schemes *183*
8.2.4	More Detailed Analyses *184*
8.2.4.1	Influence of the Dark Conductivity *184*
8.2.4.2	Influence of Traps *184*
8.2.4.3	Minority-Carrier and Majority-Carrier Mobility-Lifetime Products *186*
8.3	Experimental Setup *187*
8.4	Data Analysis *189*
8.5	Results *192*
8.5.1	Hydrogenated Amorphous Silicon *192*
8.5.1.1	Temperature and Generation Rate Dependence *192*
8.5.1.2	Surface Recombination *193*
8.5.1.3	Electric-Field Influence *193*
8.5.1.4	Fermi-Level Position *194*
8.5.1.5	Defects and Light-Induced Degradation *194*
8.5.1.6	Thin-Film Characterization and Deposition Methods *195*
8.5.2	Hydrogenated Amorphous Silicon Alloys *196*

8.5.3	Hydrogenated Microcrystalline Silicon	196
8.5.4	Hydrogenated Microcrystalline Germanium	197
8.5.5	Other Thin-Film Semiconductors	197
8.6	Density-of-States Determination	198
8.7	Summary	198
	References	198
9	**Time-of-Flight Analysis**	**203**
	Torsten Bronger	
9.1	Introduction	203
9.2	Fundamentals of TOF Measurements	204
9.2.1	Anomalous Dispersion	205
9.2.2	Basic Electronic Properties of Thin-Film Semiconductors	207
9.3	Experimental Details	208
9.3.1	Accompanying Measurements	210
9.3.1.1	Capacitance	210
9.3.1.2	Collection	212
9.3.1.3	Built-in Field	212
9.3.2	Current Decay	212
9.3.3	Charge Transient	215
9.3.4	Possible Problems	217
9.3.4.1	Dielectric Relaxation	217
9.3.5	Inhomogeneous Field	218
9.4	Analysis of TOF Results	219
9.4.1	Multiple Trapping	219
9.4.1.1	Overview of the Processes	219
9.4.1.2	Energetic Distribution of Carriers	220
9.4.1.3	Time Dependence of Electrical Current	223
9.4.2	Spatial Charge Distribution	223
9.4.2.1	Temperature Dependence	223
9.4.3	Density of States	225
9.4.3.1	Widths of Band Tails	225
9.4.3.2	Probing of Deep States	226
	References	228
10	**Electron-Spin Resonance (ESR) in Hydrogenated Amorphous Silicon (a-Si:H)**	**231**
	Klaus Lips, Matthias Fehr, and Jan Behrends	
10.1	Introduction	231
10.2	Basics of ESR	232
10.3	How to Measure ESR	235
10.3.1	ESR Setup and Measurement Procedure	235
10.3.2	Pulse ESR	238
10.3.3	Sample Preparation	239
10.4	The g Tensor and Hyperfine Interaction in Disordered Solids	240

10.4.1	Zeeman Energy and g Tensor	240
10.4.2	Hyperfine Interaction	243
10.4.3	Line-Broadening Mechanisms	245
10.5	Discussion of Selected Results	248
10.5.1	ESR on Undoped a-Si:H	248
10.5.2	LESR on Undoped a-Si:H	252
10.5.3	ESR on Doped a-Si:H	253
10.5.4	Light-Induced Degradation in a-Si:H	257
10.5.4.1	Excess Charge-Carrier Recombination and Weak Si–Si Bond Breaking	258
10.5.4.2	Si–H Bond Dissociation and Hydrogen Collision Model	260
10.5.4.3	Transformation of Existing Nonparamagnetic Charged Dangling-Bond Defects	260
10.6	Alternative ESR Detection	263
10.6.1	History of EDMR	264
10.6.2	EDMR on a-Si:H Solar Cells	265
10.7	Concluding Remarks	268
	References	269

11 **Scanning Probe Microscopy on Inorganic Thin Films for Solar Cells** *275*
Sascha Sadewasser and Iris Visoly-Fisher

11.1	Introduction	275
11.2	Experimental Background	276
11.2.1	Atomic Force Microscopy	276
11.2.1.1	Contact Mode	277
11.2.1.2	Noncontact Mode	278
11.2.2	Conductive Atomic Force Microscopy	279
11.2.3	Scanning Capacitance Microscopy	280
11.2.4	Kelvin Probe Force Microscopy	282
11.2.5	Scanning Tunneling Microscopy	284
11.2.6	Issues of Sample Preparation	285
11.3	Selected Applications	286
11.3.1	Surface Homogeneity	286
11.3.2	Grain Boundaries	288
11.3.3	Cross-Sectional Studies	291
11.4	Summary	294
	References	294

12 **Electron Microscopy on Thin Films for Solar Cells** *299*
Daniel Abou-Ras, Melanie Nichterwitz, Manuel J. Romero, and Sebastian S. Schmidt

12.1	Introduction	299
12.2	Scanning Electron Microscopy	299
12.2.1	Imaging Techniques	301
12.2.2	Electron Backscatter Diffraction	302

12.2.3	Energy-Dispersive and Wavelength-Dispersive X-Ray Spectrometry *305*	
12.2.4	Electron-Beam-Induced Current Measurements *307*	
12.2.4.1	Electron-Beam Generation *308*	
12.2.4.2	Charge-Carrier Collection in a Solar Cell *309*	
12.2.4.3	Experimental Setups *310*	
12.2.4.4	Critical Issues *311*	
12.2.5	Cathodoluminescence *312*	
12.2.5.1	Example: Spectrum Imaging of CdTe Thin Films *315*	
12.2.6	Scanning Probe and Scanning-Probe Microscopy Integrated Platform *318*	
12.2.7	Combination of Various Scanning Electron Microscopy Techniques *322*	
12.3	Transmission Electron Microscopy *323*	
12.3.1	Imaging Techniques *324*	
12.3.1.1	Bright-Field and Dark-Field Imaging in the Conventional Mode *324*	
12.3.1.2	High-Resolution Imaging in the Conventional Mode *325*	
12.3.1.3	Imaging in the Scanning Mode Using an Annular Dark-Field Detector *327*	
12.3.2	Electron Diffraction *327*	
12.3.2.1	Selected-Area Electron Diffraction in the Conventional Mode *327*	
12.3.2.2	Convergent-Beam Electron Diffraction in the Scanning Mode *328*	
12.3.3	Electron Energy-Loss Spectrometry and Energy-Filtered Transmission Electron Microscopy *329*	
12.3.3.1	Scattering Theory *329*	
12.3.3.2	Experiment and Setup *330*	
12.3.3.3	The Energy-Loss Spectrum *331*	
12.3.3.4	Applications and Comparison with EDX Spectroscopy *334*	
12.3.4	Off-Axis and In-Line Electron Holography *335*	
12.4	Sample Preparation Techniques *338*	
12.4.1	Preparation for Scanning Electron Microscopy *338*	
12.4.2	Preparation for Transmission Electron Microscopy *339*	
	References *341*	
13	**X-Ray and Neutron Diffraction on Materials for Thin-Film Solar Cells** *347*	
	Susan Schorr, Christiane Stephan, Tobias Törndahl, and Roland Mainz	
13.1	Introduction *347*	
13.2	Diffraction of X-Rays and Neutron by Matter *347*	
13.3	Neutron Powder Diffraction of Absorber Materials for Thin-Film Solar Cells *351*	
13.3.1	Example: Investigation of Intrinsic Point Defects in Nonstoichiometric $CuInSe_2$ by Neutron Diffraction *351*	
13.4	Grazing Incidence X-Ray Diffraction (GIXRD) *354*	

13.5	Energy Dispersive X-Ray Diffraction (EDXRD)	357
	References 362	

14 Raman Spectroscopy on Thin Films for Solar Cells 365
Jacobo Álvarez-García, Víctor Izquierdo-Roca, and Alejandro Pérez-Rodríguez

14.1	Introduction 365	
14.2	Fundamentals of Raman Spectroscopy 366	
14.3	Vibrational Modes in Crystalline Materials 368	
14.4	Experimental Considerations 370	
14.4.1	Laser Source 370	
14.4.2	Light Collection and Focusing Optics 372	
14.4.3	Spectroscopic Module 372	
14.5	Characterization of Thin-Film Photovoltaic Materials 373	
14.5.1	Identification of Crystalline Structures 373	
14.5.2	Evaluation of Film Crystallinity 377	
14.5.3	Chemical Analysis of Semiconducting Alloys 378	
14.5.4	Nanocrystalline and Amorphous Materials 380	
14.5.5	Evaluation of Stress 382	
14.6	Conclusions 383	
	References 384	

15 Soft X-Ray and Electron Spectroscopy: A Unique "Tool Chest" to Characterize the Chemical and Electronic Properties of Surfaces and Interfaces 387
Marcus Bär, Lothar Weinhardt, and Clemens Heske

15.1	Introduction 387
15.2	Characterization Techniques 388
15.3	Probing the Chemical Surface Structure: Impact of Wet Chemical Treatments on Thin-Film Solar Cell Absorbers 394
15.4	Probing the Electronic Surface and Interface Structure: Band Alignment in Thin-Film Solar Cells 399
15.5	Summary 405
	References 405

16 Elemental Distribution Profiling of Thin Films for Solar Cells 411
Volker Hoffmann, Denis Klemm, Varvara Efimova, Cornel Venzago, Angus A. Rockett, Thomas Wirth, Tim Nunney, Christian A. Kaufmann, and Raquel Caballero

16.1	Introduction 411
16.2	Glow Discharge-Optical Emission (GD-OES) and Glow Discharge-Mass Spectroscopy (GD-MS) 413
16.2.1	Principles 413
16.2.2	Instrumentation 413
16.2.2.1	Plasma Sources 413

16.2.2.2	Plasma Conditions	*415*
16.2.2.3	Detection of Optical Emission	*415*
16.2.2.4	Mass Spectroscopy	*416*
16.2.3	Quantification	*416*
16.2.3.1	Glow Discharge-Optical Emission Spectroscopy	*416*
16.2.3.2	Glow Discharge-Mass Spectroscopy	*417*
16.2.4	Applications	*418*
16.2.4.1	Glow Discharge-Optical Emission Spectroscopy	*418*
16.2.4.2	Glow Discharge-Mass Spectroscopy	*418*
16.3	Secondary Ion Mass Spectrometry (SIMS)	*420*
16.3.1	Principle of the Method	*420*
16.3.2	Data Analysis	*423*
16.3.3	Quantification	*425*
16.3.4	Applications for Solar Cells	*426*
16.4	Auger Electron Spectroscopy (AES)	*427*
16.4.1	Introduction	*427*
16.4.2	The Auger Process	*427*
16.4.3	Auger Electron Signals	*428*
16.4.4	Instrumentation	*429*
16.4.5	Auger Electron Signal Intensities and Quantification	*431*
16.4.6	Quantification	*432*
16.4.7	Application	*433*
16.5	X-Ray Photoelectron Spectroscopy (XPS)	*435*
16.5.1	Theoretical Principles	*435*
16.5.2	Instrumentation	*437*
16.5.3	Application to Thin Film Solar Cells	*438*
16.6	Energy-Dispersive X-Ray Analysis on Fractured Cross Sections	*440*
16.6.1	Basics on Energy-Dispersive X-Ray Spectrometry in a Scanning Electron Microscope	*440*
16.6.2	Spatial Resolutions	*442*
16.6.3	Applications	*442*
16.6.3.1	Sample Preparation	*445*
	References	*445*

17	**Hydrogen Effusion Experiments**	*449*
	Wolfhard Beyer and Florian Einsele	
17.1	Introduction	*449*
17.2	Experimental Setup	*450*
17.3	Data Analysis	*454*
17.3.1	Identification of Rate-Limiting Process	*455*
17.3.2	Analysis of Diffusing Hydrogen Species from Hydrogen Effusion Measurements	*458*
17.3.3	Analysis of H_2 Surface Desorption	*459*
17.3.4	Analysis of Diffusion-Limited Effusion	*460*

17.3.5	Analysis of Effusion Spectra in Terms of Hydrogen Density of States *462*	
17.3.6	Analysis of Film Microstructure by Effusion of Implanted Rare Gases *463*	
17.4	Discussion of Selected Results *467*	
17.4.1	Amorphous Silicon and Germanium Films *467*	
17.4.1.1	Material Density versus Annealing and Hydrogen Content *467*	
17.4.1.2	Effect of Doping on H Effusion *468*	
17.4.2	Amorphous Silicon Alloys: Si-C *469*	
17.4.3	Microcrystalline Silicon *470*	
17.4.4	Zinc Oxide Films *471*	
17.5	Comparison with Other Experiments *471*	
17.6	Concluding Remarks *472*	
	References *473*	
Part Four	**Materials and Device Modeling** *477*	
18	***Ab-Initio* Modeling of Defects in Semiconductors** *479*	
	Karsten Albe, Péter Ágoston, and Johan Pohl	
18.1	Introduction *479*	
18.2	Density Functional Theory and Methods *480*	
18.2.1	Basis Sets *480*	
18.2.2	Functionals for Exchange and Correlation *481*	
18.2.2.1	Local Approximations *481*	
18.2.2.2	Functionals Beyond LDA/GGA *481*	
18.3	Methods Beyond DFT *483*	
18.4	From Total Energies to Materials' Properties *485*	
18.5	*Ab-initio* Characterization of Point Defects *486*	
18.5.1	Thermodynamics of Point Defects *488*	
18.5.2	Formation Energies from *Ab-Initio* Calculations *493*	
18.5.3	Case study: Point Defects in ZnO *494*	
18.6	Conclusions *497*	
	References *497*	
19	**One-Dimensional Electro-Optical Simulations of Thin-Film Solar Cells** *501*	
	Bart E. Pieters, Koen Decock, Marc Burgelman, Rolf Stangl, and Thomas Kirchartz	
19.1	Introduction *501*	
19.2	Fundamentals *501*	
19.3	Modeling Hydrogenated Amorphous and Microcrystalline Silicon *503*	
19.3.1	Density of States and Transport Hydrogenated Amorphous Silicon *503*	
19.3.2	Density of States and Transport Hydrogenated Microcrystalline Silicon *507*	

19.3.3	Modeling Recombination in a-Si:H and μc-Si: H *508*	
19.3.3.1	Recombination Statistics for Single-Electron States: Shockley–Read–Hall Recombination *508*	
19.3.3.2	Recombination Statistics for Amphoteric States *510*	
19.3.4	Modeling Cu(In,Ga)Se$_2$ Solar Cells *512*	
19.3.4.1	Graded Band-Gap Devices *512*	
19.3.4.2	Issues when Modeling Graded Band-Gap Devices *513*	
19.3.4.3	Example *514*	
19.3.5	Modeling of CdTe Solar Cells *514*	
19.3.5.1	Baseline *516*	
19.3.5.2	The $\Phi_b - N_{Ac}$ (Barrier–Doping) Trade-Off *516*	
19.3.5.3	C–V Analysis as an Interpretation Aid of I–V Curves *518*	
19.4	Optical Modeling of Thin Solar Cells *519*	
19.4.1	Coherent Modeling of Flat Interfaces *519*	
19.4.2	Modeling of Rough Interfaces *519*	
19.5	Tools *521*	
19.5.1	AFORS-HET *521*	
19.5.2	AMPS-1D *522*	
19.5.3	ASA *523*	
19.5.4	PC1D *523*	
19.5.5	SCAPS *523*	
19.5.6	SC-SIMUL *524*	
	References *524*	
20	**Two- and Three-Dimensional Electronic Modeling of Thin-Film Solar Cells** *529*	
	Ana Kanevce and Wyatt K. Metzger	
20.1	Introduction *529*	
20.2	Applications *529*	
20.3	Methods *531*	
20.3.1	Equivalent-Circuit Modeling *531*	
20.3.2	Solving Semiconductor Equations *532*	
20.3.2.1	Creating a Semiconductor Model *533*	
20.4	Examples *534*	
20.4.1	Equivalent-Circuit Modeling Examples *534*	
20.4.2	Semiconductor Modeling Examples *535*	
20.5	Summary *539*	
	References *539*	

Index *541*

Preface

Inorganic thin-film photovoltaics is a very old research topic with a scientific record of more than 30 years and tens of thousands of published papers. At the same time, thin-film photovoltaics is an emerging research field due to technological progress and the subsequent tremendous growth of the photovoltaic industry during recent years. As a consequence, many young scientists and engineers enter the field not only because of the growing demand for skilled scientific personal but also because of the many interesting scientific and technological questions that are still to be solved. As a consequence, there is a growing demand for skilled scientific staff entering the field who will face a multitude of challenging scientific and technological questions. Thin-film photovoltaics aims for the highest conversion efficiencies and at the same time for the lowest possible cost. Therefore, a profound understanding of corresponding solar-cell devices and the photovoltaic materials applied is a major prerequisite for any further progress in this challenging field.

In recent years, a wide and continuously increasing variety of sophisticated and rather specialized analysis techniques originating from very different directions of physics, chemistry, or materials science has been applied in order to extend the scientific base of thin-film photovoltaics. This increasing specialization is a relatively new phenomenon in the field of photovoltaics where during the "old days" everyone was (and had to be) able to handle virtually every scientific method personally. Consequently, it becomes nowadays more and more challenging for the individual scientist to keep track with the results obtained by specialized analysis methods, the physics behind these methods, and on their implications for the devices.

The need for more communication and exchange especially among scientists and Ph.D. students working in the same field but using very different techniques was more and more rationalized during recent years. As notable consequences, very well attended "Young Scientist Tutorials on Characterization Techniques for Thin-Film Solar Cells" were established at Spring Meetings of the Materials Research Society and the European Materials Research Society. These Tutorials were especially dedicated to mutual teaching and open discussions.

The present handbook aims to follow the line defined by these Tutorials: providing concise and comprehensive lecture-like chapters on specific research methods, written by researchers who use these methods as the core of their scientific work and who at the same time have a precise idea of what is relevant for photovoltaic devices. The chapters are intended to focus on the specific methods more than on significant results. This is because these results, especially in innovative research areas, are subject to rapid change and are better dealt with by review articles. The basic message of the chapters in the present handbook focuses more on how to use the specific methods, on their physical background and especially on their implications for the final purpose of the research, that is, improving the quality of photovoltaic materials and devices.

Therefore, the present handbook is not thought as a textbook on established standard (canonical) methods. Rather, we focus on emerging, specialized methods that are relatively new in the field but have a given relevance. This is why the title of the book addresses "advanced" techniques. However, also new methods need to be judged by their implication for photovoltaic devices. For this reason, an introductory chapter (Chapter 1) will describe the basic physics of thin-film solar cells and modules and also guide to the specific advantages that are provided by the individual methods. In addition, we have made sure that the selected authors are not only established specialists concerning a specific method but also have long-time experience dealing with solar cells. This ensures that in each chapter, the aim of the analysis work is kept on the purpose of improving solar cells.

The choice of characterization techniques is not intended for completeness but should be a representative cross section through the scientific methods that have a high level of visibility in the recent scientific literature. Electrical device characterization (Chapter 2), electroluminescence (Chapter 3), photoluminescence (Chapter 7), and capacitance spectroscopy (Chapter 4) are standard optoelectronic analysis techniques for solid-state materials and devices but are also well-established and of common use in their specific photovoltaic context. In contrast, characterization of light trapping (Chapter 5) is an emerging research topic very specific to the photovoltaic field. Chapters 6, 8 and 9 deal with ellipsometry, the steady-state photocarrier grating method, and time-of-flight analysis, which are dedicated thin-film characterization methods. Steady-state photocarrier grating (Chapter 8) and time-of flight measurements (Chapter 9) specifically target the carrier transport properties of disordered thin-film semiconductors. Electron spin resonance (Chapter 10) is a traditional method in solid-state and molecule physics, which is of particular use for analyzing dangling bonds in disordered semiconductors.

The disordered nature of thin-film photovoltaic materials requires analysis of electronic, structural, and compositional properties at the nanometer scale. This is why methods such as scanning probe techniques (Chapter 11) as well as electron microscopy and its related techniques (Chapter 12) gain increasing importance in the field. X-ray and neutron diffraction (Chapter 13) as well as Raman spectroscopy (Chapter 14) contribute to the analysis of structural properties of photovoltaic materials. Since thin-film solar cells consist of layer stacks with interfaces and surfaces, important issues are addressed by understanding their chemical and electronic properties, which may be studied by means of soft X-ray and electron spectroscopy (Chapter 15). Important information for thin-film solar cell research and development are the elemental distributions in the layer stacks, analyzed by various techniques presented in Chapter 16. Specifically for silicon thin-film solar cells, knowledge about hydrogen incorporation and stability is obtained from hydrogen effusion experiments (Chapter 17).

For designing photovoltaic materials with specific electrical and optoelectronic properties, it is important to predict these properties for a given compound. Combining experimental results from materials analysis with those from *ab-initio* calculations based on density-functional theory provides the means to study point defects in photovoltaic materials (Chapter 18). Finally, in order to come full circle regarding the solar-cell devices treated in the first chapters of the handbook, the information gained from the various materials analyses and calculations may now be introduced into one-dimensional (Chapter 19) or multidimensional device simulations (Chapter 20). By means of carefully designed optical and electronic simulations, photovoltaic performances of specific devices may be studied even before their manufacture.

We believe that the overview of these various characterization techniques is not only useful for colleagues engaged in the research and development of inorganic thin-film solar cells, from which the examples in the present handbook are given, but also to those working with other types of solar cells as well as with other optoelectronic, thin-film devices.

The editors would like to thank all authors of this handbook for their excellent and (almost) punctual contributions. We are especially grateful to Ulrike Fuchs and Anja Tschörtner, WILEY-VCH, for helping in realizing this book project.

August 2010

Daniel Abou-Ras, Berlin
Thomas Kirchartz, London
and Uwe Rau, Jülich

List of Contributors

Daniel Abou-Ras
Helmholtz-Zentrum Berlin für
Materialien und Energie (HZB)
Hahn-Meitner-Platz 1
14109 Berlin
Germany

Péter Ágoston
Technische Universität Darmstadt
Institut für Materialwissenschaft
Fachgebiet Materialmodellierung
Petersenstr. 23
64287 Darmstadt
Germany

Karsten Albe
Technische Universität Darmstadt
Institut für Materialwissenschaft
Fachgebiet Materialmodellierung
Petersenstr. 23
64287 Darmstadt
Germany

Jacobo Álvarez-García
Universitat de Barcelona
Facultat de Física
Department Electrònica
C. Martí i Franquès 1
08028 Barcelona
Spain

Marcus Bär
Helmholtz-Zentrum Berlin für
Materialien und Energie (HZB)
Hahn-Meitner-Platz 1
14109 Berlin
Germany

Jan Behrends
Helmholtz-Zentrum Berlin für
Materialien und Energie (HZB)
Institut für Silizium-Photovoltaik
Kekuléstr. 5
12489 Berlin
Germany

Wolfhard Beyer
Forschungszentrum Jülich
Institut für Energieforschung (IEF-5),
Photovoltaik
Leo-Brandt-Straße
52428 Jülich
Germany

Karsten Bittkau
Forschungszentrum Jülich
Institut für Energieforschung (IEF-5),
Photovoltaik
Leo-Brandt-Straße
52428 Jülich
Germany

List of Contributors

Torsten Bronger
Forschungszentrum Jülich
Institut für Energieforschung (IEF-5), Photovoltaik
Leo-Brandt-Straße
52428 Jülich
Germany

Rudolf Brüggemann
Carl von Ossietzky Universität Oldenburg
Fakultät V – Institut für Physik
AG Greco
Carl-von-Ossietzky-Straße 9-11
26129 Oldenburg
Germany

Marc Burgelman
Universiteit Gent
Vakgroep Elektronica en Informatiesystemen (ELIS)
St.- Pietersnieuwstraat 41
9000 Gent
Belgium

Raquel Caballero
Helmholtz-Zentrum Berlin für Materialien und Energie (HZB)
Hahn-Meitner-Platz 1
14109 Berlin
Germany

Robert W. Collins
University of Toledo
Department of Physics and Astronomy
2801 W. Bancroft Street
Toledo, OH 43606
USA

Koen Decock
Universiteit Gent
Vakgroep Elektronica en Informatiesystemen (ELIS)
St.- Pietersnieuwstraat 41
9000 Gent
Belgium

Kaining Ding
Forschungszentrum Jülich
Institut für Energieforschung (IEF-5), Photovoltaik
Leo-Brandt-Straße
52428 Jülich
Germany

Varvara Efimova
Leibniz Institute for Solid State and Materials Research (IFW) Dresden
Institute for Complex Materials
Helmholtzstraße 20
01069 Dresden
Germany

Florian Einsele
Forschungszentrum Jülich
Institut für Energieforschung (IEF-5), Photovoltaik
Leo-Brandt-Straße
52428 Jülich
Germany

Matthias Fehr
Helmholtz-Zentrum Berlin für Materialien und Energie (HZB)
Institut für Silizium-Photovoltaik
Kekuléstr. 5
12489 Berlin
Germany

Levent Gütay
University of Luxembourg
Faculté des Sciences, de la Technologie
et de la Communication
41, rue du Brill
4422 Belvaux
Luxembourg

Jennifer Heath
Linfield College
Department of Physics
900 SE Baker Street
McMinnville, OR 97128
USA

Anke Helbig
University of Stuttgart
Institut für Physikalische Elektronik
Pfaffenwaldring 47
70569 Stuttgart
Germany

Clemens Heske
University of Nevada Las Vegas (UNLV)
Department of Chemistry
4505 Maryland Parkway, Box 454003
Las Vegas, NV 89154-4003
USA

Volker Hoffmann
Leibniz Institute for Solid State and
Materials Research (IFW) Dresden
Institute for Complex Materials
Helmholtzstraße 20
01069 Dresden
Germany

Víctor Izquierdo-Roca
Universitat de Barcelona
Facultat de Física
Department Electrònica
C. Martí i Franquès 1
08028 Barcelona
Spain

Ana Kanevce
Colorado State University
Department of Physics
1875 Campus Delivery
Fort Collins, CO 80523-1875
USA

and

National Renewable Energy Laboratory
1617 Cole Blvd.
Golden, CO 80401-3305
USA

Christian A. Kaufmann
Helmholtz-Zentrum Berlin für
Materialien und Energie (HZB)
Hahn-Meitner-Platz 1
14109 Berlin
Germany

Thomas Kirchartz
Imperial College London
Blackett Laboratory of Physics
Experimental Solid State Physics
Prince Consort Road
London SW7 2AZ
UK

Denis Klemm
Leibniz Institute for Solid State and
Materials Research (IFW) Dresden
Institute for Complex Materials
Helmholtzstraße 20
01069 Dresden
Germany

Jian Li
University of Toledo
Department of Physics and Astronomy
2801 W. Bancroft Street
Toledo, OH 43606
USA

List of Contributors

Klaus Lips
Helmholtz-Zentrum Berlin für
Materialien und Energie (HZB)
Institut für Silizium-Photovoltaik
Kekuléstr. 5
12489 Berlin
Germany

Roland Mainz
Helmholtz-Zentrum Berlin für
Materialien und Energie (HZB)
Hahn-Meitner-Platz 1
14109 Berlin
Germany

Sylvain Marsillac
University of Toledo
Department of Physics and Astronomy
2801 W. Bancroft Street
Toledo, OH 43606
USA

Wyatt K. Metzger
PrimeStar Solar
13100 West 43rd Drive
Golden, CO 80403
USA

Melanie Nichterwitz
Helmholtz-Zentrum Berlin für
Materialien und Energie (HZB)
Hahn-Meitner-Platz 1
14109 Berlin
Germany

Tim Nunney
Thermo Fisher Scientific
The Birches Industrial Estate
Imberhorne Lane
East Grinstead
West Sussex RH19 1UB
UK

Alejandro Pérez-Rodríguez
University of Barcelona
Catalonia Institute for Energy Research
(IREC)
C. Josep Pla 2, B2
08019 Barcelona
Spain

Bart E. Pieters
Forschungszentrum Jülich
Institut für Energieforschung (IEF-5),
Photovoltaik
Leo-Brandt-Straße
52428 Jülich
Germany

Johan Pohl
Technische Universität Darmstadt
Institut für Materialwissenschaft
Fachgebiet Materialmodellierung
Petersenstr. 23
64287 Darmstadt
Germany

Uwe Rau
Forschungszentrum Jülich
Institut für Energieforschung (IEF-5),
Photovoltaik
Leo-Brandt-Straße
52428 Jülich
Germany

Angus A. Rockett
University of Illinois
Department of Materials Science and
Engineering
1304 W. Green Street
Urbana, IL 61801
USA

Manuel J. Romero
National Renewable Energy Laboratory
1617 Cole Blvd.
Golden, CO 80401-3305
USA

Sascha Sadewasser
Helmholtz-Zentrum Berlin für
Materialien und Energie (HZB)
Hahn-Meitner-Platz 1
14109 Berlin
Germany

Sebastian Schmidt
Helmholtz-Zentrum Berlin für
Materialien und Energie (HZB)
Hahn-Meitner-Platz 1
14109 Berlin
Germany

Susan Schorr
Free University Berlin
Department for Geosciences
Malteserstr. 74-100
12249 Berlin
Germany

Michelle N. Sestak
University of Toledo
Department of Physics and Astronomy
2801 W. Bancroft Street
Toledo, OH 43606
USA

Rolf Stangl
Helmholtz-Zentrum Berlin für
Materialien und Energie (HZB)
Kekuléstraße 5
12489 Berlin
Germany

Christiane Stephan
Helmholtz-Zentrum Berlin für
Materialien und Energie (HZB)
Hahn-Meitner-Platz 1
14109 Berlin
Germany

Tobias Törndahl
Uppsala University
Solid State Electronics
PO Box 534
751 21 Uppsala
Sweden

Thomas Unold
Helmholtz-Zentrum Berlin für
Materialien und Energie (HZB)
Hahn-Meitner-Platz 1
14109 Berlin
Germany

Cornel Venzago
AQura GmbH
Rodenbacher Chaussee 4
63457 Hanau
Germany

Iris Visoly-Fisher
Ben Gurion University of the Negev
Department of Chemistry
Be'er Sheva 84105
Israel

Lothar Weinhardt
Universität Würzburg
Physikalisches Institut
Experimentelle Physik VII
Am Hubland
97074 Würzburg
Germany

Thomas Wirth
Bundesanstalt für Materialforschung
und -prüfung
Unter den Eichen 87
12205 Berlin
Germany

Pawel Zabierowski
Warsaw University of Technology
Faculty of Physics
Koszykowa 75
00-662 Warsaw
Poland

Acknowledgments

Chapter 1: The authors would like to thank Dorothea Lennartz for help with the figures. Special thanks are due to Bart Pieters for discussions on thin-film silicon solar cells.

Chapters 2 and 3: Also for these chapters, Dorothea Lennartz is gratefully acknowledged for the help with the figures.

Chapter 4: The authors gratefully acknowledge Steven W. Johnston and Jian V. Li for valuable discussions of the manuscript, as well as for assistance with the figures.

Chapter 5: The author thanks Thomas Beckers for parts of the measurements and Reinhard Carius for the helpful discussions. The Deutsche Forschungsgemeinschaft is acknowledged for the partial financial support through Grant No. PAK88.

Chapter 6: The authors gratefully acknowledge support from DOE Grants No. DE-FG36-08GO18067 and DE-FG36-08GO18073 and from the State of Ohio Third Frontier's Wright Centers of Innovation Program.

Chapter 7: The authors would like to thank Jes Larsen (University of Luxembourg) and Steffen Kretzschmar (Helmholtz-Zentrum Berlin) for additional PL measurements and Raquel Caballero and Tim Münchenberg for preparation of samples.

Chapter 8: The author is grateful to M. Bayrak and O. Neumann for some measurements.

Chapter 10: The authors greatly acknowledge Alexander Schnegg for helpful discussions, suggestions, and proofreading the manuscript. The support from Christian Gutsche for updating our literature database and designing some of the graphs of this article is also greatly appreciated. Matthias Fehr is indebted to the German Federal Ministry of Research and Education (BMBF) for financial support through the Network project *EPR-Solar*, Contract No. 03SF0328A.

Chapter 11: Iris Visoly-Fisher is grateful to David Cahen and Sidney R. Cohen for their contribution to results presented in this chapter. Sascha Sadewasser acknowledges support from Thilo Glatzel, David Fuertes Marrón, Marin Rusu, Roland Mainz, and Martha Ch. Lux-Steiner.

Chapter 12: The authors are grateful to Jaison Kavalakkatt for designing various figures and to Jürgen Bundesmann for technical support. Special thanks are due to Heiner Jaksch (Carl Zeiss NTS) and to Michael Lehmann (TU Berlin) for fruitful discussions and critical reading of the manuscript. This work was supported by the U.S. Department of Energy under Contract No. DE-AC36-08-GO28308.

Chapter 13: The authors are gratefully acknowledge H. Rodriguez-Alvarez for his valuable contributions to the in-situ EDXRD results, the support in the neutron diffraction experiments by Michael Tovar and the support in the synchrotron X-ray diffraction experiments by Christoph Genzel and the team at the EDDI beamline. Moreover, Mikael Ottosson is acknowledged for measurements in the GIXRD section.

Chapter 14: The authors are grateful to Tariq Jawhari and Lorenzo Calvo-Barrio from the Scientific-Technical Services of the University of Barcelona as well as to Edgardo Saucedo and Xavier Fontané from IREC for fruitful discussions and suggestions. A. Pérez-Rodríguez and V. Izquierdo-Roca belong to the M-2E (Electronic Materials for Energy) Consolidated Research Group and the XaRMAE Network of Excellence on Materials for Energy of the "Generalitat de Catalunya."

Chapter 15: The authors gratefully acknowledge (in alphabetically order) M. Blum, J.D. Denlinger, N. Dhere, C.-H. Fischer, O. Fuchs, T. Gleim, D. Gross, A. Kadam, F. Karg, S. Kulkarni, B. Lohmüller, M.C. Lux-Steiner, M. Morkel, H.-J. Muffler, T. Niesen, S. Nishiwaki, S. Pookpanratana, W. Riedl, W. Shafarman, G. Storch, E. Umbach, W. Yang, Y. Zubavichus, and S. Zweigart for their contributions to the results presented in this chapter. Valuable discussions with L. Kronik and J. Sites are also acknowledged. The research was funded through the Deutsche Forschungsgemeinschaft (DFG) through SFB 410 (TP B3), the National Renewable Energy Laboratory through Subcontract Nos. XXL-5-44205-12 and ADJ-1-30630-12, the DFG Emmy Noether program, and the German BMWA (FKZ 0329218C). The Advanced Light Source is supported by the Ofce of Basic Energy Sciences of the US Department of Energy under Contract Nos. DE-AC02-05CH11231 and DE-AC03-76SF00098.

Chapter 16: Volker Hoffmann, Denis Klemm, Varvara Efimova (IFW Dresden), and Cornel Venzago from AQura GmbH gratefully acknowledge the financial support from the FP6 Research Training Network GLADNET (No. MRTN-CT-2006-035459). The group from IFW Dresden thanks the Spectruma Analytik GmbH and HZB, Berlin for good collaborations. Christian A. Kaufmann and Raquel Caballero are grateful to Jürgen Bundesmann for technical support.

Chapter 17: The authors wish to thank Dorothea Lennartz and Pavel Prunici for valuable technical support. Interest and support by Uwe Rau is kindly acknowledged.

Chapter 18: The authors are grateful for the support by the Sonderforschungsbereich 595 "Ermüdung von Funktionsmaterialien" of the Deutsche Forschungsgemeinschaft (DFG).

Chapter 19: The authors are grateful to Rudi Brüggemann for discussions on solar-cell simulations. Marc Burgelman and Koen Decock acknowledge the support of the Research Foundation – Flanders (FWO; Ph.D. fellowship).

Chapter 20: This work was supported by the US Department of Energy under Contract Number DE-AC36-08GO28308 to NREL.

Abbreviations

1D	One-dimensional
2D	Two-dimensional
3D	Three-dimensional
A°X	Excitons bound to neutral acceptor
ac	Alternating current
ADF	Annular dark field
ADXRD	Angle-dispersive X-ray diffraction
AES	Auger electron spectroscopy
AEY	Auger electron yield
AFM	Atomic force microscopy
AFORS-HET	Automat for simulation of heterostructures
ALDA	Adiabatic local density approximation
AM	Amplitude modulation
AM	Air mass
AMU	Atomic mass units
ARS	Angularly resolved light scattering
AS	Admittance spectroscopy
ASA	Advanced semiconductor analysis
ASCII	American Standard Code for Information Interchange
a-Si	Amorphous silicon
A-X	Excitons bound to ionized acceptor
BF	Bright field
BS	Beam splitter
BSE	Bethe–Salpeter equation
BSE	Backscattered electrons
c-AFM	Conductive AFM
CBD	Chemical bath deposition
CBED	Convergent-beam electron diffraction
CBM	Conduction-band minimum
CBO	Conduction-band offset
CC	coupled cluster

CCD	Charge-coupled device
CHA	Concentric hemispherical analyzer
CI	configuration interaction
CIGS	$Cu(In,Ga)Se_2$
CIGSe	$Cu(In,Ga)Se_2$
CIGSSe	$Cu(In,Ga)(S;Se)_2$
CIS	$CuInSe_2$
CIS	$CuInS_2$
CISe	$CuInSe_2$
CL	Cathodoluminescence
CL	Core level
CMA	Cylindrical mirror analyzer
CN	Charge neutrality
CP	Critical point
CPD	Contact-potential difference
CSL	Coincidence-site lattice
CSS	Closed-space sublimation
CTEM	Conventional transmission electron microscopy
CV	Capacitance–voltage
cw	Continuous wave
D°h	Optical transitions between donor and free hole
D°X	Excitons bound to neutral donor
DAP	Donor–acceptor pair
DB	Dangling bond
dc	Direct current
DF	Dark field
DFPT	Density functional perturbation theory
DFT	Density functional theory
DLCP	Drive-level capacitance profiling
DLOS	Deep-level optical spectroscopy
DLTS	Deep-level transient spectroscopy
DOS	Density of states
DSR	Differential spectral response
DT	Digital
D-X	Excitons bound to ionized donor
eA°	Optical transitions between acceptor and free electron
EBIC	Electron-beam-induced current
EBSD	Electron backscatter diffraction
EDMR	Electrically detected magnetic resonance
EDX	Energy-dispersive X-ray spectrometry
EDXRD	Energy-dispersive X-ray diffraction
EELS	Electron energy-loss spectrometry
EFTEM	Energy-filtered transmission electron microscopy
EL	Electroluminescence
ELNES	Energy-loss near-edge structure

EMPA	Eidgenössische Materialprüfungsanstalt
ENDOR	electron-nuclear double resonance
EPR	Electron paramagnetic resonance
ESCA	Electron spectroscopy for chemical analysis
ESEEM	Electron-spin echo envelope modulation
ESI	Energy-selective imaging
ESR	Electron spin resonance
EXC	Free excition transition
EXELFS	Extended energy-loss fine structure
FFT	Fast Fourier transformation
FIB	Focused ion beam
FM	Frequency modulation
FP-LAPW	full potential-linearized augmented plane wave
FWHM	Full width at half maximum
FX	Free excitons
FY	Fluorescence yield
GB	Grain boundary
GD-MS	Glow discharge-mass spectroscopy
GD-OES	Glow discharge-optical emission spectroscopy
GGA	Generalized gradient approximation
GIXRD	Grazing-incidence X-ray diffraction
GNU	Is not Unix (recursive acronym)
GPL	General public licence
GW	G for Green's function and W for the screened Coulomb interaction
HAADF	High-angle annular dark field
HFI	Hyperfine interaction
HOPG	Highly oriented pyrolytic graphite
HR	High resistance
HR	High resolution
HT	High-temperature
HWCVD	Hot-wire plasma-enhanced chemical vapor deposition
HZB	Helmholtz-Zentrum Berlin
IBB	Interface-induced band bending
IPES	Inverse photoelectron spectroscopy
IR	Infrared
JEBIC	Junction electron-beam-induced current
KPFM	Kelvin-probe force microscopy
KS	Kohn–Sham
KSM	Kaplan–Solomon–Mott (model)
LBIC	Laser-beam-induced current
LCR meter	Induction, capacitance, resistance - impedance analyzer
LDA	local density approximation
LED	Light-emitting diode
LESR	Light-induced ESR

LIA	Lock-in amplifier	
LO	Longitudinal optical	
LR	Low resistance	
LT	Low-temperature	
LVM	Localized vibrational modes	
MBPT	Many-body perturbation theory	
MD	Molecular dynamics	
MIP	Mean-inner potential	
MIS	Metal-insulator-semiconductor	
ML	Monolayer	
MO	Metal oxide	
MOS	Metal-oxide-semiconductor	
MSE	Mean-square error	
mw	Microwave	
nc-AFM	Non-contact atomic force microscopy	
NIR	Near-infrared	
NIST	National Institute of Standards and Technology	
NSOM	Near-field scanning optical microscopy	
OBIC	Optical-beam-induced current	
OVC	Ordered vacancy compound	
PBE-GGA	generalized gradient approximation by Perdew, Burke, and Ernzerhof	
PCSA	Polarizer-compensator-sample-analyzer; instrument configuration for spectroscopic ellipsometry	
PDA	Photodetector array	
PDE	Partial differential equations	
PECVD	Plasma-enhanced chemical vapor deposition	
PES	Photoelectron spectroscopy	
pESR	pulsed electron spin resonance	
PEY	Partial electron yield	
PIPO	Photon-in photon-out	
PL	Photoluminescence	
PLL	Phase-locked loop	
PMT	Photomultiplier tube	
pp	Peak-to-peak	
PV	Photovoltaic	
PVD	Physical vapor deposition	
QE	Quantum efficiency	
QMA	Quadrupole mass analyzer	
QMC	Quantum Monte Carlo	
RDLTS	Reverse-bias deep-level transient spectroscopy	
REBIC	Remote electron-beam-induced current	
rf	Radio frequency	
RGB	Red-green-blue, color space	
RIXS	Resonant inelastic (soft) x-ray scattering	

RS	Raman spectroscopy
RSF	Relative sensitivity factor
RTP	Rapid thermal process
RTSE	Real-time spectroscopic ellipsometry
RZW	Ritter, Zeldow, Weiser analysis
S/N	Signal-to-noise (ratio)
SAED	Selected-area electron diffraction
SCAPS	Solar-cell capacitance simulator
SCM	Scanning capacitance microscopy
SE	Spectroscopic ellipsometry
SE	Secondary electron
SEM	Scanning electron microscopy
SIMS	Secondary-ion mass spectroscopy
SNMS	Sputtered neutral mass spectroscopy
SNOM, see also NSOM	Scanning near-field optical microscopy
SPICE	Simulation Program with Integrated Circuit Emphasis
SPM	Scanning probe microscopy
SQ	Shockley–Queisser (limit)
SR	Spectral response
SSPG	Steady-state photocarrier grating
SSRM	Scanning spreading-resistance microscopy
STEM	Scanning transmission electron microscopy
STM	Scanning tunneling microscopy
SWE	Staebler–Wronski effect
TCO	Transparent conductive oxide
TD	Trigger diode
TD-DFT	Time-dependent density functional theory
TDS	Thermal desorption spectroscopy
TEM	Transmission electron microscopy
TEY	Total electron yield
TF	Tuning fork
TO	Transversal optical
TOF	Time of flight
TPC	Transient photocapacitance spectroscopy
TPD	Temperature-programmed desorption
TU	Technical University
UHV	Ultrahigh vacuum
UPS	Ultraviolet photoelectron spectroscopy
UV	Ultraviolet
VBM	Valence-band maximum
VBO	Valence-band offset
Vis	Visible
WDX	Wavelength-dispersive X-ray spectrometry
XAES	X-ray Auger electron spectroscopy
XAS	X-ray absorption spectroscopy

XES	X-ray emission spectroscopy
XPS	X-ray photoelectron spectroscopy
XRD	X-ray diffraction
XRF	X-ray fluorescence
µc-Si	Microcrystalline silicon

Part one
Introduction

1
Introduction to Thin-Film Photovoltaics
Thomas Kirchartz and Uwe Rau

1.1
Introduction

From the early days of photovoltaics until today, thin-film solar cells have always competed with technologies based on single-crystal materials such as Si and GaAs. Owing to their amorphous or polycrystalline nature, thin-film solar cells always suffered from power conversion efficiencies lower than those of the bulk technologies. This drawback was and still is counterbalanced by several inherent advantages of thin-film technologies. Since in the early years of photovoltaics space applications were the driving force for the development of solar cells, the argument in favor of thin films was their potential lighter weight as compared with bulk materials.

An extended interest in solar cells as a source of renewable energy emerged in the mid-seventies as the limitations of fossil energy resources were widely recognized. For terrestrial power applications, the cost arguments and the superior energy balance strongly favored thin films. However, from the various materials under consideration in the fifties and sixties, only four thin-film technologies, namely amorphous hydrogen alloyed (a-)Si:H and the polycrystalline heterojunction systems CdS/Cu_xS, $CdS/CdTe$, and $CdS/CuInSe_2$, entered pilot production. Activities in the CdS/Cu_xS system stopped at the beginning of the eighties because of stability problems. At that time, amorphous silicon became the front runner in thin-film technologies keeping almost constantly a share of about 10% in a constantly growing photovoltaic market, the remaining 90% kept by crystalline Si. Despite their high-efficiency potential, polycrystalline heterojunction solar cells based on CdTe and $CuInSe_2$ did not play an economic role until to the turn of the century.

During the accelerated growth of the worldwide photovoltaic market in the first decade of new century, the three inorganic thin-film technologies increased their market share to 14%, where approximately 9% are covered by CdTe modules (numbers from 2008). With annual production figures in the GW range, inorganic thin-film photovoltaics has become a multibillion dollar business. In order to expand this position, further dramatic cost-reduction is required combined with a substantial increase in module efficiency. In this context, material and device

Advanced Characterization Techniques for Thin Film Solar Cells,
Edited by Daniel Abou-Ras, Thomas Kirchartz and Uwe Rau.
© 2011 Wiley-VCH Verlag GmbH & Co. KGaA. Published 2011 by Wiley-VCH Verlag GmbH & Co. KGaA.

characterization becomes an important task not only for quality control in an expanding industry but also remains at the very heart of further technological progress.

This book concentrates on the three inorganic thin-film technologies – thin-film Si (a-Si:H combined with microcrystalline μc-Si:H to a tandem solar cell), and the two heterojunction systems CdS/CdTe as well as CdS/Cu(In,Ga)(Se,S)$_2$. These thin-film technologies have in common that they consist of layer sequences from disordered semiconductor materials that are deposited onto a supporting substrate or superstrate with the help of vacuum technologies. This layer structure and the use of disordered materials defines a fundamental difference to devices based on crystalline c-Si where a self-supporting Si wafer is transformed into a solar cell via a solid-state diffusion of dopant atoms. Thus, there are only the front and the back surface as critical interfaces in the classical wafer solar cell (with the notable exception of the a-Si:H/c-Si heterojunction solar cell). In thin-film solar cells, the number of functional layers can amount to up to eight and more. Some of these layers have thicknesses as low as 10 nm. In large-area modules, these layers homogenously cover areas of up to 6 m^2. These special features of the inorganic thin-film photovoltaic technologies define the field for the characterization techniques discussed in this book.

Electrical characterization, electroluminescence and photoluminescence, capacitance spectroscopy, and characterization of light trapping as considered in Chapters 2–5 and 7 are common photovoltaic analysis techniques. However, the specific properties of the thin-film systems like the disordered nature of the materials, the importance of features in the nm scale, and the fact that the film thicknesses are of the order or even in some cases much below the wavelength of visible light account for the special aspects that must be considered when using these techniques. Chapters 6, 8 and 9 deal with techniques like ellipsometry, the steady-state photocarrier grating method, and time-of-flight analysis that are specific thin-film methods some even invented within the field of thin-film photovoltaics. The following Chapters 10–17 discuss classical methods for material characterization, each of them having special importance for at least one of three technologies. Again, the specific features of photovoltaic thin films like the importance of dangling bonds and hydrogen passivation in disordered Si, the need for physical and chemical material analysis on the nanometer scale, or the prominence of interface chemistry and physics in thin-film solar cells define the focus of these chapters. Chapters 18–20 at the end of this handbook deal with the theoretical description of materials and devices. *Ab-initio* modeling of semiconductor materials is indispensable, because even the basic physical properties of some of the wide variety of compounds and alloys used in thin-film photovoltaics are not satisfactorily known. Finally, successful modeling of the finished devices may be looked at as the definitive proof of our understanding of materials and interfaces.

This introductory chapter yields a brief general introduction into the basic principles of photovoltaics highlighting the specific material and device properties that are relevant for the three thin-film technologies – a-Si:H/μc-Si:H, CdS/CdTe, and CdS/Cu(In,Ga)(Se,S)$_2$.

1.2
The Photovoltaic Principle

The temperature difference between the surface of the sun with a temperature of $T = 5800$ K and the surface of the earth ($T = 300$ K) is the driving force of any solar-energy conversion. Solar cells and solar modules directly convert the solar light into electricity using the internal photoelectric effect. Thus, any solar cell needs a photovoltaic absorber material that is not only able to absorb the incoming light efficiently but also to create mobile charge carriers, electrons, and holes, that are separated at the terminals of the device without significant loss of energy. Note that in *organic* absorber materials, most light-absorption processes generate excitons and a first step of charge separation is necessary in order to dissociate the exciton into free carriers. In contrast, the low binding energy of excitons in *inorganic* semiconductors makes absorption and generation of mobile charge carriers virtually identical in appropriate absorber materials of this type. Thus, after light absorption electrons and holes are present in the absorber and must be directed toward the two different contacts to the absorber, that is, the final charge carrier separation step.

For a semiconductor acting as a photovoltaic absorber, its band-gap energy E_g is the primary quantity defining how many charge carriers are generated from solar photons with energy $E \geq E_g$. Maximizing the number of photons contributing to the short-circuit current density of a solar cell would require minimizing E_g. Since photogenerated electron hole pairs thermalize to the conduction-band and valence-band edges after light absorption, the generated energy per absorbed photon corresponds to E_g regardless of the initial photon energy E. Thus, maximizing the band-gap energy E_g maximizes the available energy per absorbed photon. Therefore, one intuitively expects that an optimum band-gap energy exists between $E_g = 0$, maximizing the generated electron–hole pairs, and $E_g \to \infty$, maximizing the generated energy contained in a single electron–hole pair. Quantitatively, this consideration is reflected in the dependence of the maximum achievable conversion efficiency of a single band-gap photovoltaic absorber material as discussed in the following section.

1.2.1
The Shockley–Queisser Theory

The maximum power conversion efficiency of a solar cell consisting of single semiconducting absorber material with band-gap energy E_g is described by the Shockley–Queisser [1] (SQ) limit. In its simplest form, the SQ limit relies on four basic assumptions: (i) the probability for the absorption of solar light by the generation of a single electron–hole pair in the photovoltaic absorber material is unity for all photon energies $E \geq E_g$ and zero for $E < E_g$. (ii) All photogenerated charge carriers thermalize to the band edges. (iii) The collection probability for all photo-generated electron–hole pairs at short-circuit is unity. (iv) The only loss mechanism in excess of the nonabsorbed photons of (i) and the thermalization losses in (v) is the spontaneous emission of photons by radiative recombination of electron–hole pairs as required by the principle of detailed balance.

In order to calculate the maximum available short-circuit current $J_{sc,SQ}$ as defined by (iii), we need the incoming photon flux ϕ_{inc} and the absorptance $A(E)$ defining the percentage of the incoming light at a certain photon energy E that is absorbed and not reflected or transmitted. The simplest approximation defined for an ideal absorber by condition (i) is a step-function, that is, $A(E) = 1$ (for $E > E_g$) and $A(E) = 0$ (for $E < E_g$). Then we have under short-circuit conditions (i.e., applied voltage $V = 0$ V)

$$J_{sc,SQ} = q \int_0^\infty A(E)\phi_{inc}(E)dE = q \int_{E_g}^\infty \phi_{inc}(E)dE \qquad (1.1)$$

where q denotes the elementary charge.

Figure 1.1a compares the spectral photon flux corresponding to the terrestrial AM1.5G norm spectrum with the black body spectrum at $T = 5800$ K, both spectra normalized to a power density of 100 mW/cm². Figure 1.1b illustrates the maximum short-circuit current density that is possible for a given band-gap energy E_g according to Eq. (1.2).

Since light absorption by generation of free carriers and light emission by recombination of electron–hole pairs is interconnected by the principle of detailed balance, in thermodynamic equilibrium the emissivity ϕ_{em} is connected to the absorptance via Kirchhoff's law $\phi_{em} = A(E)\phi_{bb}(E, T)$, where $\phi_{bb}(E, T)$ is the black body spectrum at temperature T.

In a ideal solar cell under applied voltage bias, we use Würfel's generalization [2] of Kirchhoff's law to describe the recombination current $J_{rec,SQ}$ for radiative recombination according to

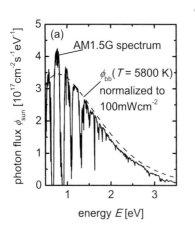
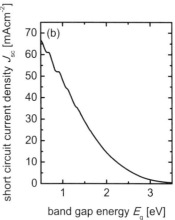

Figure 1.1 (a) Comparison of the AM1.5G spectrum with the black body spectrum of a body with a temperature $T = 5800$ K. Both spectra are normalized such that the power density is 100 mW/cm². (b) Using the AM1.5G spectrum and Eq. (2.1), we obtain the short-circuit current density $J_{sc,SQ}$ in the Shockley–Queisser limit as a function of the band-gap energy E_g of the solar absorber.

$$J_{\text{rec,SQ}} = q \int_0^\infty A(E)\phi_{\text{bb}}(E,T)\exp\left(\frac{qV}{kT}\right) dE = q \int_{E_g}^\infty \phi_{\text{bb}}(E,T)\exp\left(\frac{qV}{kT}\right) dE \quad (1.2)$$

where the second equality again results from the assumption of a sharp band-gap energy E_g. Thus, Eq. (1.2) describes the current density of a solar cell in the dark if only radiative recombination of carriers is considered corresponding to condition (iv) and the carriers have the temperature T of the solar cell according to condition (ii). The total current density J under illumination is a superposition of this radiative recombination current density and the short-circuit current density defined in Eq. (1.1). Thus, we can write

$$J(V) = J_{\text{rec,SQ}}(V) - J_{\text{sc,SQ}} = q \int_{E_g}^\infty \phi_{\text{bb}}(E) dE \exp\left(\frac{qV}{kT}\right) - q \int_{E_g}^\infty \phi_{\text{inc}}(E) dE \quad (1.3)$$

There are two contributions to the incoming photon flux ϕ_{inc}, that is, the spectrum ϕ_{sun} of the sun and the photon flux ϕ_{bb} from the environment, which has the same temperature as the sample. When we replace the incoming photon flux ϕ_{inc} with the sum $\phi_{\text{sun}} + \phi_{\text{bb}}$, Eq. (1.3) simplifies to

$$J(V) = q \int_{E_g}^\infty \phi_{\text{bb}}(E) dE \left[\exp\left(\frac{qV}{kT}\right) - 1\right] - q \int_{E_g}^\infty \phi_{\text{sun}}(E) dE \quad (1.4)$$

which is a typical diode equation with an additional photocurrent only due to the extra illumination from the sun. Now it is obvious that for zero excess illumination and zero volts applied, the current becomes zero.

Figure 1.2 shows the current density/voltage (J/V) curves of an ideal solar cell according to Eq. (1.4) for three different band-gap energies $E_g = 0.8, 1.4,$ and 2.0 eV. If we evaluate Eq. (1.4) under open-circuit conditions, that is, at $J = 0$, we find the maximum possible voltage in the fourth quadrant of the coordinate system in Figure 1.2. This voltage is called the open-circuit voltage V_{oc} and follows from Eq. (1.4) as

$$V_{\text{oc}} = \frac{kT}{q} \ln\left(\frac{\int_{E_g}^\infty \phi_{\text{sun}}(E) dE}{\int_{E_g}^\infty \phi_{\text{bb}}(E) dE} + 1\right) = \frac{kT}{q} \ln\left(\frac{J_{\text{sc,SQ}}}{J_{0,\text{SQ}}} + 1\right) \quad (1.5)$$

Here, $J_{0,\text{SQ}}$ is the saturation current density in the SQ limit, that is, the smallest possible saturation current density for a semiconductor of a given band gap. The open-circuit voltage increases nearly linearly with increasing band gap as shown in Figure 1.3a.

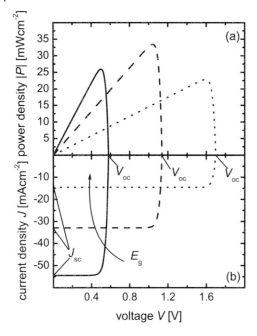

Figure 1.2 (a) Power density/voltage curves and (b) current density/voltage (J/V) curves of three ideal solar cells with band gaps $E_g = 0.8$, 1.4, and 2.0 eV, respectively. The higher the band gap E_g, the higher the open-circuit voltage V_{oc}, that is, the intercept of both power density and current density with the voltage axis. However, a higher band gap also leads to a decreased short-circuit current J_{sc} (cf. Figure 1.1b). The curves are calculated using Eq. (1.4).

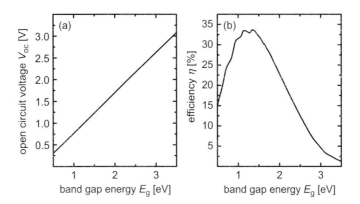

Figure 1.3 (a) Open-circuit voltage and (b) conversion efficiency as a function of the band-gap energy E_g in the Shockley–Queisser limit using an AM1.5G spectrum as illumination. The optimum band-gap energies for single junction solar cells are in the range of $1.1\,\text{eV} < E_g < 1.4\,\text{eV}$ with maximum conversion efficiencies around $\eta = 33\%$ under unconcentrated sunlight.

1.2 The Photovoltaic Principle

From Eq. (1.4), the power density follows by multiplication with the voltage. The efficiency η is then the maximum of the negative power density,[1] that is,

$$\eta = \frac{-\max(J(V)V)}{P_{opt}} = \frac{-\max(J(V)V)}{\int_{E_g}^{\infty} E\phi_{sun}(E)dE} \tag{1.6}$$

Figure 1.3b shows the final result of the SQ theory: the efficiency as a function of the band-gap energy for illumination with the AM1.5G spectrum depicted in Figure 1.1a.

1.2.2
From the Ideal Solar Cell to Real Solar Cells

The universality and simplicity of the SQ limit is due to the fact that all internal details of the solar cell are irrelevant for its derivation. However, these hidden details are the practical subjects of research on real solar cells, and especially on thin-film solar cells. It is important to understand that some of these details idealized (or neglected) by the original SQ theory [1] are not in conflict with the detailed balance principle [3].

First, starting from a step-function like absorptance toward a more complex spectral dependence of $A(E)$ is not in conflict with the radiative recombination limit (cf. Eqs. (2.1) and (2.2)). A continuous transition from zero to unity is expected from any semiconductor material with finite thickness. Especially for thin-film absorbers, maximizing light absorption is an important task requiring additional means to confine the light as discussed in Section 1.2.3. Moreover, the disorder in thin-film absorbers may lead to additional electronic states close to the band gap (so-called band tails or band-gap fluctuations) with a considerable contribution to light absorption and emission. In consequence, the achievable conversion efficiency is reduced even in the radiative limit [4].

Second, proper extraction of the photogenerated electrons and holes requires sufficiently high carrier mobilities and selectivity of the contacts to make sure that all electrons and holes are collected in the n-type and in the p-type contact. Again, these requirements are valid even when restricting the situation to radiative recombination [5]. Since mobilities in disordered thin-film materials are generally lower than in mono-crystalline absorbers, charge carrier extraction is an issue to be discussed with especial care (Section 1.2.4).

Finally, recombination in thin-film solar cells is dominated by nonradiative processes. Thus, especially the achieved open-circuit voltages are far below the radiative limit. Section 1.2.5 and the major part of Chapter 2 will deal with understanding the efficiency limits resulting from all sorts of nonradiative recombination. It is important

---

1) Negative current density means here that the current density is opposite to the current density any passive element would have. A negative power density means then that energy is extracted from the device and not dissipated in the device as it would happen in a diode, which is not illuminated.

to note that even when considering nonradiative recombination, we must not necessarily abandon a detailed balance approach [6, 7] (cf. Chapter 3).

1.2.3
Light Absorption and Light Trapping

The first requirement for any solar cell is to absorb light as efficiently as possible. Solar-cell absorbers should, therefore, be nontransparent for photons with energy $E > E_g$. For any solar cell but especially for thin-film solar cells, this requirement is in conflict with the goal of using as little absorber material as possible. Additionally, thinner absorbers facilitate charge extraction for materials with low mobilities and/or lifetimes of the photogenerated carriers. This is why light trapping in photovoltaic devices is of major importance. Light trapping exploits randomization of light at textured surfaces or interfaces in combination with the fact that semiconductor absorber layers have typical refractive indices n that are much higher than that of air ($n = 1$) or glass ($n \approx 1.5$). Typical values for the real part of the refractive index are $n > 3.5$. But beforehand, the light has to enter the solar cell, and for the reflection at the front surface, a high refractive index is a disadvantage. The reflectance

$$R = \left(\frac{n-1}{n+1}\right)^2 \tag{1.7}$$

at the interface between air and the semiconductor will become higher when the refractive index gets higher. However, the high reflection at the front surface is reduced by using several layers between air and absorber layer. The refractive indices of these layers increase gradually, and any large refractive index contrast is avoided.

For light trapping, however, a high refractive index has an advantage. When the direction of the incoming light is randomized by a scattering interface somewhere in the layer stack of the thin-film solar cell, part of the light will be guided in the solar cell absorber by total internal reflection. The percentage of light kept in the solar cell by total internal reflection increases with the refractive index, since the critical angle $\theta_c = \arcsin(1/n)$ becomes smaller. For light with a Lambertian distribution of angles, the reflectance of the front surface for light from the inside is

$$R_i = 1 - \frac{(1-R_f)\int_0^{\theta_c} \cos\theta \sin\theta\, d\theta}{\int_0^{\pi/2} \cos\theta \sin\theta\, d\theta} = 1 - \frac{(1-R_f)}{n^2} \tag{1.8}$$

Here, R_f is the reflectance at the front side of the absorber for normal incidence.

To visualize the effect, the absorption coefficient and the light trapping has on the absorptance of a solar cell, we present some calculations for a model system. Let us assume a direct semiconductor, which have absorption coefficients of the typical form

$$\alpha = \alpha_0 \sqrt{(E-E_g)/1\text{ eV}} \tag{1.9}$$

Then, the absorptance $A(E)$, that is, the percentage of photons that are absorbed and not reflected or transmitted at a certain photon energy, is calculated for flat surfaces and for an absorber thickness much larger than the wavelength of light with

$$A = (1 - R_f) \frac{\left(1 - e^{-\alpha d}\right)\left(1 + R_b e^{-\alpha d}\right)}{1 - R_f R_b e^{-2\alpha d}} \tag{1.10}$$

Here, R_b is the reflectance at the backside. Equation (1.10) assumes an infinite number of reflections at the front and the back of the absorber layer. To calculate the real absorptance of any thin-film solar cell, it is rather useless for two reasons: (i) thin-film solar cells usually consist of not only one but several layers and (ii) the layer thicknesses are of the same order than the wavelength of light and interference cannot be neglected any more. Nevertheless Eq. (1.10) is useful to test the influence of the absorption coefficient on the absorptance. Figure 1.4a compares the absorptance calculated according to Eq. (1.10) for three different values of α_0, namely $\alpha_0 = 10^4$, $\sqrt{10} \times 10^4$, and 10^5 cm^{-1}, and for a constant thickness d of the absorber of $d = 1$ μm. The reflectance at the front side is assumed to be $R_f = 0$ and the reflectance at the backside is $R_b = 1$.

To calculate the absorptance of textured cell with light trapping, it is necessary to integrate over all angles. The resulting equations are rather complicated [8, 9]; however, a simple and useful approximation exists for the case $R_b = 1$, namely [10]

$$A = \frac{1 - R_f}{1 + \frac{(1 - R_f)}{4n^2 \alpha d}} \tag{1.11}$$

Figure 1.4b shows the result of applying the absorption coefficient defined in Eqs. (1.9)–(1.11). Again, the absorptance for the case of perfect light trapping is calculated for $\alpha_0 = 10^4$, $\sqrt{10} \times 10^4$, and 10^5 cm^{-1}, $d = 1$ μm, and $R_f = 0$. The refractive index is assumed to be $n = 3.5$. It is obvious that for a given value of α_0, the

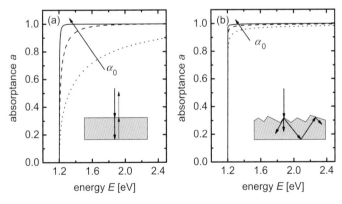

Figure 1.4 Absorptance as a function of photon energy for (a) a flat solar cell and (b) a textured solar cell with perfect light trapping. In both cases, the absorption coefficient α_0 from Eq. (1.9) is varied. The values are for both subfigures $\alpha_0 = 10^4$, $\sqrt{10} \times 10^4$, 10^5 cm^{-1}. For the same absorption coefficient, the textured solar cell has absorptances that are much closer to the perfect step function than the flat solar cell.

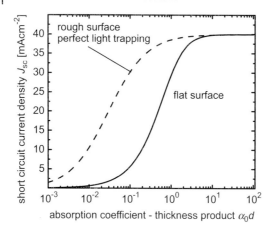

Figure 1.5 Comparison of the short-circuit current density of a flat (solid line) and a textured solar cell (dashed line) as a function of the product α_0 and the thickness d assuming an absorption coefficient according to Eq. (1.9). Especially for low absorption coefficients relative to the device thickness (low $\alpha_0 d$), light trapping increases the short-circuit current density drastically. The refractive index used for these simulations is $n = 3.5$ independent of photon energy.

absorptance of the textured solar cell comes much closer to the perfect step-function like absorptance of the SQ limit.

To visualize the effect of light trapping on the short-circuit current density J_{sc}, Figure 1.5 compares the J_{sc} as a function of the product $\alpha_0 d$ for a flat and a Lambertian surface, that is, for absorptances calculated with Eqs. (1.10) and (1.11). The band gap is chosen to be $E_g = 1.2$ eV as in Figure 1.4 so the maximum J_{sc} for high $\alpha_0 d$ is the same as in the SQ limit (cf. Figure 1.1b), namely $J_{sc,max} = 40$ mA/cm². However, for lower $\alpha_0 d$, the J_{sc} with and without light trapping differ considerably and show the benefit from structuring the surface to enhance the scattering in the absorber layer. In reality, the benefit from light trapping will be smaller since the light has to be reflected several times at the front and especially at the back surface, where we assumed the reflection to be perfect. In reality any back reflector will absorb part of the light parasitically, that is, the light is absorbed but no electron–hole pairs are created, which could contribute to the photocurrent.

1.2.4
Charge Extraction

After an electron–hole pair is generated, the charge carriers must be extracted from the absorber layer. To get a net photocurrent, the electron must leave the device at the opposite contact than the hole. This requires a built-in asymmetry that makes electrons leave the device preferentially at the electron contact and holes at the hole contact.

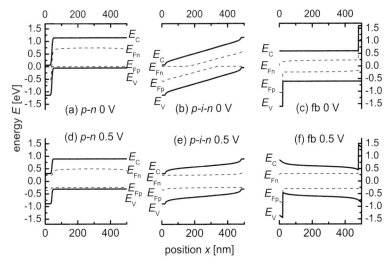

Figure 1.6 Simulation of the band diagrams of a (a, d) p–n-junction, a (b, e) p–i–n-junction, and a (c, f) flat-band (fb) solar cell under illumination. Every type of geometry is depicted under short-circuit conditions and under an applied forward bias $V = 0.5$ V.

Figure 1.6 introduces three device geometries that induce a built-in asymmetry that helps to extract oppositely charged carriers at opposite contacts. Figure 1.6a shows the band diagram of a p–n-junction solar cell under illumination and Figure 1.6d shows the same cell with an applied voltage $V = 0.5$ V. The simulations were done by solving the Poisson equation and the continuity equations with the software ASA, which is described in Chapter 19. As typical for most solar cells with a p–n-junction, the space charge region, where the bands are steep and the electric field is high, is at the very edge of the device. Most of the device consists, in our example, of a p-type base layer, where the field is practically zero. The transport of minority carriers (here electrons) to the space charge region is purely diffusive and independent from the applied voltage. That means application of a voltage does not change the electrical potential in the device, except for the space charge region.

The band bending at the junction leads to an asymmetry that separates the charges. Electrons are able to diffuse to the junction and then further to the n-type region and the electron contact. In addition, the p–n-junction serves as a barrier for holes which are in turn extracted by the back contact. Note that in the band diagram in Figure 1.6a and d, this back contact is not selective as is the p–n-junction. Therefore, also electrons can leave the device at this contact, a fact that is usually considered as contact recombination (cf. Chapter 2). $Cu(In,Ga)Se_2$ and CdTe solar cells are examples for p–n-(hetero)junctions.

For some disordered semiconductors like amorphous silicon, the electronic quality of doped layers is very poor. In addition, the mobilities and diffusion lengths are small, and thus purely diffusive transport would not lead to efficient charge extraction. The solution to this problem is the so-called p–i–n-junction diode. Here the doped layers are very thin compared to the complete thickness of the diode. The largest share of the complete absorber thickness is occupied by an intrinsic, that is, undoped layer, in between the n and p-type regions. Figure 1.6a shows the band diagram of such a p–i–n-junction solar cell under illumination and Figure 1.6d shows the same cell with an applied voltage $V = 0.5$ V. Under short-circuit conditions, the region with a nonzero electric field extends over the complete intrinsic layer. Only directly at the contacts, the field is relatively small. When a forward voltage is applied to the cell, the electric field becomes smaller as shown in Figure 1.6d. Solar cells made from a-Si:H as well as a-Si:H/μc-SiH tandem cells use the p–i–n configuration.

Both p–n-junction and p–i–n-junction solar cells have a built-in field, meaning that the bands are bended due to the different conductivity type of the layers. Theoretically, such a band bending is not necessary to separate charges as can be shown by a gedanken experiment [11]. Figure 1.6c shows the band diagram of a hypothetical flat-band solar cell under short-circuit conditions. Like the p–i–n-junction solar cell, the flat-band solar cell has an intrinsic layer sandwiched between two other layers that induce the asymmetry for charge separation. In this case, the asymmetry is not due to band bending and differently doped layers but instead due to band offsets at the heterojunction between two materials with different band gaps. Let us assume we find one contact materials with zero band offset for the electrons and a high (in this case 1 eV) band offset for the holes and another material with the exact inverse properties. In this case, the band diagram is completely flat apart from the two band offsets. Like in the p–n-junction solar cell, the charge separation at short-circuit is arranged diffusive transport that is effective, when the diffusion length is high enough.

Under applied voltage, the drawback of the flat-band solar cell becomes obvious. The voltage has to drop somewhere over the absorber layer leading to an electric field, which is opposite to the direction the charge carriers should travel. While for a p–i–n-junction solar cell the field-assisted charge extraction becomes weaker with applied voltage, in a flat-band solar cell the field hinders charge separation. This is why we consider in the following the flat-band solar cell as a paradigmatic example for a device that exhibits poor charge separation properties. In fact, some typical features that show up in the numerical simulations below are indicative in practical (but faulty) devices for problems due to insufficient contact properties.

To illustrate the basic properties of the solar-cell structures introduced in Figure 1.6, we simulated the current/voltage curves for two different mobilities μ of electrons and holes. The recombination in the device was assumed to be dominated by one defect in the middle of the device with a Shockley–Read–Hall lifetime (see Section 1.2.5) $\tau = 100$ ns for electrons and holes. In addition, we assumed a surface recombination velocity $S = 10^5$ cm/s for the holes at the electron contact ($x = 0$) and the electrons at the hole contact ($x = 500$ nm). The results are presented in Figure 1.7a ($\mu = 10^{-1}$ cm^2/Vs) and Figure 1.7b ($\mu = 10^1$ cm^2/Vs) demonstrating that short-circuit current density is substantially decreased when turning from the high to the low mobility. The fill factor FF, that is,

$$FF = \frac{P_{mpp}}{J_{sc} V_{oc}} \tag{1.12}$$

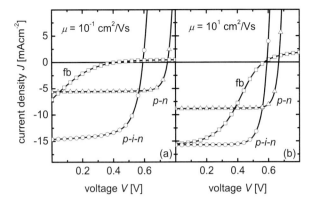

Figure 1.7 Simulated current/voltage curves of the three solar-cell geometries introduced in Figure 1.6 for two charge carrier mobilities, namely (a) $\mu = 10^{-1}\,\text{cm}^2/\text{V s}$ and (b) $\mu = 10^{1}\,\text{cm}^2/\text{V s}$. The main influence of a decreased mobility is a lower short-circuit current for the p–n-junction solar cell and a lower fill factor for the p–i–n-junction and the flat-band solar cell, which feature voltage-dependent charge carrier collection.

is for both cases relatively high. Here, P_{mpp} is the maximum power density that can be extracted from the device. Thus, the fill factor can be understood as the largest rectangle that fits between a J/V curve and the axis divided by the rectangle with the sides J_{sc} and V_{oc}.

For the p–n-junction solar cell, the open-circuit voltage also changes with mobility, which is due to increased surface recombination at high mobilities. This effect is relatively pronounced in this simulation since the complete thickness of the absorber is rather thin (500 nm) and the surface recombination velocity is assumed to be rather high ($S = 10^5\,\text{cm/s}$). The same effect also explains the relatively low short-circuit current density of the p–n-junction geometry since there is no built-in field or heterojunction that keeps the minorities away from the "wrong" contact (at $x = 500$ nm in Figure 1.6). Thus, the p–n-junctions solar cell is relatively sensitive to the lack of selectivity of the back contact, that is, to surface recombination.

The p–i–n-junction has a much higher short-circuit current density changing also very little upon decrease in mobility from $\mu = 10^1\,\text{cm}^2/\text{V s}$ (Figure 1.7b) to $\mu = 10^{-1}\,\text{cm}^2/\text{V s}$ (Figure 1.7a). However, the fill factor decreases because of the reduced capability of the device to collect all charge carriers when under forward voltage bias the built-in field is reduced (cf. Figure 1.6e). This phenomenon is called bias-dependent carrier collection. Furthermore, the open-circuit voltage of the p–i–n-cell is lower than that of its p–n-type counterpart. Nevertheless, the p–i–n-structure delivers the highest output power under the assumed, unfavorable conditions, namely relative low carrier mobilities and high surface recombination velocities.

The flat-band solar cell has the most remarkable J/V curves. The J/V curves in both mobility cases are partly bended, leading to extremely low fill factors. This so-called S-shaped characteristic becomes more pronounced in the low-mobility case. Note that, in practice, such behavior is common to devices with faulty contacts and consequent insufficient carrier separation capabilities.

1.2.5
Nonradiative Recombination

The open-circuit voltage V_{oc} of any solar cell is considerably lower than its radiative limit, implying that nonradiative recombination mechanisms like Auger recombination [12] or recombination via defects, which is usually called Shockley–Read–Hall recombination [13, 14], dominate real-world devices. Figure 1.8 compares the three main recombination mechanisms. In case of radiative recombination (a), the excess energy of the recombining electron–hole pair is transferred to a photon. In case of (b) Auger recombination [15, 16], the excess energy serves to accelerate a third charge carrier (electron or hole), which thermalizes rapidly by emitting phonons. The third recombination mechanism is Shockley–Read–Hall recombination via states in the forbidden gap. Here, the excess energy is also transferred to phonons leading to an increase in the lattice temperature of the absorber.

In very high-quality devices from monocrystalline silicon, the recombination will be limited by Auger recombination and surface recombination. That means, even with a perfect bulk material without any defects, recombination in an indirect semiconductor like silicon will most likely not be limited by radiative recombination. However, typical thin-film solar cells are made from amorphous or microcrystalline semiconductors that are far from defect-free. Here, the most important recombination mechanism is recombination via states in the forbidden gap. These states can be for instance due to defects like dangling bonds [17] or due to band tails [18–20] arising from disorder in the material. Especially in amorphous Si, there is not only a single state in the band gap as indicated in Figure 1.8c but a complete distribution of

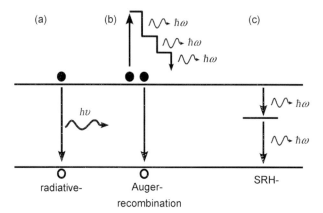

Figure 1.8 Overview over the three basic recombination mechanisms for photogenerated excess carriers in a semiconductor. The excess energy is either transferred to (a) a photon, (b) kinetic energy of an excess electron or hole, or (c) phonons. For case (b), the so-called Auger recombination, the kinetic energy of the electron is lost by collisions with the lattice, which heats up. In case (c), the emission of phonons becomes possible by the existence of states in the forbidden gap. This recombination mechanism is called Shockley–Read–Hall recombination.

states. The theory and modeling of such distributions of defects will be described later in Chapter 19, while we want to restrict ourselves here to some simple examples with a single defect state.

To visualize the influence of increased recombination rates on the current/voltage curve of solar cells, we made some numerical simulations using a very simple model for the recombination. This model assumes recombination via a defect in the middle of the forbidden gap, assuming capture cross sections σ for electrons and holes to be the same. Then the recombination rate according to Shockley–Read–Hall statistics is

$$R = \frac{np - n_i^2}{(n+p)\tau} \quad (1.13)$$

where τ is called the lifetime of the charge carrier. This lifetime depends on the density N_T of defect states, the capture cross section σ, and the thermal velocity v_{th} via

$$\tau = (v_{th}\sigma N_T)^{-1} \quad (1.14)$$

Figure 1.9 shows the current/voltage curves of a (a) p–i–n-junction solar cell and (b) a p–n-junction solar cell for a constant mobility $\mu = 1\,\text{cm}^2/\text{Vs}$ (for electrons and holes) and with a varying lifetime τ = 1 ns, 10 ns, 100 ns, 1 μs, and 10 μs. All other parameters are defined in Table 1.1. It is important to note that a reduction in the lifetime has a different influence on the two geometries, which is in accordance with what we already observed when varying the mobility. For p–i–n-junction solar cells, a decrease in the lifetime leads to a decrease in open-circuit voltage, in fill factor, and in short-circuit current density. In contrast, the p–n-junction solar cell does not suffer from a decreased fill factor. The shape of the J/V curves stays practically the same. For low lifetimes (and/or low mobilities), the charge carrier collection in p–i–n-junction solar cells is voltage dependent. For p–n-junction solar cells, this is not the case. But

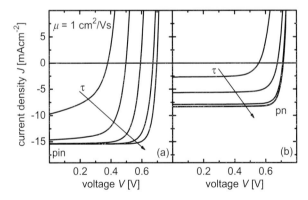

Figure 1.9 Current/voltage curves of (a) a p–i–n-junction solar cell and (b) a p–n-junction solar cell for a constant mobility $\mu = 1\,\text{cm}^2/\text{Vs}$ (for electrons and holes) and with a varying lifetime τ = 1 ns, 10 ns, 100 ns, 1 μs, and 10 μs. All other parameters are defined in Table 1.1. An increasing lifetime helps to increase V_{oc} in both cases up to the level defined by the surface recombination alone. In case of the p–i–n-junction solar cell, the FF increases as well.

Table 1.1 Summary of all parameters for the simulations in this chapter that are not changed for the simulation[a].

Parameters for all simulations in this chapter	Values
Band gap E_g	1.2 eV
Effective density of states N_C, N_V for conduction and valence band, respectively	10^{20} cm^{-3}
Doping concentrations N_D, N_A in all doped layers of p–n- and p–i–n-junction solar cells	10^{19} cm^{-3}
Total thickness d	500 nm
Generation rate G	2×10^{21} cm^{-3} s^{-1}
Surface recombination velocity S	10^5 cm/s

a) The mobilities and lifetimes, which are changed, are always given in the respective figure captions.

apart from the influence, the carrier lifetime has on charge extraction, which is very similar to the effect of the mobility; the lifetime has a pronounced influence on the open-circuit voltage. The increase in V_{oc} with increasing lifetime τ, however, seems not to follow a simple relation. For high values of τ, V_{oc} saturates for both p–i–n- and p–n-junction solar cells. This saturation is due to surface recombination, which limits the maximum attainable open-circuit voltage V_{oc}.

1.3
Functional Layers in Thin-Film Solar Cells

Until now, we have discussed the photovoltaic effect, the requirements for the material properties to come close to a perfect solar cell and the possible geometries to separate and extract charge carriers. In typical crystalline silicon solar cells, nearly all these requirements and tasks have to be fulfilled by the silicon wafer itself. Charge extraction is guaranteed by diffusing phosphorus into the first several hundred nanometers of the p-type wafer to create a p–n-junction. The wafer is texture etched to obtain a light trapping effect and to decrease the reflection at the front surface. The only additional layers that are necessary are the metal grid at the front, an antireflective coating (typically from SiN_x) and the metallization at the back.

Thin-film solar cells are usually more complex devices with a higher number of layers that are optimized for one or several purposes. In general, there are two configurations possible for any thin-film solar cell as shown in Figure 1.10. The first possibility is that light enters the device through a transparent superstrate. The superstrate has to maintain the mechanical stability of the device, while at the same time being extremely transparent. The superstrate is followed by layers which are part of the front contact, followed by the absorber layer and the layers that form the back contact. The second possibility is to inverse the layer stack, starting with the front contact, the absorber, and the back contact. These layers are all deposited on top of a substrate that is now not at the illuminated side of the device. Thus, the substrate can be transparent or opaque.

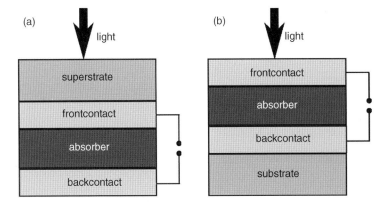

Figure 1.10 Sketch of the layer sequences to build up the system for thin-film solar cells in (a) superstrate and (b) substrate configuration. The minimum number of layers in excess of the supporting sub- or superstrate consists of the transparent and conductive front contact the absorber layer and the back contact.

Table 1.2 summarizes the roles and the requirement for the three functional layers and the sub/superstrate of thin-film solar cells. The substrate or superstrate provides mechanical stability. The functional layers are deposited onto the substrate or superstrate; thus it has to be thermally stable up to the highest temperature reached during the complete deposition process.

The front contact and back contact layers have to provide the electrical contact of the solar cell to the outside world, that is, the layers need high conductivities and must make a good electric contact to the absorber layers. In addition, the built-in field required for efficient charge extraction (especially at higher voltage bias) of a p–i–n-junction as depicted in Figure 1.6 requires doped contact layers. In devices that require efficient light trapping, usually the front and/or back contact layers are textured and have a lower refractive index than the absorber layer. Thus, the front contact layer additionally serves as an internal antireflective coating. In addition, a possible texture of the contact/absorber interface will lead to scattering of light and to increased path lengths of weakly absorbed light in the absorber layer. The back contact should have a high reflectivity so that weakly absorbed light is reflected multiple times.

Table 1.2 List of the four types of layers in a thin-film solar cell together with their specific tasks and requirements necessary for an efficient solar cell.

Layer type	Possible tasks and requirements
Substrate/superstrate	Mechanical and thermal stability, transparency (superstrate)
Front contact	Light trapping, antireflection, electrical contact, charge extraction
Absorber	Absorb light, charge extraction, low recombination
Back contact	Light trapping, high reflection, electrical contact, charge extraction

The absorber layer is central to the energy conversion process, requiring a steep rise of the absorption coefficient above the band gap, a high mobility and low recombination rates for efficient charge collection, and a high open-circuit voltage potential. In case of a p–n-junction device, the absorber layer must be moderately doped either intentionally or by intrinsic doping due to defects. In case of a p–i–n-junction device, the main absorber layer, the i-layer, should be undoped.

1.4
Comparison of Various Thin-Film Solar-Cell Types

The basic schemes of the layer stack of a thin-film solar cell, as presented in Figure 1.10, are implemented in different ways in the three most common inorganic thin-film technologies to date. These technologies are the $Cu(In,Ga)Se_2$ solar cell, the CdTe-based solar cell, and the thin-film silicon solar cell with amorphous and microcrystalline silicon absorbers. In the following, we will briefly discuss the main characteristics of these three technologies as well as the main challenges in future developments and how characterization of materials and devices can help to improve the devices. For those readers who desire a more detailed insight in the physics and technology of the different thin-film solar cells, we refer to a number of books and review articles on the topic. The physics and particularly the fabrication of all types of thin-film solar cells are discussed in Refs. [21–23], the physics of Cu-chalcopyrite solar cells in Ref. [24], the interfaces of CdS/CdTe solar cells in Ref. [25], the physics of amorphous hydrogenated silicon in Ref. [26], the physics and technology of thin-film silicon solar cells in Refs. [27–30], and the aspect of charge transport in disordered solids in Ref. [31].

1.4.1
$Cu(In,Ga)Se_2$

1.4.1.1 Basic Properties and Technology
Solar cells with an absorber layer made from $Cu(In,Ga)Se_2$ are currently the state of the art of the evolution of Cu-based chalcopyrites for use as solar cells. Heterojunctions between CdS and Cu_2S were the basis for first approaches for thin-film solar cells since the 1950s [32–35]. In 1974 first work on the light emission and light absorption of $CdS/CuInSe_2$ diodes was published [36–38]. While $CuInSe_2$ was not further considered for applications as a near-infrared light-emitting diode, its high absorption coefficient and electronically rather passive defects make it a perfect choice for use as a microcrystalline absorber material. Inclusion of Ga atoms on the In lattice site such that the ratio of $Ga/(Ga + In)$ becomes around 20% shifts the band gap from 1.04 eV to around 1.15 eV, which is nearly perfect for a single-junction cell (cf. Figure 1.3). Today, thin-film solar cells with a $Cu(In,Ga)Se_2$ absorber layer are the most efficient thin-film technology with laboratory efficiencies up to 20% [39].

The classical layer stack for this type of solar cell is shown in Figure 1.11a. It consists of a typically 1 μm thick Mo layer deposited on a soda-lime glass substrate

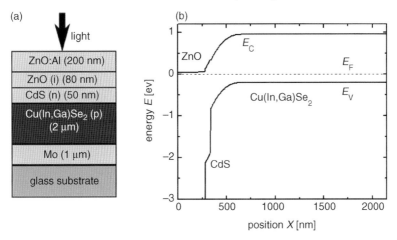

Figure 1.11 (a) Layer-stacking sequence and (b) energy band diagram of a typical ZnO/CdS/Cu(In, Ga)Se$_2$ heterojunction solar cell.

and serving as the back contact for the solar cell. Then, Cu(In,Ga)Se$_2$ is deposited on top of the Mo back electrode as the photovoltaic absorber material. This layer has a thickness of 1–2 μm. The heterojunction is then completed by chemical bath deposition (CBD) of CdS (typically 50 nm) and by the sputter deposition of a nominally undoped (intrinsic) i-ZnO layer (usually of thickness 50–70 nm) and then a heavily doped ZnO:Al window layer.

The Cu(In,Ga)Se$_2$ absorber material yielding the highest efficiencies is prepared by coevaporation from elemental sources. The process requires a maximum substrate temperature of ∼550 °C for a certain time during film growth, preferably toward the end of growth. Advanced preparation sequences always include a Cu-rich stage during the growth process and end up with an In-rich overall composition in order to combine the large grains of the Cu-rich stage with the otherwise more favorable electronic properties of the In-rich composition. The first example of this kind of procedure is the so-called Boeing or *bilayer process* [40], which starts with the deposition of Cu-rich Cu(In,Ga)Se$_2$ and ends with an excess In rate to achieve a final composition that is slightly In-poor. The most successful coevaporation process is the so-called *three-stage process* [41] where first (In,Ga)$_2$Se$_3$ (likewise In, Ga, and Se from elemental sources to form that compound) is deposited at a lower temperatures (typically around 300 °C). Then Cu and Se are evaporated at an elevated temperature and finally again In, Ga, and Se to ensure the overall In-rich composition of the film even if the material is Cu-rich during the second stage.

The second class of absorber preparation routes is based on the separation of deposition and compound formation into two different processing steps. High efficiencies are obtained from absorber prepared by selenization of metal precursors in H$_2$Se [42] and by rapid thermal processing of stacked elemental layers in an Se atmosphere [43]. These sequential processes have the advantage that approved large-area deposition techniques such as sputtering can be used for the deposition of the materials. The Cu(In,Ga)Se$_2$ film formation then requires a second step, the selenization step typically performed at similar temperatures as the coevaporation process. Both absorber preparation routes are now used in industrial application.

Important for the growth of the Cu(In,Ga)Se$_2$ absorber is the active role of Na during absorber growth. In most cases, the Na comes from the glass substrate and diffuses into the absorber [44]. But there are also approaches where Na is incorporated by the use of Na-containing precursors [45, 46]. The explanations for the beneficial impact of Na are manifold, and it is most likely that the incorporation of Na in fact results in a *variety* of consequences (for a review see Ref. [47]).

1.4.1.2 Layer-Stacking Sequence and Band Diagram of the Heterostructure

Figure 1.11 displays the layer-stacking sequence (a) and the band diagram of the ZnO/CdS/Cu(In,Ga)Se$_2$ heterojunction (b). The back contact consists of a sputtered Mo layer. In excess of producing a functional, conductive contact, proper preparation of this layer is also important for adhesion of the absorber film and, especially, for the transport of Na from the glass substrate through the Mo layer into the growing absorber. A homogeneous and sufficient supply of Na depends much on the microstructure of this layer. In contrast, if Na is supplied from a precursor additional blocking layers prevent out-diffusion of Na from the glass. Quantitative chemical depth profiling as described in Chapter 16 is a decisive tool to shed more light into the role of Na and on its way how it is functional during absorber growth.

The Cu(In,Ga)Se$_2$ absorber material grown on top of the Mo contact is slightly p-type doped by native, intrinsic defects, most likely Cu-vacancies [48]. However, the net doping is a result of the difference between the acceptors and an almost equally high number of intrinsic donors [49, 50]. Thus, the absorber material is a highly compensated semiconductor. Furthermore, the material features electronic metastabilities like persistent photoconductivity [51], which are theoretically explained by different light-induced defect relaxations [52]. However, final agreement on the observed metastability phenomena has not yet achieved, leaving an urgent need for further theoretical and experimental access to the complex defect physics of Cu(In,Ga)Se$_2$ (for a review of the present status, see Ref. [48]). Some experimental and theoretical methods helpful for further research are outlined in Chapters 7 and 18 of this book.

Another puzzle is the virtual electronic inactivity of most grain boundaries in properly prepared polycrystalline Cu(In,Ga)Se$_2$ absorbers being one essential ingredient for the high photovoltaic efficiencies delivered by this material. A discussion of the present status is given in Ref. [53]. A great part of the structural analysis methods discussed in this book (Chapters 11–14) describes tools indispensable for a better understanding of the microstructure of the Cu(In,Ga)Se$_2$ absorber material.

The surface properties of Cu(In,Ga)Se$_2$ thin films are especially important, as this surface becomes the active interface of the completed solar cell. The free surface of as-grown Cu(In,Ga)Se$_2$ films exhibits a very unique feature, namely a widening of the band gap with respect to the bulk of the absorber material [54, 55]. This band-gap widening results from a lowering of the valence band and is effective in preventing interface recombination at the absorber buffer interface [56, 57]. This surface layer has an overall Cu-poor composition and a thickness of 10–30 nm [58]. Understanding the interplay between this surface layer and the subsequently deposited buffer layer is one of the decisive challenges for the present and future research.

The 50 nm thick CdS buffer layer is in principle too thin to complete the heterojunction. In fact the role of the CdS buffer in the layer system is still somewhat obscure. It is however clear that the undoped (i) ZnO layer is also a vital part of a successful buffer/window combination. Furthermore, both interfaces of the CdS interlayer to the Cu(In,Ga)Se$_2$ absorber and to the (i) ZnO play a vital role [59]. Under standard preparation conditions, the alignment of the conduction bands at both interfaces is almost flat [60] such that neither barrier for electron transport occurs nor is the band diagram distorted in a way to enhance interface recombination. However, it turns out that a replacement of CdS by a less cumbersome layer is not straightforward. Although while promising materials like In(OH,S), Zn(OH,S), In$_2$Se$_3$, ZnSe, ZnS (for an overview see Ref. [61]) mostly in combination with standard ZnO double layer have been investigated in some detail, no conclusive solution has been found despite reported efficiencies of 18% using ZnS buffer layers [62]. Recent research [63] focuses at combinations of Zn(S,O,OH)/ZnMgO replacing the traditional CdS/(i) ZnO combination. Alternative buffer layers like ZnS also have the advantage of a higher band-gap energy $E_g = 3.6$ eV compared to that of CdS $E_g = 2.4$ eV. By the higher E_g, parasitic absorption in the buffer layer is restricted to a much narrower range and the short circuit current density in Cd-free cells can exceed that of standard devices by up to 3 mA/cm^2 [63]. However, all technological improvements rely on our scientific understanding of the physics, chemistry, and microstructure of the heterointerfaces involved in the solar cell. Surface analysis methods as those discussed in Chapters 13 and 15 have already contributed much to our present knowledge and provide the promise to deepen it further.

1.4.2
CdTe

1.4.2.1 Basic Properties and Technology

Just as the CdS/Cu(In,Ga)Se$_2$ solar cell, also the CdS/CdTe devices are descendants of the first CdS/Cu$_2$S solar cells. In the mid-1960s, first experiments with tellurides were performed. Efficiencies between 5% and 6% were obtained for CdTe/CuTe$_2$ devices [64, 65]. Since Cu diffusion led to instabilities in these devices, instead CdS and CdTe were combined to form a p–n-heterojunction with efficiencies around 6% [66]. Thirty years later, the efficiency has increased to above 16% [67]. In addition, CdTe solar modules represent the by far most successful photovoltaic thin-film technology with a share of almost 10% in the global photovoltaic market (data from 2008).

One decisive reason for this success is the relative ease which with CdTe solar cells and modules are prepared. Several types of transparent conductive oxides (TCO) are used as front contact materials for the preparation of CdTe solar cells, SnO$_2$:F and In$_2$O$_3$:F being the most common ones. Both materials, CdS and CdTe, forming the heterojunction of the solar cell, are grown with similarly fast and reliable methods, including closed-space sublimation, spraying, screen printing followed by sintering, and electrodeposition. Since CdS grows natively as n-type and CdTe as p-type material, the p–n-heterojunction forms automatically.

However, in order to improve the device efficiency substantially, an additional step, the $CdCl_2$-activation, is necessary. A vapor-based approach is most useful with regard to industrial applications [68]. The activation step leads to an intermixing of CdS and CdTe close to the heterointerface and to the formation of a Cu(Te,S) compound. In some cases, recrystallization of the CdTe film was observed after $CdCl_2$ treatment [69]. In any case, the intermixing process is decisive for the improvement of the device performance.

The major challenge for reliable manufacture of efficient devices is to produce a stable and ohmic back contact to the CdTe absorber with its high electron affinity. Often, back contacts are made with materials that contain Cu, such as Cu_2Te, ZnTe:Cu, or HgTe:Cu, enabling a relatively low contact resistance. However, Cu-diffusion in CdTe is fast and extends deeply into the absorber, thereby affecting considerably the stability of the device [70]. Cu-free alternative contact materials embrace, for example, Sb_2Te_3 [71]. Often, an etching step is used to produce a Te-rich interlayer, providing higher p-type doping and, consequently, a reasonably low-ohmic contact [72].

1.4.2.2 Layer-Stacking Sequence and Band Diagram of the Heterostructure

From the point of view of the layer-stacking sequence and the band diagram shown in Figure 1.12a and b, the CdS/CdTe heterostructure is quite similar to those of the CdS/Cu(In,Ga)Se$_2$ heterostructure given in Figure 1.11. One obvious difference is the low doping density of the CdTe absorber, making the device somewhat a hybrid between a p–i–n- and a p–n-junction. The built-in field almost stretches from the heterointerface toward the back contact. As we have already seen, such a configuration is helpful

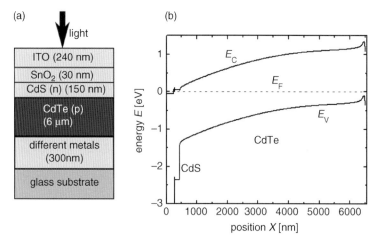

Figure 1.12 (a) Layer stacking sequence and (b) energy-band diagram of a typical CdTe-based solar cell following Ref. [73].

for carrier collection but has the drawback of delivering lower open circuit voltages than a p–n-type device with a relatively narrow space charge region.

The band diagram at the back contact features a highly p-doped region due to Cu-indiffusion or due to the formation of a Te-rich interlayer. This leads to a relatively thin yet high barrier for holes. Thus, the electrical contact is achieved via tunneling from the absorber into the back metal. The modeling of CdTe solar cells, including a proper approach to the back contact, which usually is by far not perfectly ohmic, represents a major challenge as discussed in Chapter 19.

The average grain sizes in the polycrystalline CdTe absorbers range from 1 to 2 mm, thus somewhat larger than in $Cu(In,Ga)Se_2$. However, the grain boundaries are considerably more electronically active than in $Cu(In,Ga)Se_2$. For instance, photocurrent concentration along grain boundaries [74] indicates type inversion of grain boundaries in CdTe. This could be helpful for current collection (along the grain boundaries) but also implies losses for the open circuit voltage. Again, connecting microstructural analysis with highly resolved measurements of electronic properties by scanning techniques as described in Chapters 11 and 12 will clarify the picture in the future.

The favorable, flat conduction band alignment at the CdTe/CdS as well as at the CdS/TCO interface as featured by Figure 1.12b is similar to the situation for $Cu(In,Ga)Se_2$ devices. In CdTe solar cells this is basically a result of the $CdCl_2$ activation process and of intermixing [73].

1.4.3
Thin-Film Silicon Solar Cells

1.4.3.1 Hydrogenated Amorphous Si (a-Si:H)
Central to the working principle of semiconductors is the forbidden energy gap derived from the periodicity of the crystal lattice. However, it is exactly this strict periodicity that is lacking in amorphous semiconductors, which have a short-range order but no long-range order as their crystalline counterparts. The structural disorder caused by variations in bond lengths and angles has several implications for the electronic and optical properties of amorphous materials. The most important feature is the peculiar density of electronic states in amorphous silicon featuring localized states close to the band edges that arise from disorder and a distribution of deep states due to unpassivated, that is, dangling, bonds. In addition the word band gap is no longer adequate in amorphous semiconductors. Instead, an optical gap is defined from the onset of absorption, while a mobility gap is defined as the approximate demarcation line between localized and extended states [75]. Despite the fact that the mobility gap is not a forbidden zone for electrons but instead full of localized states, amorphous silicon still proves to be a useful material for thin-film devices like solar cells, photodetectors, and transistors [76].

While first crystalline silicon solar cells with reasonable efficiencies of about $\eta = 6\%$ were already developed in 1954 [77], the research on amorphous silicon first needed two breakthroughs before the fabrication of the first amorphous silicon solar cells in 1976 became possible [78]. The first breakthrough was the realization that the addition of considerable amounts of hydrogen helped to passivate dangling bonds in the amorphous material thereby leading to sufficiently low defect densities that hydrogenated amorphous silicon showed some of the important characteristics of useful semiconductors like dopability and photoconductivity [79]. The second breakthrough was the successful doping of amorphous silicon [80].

Despite the defect passivation with hydrogen, the defect densities in a-Si:H are still relatively high with diffusion lengths between 100 and 300 nm [81]. In doped a-Si:H layers, the defect density is two or three orders of magnitude higher and the diffusion length is accordingly even lower. Thus, a p–n-junction as used in crystalline silicon but also in Cu(In,Ga)Se$_2$ as well as CdTe solar cells would not work for a-Si:H, since the diffusion length is too low. Since the absorber thickness cannot be made much thinner than the diffusion length due to the large losses because of insufficient light absorption, a p–i–n-junction configuration has to be used. The first advantage is that most of the absorber layer consists of intrinsic a-Si:H with its higher carrier lifetime than doped a-Si:H. The second advantage is that the built-in field helps with extracting charge carriers as shown in Figure 1.13. The advantage of the p–i–n-configuration is that the electron and hole concentrations are similar in a relatively large portion of the absorber volume, which increases defect recombination, which is automatically highest, when electron and hole concentrations are equal.

Figure 1.13 shows the typical layer stack and band diagram of an a-Si:H p–i–n-type solar cell. Usually a superstrate configuration is used, although a substrate configuration is also possible. In the latter case, the solar cell is deposited on the substrate starting with the back contact and the n-type layer. Thus, such a solar cell is called nip

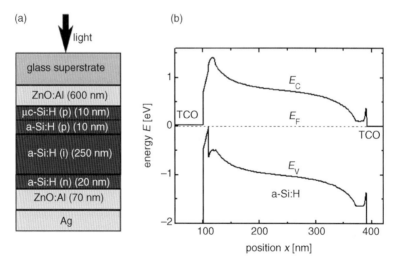

Figure 1.13 (a) Stacking sequence and (b) band diagram of a typical a-Si:H p–i–n solar cell. The main absorber layer is intrinsic while the built-in field is due to the thin doped silicon layers. Due to the asymmetric mobilities between electrons and holes, the p-type layers will always be on the illuminated side, ensuring that the holes with their lower mobility have the shorter way to the contacts.

solar cell, where n–i–p represents the deposition order. In both cases of a substrate or superstrate configuration, the illumination is always from the p-side. This is due to the lower mobility of holes in a-Si:H. It is therefore beneficial to have the hole contact on the illuminated side, where the generation rate is higher. With the hole contact on the illuminated side, the distance the slowest carrier has to travel to the contacts is minimized.

1.4.3.2 Metastability in a-Si:H: The Staebler–Wronski Effect

Shortly after the first reports on a-Si:H solar cells, Staebler and Wronski published their findings on metastability in a-Si:H [82]. Under illumination, the conductivity of a-Si:H degrades but can be restored by annealing at temperatures of about 425 K. This degradation effect is known as the Staebler–Wronski effect (SWE). The metastable behavior is ascribed to the light-induced creation of additional defects. It is generally accepted that these metastable defects are additional dangling bonds that act like recombination centers in the material and that these dangling bonds are created by the breaking of weak or strained Si–Si bonds. Hydrogen plays an important role in the metastable behavior of a-Si:H; however, there is no consensus on the exact mechanisms involved and the role of hydrogen in the SWE [83–86]. The creation of additional recombination centers affects a-Si:H-based solar cells to such a degree that the SWE is a severe limitation for the application of a-Si:H in single-junction solar cells.

1.4.3.3 Hydrogenated Microcrystalline Silicon (μc-Si:H)

As can be seen by comparison with Figure 1.3b, the high optical gap of a-Si:H of approximately 1.75 eV (the exact value depending on the definition and on the hydrogen content) is too high for a single-junction solar cell. It was, therefore, an important discovery in thin-film silicon solar-cell research to find a way to prepare hydrogenated microcrystalline silicon (μc-Si:H) with approximately the same band gap as crystalline silicon (E_g (c-Si) = 1.12 eV) that had a sufficient quality for use in solar cells. First μc-Si:H layers were deposited in the 1960s [87] and successful doping in the 1970s [88]. However, the material had insufficient electronic quality for use in solar cells. The use of gas purifiers in the 1990s by the Neuchâtel group made the fabrication of μc-Si:H layers with sufficiently low oxygen contents [89–91] and the successful fabrication of first μc-Si:H solar cells with reasonable efficiencies possible [92, 93].

1.4.3.4 Micromorph Tandem Solar Cells

One possibility to overcome [94] the efficiency limit of SQ is the use of a multijunction solar cell with absorber layers having different band gaps. The highest band-gap absorber should be on the illuminated side such that all high-energy photons are absorbed by the absorber with the higher band gap and the low-energy photons are absorbed by the cell or the cells with the lower band gap(s). If every absorber has its own p–n- or p–i–n-junction, then they can be deposited on top of each other such that one obtains two or more series connected solar cells on top of each other. This approach minimizes the losses due to thermalization of carriers and due to the transparency of any solar cell for photons with energies below the band gap of the absorber. With a similar approach as discussed in Section 1.1, the efficiency [95] and the optimal band-gap combinations can be calculated for multijunction solar cells in general and tandem (i.e., two-junction solar cells) in particular [96, 97]. It is a fortunate coincidence that the optimum combination for a tandem solar cell is close to the actual band gaps of amorphous (E_g (a-Si:H) ≈ 1.75 eV) and microcrystalline silicon (E_g (μc-Si:H) ≈ 1.2 eV). Although in principle efficiencies above the SQ limit for single-junction solar cells are possible with such a configuration, in reality the efficiencies are much lower than the SQ limit and even lower than efficiencies of real crystalline Si single-junction solar cells. Nevertheless, the tandem cell made from a-Si:H and μc-Si:H has achieved slightly higher efficiencies than either of the single-junction devices (see Ref. [29] for an overview). For these thin-film tandem cells with their relatively low mobilities, a second motivation arises for the use of tandem solar cells. Since the built-in field decreases for increasing thickness and since the charge collection becomes increasingly difficult with increasing distance to the contacts, thin solar cells have higher fill factors than thicker solar cells. The tandem approach is a useful way to keep individual cell thicknesses low and at the same time have a higher total thickness and a better absorptance.

1.5
Conclusions

Despite more than 30 years of research invested in each of the three thin-film solar-cell technologies considered here, a large series of questions has still to be answered. The need for more "know-why" in addition to the available "know-how" is urged by the responsibility of scientists toward a steadily growing industry and toward a world in need for clean energy. Fortunately, more and more specialists for sophisticated physical and chemical analysis methods enter the field and help improving our common understanding as well as improving our technology. The most satisfying answers always will arise from a combination of a solid understanding of the photovoltaic principles with the results from various methods analyzing the electronic, chemical, and structural properties of all the layers and interfaces in the device.

References

1 Shockley, W. and Queisser, H.J. (1961) Detailed balance limit of efficiency of p–n junction solar cells. *J. Appl. Phys.*, **32**, 510.
2 Würfel, P. (1982) The chemical potential of radiation. *J. Phys. C*, **15**, 3967–3985.
3 Bridgman, P.W. (1928) Note on the principle of detailed balancing. *Phys. Rev.*, **31**, 101.
4 Rau, U. and Werner, J.H. (2004) Radiative efficiency limits of solar cells with lateral band-gap fluctuations. *Appl. Phys. Lett.*, **84**, 3735–3737.
5 Mattheis, J., Werner, J.H., and Rau, U. (2008) Finite mobility effects on the radiative efficiency limit of pn-junction solar cells. *Phys. Rev. B*, **77**, 085203.

6 Rau, U. (2007) Reciprocity relation between photovoltaic quantum efficiency and electroluminescent emission of solar cells. *Phys. Rev. B*, **76**, 085303.

7 Kirchartz, T. and Rau, U. (2008) Detailed balance and reciprocity in solar cells. *phys. status solidi (a)*, **205**, 2737–2751.

8 Green, M.A. (2002) Lambertian light trapping in textured solar cells and light-emitting diodes: analytical solutions. *Prog. Photovolt. Res. Appl.*, **10**, 235.

9 Mattheis, J. (2008) Mobility and homogeneity effects on the power conversion efficiency of solar cells, Dissertation, Universität Stuttgart, http://elib.uni-stuttgart.de/opus/volltexte/2008/3697/ (23.2.2010).

10 Tiedje, T., Yablonovitch, E., Cody, G.D., and Brooks, B.G. (1984) Limiting efficiency of silicon solar cells. *IEEE Trans. Electron Devices*, **ED-31**, 711.

11 Würfel, P. (2002) *Physica E*, **14**, 18–26.

12 Green, M.A., Limits on the open-circuit voltage and efficiency of silicon solar cells imposed by intrinsic auger processes. *IEEE Trans. Electron Devices*, **ED-31**, 671.

13 Hall, R.N. (1952) Electron–hole recombination in germanium. *Phys. Rev.*, **87**, 387.

14 Shockley, W. and Read, W.T. (1952) Statistics of the recombinations of holes and electrons. *Phys. Rev.*, **87**, 835.

15 Auger, P. (1925) Sur L'effet photoélectrique composé. *J. Physique et Le Radium*, **6**, 205.

16 Meitner, L. (1922) Über die β-Strahl-Spektra und ihren Zusammenhang mit der γ-Strahlung. *Zeitschrift für Physik A Hadrons and Nuclei.*, **11**, 35.

17 Dersch, H., Stuke, J., and Beichler, J. (1981) Light-induced dangling bonds in hydrogenated amorphous silicon. *Appl. Phys. Lett.*, **38**, 456.

18 Tiedje, T., Cebulka, J.M., Morel, D.L., and Abeles, B. (1981) Evidence for exponential band tails in amorphous silicon hydride. *Phys. Rev. Lett.*, **46**, 1425.

19 Schiff, E.A. (1981) Trap-controlled dispersive transport and exponential band tails in amorphous silicon. *Phys. Rev. B*, **24**, 6189.

20 Fedders, P.A., Drabold, D.A., and Nakhmanson, S. (1998) Theoretical study on the nature of band tail states in amorphous Si. *Phys. Rev. B*, **58**, 15624.

21 Poortmans, J. and Arkhipov, V. (eds) (2006) *Thin Film Solar Cells – Fabrication, Characterization and Applications*, John Wiley & Sons, Ltd, Chichester, UK.

22 Hamakawa, Y. (ed.) (2004) *Thin-Film Cells: Next Generation Photovoltaics and Its Applications*, Springer, Berlin.

23 Chopra, K.L., Paulson, P.D., and Dutta, V. (2004) Thin-film solar cells: An overview. *Prog. Photovolt. Res. Appl.*, **12**, 69.

24 Siebentritt, S. and Rau, U. (eds) (2006) *Wide-Gap Chalcopyrites*, Springer, Berlin.

25 Jaegermann, W., Klein, A., and Mayer, T. (2009) Interface engineering of inorganic thin-film solar cells – materials-science challenges for advanced physical concepts. *Adv. Mater.*, **21**, 4196.

26 Street, R.A. (1991) *Hydrogenated Amorphous Silicon*, Cambridge University Press, Cambridge.

27 Zeman, M. and Schropp, R.E.I. (1998) *Amorphous and Microcrystalline Silicon Solar Cells: Modeling, Materials and Device Technology*, Kluwer Academic Publishers, Norwell.

28 Shah, A.V., Schade, H., Vanecek, M., Meier, J., Vallat-Sauvain, E., Wyrsch, N., Kroll, U., Droz, C., and Bailat, J. (2004) Thin-film silicon solar cell technology. *Prog. Photovolt. Res. Appl.*, **12**, 113.

29 Schropp, R.E.I., Carius, R., and Beaucarne, G. (2007) Amorphous silicon, microcrystalline silicon, and thin-film polycrystalline silicon solar cells. *MRS Bull.*, **32**, 219.

30 Deng, X. and Schiff, E.A. (2003) Amorphous Silicon-based solar cells, in *Handbook of Photovoltaic Science and Engineering* (eds A. Luque and S. Hegedus) John Wiley & Sons Ltd, Chichester, UK, p. 505.

31 Baranovski, S. (ed.) (2006) *Charge Transport in Disordered Solids – with Applications in Electronics*, John Wiley & Sons, Ltd, Chichester, UK.

32 Reynolds, D.C., Leies, G., Antes, L.L., and Marburger, R.E. (1954) Photovoltaic effect in cadmium sulfide. *Phys. Rev.*, **96**, 533.

33 Böer, K.W. (1976) Photovoltaic effect in CdS–Cu$_2$S heterojunctions. *Phys. Rev. B*, **13**, 5373.

34 Böer, K.W. and Rothwarf, A. (1976) Materials for solar photovoltaic energy conversion. *Annu. Rev. Mater. Sci.*, **6**, 303.

35 Pfisterer, F. (2003) The wet-topotaxial process of junction formation and surface treatments of Cu_2S–CdS thin-film solar cells. *Thin Solid Film.*, **431–432**, 470.

36 Migliorato, P., Tell, B., Shay, J.L., and Kasper, H.M. (1974) Junction electroluminescence in $CuInSe_2$. *Appl. Phys. Lett.*, **24**, 227.

37 Wagner, S., Shay, J.L., Migliorato, P., and Kasper, H.M. (1974) $CuInSe_2$/CdS heterojunction photovoltaic detectors. *Appl. Phys. Lett.*, **25**, 434.

38 Shay, J.L., Wagner, S., and Kasper, H.M. (1975) Efficient $CuInSe_2$/CdS Solar Cells. *Appl. Phys. Lett.*, **27**, 89.

39 Repins, I., Contreras, M.A., Egaas, B., DeHart, C., Scharf, J., Perkins, C.L., To, B., and Noufi, R. (2008) 19.9%-efficient ZnO/CdS/CuInGaSe$_2$ solar cell with 81.2% fill factor. *Prog. Photovolt.: Res. Appl.*, **16**, 235.

40 Mickelsen, R.A. and Chen, W.S. (1980) High photocurrent polycrystalline thin-film CdS/$CuInSe_2$ solar cell. *Appl. Phys. Lett.*, **36**, 371–373.

41 Gabor, A.M., Tuttle, J.R., Albin, D.S., Contreras, M.A., Noufi, R., Jensen, D.G., and Hermann, A.M. (1994) High-efficiency $CuIn_xGa_{1-x}Se_2$ solar cells from $(In_xGa_{1-x})_2Se_3$ precursors. *Appl. Phys. Lett.*, **65**, 198–200.

42 Binsma, J.J.M. and Van der Linden, H.A. (1982) Preparation of thin $CuInS_2$ films via a two-stage process. *Thin Solid Film.*, **97**, 237–243.

43 Probst, V., Karg, F., Rimmasch, J., Riedl, W., Stetter, W., Harms, H., and Eibl, O. (1996) Advanced stacked elemental layer progress for Cu(InGa)Se$_2$ thin film photovoltaic devices. *Mater. Res. Soc. Symp. Proc.*, **426**, 165–176.

44 Stolt, L., Hedström, J., Kessler, J., Ruckh, M., Velthaus, K.O., and Schock, H.W. (1993) ZnO/CdS/$CuInSe_2$ thin-film solar cells with improved performance. *Appl. Phys. Lett.*, **62**, 597–599.

45 Holz, J., Karg, F., and v. Phillipsborn, H. (1994) The effect of substrate impurities on the electronic conductivity in CIGS thin films. Proc. 12th. European Photovoltaic Solar Energy Conf., Amsterdam, H. S. Stephens & Associates, Bedford, pp. 1592–1595.

46 Nakada, T., Iga, D., Ohbo, H., and Kunioka, A. (1997) Effects of sodium on Cu(In,Ga)Se$_2$-based thin films and solar cells. *Jpn. J. Appl. Phys.*, **36**, 732–737.

47 Rocket, A. (2005) The effect of Na in polycrystalline and epitaxial single-crystal $CuIn_{1-x}Ga_xSe_2$. *Thin Solid Films*, **480**, 2–7.

48 Siebentritt, S., Igalson, M., Persson, C., and Lany, S. (2010) The electronic structure of chalcopyrite-bands, point defects and grain boundaries. *Progr. Photovolt. Res. Appl.* (published online, doi: 10.1002/pip.936)

49 Dirnstorfer, I., Wagner, M., Hofmann, D.M., Lampert, M.D., Karg, F., and Meyer, B.K. (1998) Characterization of CuIn(Ga)Se$_2$ thin films – II. In-rich layers. *phys. status solidi (a)*, **168**, 163–175.

50 Bauknecht, A., Siebentritt, S., Albert, J., and Lux-Steiner, M.C. (2001) Radiative recombination via intrinsic defects in $Cu_xGa_ySe_2$. *J. Appl. Phys.*, **89**, 4391–4400.

51 Rau, U., Schmitt, M., Parisi, J., Riedl, W., and Karg, F. (1998) Persistent photoconductivity in Cu(In,Ga)Se$_2$ heterojunctions and thin films prepared by sequential deposition. *Appl. Phys. Lett.*, **73**, 223–225.

52 Lany, S. and Zunger, A. (2008) Intrinsic DX centers in ternary chalcopyrite semiconductors. *Phys. Rev. Lett.*, **100**, 016401.

53 Rau, U., Taretto, K., and Siebentritt, S. (2009) Grain boundaries in Cu(In,Ga)(Se,S)$_2$ thin-film solar cells. *Appl. Phys. A*, **96**, 221–234.

54 Schmid, D., Ruckh, M., Grunwald, F., and Schock, H.W. (1993) Chalcopyrite/defect chalcopyrite heterojunctions on basis of $CuInSe_2$. *J. Appl. Phys.*, **73**, 2902–2909.

55 Morkel, M., Weinhardt, L., Lohmüller, B., Heske, C., Umbach, E., Riedl, W., Zweigart, S., and Karg, F. (2001) Flat conduction-band alignment at the CdS/$CuInSe_2$ thin-film solar-cell heterojunction. *Appl. Phys. Lett.*, **79**, 4482–4484.

56 Dullweber, T., Hanna, G., Rau, U., and Schock, H.W. (2001) A new approach to high-efficiency solar cells by band gap

grading in Cu(In,Ga)Se$_2$ chalcopyrite semiconductors. *Sol. Energy Mater. Sol. Cells*, **67**, 1–4.
57 Turcu, M., Pakma, O., and Rau, U. (2002) Interdependence of absorber composition and recombination mechanism in Cu(In, Ga)(Se,S)$_2$ heterojunction solar cells. *Appl. Phys. Lett.*, **80**, 2598–2600.
58 Kötschau, I.M. and Schock, H.W. (2003) Depth profile of the lattice constant of the Cu-poor surface layer in (Cu$_2$Se)$_{1-x}$(In$_2$Se$_3$)$_x$, evidenced by grazing incidence X-ray diffraction. *J. Phys. Chem. Sol.*, **64**, 1559–1563.
59 Nguyen, Q., Orgassa, K., Koetschau, I., Rau, U., and Schock, H.W. (2003) Influence of heterointerfaces on the performance of Cu(In,Ga)Se$_2$ solar cells with CdS and In(OH$_x$,S$_y$) buffer layers. *Thin Solid Film.*, **431**, 330–334.
60 Weinhardt, L., Heske, C., Umbach, E., Niesen, T.P., Visbeck, S., and Karg, F. (2004) Band alignment at the i-ZnO/CdS interface in Cu(In,Ga)(S,Se)$_2$ thin-film solar cells. *Appl. Phys. Lett.*, **84**, 3175–3177.
61 Hariskos, D., Spiering, S., and Powalla, M. (2005) Buffer layers in Cu(In,Ga)Se$_2$ solar cells and modules. *Thin Solid Film.*, **480**, 99–109.
62 Nakada, T. and Mizutani, M. (2002) 18% efficiency Cd-free Cu(In, Ga)Se$_2$ thin-film solar cells fabricated using chemical bath deposition (CBD)-ZnS buffer layers. *Jpn. J. Appl. Phys.*, **41**, L165–L167.
63 Hariskos, D., Fuchs, B., Menner, R., Naghavi, N., Hubert, C., Lincot, D., and Powalla, M. (2009) The Zn(S,O,OH)/ZnMgO buffer in thin-film Cu(In,Ga)(Se,S)$_2$-based solar cells part II: Magnetron sputtering of the ZnMgO buffer layer for in-line co-evaporated Cu(In,Ga)Se$_2$ solar cells. *Prog. Photovolt. Res. Appl.*, **17**, 479–488.
64 Cusano, D.A. (2010) CdTe solar cells and photovoltaic heterojunctions in II–VI compounds. *Solid State Electron.*, **6**, 217.
65 Bonnet, D. (2004) The evolution of the CdTe thin film solar cell. Proc. Europ. Photov. Sol. Ener. Conf., Paris, June 2004. WIP Renewable Energies, Munich, p. 1657.
66 Bonnet, D. and Rabenhorst, H. (1972) New results on the development of a thin film p-CdTe/n-CdS heterojunction solar cell. Proc. 9th IEEE Photov. Spec. Conf., Silver Springs, MD, May 1972. IEEE, New York.
67 Wu, X., Keane, J.C., Dhere, R.G., DeHart, C., Duda, A., Gessert, T.A., Asher, S., Levi, D.H., and Sheldon, P. (2001) 16.5%-efficient CdS/CdTe polycrystalline thin-film solar cell. Proc. 17th Europ. Photov. Sol. Energy Conf. Munich, October 2001. WIP Renewable Energies, Munich, p. 995.
68 McCandless, B.E., Qu, Y., and Birkmire, R.W. (1994) A treatment to allow contacting CdTe with different conductors. Proc. 24th IEEE Photov. Spec. Conf., pp. 107–110.
69 Moutinho, H.R., Al-Jassim, M.M., Levi, D.H., Dippo, P.C., and Kazmerski, L.L. (1998) Effects of CdCl$_2$ treatment on the recrystallization and electro-optical properties of CdTe thin films. *J. Vac. Sci. Technnol.*, **16**, 1251–1257.
70 Dobson, K.D., Visoly-Fischer, I., Hodes, G., and Cahen, D. (2000) Stability of CdTe/CdS thin-film solar cells. *Sol. Energy Mater. Sol. Cells*, **62**, 295–325.
71 Romeo, N., Bosio, A., Canevari, V., and Podesta, A. (2005) Recent progress on CdTe/CdS thin film solar cells. *Sol. Energy*, **77**, 795–801.
72 Bätzner, D.L., Romeo, A., Zogg, H., Tiwari, A.N., and Wendt, R. (2000) *Thin Solid Films*, **361–362**, 463–467.
73 Fritsche, J., Kraft, D., Thissen, A., Mayer, T., Klein, A., and Jaegermann, W. (2002) Band energy diagram of CdTe thin film solar cells. *Thin Solid Film.*, **403–404**, 252–257.
74 Visoly-Fisher, I., Cohen, S.R., Gartsman, K., Ruzin, A., and Cahen, D. (2006) Understanding the beneficial role of grain boundaries in polycrystalline solar cells from single-grain-boundary scanning probe microscopy. *Adv. Funct. Mater.*, **16**, 649–660.
75 Pieters, B.E., Stiebig, H., Zeman, M., and van Swaaij, R.A.C.M. (2009) Determination of the mobility gap of µc-Si:H in pin solar cells. *J. Appl. Phys.*, **105**, 044502.
76 Street, R.A. (1991) *Hydrogenated Amorphous Silicon*, Cambridge University Press, Cambridge, pp. 363–403.

77 Chapin, D.M., Fuller, C.S., and Pearson, G.L. (1954) A new silicon p–n junction photocell for converting solar radiation into electrical power. *J. Appl. Phys.*, **25**, 676.

78 Carlson, D.E. and Wronski, C.R. (1976) Amorphous silicon solar cell. *Appl. Phys. Lett.*, **28**, 671.

79 Chittick, R.C., Alexander, J.H., and Sterling, H.F. (1969) The preparation and properties of amorphous silicon. *J. Electrochem. Soc.*, **116**, 77.

80 Spear, W.E. and LeComber, P.G. (1975) Substitutional doping of amorphous silicon. *Solid State Commun.*, **17**, 1193.

81 Zeman, M. (2006) Advanced amorphous silicon solar cell technology, in *Thin Film Solar Cells – Fabrication, Characterization and Applications* (eds J. Poortmans and V. Arkhipov) John Wiley & Sons, Ltd, Chichester, UK, p. 204.

82 Staebler, D.L. and Wronski, C.R. (1977) Reversible conductivity changes in discharge-produced amorphous Si. *Appl. Phys. Lett.*, **31**, 292.

83 de Walle, C.G.V. and Street, R.A. (1995) Silicon–hydrogen bonding and hydrogen diffusion in amorphous silicon. *Phys. Rev. B.*, **51**, 10615.

84 Stutzmann, M., Jackson, W.B., and Tsai, C.C. (1986) Annealing of metastable defects in hydrogenated amorphous silicon. *Phys. Rev. B.*, **34**, 63.

85 Powell, M.J., Deane, S.C., and Wehrspohn, R.B. (2002) Microscopic mechanisms for creation and removal of metastable dangling bonds in hydrogenated amorphous silicon. *Phys. Rev. B*, **66**, 155212.

86 Branz, H. (1999) Hydrogen collision model: Quantitative description of metastability in amorphous silicon. *Phys. Rev. B*, **59**, 5498.

87 Veprek, S., Marecek, V., and Anna Selvan, J.A. (1968) The preparation of thin layers of Ge and Si by chemical hydrogen plasma transport. *Solid State Electron.*, **11**, 683.

88 Usui, S. and Kikuchi, M. (1979) Properties of heavily doped GD-Si with low resistivity. *J. Non-Cryst. Sol.*, **34**, 1.

89 Kroll, U., Meier, J., Keppner, H., Littlewood, S.D., Kelly, I.E., Giannoulès, P., and Shah, A. (1995) Origin and incorporation mechanism for oxygen contaminants in a-Si:H and mc-Si:H films prepared by the very high frequency (70MHz) glow discharge technique. *Mater. Res. Soc. Symp. Proc.*, **377**, 39.

90 Kroll, U., Meier, J., Keppner, H., Littlewood, S.D., Kelly, I.E., Giannoulès, P., and Shah, A. (1995) Origins of atmospheric contamination in amorphous silicon prepared by very high frequency (70MHz) glow discharge. *J. Vac. Sci. Technol. A*, **13**, 2742.

91 Torres, P., Meier, J., Flückiger, R., Kroll, U., Selvan, J.A.A., Keppner, H., Shah, A., Littlewood, S.D., Kelly, I.E., and Giannoulès, P. (1996) Device grade microcrystalline silicon owing to reduced oxygen contamination. *Appl. Phys. Lett.*, **69**, 1373.

92 Meier, J., Dubail, S., Flückiger, R., Fischer, D., Keppner, H., and Shah, A. (1994) Intrinsic microcrystalline silicon – a promising new thin film solar cell material. Proc. 1st World Conf. Photov. Energy Conv., Hawai, p. 409.

93 Flückiger, R. (1995) Microcrystalline silicon thin-films deposited by VHF Plasma for solar cell applications. Ph.D. thesis, Institute of Microtechnology, University of Neuchatel.

94 Green, M.A. (2001) Third generation photovoltaics: ultra-high conversion efficiency at low cost. *Prog. Photovolt. Res. Appl.*, **9**, 123.

95 Henry, C.H. (1980) Limiting efficiencies of ideal single and multiple energy gap terrestrial solar cells. *J. Appl. Phys.*, **51**, 4494.

96 de Vos, A. (1980) Detailed balance limit of the efficiency of tandem solar cells. *J. Phys. D: Appl. Phys.*, **13**, 839.

97 Coutts, T.J., Ward, J.S., Young, D.L., Emery, K.A., Gessert, T.A., and Noufi, R. (2003) Critical issues in the design of polycrystalline, thin-film tandem solar cells. *Prog. Photovolt. Res. Appl.*, **11**, 359.

Part Two
Device Characterization

2
Fundamental Electrical Characterization of Thin-Film Solar Cells

Thomas Kirchartz, Kaining Ding, and Uwe Rau

2.1
Introduction

This chapter discusses device characterization methods, that is, methods to determine the response of a solar cell to optical and electrical excitation. Thus, device characterization deals directly with the finished product and is, thus, directly related to the final goal of all research efforts, namely to produce an efficient solar cell. However, since the device is characterized as a whole, the interpretation of device measurements is complex because a large number of optical and electronic effects contribute to a relatively featureless result like the current voltage characteristics. Thus, the challenge of device characterization lies not only in the measurement of the samples but mostly in the interpretation of data, which often goes hand in hand with device simulation as described in the fourth part of this book.

A variety of device characterization methods are conceivable. The current of a solar cell can be measured as a function of applied voltage, illumination intensity, wavelength of monochromatic illumination, illumination position, and sample temperature. In addition, the solar cell can be used as a light-emitting diode and its emission can be detected as a function of position, temperature, and photon wavelength. Finally, the impedance of the device applying alternating voltages can be measured as a function of frequency or voltage. Part Two of this book will deal with all these characterization methods in three chapters, while this chapter will restrict itself to the first group of methods. The most fundamental characterization methods are those where the current of a solar cell is measured in response to a variety of illumination conditions and as a function of the applied voltage. Especially the measurement of the illuminated current/voltage (J/V) curves under standard measuring conditions is of crucial importance for the determination and comparison of the efficiency of the optoelectrical energy conversion process.

Advanced Characterization Techniques for Thin Film Solar Cells,
Edited by Daniel Abou-Ras, Thomas Kirchartz and Uwe Rau.
© 2011 Wiley-VCH Verlag GmbH & Co. KGaA. Published 2011 by Wiley-VCH Verlag GmbH & Co. KGaA.

2.2
Current/Voltage Curves

2.2.1
Shape of Current/Voltage Curves and their Description with Equivalent Circuit Models

In Chapter 1, the concept of the illuminated J/V curve and its basic features have already been discussed. We started our discussion on J/V curves with the ideal solar cell in the Shockley–Queisser limit [1]. In this case, the thermodynamic argumentation gave a J/V curve of the form

$$J = J_0 \left[\exp\left(\frac{qV}{kT}\right) - 1\right] - J_{sc} \qquad (2.1)$$

where kT/q is the thermal voltage. A J/V curve of such a shape – however with different values for the saturation current density J_0 and the short-circuit current density J_{sc} – follows also from a device simulation of a p–n- or p–i–n-junction solar cell as described in Chapter 19 as long as the recombination rate R is proportional to the product of electron and hole concentration ($R \propto np$) and as long as the carrier mobilities are sufficiently high to extract all charge carriers under all voltages under consideration. In this case, the recombination current density J_{rec}, which is equal per definition to the dark current density J_d, is simply

$$\begin{aligned}J_{rec} = J_d &= q\int_0^d R dx = q\int_0^d B(np - n_i^2)dx = qBd\left[\exp\left(\frac{qV}{kT}\right) - 1\right] \\ &= J_0\left[\exp\left(\frac{qV}{kT}\right) - 1\right]\end{aligned} \qquad (2.2)$$

The requirement of sufficiently high mobilities means also that it is reasonable to assume $np = n_i^2 \exp[qV/(kT)]$ throughout the whole device thickness d, where n_i is the intrinsic carrier concentration. In the following, this will be illustrated by numerical simulations. Figure 2.1a and b shows the simulated dark and illuminated J/V curve of such an idealized device in the case when only radiative recombination is considered. The simulated device has a thick (2.5 μm) p-type layer and a thin (200 nm) n-type layer as is typical for most p–n-junction solar cells. The dark current density J_d is a simple exponential function with a slope of q/kT as rationalized from Eq. (2.2) and the illuminated current density J_{il} is shifted by the value J_{sc} into the fourth quadrant of the coordinate system. Thus, the difference $J_{ph} = J_{il} - J_d$ between illuminated and dark J/V curve, the photocurrent, equals always the short-circuit current density J_{sc}.

Although a simple description of the J/V curve as in Eq. (2.1) represents the main characteristics of most solar-cell J/V curves, the reality looks slightly different in several respects. The first typical feature of any p–n-junction solar cell is shown in Figure 2.1c and d. Typically, the dominant recombination mechanism is not radiative

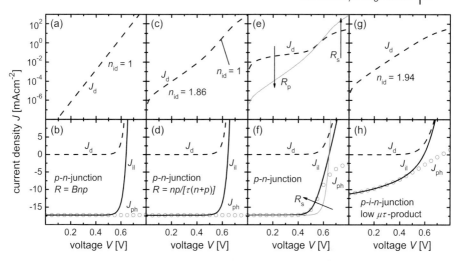

Figure 2.1 Semilogarithmic plots (a,c,e,g) of dark J/V characteristics and linear plots (b, d, f, h) of dark (dashed lines) and illuminated (full lines) J/V characteristics as well as of the difference $J_{ph} = J_{il} - J_d$ (open circles). In (a, b) the characteristics of a p–n-junction diode resulting from radiative recombination is shown leading to an ideal slope of the dark J/V with an ideality $n_{id} = 1$ and a voltage-independent photocurrent. (c, d) The departure from an ideal diode law in case of typical p–n-junction solar cells, where the low-energy part of the dark J/V features a second slope with a higher ideality factor $n_{id} = 1.86$, which originates from SRH recombination in the space-charge region. (e, f) The addition of a series and parallel resistance with the gray line representing the case with $R_s = 0$ and $R_p = \infty$ for reference. Note that J_{ph} is voltage-dependent despite the fact that carriers are efficiently collected. This can be used to determine the series resistance. (g, h) A p–i–n-junction with a low-mobility lifetime product and a subsequently strongly voltage-dependent photocurrent J_{ph}.

recombination and thus, the recombination rate does not scale directly with the np-product. Instead, the typical recombination mechanism in thin-film solar cells is Shockley–Read–Hall recombination [2, 3] via defects in the band gap. Assuming a defect in the middle of the band gap, the recombination rate scales with

$$R = \frac{np - n_i^2}{(n+p)\tau} \tag{2.3}$$

where τ is the lifetime of electrons and holes, which we assume to be equal. Assuming such a recombination rate and a p–n-junction solar cell, the dark J/V curve as shown in Figure 2.1c differs considerably from the one for radiative recombination (Figure 2.1a). In addition to the voltage range $0.5\,\text{V} < V < 0.7\,\text{V}$, where the slope of the curve is q/kT, the slope is considerably smaller at lower voltages $V < 0.5\,\text{V}$. This difference comes from the various positions where the recombination takes place dominantly at lower or higher voltages, respectively. At lower voltages, the recombination in the space-charge region is dominant. Within the space-charge region of a p–n-junction (and similar within the intrinsic layer of a p–i–n-junction), the carrier concentrations change from $p \gg n$ (toward the p-side) to $n \gg p$ (toward the n-side).

Thus, the maximum of the recombination rate R in Eq. (2.3) is always found within the junction for $n = p$, leading to

$$R(n = p) = \frac{np - n_i^2}{(n+p)\tau} \approx \frac{n}{2\tau} = \frac{\sqrt{np}}{2\tau} \tag{2.4}$$

Assuming flat quasi-Fermi levels throughout the space-charge region, the product np is proportional to $\exp(qV/kT)$, where the internal voltage V is the quasi-Fermi level splitting (divided by the elementary charge). Thus, the recombination rate scales via

$$R(n = p) \propto \sqrt{\exp\left(\frac{qV}{kT}\right)} = \exp\left(\frac{qV}{2kT}\right) \tag{2.5}$$

with the internal voltage V. Obviously, the slope of the recombination rate and subsequently also the slope of the integral recombination rate, that is, the recombination current, is now around $q/2kT$. We call this factor 2 in the denominator of the exponent in Eq. (2.5) the ideality factor n_{id}, sometimes also the diode quality factor. When comparing this value with the according simulation in Figure 2.1c, we see that the actual value of the ideality factor is slightly smaller, that is, $n_{id} = 1.86$ at lower voltages, which is due to the fact that $n = p$ is only a limiting situation, which is exactly valid only at one position in the space-charge region. The integration over the recombination in the space-charge region then gives an ideality factor slightly smaller than 2 with an exact value that depends on the details of thicknesses, doping concentrations, and electrical parameters of the device.

While the above reasoning simply replaced the integral of the recombination rate R through the space-charge region by the maximum of the integrand, more precise analytical approximations are possible that take into account different lifetimes of holes and electrons as well as the recombination through the entire space-charge region [4]. This theoretical approach predicts diode ideality factors in a range $1 \leq n_{id} \leq 2$ dependent on the energy of the recombination center and the respective capture cross sections for electrons and holes. The explanation of $n_{id} > 2$ is not possible by considering recombination via a single recombination center. A multiple step recombination process via a series of trap states distributed in space and energy would explain such large ideality factors for a recombination process situated in the space-charge region [5]. This model is especially suited to describe the dependence of the ideality factor of CdS/CdTe heterojunction solar cells. Also the enhancement of recombination by tunneling – a process that makes the recombination rate dependent on the local electrical field and, in consequence, on the applied bias voltage – predicts ideality factors that may exceed 2 [6]. This theory of tunneling-enhanced recombination especially applies to $Cu(In,Ga)Se_2$ with high Ga content or $CuGaSe_2$ solar cells [7].

The preceding discussion made clear that the ideality factor is strongly influenced by the recombination mechanism. However, the unique identification of recombination mechanisms via the ideality factor alone is not possible. Turning back to the simulations, we notice that at higher voltages, the ideality factor of the dark J/V curve in Figure 2.1c is again 1, as in Figure 2.1a, although the recombination mechanism is different. This is due to the fact that for the example in Figure 2.1c the recombination for voltages $V > 0.5\,\text{V}$ is dominated by recombination in the p-type layer of the p–n-junction (which is assumed to be much thicker than the n-type layer). In the p-type layer in the dark and for not too high voltages, the electron concentration is much smaller than the hole concentration ($n \ll p$). Simplifying Eq. (2.3) yields

$$R(n \ll p) = \frac{np - n_i^2}{(n+p)\tau} \approx \frac{n}{\tau} \tag{2.6}$$

Since, the whole internal voltage V is now used to increase the minority carrier concentration, it holds

$$R(n \ll p) \approx \frac{n}{\tau} \propto \exp\left(\frac{qV}{kT}\right) \tag{2.7}$$

Obviously, the ideality factor is again unity, also for recombination via defects, as long as the concentration of one type of carrier is much smaller than the concentration of the other. With the same argument, also the recombination at surfaces of p–n- or p–i–n-junctions will lead to ideality factors of 1, since at the contacts usually $n \gg p$ or $p \gg n$ is fulfilled.

For p–n-junction solar cells, there are typically two voltage ranges with different ideality factors as shown in Figure 2.1c. At lower voltages, the ideality factor is close to but smaller than 2 (if multistep recombination or tunneling is absent), and the recombination is dominated by the space-charge region. For higher voltages, the ideality factor is close to 1 indicating defect recombination in the volume or the surfaces of the absorber away from the space-charge region. For this reason, determination of ideality factors in p–n-junction solar cells is usually done by fitting two-diode models to experimental data. The two-diode model in the dark has the form

$$J = J_{01}\left[\exp\left(\frac{qV}{n_{id1}kT}\right) - 1\right] + J_{02}\left[\exp\left(\frac{qV}{n_{id2}kT}\right) - 1\right] \tag{2.8}$$

where n_{id1} and n_{id2} are the two ideality factors and J_{01} and J_{02} the two corresponding saturation current densities.

Up to now, we discussed the dependence of the recombination rate and the recombination current on the internal voltage V, defined as the quasi-Fermi level splitting in the space-charge region. We implicitly assumed that the voltage measured is equal to this internal voltage. Now, we have to distinguish the externally measured voltage V_{ext} from the internal voltage V_i at the p–n-junction. The external voltage is larger than the internal voltage by a term $\Delta V = V_{\text{ext}} - V_i$ that scales roughly in a linear way with current density J. The proportionality factor between ΔV and J is the series resistance R_s. The series resistance may originate from the finite conductivity of the absorber layers themselves or from the front and back contacts. For $V > 0.7\,\text{V}$ in Figure 2.1c, we already see that the simulated dark J/V curve (dashed line) has a lower current density at a given voltage than the solid line indicating the two-diode fit according to Eq. (2.8). For this simulation, this is due to the finite mobility and thus finite conductivity of the p-type layer of the p–n-junction.

For Figure 2.1e and f, we explicitly included an external series resistance and a finite shunt resistance to show its influence on the J/V curves under illumination and in the dark. We add an external series resistance of $R_s = 5\,\Omega\,cm^2$ and an external shunt resistance of $R_p = 5\,k\Omega\,cm^2$. The series resistance leads to an increased voltage at constant current density in case of the dark J/V curve, that is, more voltage has to be applied for the same current to flow since part of the voltage drops over the external resistance and not over the p–n-junction. Under illumination, the curve with series resistance is shifted in the opposite direction. Now, the voltage at the junction must be higher than the voltage at the external contacts to drive a current through an external load resistance. Now, the photocurrent J_{ph} defined as the difference between dark and illuminated J/V curve is no longer constant, since the voltage drop depends on the current, which is – at a given voltage – different under illumination and in the dark.

The shunt resistance R_p has a large influence on the dark J/V curve in the lower voltage range where the current increases drastically when compared to the characteristics without R_p. In contrast at higher voltages, the differential conductivity $G_d = \partial J/\partial V$ of the diode itself increases exponentially, while the conductivity of the shunt stays constant. Thus, for higher voltages, the shunt disappears and is also nearly invisible in the linear plot of dark and illuminated J/V curve.

Including both shunt and series resistance in the J/V curve leads to

$$J = J_{01}\left[\exp\left(\frac{q(V-JR_s)}{n_{id1}kT}\right)-1\right] + J_{02}\left[\exp\left(\frac{q(V-JR_s)}{n_{id2}kT}\right)-1\right] + \frac{V-JR_s}{R_p} - J_{sc} \quad (2.9)$$

This equation is frequently used to analyze J/V measurements of p–n-junction solar cells. A useful property of this description of the J/V curve is that it consists entirely of basic circuit elements, like ideal diodes (defined by n_{id} and J_0), resistances (R_s, R_p), and a current source (J_{sc}) as depicted in Figure 2.2. Thus, such a description is particularly useful for two-dimensional modeling of solar cells and modules by solving networks of diodes, resistances, and current sources with appropriate software tools like SPICE (cf. Chapter 20).

Especially for disordered, low-mobility solar cells that are fabricated as p–i–n-diodes such as amorphous and microcrystalline thin-film solar cells, an expression like Eq. (2.9) is still insufficient to describe the measured results well. As shown in Chapter 19, the charge carrier collection of p–i–n-type devices is inherently voltage-dependent. Thus, the short-circuit current density is not a voltage-independent constant anymore. Figure 2.1g and h shows the results of a simulation of a p–i–n-junction solar cell with relatively low mobility-lifetime product $\mu\tau = 10^{-9}\,cm^2\,V^{-1}$. Obviously, although there are no external series resistances assumed, the photocurrent is strongly voltage dependent and decays rapidly for increasing voltages, even changing its sign slightly below 0.8 V. The exact shape of the voltage-dependent photocurrent cannot be reproduced with analytical equations but it must be calculated by numerical simulations as presented in Chapter 19. However, there are some analytical approximations, as for example, that described in Refs. [8, 9]. Thus, for low-mobility p–i–n-type diodes,

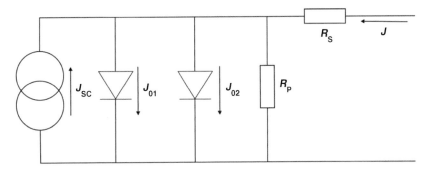

Figure 2.2 Equivalent circuit useful for the description of p–n-junction solar cells consisting of a current source representing the short-circuit current, two diodes for the recombination in the space-charge region and one series and parallel resistance. Note that a representation with an equivalent circuit is difficult for p–i–n-type solar cells, since the photocurrent there is inherently voltage dependent.

the use of equivalent circuit models is difficult or impossible, depending on the specific needs of the given application.

2.2.2
Measurement of Current/Voltage Curves

Details concerning the correct measurement of J/V curves of solar cells under illumination have been discussed in Refs. [10, 11]. Thus, we restrict ourselves here to a brief overview of the measurement itself. Illuminated J/V curves are usually measured under standard testing conditions, that is, a sample temperature of 25 °C and a predefined spectrum, which is for nonconcentrating solar cells usually the AM1.5G spectrum [12]. Figure 2.3 shows a typical setup for exact measurements of the illuminated J/V curve. The biggest challenge for this measurement is to have a light source generating a spectrum that resembles the solar spectrum as much as possible. Since the terrestrial solar spectrum is close to that of a black body with a temperature of about 5800 K, any light bulb will have a black body spectrum with a much lower temperature since no element can withstand these temperatures without melting. The pure metal with the highest melting point is W (approximately 3700 K), which is therefore commonly used in light bulbs. Since any black body source will not reach the temperature of the sun and thus the spectrum of the sun without melting, a W halogen lamp is usually combined with a Xe-lamp to get a close match to the solar spectrum.

Assuming that the device under test is illuminated with a spectrum resembling the standard AM1.5G spectrum, the device is contacted and the load resistance is varied such that the voltage changes. The voltage and current are usually measured during the voltage sweep with a four-point probe technique, that is, the current measurement is connected in series with the load resistance, while the voltage measurement requires two separate probes. The circuit containing the voltage measurement has a high resistance and avoiding any resistive voltage drop affecting the measurement result.

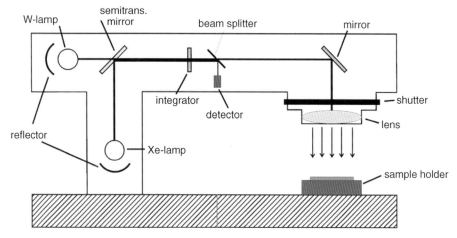

Figure 2.3 Schematic of a solar simulator for J/V measurements under illumination with a spectrum resembling the standard AM1.5G. To better approximate the solar spectrum a W-lamp and a Xe-lamp are combined.

2.2.3
Determination of Ideality Factors and Series Resistances

One part of the analysis of J/V curves is the determination of characteristic properties of the curve, as for instance the series resistance and the diode quality factor at various voltages. One method is to fit a one- or two-diode model (Eq. (2.8)) to the experiment. An alternative method for the determination of the series resistance in p–n-junction solar cells with voltage-independent charge carrier collection is the comparison of dark and illuminated J/V curves. As we have already seen in Figure 2.1f, the difference between dark and illuminated current voltage curves is not a constant, when the series resistance $R_s > 0$ even when the actual charge carrier collection process is not voltage dependent. For a better understanding of this difference, let us shift the current density J_{il} under illumination by, for example, the AM1.5G spectrum by the short-circuit current density $J_{sc}^{AM1.5}$ at exactly this AM1.5G illumination to obtain the current density $J_{il} + J_{sc}^{AM1.5}$. Then the J/V curve reads

$$J_{il} + J_{sc}^{AM1.5} = J_0 \exp\left(\frac{q(V_{il} - J_{il} R_s)}{n_{id} kT} - 1\right) \qquad (2.10)$$

if parallel resistances are neglected, and assuming that the solar-cell characteristic is well described by a one-diode model in the relevant voltage range. The voltage V_{il} at a given current density is consequently

$$V_{il} = \frac{n_{id} kT}{q} \ln\left(\frac{J_{il} + J_{sc}^{AM1.5}}{J_0} + 1\right) + J_{il} R_s \qquad (2.11)$$

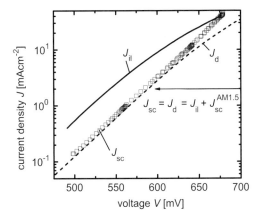

Figure 2.4 Dark current density J_d (dashed line) illuminated current density J_{il} (solid line) and illumination-dependent short-circuit current density J_{sc} (open squares) as a function of voltage or open-circuit voltage V_{oc} in case of the J_{sc} on a semilogarithmic scale. The J_{il}/V curve is shifted by the J_{sc} at AM1.5G conditions at which the J_{il}/V curve was measured. From the voltage differences at constant current densities in this plot, the series resistance of a p–n-junction solar cell can be calculated.

With similar arguments, the voltage V_d in the dark is

$$V_d = \frac{n_{id} kT}{q} \ln\left(\frac{J_d}{J_0} + 1\right) + J_d R_s \qquad (2.12)$$

If we now calculate the difference between the voltage in the dark and the voltage under illumination at a given current density $J_d = J_{il} + J_{sc}^{AM1.5}$, we obtain

$$V_d - V_{il} = J_d R_s - J_{il} R_s = J_{sc}^{AM1.5} R_s \qquad (2.13)$$

Consequently, the series resistance can easily be determined from the voltage difference in the dark and under illumination and the known short-circuit current density. Figure 2.4 shows the plot of the illuminated and dark J/V curves of a crystalline Si solar cell shifted into the first quadrant of the coordinate system and plotted with a logarithmic current axis. It can be seen that the voltage difference between dark and illuminated curves is relatively constant. At a current density $J_d = J_{il} + J_{sc}^{AM1.5} = 2$ mAcm^{-2} (indicated by the black arrow), the series resistance is $R_s = \Delta V/J_{sc}^{AM1.5} = 45$ mV$/40.4$ mAcm$^{-2} \approx 1.1 \, \Omega$ cm^2. With this method, the series resistance is determined as a function of the current density.

However, the method is only suitable as long as the series resistance does not depend on illumination conditions. For this situation, Aberle et al. [13] proposed to use the open-circuit voltage measured at different illumination conditions, yielding a so-called J_{sc}/V_{oc} curve. This curve follows the equation

$$V_{oc} = \frac{n_{id} kT}{q} \ln\left(\frac{J_{sc}}{J_0} + 1\right) \qquad (2.14)$$

where J_{sc} denotes the variable short-circuit current density at various illumination intensities. From Eq. (2.14) it follows that V_{oc} is completely independent of series resistance because no current flows at open-circuit conditions. The J_{sc}/V_{oc} curve is thus a series-resistance-free J/V curve. Consequently, the series resistance under illumination follows from

$$V_{oc} - V_{il} = -J_{il}R_s = (J_{sc} - J_{sc}^{AM1.5})R_s \qquad (2.15)$$

and in the dark from

$$V_d - V_{oc} = J_d R_s = J_{sc} R_s \qquad (2.16)$$

is obtained from comparison of the illuminated and the dark J/V curves with the J_{sc}/V_{oc} curve, which is shown in Figure 2.4 (open symbols).

A second useful property of the J_{sc}/V_{oc} representation is the fact that the voltage range, where the curve follows an exponential relation, is enlarged. In Figure 2.1e, the voltage range, where the dark J/V is not heavily affected by either shunt or series-resistance effects, is very small and it would be difficult to determine an ideality factor from the dark J/V curve. Since the J_{sc}/V_{oc} curve is independent of external series resistances, the determination of ideality factors from J_{sc}/V_{oc} curves becomes much easier using the slope of the J_{sc}/V_{oc} curve via

$$n_{id} = \frac{q}{kT}\frac{dV_{oc}}{d\ln(J_{sc})} = \frac{q}{kT}\frac{dV_{oc}}{dJ_{sc}}J_{sc} \qquad (2.17)$$

2.2.4
Temperature-Dependent Current/Voltage Measurements

The temperature is a useful parameter to vary, when analyzing a solar cell by means of J/V measurements. The temperature dependence of J/V curves is mainly due to the temperature dependence of the equilibrium concentrations of the charge carriers, either electrons, holes or electrons, *and* holes. To illustrate this statement, let us go back to the simple case of volume recombination described in Eq. (2.6). The recombination rate R scales with the minority carrier concentration n in this case. At a given voltage, $n \sim n_i^2/N_A$, where N_A is the acceptor concentration. Since the intrinsic carrier concentration n_i is given by

$$n_i^2 = N_C N_V \exp\left(\frac{-E_g}{kT}\right) \qquad (2.18)$$

where N_C and N_V are the effective density of states of conduction and valence band, respectively, also the recombination rate in the volume should scale with a term $\exp(-E_g/kT)$. However, the activation energy E_a of the recombination rate is not always identical to the band gap E_g as in this example.

A typical task that is accomplished by temperature-dependent J/V analysis is the distinction between bulk and interface recombination. Figure 2.5 uses the band diagram to illustrate the different recombination mechanisms that can occur in a

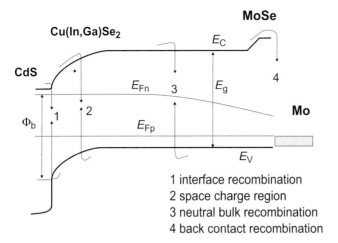

Figure 2.5 Schematic band diagram of a CdS/Cu(In,Ga)Se$_2$ solar cell showing the four main recombination mechanisms that can occur. The four different locations, where recombination with different ideality factors and activation energies can take place are (1) the CdS/Cu(In,Ga)Se$_2$ interface, (2) the space-charge region, (3) the neutral bulk and (4) the back contact (interface between Cu(In,Ga)Se$_2$ and Mo). The quantities E_C and E_V stand for the conduction and valence band, E_{fn} and E_{fp} for the quasi-Fermi-levels of electrons and holes, Φ_b is the interface barrier.

CdS/Cu(In,Ga)Se$_2$ solar cell. In case of interface recombination at the CdS-Cu(In,Ga)Se$_2$ interface (mechanism 1 in Figure 2.5), the respective activation energy is the potential barrier height $\Phi_b = E_{fn} - E_V$ at the interface (E_{fn} being the electron Fermi level at the interface). This is due to the case that holes are the minorities at the interface and that subsequently the recombination rate at the interface scales with hole concentration [14]. The hole concentration in turn is given by

$$p = N_V \exp\left(\frac{qV - \Phi_b}{kT}\right) \tag{2.19}$$

which directly demonstrates that Φ_b is the activation energy of interface recombination. In contrast the bulk recombination mechanisms 2 and 3 in Figure 2.5, that is, recombination in the space-charge region and in the neutral region are thermally activated by the band-gap energy E_g. Due to the band offset between the Cu(In,Ga)Se$_2$ absorber and a thin MoSe$_2$ layer forming between the absorber and the metallic Mo back contact [15], the activation energy for this process 4 is even larger by this amount than E_g.

Obviously, a temperature-dependent study of the recombination current should be capable of determining the activation energy and subsequently discriminate between limitations by interface or volume recombination. Since the open-circuit voltage V_{oc} is the photovoltaic parameter, which is most directly affected by recombination, we now study the temperature dependence of V_{oc} at a given illumination intensity and thus a given short-circuit current density J_{sc}. The relation between the two is given by [16]

$$J_{sc} = J_0 \left[\exp\left(\frac{qV_{oc}}{n_{id}kT}\right) - 1\right] = J_{00} \exp\left(\frac{-E_a}{n_{id}kT}\right) \left[\exp\left(\frac{qV_{oc}}{n_{id}kT}\right) - 1\right] \quad (2.20)$$

where J_{00} is a weakly temperature-dependent prefactor of the saturation current density J_0. The usefulness of such a description becomes obvious when resolving for the open-circuit voltage. Now, we obtain

$$qV_{oc} = E_a - n_{id}kT\ln\left(\frac{J_{00}}{J_{sc}}\right) \quad (2.21)$$

that is, a direct relation between the open-circuit voltage V_{oc} and the activation energy E_a. The prefactor J_{00} is typically much larger than J_{sc}, thus qV_{oc} is smaller than the activation energy by the roughly linearly temperature-dependent second term in Eq. (2.21). A temperature-dependent measurement of V_{oc} should yield the activation energy as the extrapolated value of $V_{oc}(T=0\,\text{K})$ as long as J_{00} and J_{sc} are reasonably independent of temperature. As an example, Figure 2.6a shows the result of such a measurement for four different Cu(In,Ga)(Se,S)$_2$ solar cells [17].

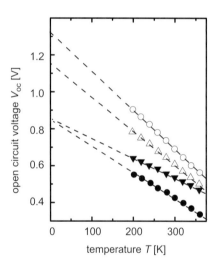

Figure 2.6 Temperature dependence of the open-circuit voltage V_{oc} for different Cu(In,Ga)(Se,S)$_2$ solar cells with different band-gap energies due to different In/Ga and Se/S ratios. The open symbols correspond to devices that are grown with a Cu-poor final composition and have a band-gap energy (as calculated from the stochiometry) of $E_g = 1.49\,\text{eV}$ (circles) and 1.22 eV (triangles). The extrapolated open-circuit voltage V_{oc} ($T=0\,\text{K}$) roughly follow E_g, whereas for the devices grown under Cu-rich conditions V_{oc} ($T=0\,\text{K}$) is independent of E_g (1.15 eV, circles, and 1.43 eV, triangles). The latter finding points to the fact that recombination in such devices has an activation energy given by the height of the interface barrier Φ_b.

The data with open symbols stem from samples with a Cu-poor composition. The extrapolated open-circuit voltage $V_{oc}(T=0\,K)$ roughly follow their band-gap energy E_g (as determined independently from their stoichiometry). The dominant recombination path for these devices, therefore, is bulk recombination. In contrast, samples with absorbers that are grown under Cu-rich conditions have low activation energies that are independent from the absorber's band-gap energy. This finding points to the fact that their V_{oc} is limited by interface recombination and the activation energy corresponds to the interfacial barrier Φ_b.

Obviously such an analysis of the dominant recombination mechanism requires investigation of a large series of samples (as done in Ref. [17]) and Figure 2.6 just illustrates exemplarily the experimental procedure. More examples covering different types of solar cells ($CuInSe_2$, CdTe, and a-Si:H) are found in Ref. [18].

2.3
Quantum Efficiency Measurements

2.3.1
Definition

A J/V measurement yields information on the absolute value of the short-circuit current density J_{sc} produced in a solar cell. However, this simple measurement does not yield information on the origin of the loss mechanisms that are responsible for the fact that not every photon in the solar spectrum contributes to J_{sc}. In an ideal solar cell, corresponding to the Shockley–Queisser limit as discussed in Chapter 1, every photon with a suitable energy $E > E_g$ leads to one electron–hole pair that is collected at the terminals of the solar cell. In real solar cells, this is not the case and we are interested in knowing the reasons for these losses. An appropriate method to literally shed light onto this problem is the spectrally resolved measurement of the short-circuit current, that is, $J_{sc}(E)$ likewise $J_{sc}(\lambda)$, depending on whether the result is plotted versus photon energy E or versus wavelength λ. The external quantum efficiency Q_e is defined as the number of electrons collected per photon incident on the solar cell according to

$$Q_e(E) = \frac{1}{q}\frac{dJ_{sc}(E)}{d\Phi(E)} \qquad (2.22)$$

where $d\Phi(E)$ is the incident photon flux in units of $[\Phi] = cm^{-2}s^{-1}$ in the (photon) energy interval dE that leads to the short-circuit current density dJ_{sc}.

A second frequently used quantity is the spectral response SR defined as the current produced per unit optical power incident on the solar cell. Consequently, the spectral response has the unit $[SR] = A/W$ and relates to the quantum efficiency via

$$SR = \frac{dJ_{sc}(E)}{d\Phi(E)}\frac{1}{E} = \frac{qQ_e}{E} \qquad (2.23)$$

In the ideal Shockley–Queisser case, we would have $Q_e(E) = 1$ for $E \geq E_g$ and $Q_e(E) = 0$, otherwise. In real solar cells, we have $Q_e(E) < 1$ (even for $E \geq E_g$) resulting either from (i) optical or (ii) recombination losses. The optical losses can be further broken down to losses due to reflection and due to parasitic absorption within the device.

The reflection losses can be assessed by an additional measurement using a spectrometer, equipped with an integrating sphere determining the reflectance R, thus allowing us to quantify this loss mechanism separately. For an opaque solar cell we know that all photons that are not reflected are absorbed in the device, that is, the absorptance A is given by $A = 1 - R$. The internal quantum efficiency Q_i is then defined as the number of collected electrons per number of photon *absorbed* in the solar cell according to

$$Q_i(E) = \frac{Q_e(E)}{1 - R(E)} \qquad (2.24)$$

Note that sometimes Q_i is defined as the number of collected electrons per number of photons *entering* in the solar cell. This definition includes weakly absorbed light that enters the solar cell and leaves it after reflection at internal interfaces or surfaces. In this case, the overall reflectance R must be replaced by the front reflectance R_f.

In internal quantum-efficiency spectra, the influence of (front surface) reflection is eliminated. However, it still contains information on optical as well as electrical properties of the device. For instance the effect of a surface texture on the path-length enhancement of weakly absorbed light (cf. Chapters 1 and 5) influences both internal and external quantum efficiency. In addition in a typical thin-film layer system, each layer, except for the photovoltaically active one(s), will lead to parasitic absorption. Therefore, the first step for the analysis of Q_i requires an *optical* model for the device to determine the absorptance A_i of each layer i in the system. Obviously, such a model requires the knowledge of thicknesses, refractive indices, and absorption coefficients determined, for example, with spectroscopic ellipsometry for reference layers from the respective materials. For thin layers, reflection, absorption, and transmission feature interference effects, which allow checking the accuracy of the simulation of the absorptance by comparison of measured and calculated reflectance. In case the absorptance A_i of a photovoltaically active layer is known, one may define an internal quantum efficiency Q_i^* for this layer via [19–21]

$$Q_i^*(E) = \frac{Q_e(E)}{A_i(E)} \qquad (2.25)$$

In the photovoltaically active layer, absorption is exclusively due to generation of electron–hole pairs (neglecting a possible contribution from free carrier absorption which may occur at long wavelengths in doped absorber materials). Under above assumption, the absorptance $A_i(E)$ is given by integrating the generation function $g(x,E)$ over the thickness d of the layer. Losses within the active layer are then only due to recombination and only starting from this point an *electronic* model must be applied. Typically, such a model calculates the collection probability $f_c(x)$ for electrons and holes such that at the very end of the analysis, we may use

$$Q_i^*(E) = \frac{Q_e(E)}{A_i(E)} = \frac{\int_0^d g(x, E) f_c(x) dx}{\int_0^d g(x, E) dx} \qquad (2.26)$$

explaining the external quantum efficiency layer by layer.

2.3.2
Measurement Principle and Calibration

Figure 2.7 shows two typical quantum efficiency measurement setups: (a) a monochromator-based setup and (b) a setup with a filter wheel equipped with interference

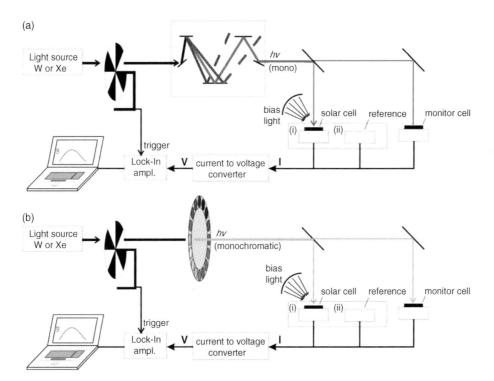

Figure 2.7 Scheme of two quantum efficiency setups – (a) a monochromator-based setup and (b) a setup with a filter wheel. In both cases, chopped monochromatic light illuminates first the reference (during calibration) and then the sample (during measurement). The current output of reference or sample is converted to voltage and then amplified with a lock-in amplifier triggered by the chopper wheel synchronization output. Temporal variations in intensity of the monochromatic light can be monitored with a monitor diode measuring intensity during calibration and measurement.

filters. The advantage of using monochromators is the high wavelength resolution and the broad spectral range. The wavelength resolution is especially high when double-stage monochromators are used. The disadvantage of double-stage monochromators compared with single-stage monochromators is the lower light throughput combined with a higher price and larger size. The advantage of filter wheels is the high optical throughput and the larger illuminated area. While monochromator-based setups illuminate only a small spot of few millimeters diameter, a filter wheel completely illuminates larger solar cells and mini-modules of several centimeters edge length. In the following, we will first describe the operation and calibration of a monochromator-based system. Subsequently, we will discuss the aspects that are different for a setup with filter wheel.

In case of the setup with grating monochromator, first white light from a W-halogen lamp or a Xe-arc lamp is chopped before entering the monochromator. The chopper is needed to obtain a periodic signal, which a lock-in amplifier can use. Care must be taken that higher modes due to shorter wavelengths (e.g., $\lambda/2$) are suppressed by the use of filters.

The monochromatic light is then focused on the (i) solar cell to be measured (during the actual measurement) or (ii) on the reference cell or detector (during the calibration). The reference used for calibration of the setup can either (i) be a pyroelectric radiometer for the relative calibration combined with a solar cell or photodiode for the absolute calibration at one wavelength or (ii) one reference solar cell for the whole spectral range. The advantage of using a pyroelectric radiometer lies in its spectrally independent sensitivity over a broad wavelength range. Thus, the calibration has a high quality for all wavelengths, where the intensity of the lamp is sufficient for a high signal to noise ratio in the radiometer. In case of the reference solar cell, not only the intensity of the lamp but also the quantum efficiency of the reference cell must be sufficiently high. Using a reference cell (without a radiometer) is particularly useful as long as the reference cell has high quantum efficiency for all wavelengths of interest of the device under test. For instance, high-efficiency crystalline silicon solar cells have high quantum efficiency ($>80\%$) in the spectral range where amorphous and microcrystalline silicon solar cells are sensitive.

In order to account for temporal variations in the absolute intensity of the light source, monitor solar cells are often used. Using a semitransparent mirror, part of the light leaving the monochromator or filter wheel is directed on a monitor solar cell, both during calibration and during measurement. For each wavelength, the ratio of intensity on the monitor cell during calibration and measurement is multiplied with the resulting quantum-efficiency value. When using monitor solar cells, it is important that the quantum efficiency of the monitor solar cell is sufficiently high in the complete spectral range of interest.

The current signal from the monitor and the test solar cell are then converted into a voltage by a current-to-voltage converter with a typical amplification ratio of 10^4–10^6 V/A. The voltage output of the converter serves as input for the lock-in amplifier that uses the synchronization output of the chopper controller as trigger input. The amplified signal of the lock-in amplifier is then read and displayed by a computer.

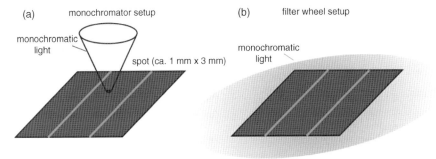

Figure 2.8 Schematic of the two possible approaches when illuminating a solar cell during a quantum-efficiency measurement. (a) With a monochromator-based setup, a typical spot size is in the millimeter range and will be smaller than most investigated cells. Thus, the spot illuminates only a small part of the cell and the quantum efficiency will be a local quantity, which may change when moving the spot. (b) In case of a filter wheel setup, it is possible to illuminate solar cells or small modules homogeneously and thus get an average quantum efficiency.

The main difference in operation of a filter wheel based setup concerns the calibration. As Figure 2.8 shows, the light from the monochromator is focused on the solar cell and illuminates only a small spot of typically 1 mm × 3 mm size. The filter wheel, however, illuminates larger areas. In order to measure the complete quantum efficiency of a solar cell of a certain size, the illuminated area must be much larger than the solar cell such that the complete cell area is illuminated homogeneously. This difference in illuminated area being either much smaller or larger than the sample requires different ways of calibration. While in the grating-based setup the current of the reference device is measured and compared to the current of the sample, for filter wheel setups the current *densities* have to be used.

2.3.3
Quantum Efficiency Measurements of Tandem Solar Cells

Especially in case of thin-film Si solar cells, multijunction solar cells are in the focus of research to make better use of the solar spectrum. Figure 2.9 shows the quantum efficiency of an a-Si:H/µc-Si:H tandem cell on textured ZnO:Al superstrate. The measurement of the two quantum efficiency spectra corresponding to both series-connected subcells requires the subsequent use of two bias light sources. In a first step, bias light with a short wavelength generates a photocurrent in the top cell that is *not* subject to the measurement. As a consequence, the bottom cell under investigation is limiting the output current of the tandem device. Now chopped monochromatic light with varied wavelength is used to produce the quantum efficiency spectrum of the bottom cell. A second measurement then uses longer wavelength light to generate a photocurrent in the bottom cell and the chopped light probes the quantum efficiency of the top cell. Care must be taken to ensure that for both

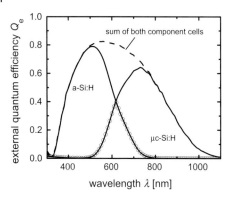

Figure 2.9 Example for the quantum efficiency of a-Si:H/μc-Si:H tandem cell on textured ZnO: Al superstrates. The quantum efficiencies of top and bottom cell are indicated as solid lines and are measured using a bias light to flood the respective other subcell. In addition, the sum of both is indicated as dashed line and the minimum value is indicated as open circles. The latter indicates the measurement result one would obtain when measuring without bias light.

measurements the bias light is in a spectral range where a pronounced difference in the quantum efficiency of top- and bottom cell exists. Also the intensity of the bias light sources must be larger than that of the chopped light for all wavelengths [22–24].

2.3.4
Differential Spectral Response (DSR) Measurements

As defined in Section 2.3.1, the quantum efficiency measures the spectrally resolved photocurrent at *short-circuit*. For many applications, and especially for thin-film solar cells, the restriction to the short-circuit situation does not give results that are representative for other operating points of the cells, too [18, 25]. A dependency of photocurrent on voltage or illumination is equivalent to a violation of the superposition principle in photovoltaics that has been discussed in Section 2.2. The main result was that the superposition principle is violated when the charge carrier collection depends on the number of charge carriers in the device and, thus, on the voltage and/or illumination bias. Such bias-dependent collection, in turn, originates either from a varying electric field in low-mobility devices as is the case for any p–i–n-type (cf. Figure 2.1h) solar cell or from a recombination rate $R \neq Bnp$. Recombination rates that are not proportional to the product np of electron and hole concentrations may result either from Shockley–Read–Hall recombination via (defect) states in the band gap or from Auger recombination (cf. Chapter 1).

In these situations, where the photocurrent is voltage and/or illumination-bias-dependent, the measurement of the quantum efficiency at various biases gives valuable information. As an example, in p–i–n-type solar cells, the comparison of the quantum efficiency at reverse voltages compared to those at short-circuit conditions or higher voltages allows distinguishing between collection losses and optical losses.

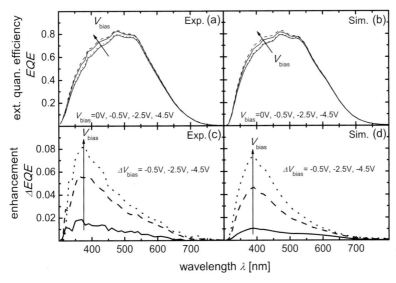

Figure 2.10 Bias-dependent quantum efficiency of the top cell in an a-Si:H/μc-Si:H tandem solar cell. (a, c) The experiment and (b, d) the simulation. (a, b) The quantum efficiency itself and (c, d) the difference in quantum efficiency compared to the short-circuit situation.

This is because the more negative the bias voltage V, the higher the electrical field F in the intrinsic layer of the p–i–n-diode, and the better the collection of carriers. If one can decrease the voltage V such that the quantum efficiency $Q_e(\lambda, V)$ does not increase any more with increasing reverse voltage, the electric field F is sufficiently high to drive all photogenerated carriers to their respective junctions and the collection efficiency $f_c = 1$. The quantum efficiency at such reverse voltages is then equal to the absorptance and, thus only determined by the optical properties of the device. Comparing the absorptance obtained from reverse-biased quantum efficiency measurements with the quantum efficiency at higher bias voltages allows eventually the calculation of losses due to carrier collection at a certain voltage.

Figure 2.10 shows an example for a bias-dependent quantum efficiency measurement of an a-Si:H/μc-Si:H tandem solar cell. Presented is only the quantum efficiency of the top cell. This quantum efficiency is measured for different negative bias voltages to show the effect of the electric field in the p–i–n-junction of the a-Si:H cell. The gain in quantum efficiency (b) is mainly in the wavelength range, where a considerable part of the light is absorbed in the p-layer, which has a poor electronic quality. Numerical simulations (c,d) as described in Chapter 19 have been used to analyze the electrical properties of the device.

2.3.5
Interpretation of Quantum Efficiency Measurements in Thin-Film Silicon Solar Cells

The interpretation of quantum efficiency measurements differs considerably depending on the solar cell type. For crystalline Si, for instance, the quantum

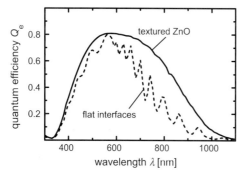

Figure 2.11 Comparison of the typical quantum efficiency of a μc-Si:H solar cell with a textured ZnO (solid line) and a flat ZnO.

efficiency in the wavelength range 780 nm $< \lambda <$ 950 nm is typically used to determine the effective diffusion length of the base layer of the device [26–29]. For even higher wavelengths, information about the optical properties of the back contact is obtained. In CdS/Cu(In,Ga)Se$_2$ solar cells, different evaluation methods yield information about charge carrier collection properties of the buffer layer [30] or the absorber layer [31, 32]. Since a complete description of all methods is beyond the scope of this chapter, we focus here mainly on thin-film silicon solar cells and discuss the main challenges in interpretation of quantum efficiency measurements.

According to Eq. (2.26), the quantum efficiency consists of contributions due to absorption of photons and generation of charge carriers as well as of contributions due to carrier collection. In order to show an example for the influence of optical properties of the device on the quantum efficiency, Figure 2.11 compares the external quantum efficiency of two μc-Si:H solar cells. The cell represented by the solid line is grown on a texture-etched ZnO:Al superstrate [33], while the cell with the dashed line is grown on the same superstrate but without etching. All thicknesses of the Si layers in the stack were the same, which was achieved by depositing the active layers at the same time during the same deposition process. There are two striking differences between the two curves. First, the textured solar cell has considerably higher quantum efficiency especially for higher wavelengths. Second, the quantum efficiency of the untextured solar cell shows interference patterns, which are absent for the textured solar cell.

The reason for the increased quantum efficiency of the textured solar cell lies in the concept of light trapping that was described in Chapters 1 and 5. Light that passes the interface between the textured ZnO and the Si is scattered by the rough interface and the refractive index difference (between 1.5 and 2 for ZnO:Al and between 3 and 5 for Si, depending on crystallinity and wavelength). The scattered light has now a high probability of being trapped in the absorber layer by total internal reflection. Thus, the weakly absorbed light, which travels in a flat cell only twice the cell thickness, is reflected multiple times in the textured cell. The increased optical path length of the light leads to the increased quantum efficiency at a given absorption coefficient and thickness. In case of the untextured solar cell, interference effects play a role since the cell thickness (thickness of all silicon layers is 1230 nm) is of the order of the wavelength of light and is smaller than the coherence length of the illumination. As a consequence, there are interferences that depend on the layer thicknesses and the refractive indices. For the textured cell, these interference patterns are destroyed by the light scattering at the ZnO/Si interface.

Obviously, a more detailed analysis of the quantum efficiencies in Figure 2.11 would be of interest. This requires numerical simulations, which are described in more detail in Chapter 19. Analytical approaches to model both the optical part as well as the electronic part of the quantum efficiency are not possible in such thin p–i–n-type devices. This is due to the fact that both electrons and holes are relevant for transport and recombination, which means that coupled differential equations for electron and hole transport have to be solved. For p–n-junction solar cells, most evaluation schemes rely on the fact that only one type of carrier dominates recombination. This is due to the fact that most p–n-junction devices have a thick base and a very thin emitter. Due to the large size difference, the minority carrier in the base is usually the relevant carrier and the solution of the continuity equation for the minority carrier in the base is sufficient to cover the electronic aspect of the problem.

The second challenge associated with p–i–n-junction thin-film solar cells is the fact that the wavelength of light and the cell thickness are comparable in magnitude and that geometric optics has to be abandoned in favor of wave optics. Calculating the absorption of a solar cell with perfectly flat interfaces is a simple numerical problem that is solved with a matrix transfer formalism [34]. The only prerequisite is a precise measurement of the optical properties of the individual layers by ellipsometry and photothermal deflection spectroscopy.

For textured surfaces, however, the calculation of the absorptance of the layers requires more sophisticated techniques. For research on light-trapping properties of a certain surface texture [35–38], Maxwell's equations have to be solved in three dimensions, which requires extensive computational resources. For a quick analysis of quantum efficiency measurements, more approximate methods [39] have to be used that do not consider interference effects for scattered light and that consider the interface properties only in the form of effective parameters such as haze (the ratio between light intensity that is scattered when being transmitted or reflected at an interface and the complete light intensity transmitted or reflected at an interface) and angular distribution function (describing the distribution of angles after the light is scattered at an interface) [40, 41]. Although the optical models for textured surfaces are far from being exact, a decent fit to the measured quantum efficiency is possible.

When all optical parameters of the layers are measured and the interface properties are determined such that the optical model allows a good description of the experiment, simulations of quantum efficiency measurements yield approximations of the major loss mechanisms affecting the short-circuit current. Figure 2.12 shows the absorptances of all relevant layers in (a) a μc-Si:H solar cell on textured ZnO:Al superstrate and (b) a CdS/Cu(In,Ga)Se$_2$ solar cell on a flat glass substrate simulated with ASA [42] and using the optical model genpro3 (a) and genpro1 (b). The major photocurrent losses for the μc-Si:H solar cell occur by absorption in the ZnO:Al layer and, especially, in the p-type Si layer. These layers are the first ones that the light has to

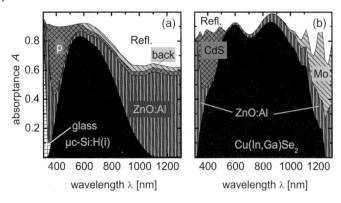

Figure 2.12 Comparison of the absorption in the different layers of (a) a typical μc-Si:H solar cell (redrawn from Ref. [43]) and (b) a typical Cu(In,Ga)Se$_2$ solar cell. The regions that are black indicate the absorption in the respective absorber layers. Since most of the photons absorbed there contribute to the short-circuit current, the black regions correspond well to the external quantum efficiency while the other areas indicate how severe losses are in the different layers.

pass after the glass. Thus, the light has a high intensity when passing through these layers and much of the light is absorbed there. While ZnO:Al exhibits a relatively low absorption coefficient but a high thickness (about 500 nm), the p-type Si layer is very thin (around 20 nm) but has a high absorption coefficient in the complete wavelength range also relevant for the absorber layer. For wavelengths higher than 900 nm, most light is absorbed again in the ZnO:Al layer since for higher wavelengths free carrier absorption in the ZnO:Al becomes more dominant. In addition, the absorber layers become transparent and, thus, light is reflected multiple times in the cell and will be absorbed primarily in the thick front contact ZnO and the three back contact layers (n-type Si, ZnO, Ag).

In case of the Cu(In,Ga)Se$_2$ solar cell, the low wavelength regime is pretty similar to μc-Si:H. Part of the high-energy photons is absorbed in the ZnO:Al, part in the CdS, which has a similar role as the p-type Si in the μc-Si:H layer stack. The main difference between μc-Si:H and Cu(In,Ga)Se$_2$ becomes apparent for higher wavelengths. Here, the absorption in the Cu(In,Ga)Se$_2$ is much higher than in the μc-Si:H because of the much higher absorption coefficient of Cu(In,Ga)Se$_2$ compared to μc-Si:H. Thus, even for wavelengths around 1100 nm, that is, just above the band gap of Cu(In,Ga)Se$_2$ and μc-Si:H, the absorptance of the Cu(In,Ga)Se$_2$ absorber is above 50%, while it is nearly zero in case of μc-Si:H. The higher absorptance of Cu(In,Ga)Se$_2$ in the infrared is possible despite the fact that no intentional texturing of the surface is required to achieve such a high absorptance in the long wavelength range as shown in Figure 2.12b. Note that the optical data for Figure 2.12 were measured [43] by means of ellipsometry and photothermal deflection spectroscopy (in case of the μc-Si:H cell) and the data for the Cu(In,Ga)Se$_2$ cell are taken from Ref. [44].

In addition to an entirely optical analysis of the quantum efficiency, the loss mechanisms of the photocurrent can also be analyzed in terms of optical losses and charge collection losses. However, this requires a careful determination of all

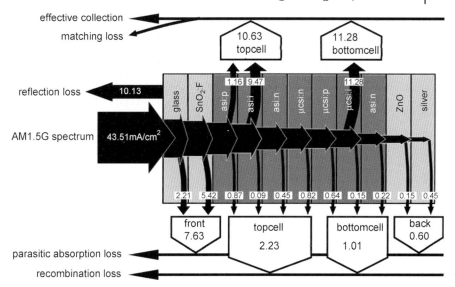

Figure 2.13 Using a simulation of a a-Si:H/μc-Si:H tandem solar cell, the losses in the different layers are calculated. The total incident photon flux of the AM1.5G spectrum between 300 and 1100 nm is taken as input leading to 43.51 mA cm^{-2} as the maximum J_{sc} for a single-junction solar cell (and 0.5×43.51 mA cm^{-2} for a tandem solar cell). Losses in the Si layers are denoted as recombination losses, while losses in the front and back contact layers are termed parasitic absorption losses although the distinction between the two is not strict.

electrical parameters, which has to be done by using a variety of different measurement techniques. For the case of a-Si:H, μc-Si:H, and Cu(In,Ga)Se$_2$, an overview over typical parameters and the measurement techniques to obtain them is given in Refs. [45–52]. For the case of a a-Si:H/μc-Si:H tandem solar cell, we performed a simulation of the losses in photocurrent in each layer as shown in Figure 2.13. We distinguished between parasitic absorption losses, recombination losses, and matching losses although this distinction does not discriminate between fundamentally different loss mechanisms. *Parasitic absorption* is defined here as the absorption in a layer where either photogenerated electron–hole pairs will not reach the junction or where absorption does not lead to photogenerated electron–hole pairs at all. These layers are the transparent conductive oxide (TCO) layers and the silver back reflector. The glass superstrate contributes to parasitic absorption as well. *Recombination losses* are the losses in all silicon layers due to finite collection efficiencies and finite mobility-lifetime products. It becomes clear that the distinction between parasitic absorption loss and recombination loss is somewhat arbitrary, when it comes to generation of electron–hole pairs in layers that are of low electronic quality such as the TCO layers or the doped silicon layers. Only the free carrier absorption, which is dominant in the TCO layers for higher wavelengths, is clearly parasitic absorption. The third loss mechanism is only relevant in multijunction solar cells. In these devices, the lowest photocurrent of the subcells limits the total photocurrent because of the series connection of the subcells. As a consequence, the additional photogenerated and collected carriers in the cell with the higher photocurrent are lost by recombination. Thus, the matching loss in a multijunction solar cell is only a special type of recombination loss.

In conclusion, we can say that a loss analysis as detailed in Figure 2.13 is a useful tool to determine the major loss mechanisms of the photocurrent for a specific solar cell. It will be based on the measurement of the quantum efficiency Q_e and the numerical simulation of the measured Q_e. From the simulation, the losses in the solar cell are obtained. The critical point for analyses underlying Figures 2.12 and 2.13 is the assumption that a good fit of the simulated quantum efficiency to the measured one implies that the loss mechanisms in the real cell are the same as in the simulation. In order to make such an analysis reliable, all electrical and optical parameters have to be determined – at least approximately – with various additional methods. In addition, results of optical simulations of thin-film solar cells grown on rough substrates have to be critically examined due to the lack of reasonably fast and at the same time correct methods to simulate the absorptance of such layer stacks.

References

1 Shockley, W. and Queisser, H.J. (1961) Detailed balance limit of efficiency of p–n junction solar cells. *J. Appl. Phys.*, **32**, 510.

2 Hall, R.N. (1952) Electron–hole recombination in germanium. *Phys. Rev.*, **87**, 387.

3 Shockley, W. and Read, W.T. (1952) Statistics of the recombinations of holes and electrons. *Phys. Rev.*, **87**, 835.

4 Sah, C.T., Noyce, R.N., and Shockley, W. (1957) Carrier generation and recombination in P–N-junctions and P–N-junction characteristics. *Proc. IRE*, **45**, 1228–1243.

5 Nardone, M., Karpov, V.G., Shvydka, D., and Attygale., M.L.C. (2009) Theory of electronic transport in noncrystalline junctions. *J. Appl. Phys.*, **106**, 074503.

6 Rau, U. (1999) Tunneling-enhanced recombination in Cu(In,Ga)Se$_2$ heterojunction solar cells. *Appl. Phys. Lett.*, **74**, 111–113.

7 Nadenau, V., Rau, U., Jasenek, A., and Schock, H.W. (1999) Electronic properties of CuGaSe$_2$-based heterojunction solar cells. Part I. Transport analysis. *J. Appl. Phys.*, **87**, 584–593.

8 Crandall, R.S. (1983) Modeling of thin film solar cells: Uniform field approximation. *J. Appl. Phys.*, **54**, 7176.

9 Hof, C. (1999) Thin Film Solar Cells of Amorphous Silicon: Influence of i-layer Material on Cell Efficiency. Université de Neuchâtel, Dissertation.

10 Osterwald, C.R. (2003) Standards, calibration and testing of PV modules and solar cells (eds T. Markvart and L. Castañer), *Practical Handbook of Photovoltaics*, Elsevier, Kidlington Oxford, p. 793.

11 Emery, K. (2003) Measurement and characterization of solar cells and modules (eds A. Luque and S. Hegedus), *Handbook of Photovoltaic Science and Engineering*, John Wiley & Sons Ltd, Chichester, UK, p. 701.

12 ASTM G173 (2010) *Standard Tables for Reference Solar Spectral Irradiances: Direct Normal and Hemispherical on 37° Tilted Surface*, American Society for Testing and Materials, West Conshocken, PA, USA, http://rredc.nrel.gov/solar/spectra/am1.5/ (12th March 2010).

13 Aberle, A.G., Wenham, S.R., and Green, M.A. (1993) A new method for accurate measurements of the lumped series resistance of solar cells. Proc. of the 23rd IEEE Photov. Spec. Conf. Louisville, Kentucky, IEEE New York, p. 133.

14 Scheer, R. (2009) Activation energy of heterojunction diode currents in the limit

of interface recombination. *J. Appl. Phys.*, **105**, 104505.
15. Wada, T., Kohara, N., Negami, T., and Nishitani, M. (1996) Chemical and structural characterization of Cu(In,Ga)Se$_2$/Mo interface in Cu(In,Ga)Se$_2$ solar cells. *Jpn. J. Appl. Phys.*, **35**, L1253–L1256.
16. Rau, U. and Schock, H.W. (1999) Electronic properties of Cu(In,Ga)Se$_2$ heterojunction solar cells – recent achievements, current understanding, and future challenges. *Appl. Phys. A*, **69**, 131.
17. Turcu, M., Pakma, O., and Rau, U. (2002) Interdependence of absorber composition and recombination mechanism in Cu(In,Ga)(Se,S)$_2$ heterojunction solar cells. *Appl. Phys. Lett.*, **80**, 2598.
18. Hegedus, S.S. and Shafarman, W.N. (2004) Thin-film solar cells: Device measurement and analysis. *Prog. Photovolt. Res. Appl.*, **12**, 155.
19. Slooff, L.H., Veenstra, S.C., Kroon, J.M., Moet, D.J.D., Sweelsen, J., and Koetse, M.M. (2007) Determining the internal quantum efficiency of highly efficient polymer solar cells through optical modeling. *Appl. Phys. Lett.*, **90**, 143506.
20. Dennler, G., Forberich, K., Scharber, M.C., Brabec, C.J., Tomis, I., Hingerl, K., and Fromherz, T. (2007) Angle dependence of external and internal quantum efficiencies in bulk-heterojunction organic solar cells. *J. Appl. Phys.*, **102**, 054516.
21. Law, M., Beard, M.C., Choi, S., Luther, J.M., Hanna, M.C., and Nozik, A.J. (2008) Determining the internal quantum efficiency of PbSe nanocrystal solar cells with the aid of an optical model. *Nano Lett.*, **8**, 3904.
22. Deng, X. and Schiff, E.A. (2003) Measurement amorphous silicon-based solar cells (eds A. Luque and S. Hegedus), *Handbook of Photovoltaic Science and Engineering*, John Wiley & Sons Ltd, Chichester, UK, p. 549.
23. Burdick, J. and Glatfelter, T. (1986) Spectral response and IV measurements of tandem amorphous-silicon alloy solar-cells. *Solar Cell.*, **18**, 310.
24. Müller, R. (1993) Spectral response measurements of 2-terminal triple junction a-Si solar-cells. *Sol. Energy Mater. Sol. C.*, **30**, 37.
25. Hegedus, S.S. and Kaplan, R. (2002) Analysis of quantum efficiency and optical enhancement in amorphous Si p–i–n solar cells. *Prog. Photovolt. Res. Appl.*, **10**, 257.
26. Arora, N.D., Chamberlain, S.G., and Roulston, D.J. (1980) Diffusion length determination in p–n junction diodes and solar cells. *Appl. Phys. Lett.*, **37**, 325.
27. Basore, P.A. (1993) Extended spectral analysis of internal quantum efficiency. Proc. of the 23rd IEEE Photov. Spec. Conf. Louisville, Kentucky, IEEE New York, p. 147.
28. Hirsch, M., Rau, U., and Werner, J.H. (1995) Analysis of internal quantum efficiency and a new graphical evaluation scheme. *Solid State Electron.*, **38**, 1009.
29. Brendel, R. and Rau, U. (1999) Injection and collection diffusion lengths of polycrystalline thin-film solar cells. *Solid State Phenom.*, **67–68**, 81.
30. Engelhardt, F., Bornemann, L., Köntges, M., Meyer, T., Parisi, J., Pschorr-Schoberer, E., Hahn, B., Gebhardt, W., Riedl, W., and Rau, U. (1999) Cu(In,Ga)Se$_2$ solar cells with a ZnSe buffer layer: Interface characterization by quantum efficiency measurements. *Prog. Photovolt. Res. Appl.*, **7**, 423.
31. Parisi, J., Hilburger, D., Schmitt, M., and Rau, U. (1998) Quantum efficiency and admittance spectroscopy on Cu(In,Ga)Se$_2$ solar cells. *Sol. Energy Mater. Sol. C.*, **50**, 79.
32. Kniese, R., Lammer, M., Rau, U., and Powalla, M. (2004) Minority carrier collection in CuGaSe$_2$ solar cells. *Thin Solid Film.*, **451–452**, 430.
33. Löffl, A., Wieder, S., Rech, B., Kluth, O., Beneking, C., and Wagner, H. (1997) Al-doped ZnO films for thin-film solar cells with very low sheet resistance and controlled texture. Proc. 14th Europ. Photovolt. Sol. En. Conf., Barcelona, p. 2089.
34. Berning, P.H. (1963) Theory and calculations of optical thin films, in *Physics of Thin Films*, vol. 1 (ed. G. Hass), Academic, New York, pp. 69–121.

35 Rockstuhl, C., Lederer, F., Bittkau, K., and Carius, R. (2007) Light localization at randomly textured surfaces for solar-cell applications. *Appl. Phys. Lett.*, **91**, 171104.

36 Rockstuhl, C., Lederer, F., Bittkau, K., Beckers, T., and Carius, R. (2009) The impact of intermediate reflectors on light absorption in tandem solar cells with randomly textured surfaces. *Appl. Phys. Lett.*, **94**, 211101.

37 Bittkau, K., Beckers, T., Fahr, S., Rockstuhl, C., Lederer, F., and Carius, R. (2008) Nanoscale investigation of light-trapping in a-Si:H solar cell structures with randomly textured interfaces. *phys. status solidi (a)*, **205**, 2766.

38 Rockstuhl, C., Fahr, S., Lederer, F., Bittkau, K., Beckers, T., and Carius, R. (2009) Local versus global absorption in thin-film solar cells with randomly textured surfaces. *Appl. Phys. Lett.*, **93**, 061105.

39 Tao, G., Zeman, M., and Metselaar, J.W. (1994) Accurate generation rate profiles in a-Si:H solar cells with textured TCO substrates. *Sol. Energy Mater. Sol. C.*, **34**, 359.

40 Krč, J., Zeman, M., Topič, M., Smole, F., and Metselaar, J.W. (2002) Analysis of light scattering in a-Si:H-based PIN solar cells with rough interfaces. *Sol. Energy Mater. Sol. C.*, **74**, 401.

41 Krč, J., Smole, F., and Topič, M. (2003) Analysis of light scattering in amorphous Si:H solar cells by a one-dimensional semi-coherent optical model. *Prog. Photovolt. Res. Appl.*, **11**, 15.

42 Pieters, B.E., Krč, J., and Zeman, M. (2006) Advanced numerical simulation tool for solar cells – ASA5. Conference Record of the 2006 IEEE 4th World Conference on Photovoltaic Energy Conversion, WCPEC-4 2 art. no. 4059936 p. 1513.

43 Ding, K. (2009) Characterisation and simulation of a-Si:H/μc-Si:H tandem solar cells, Master Thesis, RWTH Aachen.

44 Orgassa, K. (2004) Coherent optical analysis of the ZnO/CdS/Cu(In,Ga)Se$_2$ thin film solar cell, Dissertation, Universität Stuttgart.

45 Jiang, L., Lyou, J.H., Rane, S., Schiff, E.A., Wang, Q., and Yuan, Q. (2000) Open-circuit voltage physics in amorphous silicon solar cells. *Mater. Res. Soc. Symp. Proc.*, **609**, A18.3.1–A18.3.12.

46 Zhu, K., Yang, J., Wang, W., Schiff, E.A., Liang, J., and Guha, S. (2003) Bandtail limits to solar conversion efficiencies in amorphous silicon solar cells. *Mater. Res. Soc. Symp. Proc.*, **762**, 297.

47 http://physics.syr.edu/~schiff/AMPS/SU_Parameter_Suggestions.html (18 March 2010).

48 Liang, J., Schiff, E.A., Guha, S., Yan, B., and Yang, J. (2006) Hole mobility limit of amorphous silicon solar cells. *Appl. Phys. Lett.*, **88**, 063512.

49 Willemen, J.A. (1998) Modelling of Amorphous Silicon Single- and Multi-Junction Solar Cells, Dissertation, TU Delft, http://repository.tudelft.nl/assets/uuid:0771d543-af5f-4579-8a35-0b68d34a1334/emc_willemen_19981016.PDF (18 March 2010).

50 Pieters, B.E., Stiebig, H., Zeman, M., and van Swaaij, R.A.C.M.M. (2009) Determination of the mobility gap of intrinsic μc-Si:H in p–i–n solar cells. *J. Appl. Phys.*, **105**, 044502.

51 Pieters, B.E. (2008) Characterization of Thin-Film Silicon Materials and Solar Cells through Numerical Modeling, Dissertation, TU Delft. p. 105 http://repository.tudelft.nl/assets/uuid:83bc5ff4-8e33-4962-9014-68cfbad8226f/pieters_20081111.pdf. (18 March 2010).

52 Gloeckler, M. (2005) Device Physics of Cu(In,Ga)Se$_2$ Thin-Film Solar Cells, Dissertation, Colorado State University, Fort Collins, http://www.physics.colostate.edu/groups/photovoltaic/PDFs/MGloeckler_Thesis.pdf (18 March 2010).

3
Electroluminescence Analysis of Solar Cells and Solar Modules

Thomas Kirchartz, Anke Helbig, Bart E. Pieters, and Uwe Rau

3.1
Introduction

Solar cells are large-area diodes optimized for light absorption and charge carrier collection. Inversing the normal operating mode by injecting charge carriers under applied forward bias leads to recombination in the device. Although most solar cells are not particularly efficient light emitters, part of the recombination will be radiative, leading to detectable emission of photons with energies around the band gap of the solar-cell absorber. Electroluminescence (EL) characterization of solar cells has been used since the early 1990s [1–4]; however, the number of papers dealing with the field has risen substantially after a pioneering publication of Fuyuki and coworkers [5]. In 2005, the unique capability of EL imaging was shown for the first time. No longer was only the spectrally resolved EL measured as a function of temperature or injection current. Instead, the EL was detected by a charge-coupled device (CCD) camera, and spatially resolved information was obtained with measuring durations ranging from less than a second to some minutes depending on the applied voltage and the quality of the solar cell or module. Within few years after its introduction, EL imaging has become a valuable tool for fast and spatially resolved characterization of crystalline Si solar cells and modules [6–13].

Figure 3.1 depicts the two typical setups for EL characterization: (a) the EL imaging setup with a CCD camera, a power source, and a shielding against ambient light, and (b) the EL spectroscopy setup with a monochromator, a cooled Ge detector, and a lock-in amplifier. Instead of a Ge detector, also configurations with a photomultiplier or a Si CCD camera are possible for spectroscopic applications that do not require the infrared sensitivity of the low band-gap semiconductor Ge.

The attractiveness of EL imaging results on the one hand from the high spatial resolution combined with its simplicity and swiftness, which is clearly superior to alternative techniques such as the light-beam-induced current (LBIC) measurement. The measuring times of EL images are around two orders of magnitude shorter compared with LBIC. The technique can be used to survey entire modules but also to visualize microscopic defects on the micrometer scale. On the other hand, EL, that is,

---

Advanced Characterization Techniques for Thin Film Solar Cells,
Edited by Daniel Abou-Ras, Thomas Kirchartz and Uwe Rau.
© 2011 Wiley-VCH Verlag GmbH & Co. KGaA. Published 2011 by Wiley-VCH Verlag GmbH & Co. KGaA.

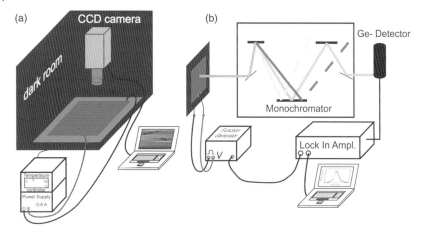

Figure 3.1 Schematic drawing of the measurement setups used for (a) spatially resolved EL imaging under forward or reverse bias and (b) EL spectroscopy.

the emission of light by application of an electrical bias, is just the complementary reciprocal action of the photovoltaic effect taking place in solar cells and modules. Therefore, all important physical processes that influence the photovoltaic performance of a solar cell – such as recombination, resistive, and optical losses – are in a complementary way reflected in the EL of the same device. This fact guarantees the relevance of the method. In addition to spatially resolved methods, also spectrally resolved EL [14–17] has proven to be a suitable tool for the analysis of solar cells. Finally, both spatially and spectrally resolved EL can be combined to a certain extent by the comparison of EL images made using different spectral filters [8, 18, 19].

This chapter starts with a brief comprehensive description of the basic theory underlying both spatially and spectrally resolved EL analysis. This description uses the convenience provided by the fundamental reciprocity between EL emission and photovoltaic action of solar cells [20]. Subsequently, Section 3.3 discusses recent results on spectrally resolved EL of different types of solar cells. Since the bulk of the literature on EL imaging studies crystalline Si solar cells, Section 3.4 reviews photographic surveying of Si wafer based solar cells. Section 3.5 introduces EL images acquired on $Cu(In,Ga)Se_2$ modules and explains the determination of the sheet resistance of front and back contact. Section 3.6 shows how to simulate EL emission from solar modules and how to determine the size and radius of shunts from such a simulation.

3.2
Basics

EL, that is, the emission of light in consequence to the application of a forward voltage bias to a diode, is the reciprocal action to the standard operation of a solar cell, namely the conversion of incident light into electricity. According to the reciprocity theorem, the EL intensity ϕ_{em} of a p–n-junction solar cell emitted at any position $\mathbf{r} = (x,y)$ of the solar-cell surface is given by [20]

$$\phi_{em}(E,\mathbf{r}) = [1-R(E,\mathbf{r})]Q_i(E,\mathbf{r})\phi_{bb}(E)\exp\left(\frac{qV(\mathbf{r})}{kT}\right)$$
$$= Q_e(E,\mathbf{r})\phi_{bb}(E)\exp\left(\frac{qV(\mathbf{r})}{kT}\right) \quad (3.1)$$

where kT/q is the thermal voltage, $V(\mathbf{r})$ is the internal junction voltage, E is the photon energy, and $Q_e(\mathbf{r}) = [1-R(\mathbf{r})]Q_i(\mathbf{r})$ is the local external quantum efficiency (QE) determined by the front surface reflectance $R(\mathbf{r})$ and the internal QE $Q_i(\mathbf{r})$. The spectral photon density ϕ_{bb} of a black body

$$\phi_{bb}(E) = \frac{2\pi E^2/(h^3 c^2)}{\exp(E/kT)-1} \quad (3.2)$$

depends on Planck's constant h and the vacuum speed c of light. Recording the EL emission of a solar cell with a CCD camera, the EL signal $S_{cam}(E,\mathbf{r})$ in each camera pixel is

$$S_{cam}(\mathbf{r}) = \int Q_{cam}(E) Q_e(E,\mathbf{r})\phi_{bb}(E) dE \exp\left(\frac{qV(\mathbf{r})}{kT}\right) \quad (3.3)$$

where Q_{cam} is the energy-dependent sensitivity of the detecting camera. Since in Eq. (3.3) $\phi_{bb}(E)$ and $Q_{cam}(E)$ depend on energy but not on the surface position $\mathbf{r}$, lateral variations in the detected EL intensity emitted from different surface positions originate only from the lateral variations of the external QE Q_e and of the internal voltage V. Hence, Eqs. (3.1) and thus (3.3) consider all losses occurring in solar cells: the external QE Q_e expresses the recombination and optical losses, while the internal voltage V reflects the resistive losses. As will be shown below, in some cases the exponential voltage-dependent term dominates the image, making EL analysis a tool that is especially suitable to analyze resistive losses.

In order to discriminate between recombination and optical losses that influence the detected and spectrally integrated EL spectrum on the one hand and resistive losses that only affect the absolute EL intensity on the other hand, a single EL image is insufficient. Measuring spectrally resolved EL at every position of the solar cell, however, would sacrifice the speed advantage of luminescent imaging. Thus, comparing the difference of images that use different spectral filters represents a viable compromise that combines spatial resolution with spectral information. Application of a spectral filter with transmittance T_{fil} changes Eq. (3.3) to

$$S_{cam}^{fil}(\mathbf{r}) = \int Q_{cam}(E) T_{fil}(E) Q_e(E,\mathbf{r})\phi_{bb}(E) dE \exp\left(\frac{qV(\mathbf{r})}{kT}\right) \quad (3.4)$$

It is easily seen that the measurement of the contrast

$$C_s(\mathbf{r}) = \frac{S_{\text{cam}}^{\text{fil},1}(\mathbf{r})}{S_{\text{cam}}^{\text{fil},2}(\mathbf{r})} = \frac{\int Q_{\text{cam}}(E) T_{\text{fil},1}(E) \phi_{\text{em}}(E,\mathbf{r}) dE}{\int Q_{\text{cam}}(E) T_{\text{fil},2}(E) \phi_{\text{em}}(E,\mathbf{r}) dE} \qquad (3.5)$$

between two EL images acquired at the same voltage V using two different filters as proposed by Würfel et al. [8] cancels out the voltage-dependent term in Eq. (3.4). Thus, the combination of suitable filters isolates the optical properties and the recombination losses from the resistive losses.

Finally, for spectrally resolved EL measured integrally over the area of a solar cell, we obtain from Eq. (3.3) for the detected signal

$$S_{\text{det}}(E) = Q_{\text{det}}(E) \phi_{\text{bb}}(E) dE \int Q_e(E,\mathbf{r}) \exp\left(\frac{qV(\mathbf{r})}{kT}\right) d\mathbf{r} \qquad (3.6)$$

where $Q_{\text{det}}(E)$ denotes the spectral sensitivity of the detector system including the detector itself as well as the monochromator. Depending on the details of the optics used in the experiment, the spatial integral in Eq. (3.6) extends over a smaller or larger part of the cell area.

It becomes clear from Eqs. (3.5) and (3.6) that the spectral sensitivities $Q_{\text{cam}}(E)$ and $Q_{\text{det}}(E)$ of the camera or the detector are critical for the analysis of the EL emission. Most photographic EL setups use a Si-CCD camera due to the lower noise and the lower price compared with an InGaAs camera. As shown in Figure 3.2 (solid line), the QE of such a Si-CCD camera is below 10^{-3} at the maximum of the EL emission. The integral EL intensity detected by the camera (open circles in Figure 3.2) is

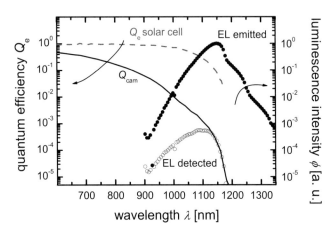

Figure 3.2 Comparison of the QE of Si solar cell and Si-CCD camera with the electroluminescence emitted by the solar cell and detected by the CCD camera. The connection between EL and QE of the solar cell is given by Eq. (3.1), and the connection between the camera sensitivity Q_{cam} and the EL detected by the camera is given by Eq. (3.3).

reduced by more than three orders of magnitude compared with the imaginary case of a perfect detector with a QE of unity over the whole spectral range of interest.

3.3
Spectrally Resolved Electroluminescence

In the past, the bulk of spectroscopic luminescence investigations of photovoltaic materials have been done using optical and not electrical excitation. These spectroscopic photoluminescence (PL) studies helped, for example, to determine the dislocation densities in multicrystalline Si wafers [21–23] or to investigate defects [24, 25] and the quasi-Fermi-level splitting in chalcopyrites [26] and thin-film Si solar cells [27]. Spectroscopic EL measurements have been rarely used for the characterization of solar cells, in spite of the fact that EL emission is the complementary action to the photovoltaic effect and is directly related to QE measurements. However, some recent publications [14–19] make use of the reciprocity between QE and EL (Eq. (3.1)) and explore the applicability of spectrally resolved EL for solar-cell characterization.

According to Eq. (3.1), the EL spectrum contains in principle the same information as the QE spectrum. However, the spectral range where useful information can be gained is quite different for both methods, EL and QE. This is because the weighting factor in Eq. (3.1), the black-body spectrum ϕ_{bb}, depends exponentially on photon energy. This exponential energy dependency of ϕ_{bb} leads to the fact that EL is only measurable close to the band gap, since all higher energies E are damped by $\exp(-E/kT)$. In general, a QE measurement contains information on optical properties as well as on the recombination behavior of the solar cell. The optical properties, including the path-length enhancement by the light-trapping scheme, the absorption coefficient of the absorber material, and the quality of the back reflector, dominate the QE especially at lower photon energies close to and below the band-gap energy. Thus, the Boltzmann factor of the black-body spectrum emphasizes these optical properties, which are therefore relatively easy to extract from luminescence spectra as compared with QE spectra. One example is the determination of the absorption coefficient. Although it is possible to determine the absorption coefficient α of crystalline Si at room temperature by QE measurements [28], the determination via PL measurements has proven to be far more sensitive [29]. In case of Si where the absorption coefficient is well known for energies below the band gap (cf. Refs. [28, 29]), EL spectra reveal information on the light-trapping properties of the measured cell as for instance the path-length enhancement factor [17, 30].

In contrast, the electronic properties of the solar cell, such as the diffusion length or the surface recombination velocity of the minority carriers, are usually extracted from the QE in an energy range slightly above the band-gap energy, where the QE is still close to its maximum. In this energy range, which corresponds to wavelength λ range around 780 nm $< \lambda <$ 920 nm for crystalline Si, the QE has a much better signal-to-noise ratio than the EL. Thus, the determination of the effective diffusion length – using for instance the method of Basore [31] – is more easily achieved by a standard QE measurement than with EL.

A useful feature of Eq. (3.1) is that it allows us to determine the so-called radiative saturation current density $J_{0,\mathrm{rad}}$ [32]. This radiative and therefore lower limit of the saturation current density J_0 is defined by the photon flux keeping a solar cell in thermodynamic equilibrium. This photon flux cannot be measured directly, since it will not create excess carriers in a detector which is at the same temperature as the device. However, the radiative recombination current

$$J_{\mathrm{rad}} = J_{0,\mathrm{rad}} \left[\exp\left(\frac{qV}{kT}\right) - 1 \right] \qquad (3.7)$$

resulting from a voltage V that drops over the rectifying junction is accessible by making use of the reciprocity relation. With a calibrated setup (absolute number of photons per area and time interval is known), $J_{0,\mathrm{rad}}$ could be derived, provided the voltage V is known accurately. However, a much simpler way to determine $J_{0,\mathrm{rad}}$ is given by Eq. (3.1). The measurement of the photovoltaic QE Q_e directly leads to the radiative saturation current density

$$J_{0,\mathrm{rad}} = q \int_0^\infty Q_e(E) \phi_{\mathrm{bb}}(E) dE \qquad (3.8)$$

In practice, the simplest way is to scale the EL emission with the calibrated QE as shown in Figure 3.3 and then to integrate the EL emission to obtain the correct $J_{0,\mathrm{rad}}$. The quantity $J_{0,\mathrm{rad}}$ defines not only the lower limit for the actual saturation current density J_0 of a real solar cell but also the upper limit for the open-circuit voltage $V_{\mathrm{oc,rad}} = kT/q \ln(J_{\mathrm{sc}}/J_{0,\mathrm{rad}} + 1)$. Comparing the radiative open-circuit voltage $V_{\mathrm{oc,rad}}$ with the actual V_{oc} determined from a current/voltage (J/V) measurement gives the loss $\Delta V_{\mathrm{oc}} = V_{\mathrm{oc,rad}} - V_{\mathrm{oc}}$ due to nonradiative recombination. This voltage loss is a useful quantity to compare the quality of different solar-cell materials in terms of their recombination. Interestingly, there is a direct relationship between ΔV_{oc} and the QE Q_{LED} of a light-emitting diode (LED) since both quantities depend only on the ratio of nonradiative to radiative recombination [20].

Figure 3.3 shows the QE (solid triangles) and the EL spectrum (solid lines) of three different solar-cell materials, that is, (a) crystalline Si, (b) Cu(In,Ga)Se$_2$ (CIGS), and (c) GaInAs. In addition, we compare the QE measured with a QE setup as described in Chapter 2 with the QE obtained from the EL spectrum using Eq. (3.1). Since the EL spectrum is measured only in relative units, the QE from EL is shifted in such a way that it fits best to the direct QE. The materials shown in Figure 3.3 are crystalline Si, an indirect semiconductor, Cu(In,Ga)Se$_2$, a polycrystalline, direct semiconductor, and GaInAs, a crystalline, direct semiconductor. While the first two samples are single-junction solar cells, the GaInAs solar cell is the middle cell of a triple-junction solar cell consisting of a GaInP/GaInAs/Ge stack [16]. Correspondingly, the shape of the absorption edge and the width of the EL spectrum are different for the three devices. In case of Si and Cu(In,Ga)Se$_2$, the peaks are broad and the absorption edge is smeared out, while for GaInAs, the spectrum is quite narrow and the QE changes abruptly from zero to its saturation value. The indirect band gap of Si explains the

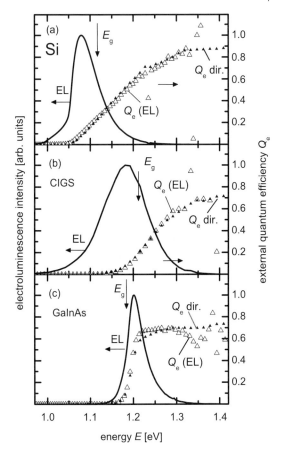

Figure 3.3 Comparison of EL spectra (lines) and quantum efficiencies (open triangles) of three different solar cells based on (a) crystalline Si, (b) Cu(In,Ga)Se$_2$ (CIGS), and (c) GaInAs. In addition to the directly measured quantum efficiencies, the filled triangles represent the quantum efficiencies that follow from the EL spectra using Eq. (3.1). A relatively broad EL peak as for the indirect semiconductor Si and the disordered polycrystalline semiconductor Cu(In,Ga)Se$_2$ correlates with a slow rise from low to high QE. In contrast, the direct and crystalline semiconductor GaInAs features a sharp peak and a small transition region between high and low QE. Data taken from Refs. [14] and [16].

differences between Si and GaInAs: the lower probability of radiative transitions and correspondingly the lower and less abrupt absorption coefficient of the indirect band-gap semiconductor Si explains the broader transition region between high and low absorption and the broader peak. In case of Cu(In,Ga)Se$_2$, the reason for the broad peak lies in the disordered nature of the polycrystalline semiconductor. Both the intentional stoichiometric inhomogeneity, due to band-gap grading normal to the cell surface, and the unintentional band-gap variation [33], due to lateral changes in stoichiometry ([In]/[Ga] ratio), lead to a clear broadening of the EL peak and of the transition region between high and low absorption in the QE spectrum.

Despite the different nature of the three investigated samples, the QE derived from the EL (open triangles) fits well to the directly measured QE (solid triangles). For all three cells, we are able to calculate the radiative saturation current density $J_{0,\mathrm{rad}}$ and subsequently the radiative open-circuit voltage

$$V_{\mathrm{oc,rad}} = \frac{kT}{q}\ln\left(\frac{J_{\mathrm{sc}}}{J_{0,\mathrm{rad}}} + 1\right) \quad (3.9)$$

By comparing $V_{\mathrm{oc,rad}}$ with the open-circuit voltage V_{oc} as derived from a J/V measurement, we obtain the values $\Delta V_{\mathrm{oc}} = V_{\mathrm{oc,rad}} - V_{\mathrm{oc}} = 185\,\mathrm{mV}$ for the Si cell, $\Delta V_{\mathrm{oc}} = 216\,\mathrm{mV}$ for the Cu(In,Ga)Se$_2$ solar cell, and $\Delta V_{\mathrm{oc}} = 132\,\mathrm{mV}$ for the GaInAs cell. Thus, since the GaInAs as a direct, crystalline semiconductor has the best ratio between radiative and nonradiative recombination, it also possesses – relative to the band-gap energy – the best open-circuit voltage and is also the best LED of the three cells. Recently, similar investigations have also been performed for organic bulk-heterojunction solar cells, revealing that the LED QE in polymer/fullerene blends is much lower than in most inorganic solar cells [34]. This implies that the ratio of radiative to nonradiative recombination is less favorable in current organic solar cells, and explains the relatively low open-circuit voltages of these devices.

Multijunction solar cells offer another particularly interesting application for spectrally resolved EL measurements. Characterization of multijunction solar cells is generally impeded by the difficulty to measure the cells in a monolithic stack independently. Monolithic means here that all layers of the cell are grown on top of each other without the possibility to contact the individual cells. Since the band-gap energies of the cells in a multijunction solar cell are usually different to make best use of the solar spectrum, the EL spectra will be at different energetic positions close to these band-gap energies. Measuring, for example, one spectrum of a triple-junction solar cell will then contain three EL peaks corresponding to the different subcells [16]. According to Eq. (3.1), the absolute amount of luminescence changes exponentially with the applied voltage. Thus, the logarithm of the EL intensity of the three peaks will then be linked to the voltage applied to the corresponding subcell at a given injection current. By varying the injection currents and measuring the EL spectra, the internal voltages of the three subcells in a triple-junction solar cell could then be determined except for an additive offset voltage if the measurement was not calibrated for absolute photon fluxes. The determination of the unknown offset voltage is possible using the open-circuit voltage V_{oc} and the corresponding short-circuit current density J_{sc} at any illumination to scale the sum of the voltages of the subcells at an injection current density $J_{\mathrm{inj}} = J_{\mathrm{sc}}$ such that it corresponds to the open-circuit voltage of the complete cell.

3.4
Spatially Resolved Electroluminescence of c-Si Solar Cells

The rapidly growing interest in EL as a characterization method for solar cells is due to the use of digital CCD cameras for image detection. Although the information

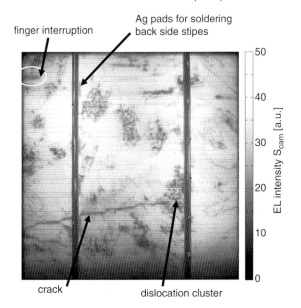

Figure 3.4 Example of an EL image of a 12.5 × 12.5 cm² multicrystalline Si solar cell at 3 A featuring both process-induced failures such as locally reduced EL emission due to broken fingers or cracks as well as variations in the EL emission, due to variations of electronic properties such as bulk diffusion lengths (e.g., reduced EL at dislocations) and the surface recombination velocity (e.g., reduced EL at unpassivated back side under Ag pads).

contained in a single pixel is spectrally integrated information, the ability to get an image in very short time frames has made EL imaging a standard tool for the quality control of finished crystalline Si solar cells. Possible reasons for spatial variations of the luminescence belong to two major categories:

1) The cell or part of the cell has a severe local failure such as a crack in the whole wafer or a broken finger that isolates a part of the cell from the rest as shown in Figure 3.4.
2) The optical or electronic properties of the solar cell vary locally such as for instance the minority carrier diffusion length is lower close to dislocations or grain boundaries. Likewise, the surface recombination velocity is higher at the metal contacts of a point-contact mask (where the metal is in direct contact with the absorber layer) as compared with the parts of the back surface covered with a passivation layer.

Failure analyses, that is, investigations of category (1), need to detect a specific process-induced problem such as a crack or a broken finger, and they have to distinguish such a problem from causes such as locally reduced diffusion lengths. This is achieved either by additional visual inspection or by using properly programmed image processing software.

Investigations of category (2), that is, the quantitative determination of physical parameters from the EL image, require methods that go beyond the inspection of a single image. As pointed out in Section 3.2, every pixel in the EL image contains information on resistive, optical, and recombination effects that can be discriminated due to their different energy and voltage dependence. To determine optical and electronic parameters such as effective diffusion length or the back-side reflectance, one has to isolate the energy-dependent parts of the EL from the voltage-dependent ones using filters as described in Refs. [8, 30] and [35]. Equation (3.5) in Section 3.2 shows that the ratio of two EL images acquired using various filters but at the same voltage is free of any resistive effect. The exponential voltage dependence of the EL signal that is sensitive to any resistive effects cancels out and the only changes in the image due to the different wavelength ranges selected by the filters remain.

Although giving absolute information on effective diffusion lengths from relative light intensities captured by a camera, the filter-method is rather sensitive to the exact filter transmission and camera sensitivity. One approach to overcome these shortcomings was presented recently by Hinken et al. using PL imaging instead of EL imaging [36]. The method makes use of one additional degree of freedom present in PL imaging. While for EL only the voltage and injection current are variable, for PL imaging of solar cells one can vary the voltage applied to the contacts of the solar cell independently from the injection of minority carriers, which is done by illuminating the sample with a light source, for example, a laser. Thus, by comparison of images acquired at open-circuit and short-circuit conditions, the effective diffusion length is calculated [36].

In case of EL imaging, changing the voltage at the contacts is equivalent to changing the injection of minority carriers since no additional illumination is present. It is not possible to change these two independently, that is, one degree of freedom is reduced compared with PL imaging. However, the variation of the voltage still provides additional information on both, resistive [10] and recombination [37] effects. Especially the locally resolved determination of the series resistance via EL and PL has recently attracted interest. The series resistance R_s at the position $\mathbf{r} = (x,y)$ is usually [38] defined by

$$R_s(\mathbf{r}) = \frac{V_{ext} - V(\mathbf{r})}{J(\mathbf{r})} \qquad (3.10)$$

where V_{ext} is the external voltage at the contacts, $V(\mathbf{r})$ the local voltage, and $J(\mathbf{r})$ the local current density. The two quantities known are the external voltage V_{ext} and the total current I or the current density $J = I/A$ averaged over the area A. This leaves the local voltage and the local current density as the remaining quantities to be determined from EL. While the local voltage can easily be calculated from the logarithm of the local EL intensity according to Eq. (3.1) (except for a constant offset), the local current density is the most critical quantity. The first attempt to determine the local series resistance by Hinken et al. [10] used an evaluation routine similar to Werner's method [39] to determine the local series resistance from the local emission and the first derivative of the EL emission for different voltages. This original method has the disadvantage that it assumes the local current density to be homogeneous. Thus, the method was only applicable to monocrystalline solar cells but not to multicrystalline Si solar cells. Attempts to improve this method were published by

Haunschild et al. [38] and Breitenstein et al. [40]. Ref. [38] assumed that the EL intensity at low voltages, where resistive voltage losses are irrelevant due to the low differential conductance of the p–n-junction at low voltages, is approximately proportional to the effective diffusion length. This assumption implies that the local current density is inversely proportional to the local EL intensity and can thus be obtained from one low-voltage EL image, again except for a constant factor. This method has still one disadvantage, namely that it requires a measurement at relatively low voltages to make sure that there are no resistive effects. Low voltages imply a lower EL signal and in turn a long integration time. To improve the speed of the method, Ref. [40] proposed an iterative computation scheme using different EL measurements taken at higher voltages with low integration times.

In addition, a solar cell with locally different current densities does no longer have a local series resistance that is a well-defined quantity. Theoretically, the series resistance according to Eq. (3.10) depends not only on local properties but also on all currents flowing everywhere in the device. This claim can be tested by assuming a network of two diodes with ohmic (voltage independent) series resistances, and the diodes themselves are connected with resistances mimicking the emitter layer and the grid. If both diodes have different diode properties, then the apparent series resistance of one of the two diodes will always depend on the current flowing through the other diode and will, therefore, not be constant when varying the voltage.

It should be mentioned that apart from the methods using EL imaging, which are presented here, there are a large number of methods developed for PL imaging as well that allow similar investigations on solar cells and wafers [41–48].

3.5
Electroluminescence Imaging of Cu(In,Ga)Se$_2$ Thin-Film Modules

EL analysis of thin-film modules is not yet as common as it is for wafer-based solar cells. Nevertheless, the suitability and the potential of EL analysis of these devices is analogous to that of Si cells as long as two requirements are fulfilled: the thin-film solar cell should be a p–n-junction device and the emission should be due to recombination of free electrons and holes in contrast to, for example, tail-to-tail emission [49]. Cu(In,Ga)Se$_2$ solar cells fulfill these requirements but not amorphous or microcrystalline solar cells. The following discusses the investigation of a Cu(In,Ga)Se$_2$ module [50, 51] as a general example for the analysis of a thin-film module. Variations of the material quality and stoichiometry in Cu(In,Ga)Se$_2$ solar cells occur on relatively small length scales below 20 μm [33, 52, 53] and would, therefore, require microscopic investigations of the luminescence [54–58]. However, prominent features in EL images on the module level are predominantly due to resistive effects, that is, either caused by series resistances or by shunts. This can be seen from Figure 3.5 presenting EL images of a Cu(In,Ga)Se$_2$ module at two different current densities (a) $J = 6.25$ mA/cm^2 and at (b) $J = 37.5$ mA/cm^2. The module consists of $N_c = 42$ cells connected in series with single cells of an area 20×0.4 cm^2. The image acquired at the lower current density (Figure 3.5a) shows

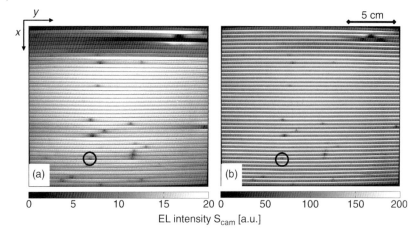

Figure 3.5 EL images at (a) $J = 6.25$ mA/cm^2 and (b) $J = 37.5$ mA/cm^2 of the same Cu(In,Ga)Se$_2$ module. Areas with quenched EL intensity are caused by shunts, which have a larger influence on the current distribution through the cell when current densities and, therefore, the differential conductance of the p–n-junction are small as in (a). Since the differential conductance of the p–n-junction increases with increasing current density, the EL intensity drop in the x direction becomes steeper as displayed in (b). The circles indicate the shunt that is used for the simulation in Figure 3.10.

dark cells at the top of the module (i.e., for low x-values) as the most striking feature. When increasing the current density to $J = 37.5$ mA/cm^2, there are no cells left that show a low emission over the whole width (i.e., extension in y direction). However, roughly circular dark spots, especially in the upper right corner, remain visible. In addition, every cell in Figure 3.5b shows a characteristic intensity gradient from high intensity at the top to low intensity at the bottom with little variation in y direction. This intensity gradient is not visible in Figure 3.5a for the image taken with the lower current density.

The macroscopic analysis discussed in the following is an example where it is reasonable to assume that $Q_e(E,\mathbf{r})$ is almost spatially independent, especially because the exponential dependence of the variations of the internal junction voltage $V(\mathbf{r})$ has a much stronger impact on the EL intensity than possible spatial variations of $Q_e(\mathbf{r})$. Thus, assuming a spatially and voltage-independent Q_e rearranges Eq. (3.3) to

$$S_{\text{cam}}(\mathbf{r}) = \int Q_{\text{cam}}(E)Q_e(E)\phi_{\text{bb}}(E)dE \exp\left(\frac{qV(\mathbf{r})}{kT}\right) \tag{3.11}$$

Consequently, we can determine from S_{cam} the voltage drop over the junction

$$V(\mathbf{r}) = \frac{kT}{q}\left[\ln\{S_{\text{cam}}(\mathbf{r})\} - \ln\left\{\int Q_{\text{cam}}(E)Q_e(E)\phi_{\text{bb}}(E)dE\right\}\right]$$
$$= \Delta V(\mathbf{r}) + V_{\text{offs}} \tag{3.12}$$

except for a spatially constant offset voltage V_{offs}.

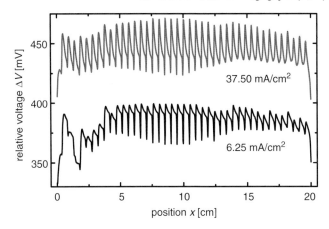

Figure 3.6 Internal voltage line scans (x direction) of the whole module of Figure 3.5 taken at different bias current densities $J_D = 37.5$ and 6.25 mA/cm². The effect of shunts on the voltage is more pronounced for smaller than for higher injection current densities. This effect is most obvious for the cells located at positions $1\,\text{cm} < x < 4\,\text{cm}$. The line scans are averaged in y direction over the whole module length l.

Figure 3.6 visualizes the application of Eq. (3.12) to the EL data from Figure 3.5 to obtain the relative voltage ΔV as a function of the coordinate x across all cells in the module. Note that we have generated the line scan by averaging over the y-coordinate, that is, over the whole length of the module. Two important features are immediately obvious from these line scans: first, the relatively low voltage drop across some cells due to the shunts, especially visible at low bias. Second, the voltage losses in x direction across individual cells due to the sheet resistance of window and back-contact layer, especially pronounced at large bias.

For a more quantitative access to the data in Figure 3.6, we need to model the voltage distribution along the whole width w of one subcell. This requires the solution of the coupled current continuity equations in the window layer and in the back contact [51]. In one dimension, we have

$$\frac{d^2}{dx^2}V_1 = -\varrho_1^{sq}\frac{d}{dx}j_1^p = \varrho_1^{sq}J(V) \tag{3.13}$$

and

$$\frac{d^2}{dx^2}V_2 = -\varrho_2^{sq}\frac{d}{dx}j_2^p = -\varrho_2^{sq}J(V) \tag{3.14}$$

where V_1, V_2 denote the voltages, j_1^p, j_2^p are the line current densities, and the ϱ_1^{sq}, ϱ_2^{sq} are the sheet resistances of the window layer and the back contact.

The solution of Eqs. (3.13) and (3.14) is given by

$$\Delta V = -\frac{j_{\max}^p \varrho_1^{sq}}{\lambda}\sinh(\lambda x) + \frac{j_{\max}^p[\varrho_1^{sq}\cosh(\lambda w) + \varrho_2^{sq}]}{\lambda \sinh(\lambda w)}\cosh(\lambda x) + \text{const} \tag{3.15}$$

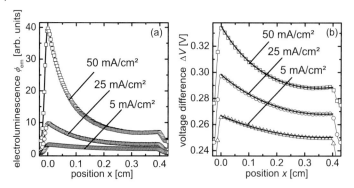

Figure 3.7 (a) Line scans of the electroluminescence intensity across one individual cell of the module shown in Figure 3.5 at different bias current densities $J_D = 50, 25, 5\,\text{mA/cm}^2$. (b) Line scans of the relative internal voltages ΔV calculated from the EL line scans in (a) according to Eq. (3.12). The solid lines represent the fits of Eq. (3.15) to the experimental voltage data (open symbols). The sheet resistances used in all fits are $\varrho^{sq}_{ZnO} = 18.2\,\Omega/\text{sq}$ and $\varrho^{sq}_{Mo} = 1.1\,\Omega/\text{sq}$ for the ZnO window layer and for the Mo back contact, respectively. Note that the experimental data are averages over 1376 lines in y direction (corresponding to a width of 3.8 cm). Data taken from Ref. [51].

with the inverse characteristic length $\lambda = [G_D(\varrho^{sq}_1 + \varrho^{sq}_2)]^{1/2}$ and $G_D = dJ/dV$ as the differential conductance at the given bias conditions.

For the investigation of the sheet resistances, EL images of a nonshunted region of the module were recorded with a higher spatial resolution. Figure 3.7 shows the calculated ΔV values calculated from the EL signals across a single nonshunted cell in x direction for three exemplary different bias current densities $J = 50, 25,$ and $5\,\text{mA/cm}^2$. Note that we have averaged the signals over 1376 lines in y direction (corresponding to a width of 3.8 cm).

To fit these experimental data, we have the choice either to determine the junction conductance G_D from an additional measurement of J_{sc}/V_{oc} independently (as in Ref. [51]) or to include the G_D values at each bias point into the fitting procedure. The latter method is based on the EL experiment alone and has the advantage of not needing an extra calibration measurement. The solid lines in Figure 3.7b show the result of a simultaneous fit of Eq. (3.15) to the experimental data obtained for the different bias current densities. Fitting parameters are the sheet resistances ϱ^{sq}_{ZnO} of the ZnO and ϱ^{sq}_{Mo} of the Mo back contact as well as the differential junction conductance G_D at each bias point. The fitting of the data presented in Figure 3.7b, and of similar data of the same module at six other voltages, yields $\varrho^{sq}_{ZnO} = 18.2\,\Omega/\text{sq}$ and $\varrho^{sq}_{Mo} = 1.1\,\Omega/\text{sq}$. Since, these values are very close to the results of the calibrated method ($\varrho^{sq}_{ZnO} = 18.0\,\Omega/\text{sq}, \varrho^{sq}_{Mo} = 1.25\,\Omega/\text{sq}$) [51], we conclude that the determination of the sheet resistances from the EL data alone is reasonably reliable. Note that the voltage curve across a cell must have a minimum to allow us to determine series resistances of both contacting layers. For the module investigated here, such a minimum is not visible until the current density $J = 50\,\text{mA/cm}^2$. The reason for that is the difference of a factor of 15 between both series resistance values. The voltage

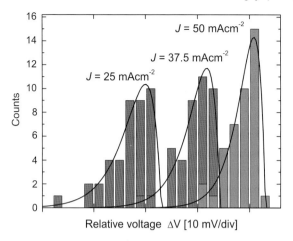

Figure 3.8 Histogram of the average relative voltage drop ΔV over the cells of the mini-module.

curve of a module with almost similar and high series resistances of both contacting layers would possess a minimum almost in the middle of the cell and that already at low current densities.

The data in Figure 3.7 allow us not only to analyze the voltage drop over individual cells but also to compare the voltages that drop over different cells. As a valid representation of such a cell voltage, we take the spatial average of the voltages determined across the cell. Again, we may calibrate the measurement with additional measurement of J_{sc}/V_{oc}, which allows us to deduce absolutely scaled J/V curves of all individual cells [51]. In practice, one might wish to avoid additional measurements. In this case, only voltages modulo an unknown offset are feasible. This possibility for a simple EL imaging based quality control method is illustrated in Figure 3.8 for three different bias current densities. As expected from our observations in Figures 3.5 and 3.7, the influence of shunts on the voltage distribution is highest at low-bias currents leading accordingly to a wide distribution of cell voltages. At higher bias currents, the discrepancy between the voltages of shunted and nonshunted cells becomes smaller and the distribution becomes narrower.

3.6
Modeling of Spatially Resolved Electroluminescence

Quantitative information about the internal voltages of the subcells and about the sheet resistance of the window layer follows directly from the EL measurements. In order to make a quantitative analysis of a shunt, we carried out spatially resolved simulations of the module, followed by fitting of the parameters of the model to the experiment. For such spatially resolved simulations of p–n-diodes, we suggest the use of an equivalent circuit composed of resistances to model the transparent conductive oxide (TCO) and metal back contact, diodes to model the p–n-junction,

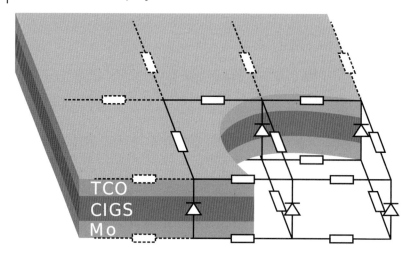

Figure 3.9 Schematic illustration of a Cu(In,Ga)Se$_2$ solar cell and the electronic network equivalent. The module is divided in rectangular meshes where each mesh cell is represented by a lumped diode in the network equivalent. The mesh cells are connected with each other via resistances, representing the TCO and the Mo contacts.

and resistances to model the shunts. Similar simulations have been done previously for c-Si solar cells [12] and the general concept of such simulations is also described in Chapter 20. To calculate the electronic network, we use the software SPICE [59].

The circuit we used to simulate modules in SPICE is illustrated in Figure 3.9. The module is divided in a rectangular mesh. Each mesh cell consists of a diode with the same area as the mesh cell, connecting the front TCO with the back contact. Resistances connecting the mesh cells with each other represent the TCO and the metal back contact. In order to obtain a good resolution in the simulations, relatively large electronic networks are required. In order to reduce simulation times, it is advisable to make use of a variable mesh along the length of the cell stripes, reducing the computation time while preserving a good resolution around the shunt. Current-crowding around a shunt can, in some cases, also extend to the adjacent cell-stripes, leading to an increased EL-intensity in the adjacent cell stripes near to the shunt. We, therefore, extended our model to simulate three series-connected cell stripes.

In the simulations, the values for the resistances making up the TCO and Mo contacts were determined from the sheet resistances of the corresponding layers ($\varrho_{ZnO}^{sq} = 18.0\ \Omega/\text{sq}$, $\varrho_{Mo}^{sq} = 1.25\ \Omega/\text{sq}$, as discussed in the previous section). The properties of the junction were determined by fitting the saturation current density and ideality factor to the J/V curves of the individual cells as determined in Ref. [51]. The ideality factor n_{id} was determined as $n_{id} = 1.17$ and the saturation current density $J_0 = 3.38 \times 10^{-8}\ \text{mA/cm}^2$. The remaining parameters needed for the simulation are the shunt resistance and radius. In principle, we can determine these parameters by fitting the simulations to experimental data. However, the shunt radius appears to be smaller than one pixel in here used EL images, preventing an accurate estimation of

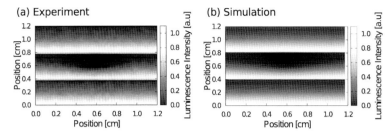

Figure 3.10 Experimental (a) and simulated (b) electroluminescence around a shunt at a current density of 25 mA/cm² through the module. This shunt corresponds to the shunt in Figure 3.5 indicated by circles.

the shunt radius. For this reason, we use a shunt with an area approximately equal to the area of one pixel in the experimental EL images. The area of one pixel in the experimental data is approximately 3.29×10^{-4} cm². In the simulations, we use a slightly higher resolution where one mesh cell is 1.0×10^{-4} cm². As we assume a roughly circular shunt, we chose a shunt area consisting of five mesh cells, that is, representing a shunt with an area of 5×10^{-4} cm². This leaves only the resistance of the shunt as a fit parameter.

Figure 3.10 shows a small section of both, the experimental (a) and simulated (b) EL image, at a current density $J = 25$ mA/cm². The shunt in Figure 3.10 is also indicated with circles in Figure 3.10. A good agreement of experiment and simulation is obtained for a shunt resistance of 300 Ω (i.e., 150 mΩ cm²), not only for the current density shown here but also for other measured and simulated currents ranging from 12.5 to 50 mA/cm².

From further simulations with varying sheet resistances of the window layer, we obtain similar results, as discussed in Ref. [60]. Higher series resistances actually reduce the influence of the shunt on the voltage drop, since the current flowing toward a shunt is limited by the distributed series resistance of the window layer.

Research on the combination of two- or three-dimensional modeling of solar modules and imaging techniques like EL imaging is just at the beginning, and first results have been discussed above. Further research will have to target for numerical tools that are fast while at the same time being accurate and able to deal with high-resolution grids. The combination of experiment and simulation has the potential to become a tool both for quantification of shunt resistances and for the calculation of the impact a shunt has on the performance of a module under arbitrary illumination conditions.

References

1 Wang, K., Silver, M., and Han, D. (1993) Electroluminescence and forward bias current in p–i–n and p–b–i–n a-Si:H solar cells. *J. Appl. Phys.*, **73**, 4567.

2 Yan, B., Han, D., and Adriaenssens, G.J. (1996) Analysis of post-transit photocurrents and electroluminescence spectra from a-Si:H solar cells. *J. Appl. Phys.*, **79**, 3597.

3 Han, D., Wang, K., and Yang, L. (1996) Recombination and metastability in amorphous silicon p–i–n solar cells made with and without hydrogen dilution studied by electroluminescence. *J. Appl. Phys.*, **80**, 2475.

4 Feldman, S.D., Collins, R.T., Kaydanov, V., and Ohno, T.R. (2004) Effects of Cu in CdS/CdTe solar cells studied with patterned doping and spatially resolved luminescence. *Appl. Phys. Lett.*, **85**, 1529.

5 Fuyuki, T., Kondo, H., Yamazaki, T., Takahashi, Y., and Uraoka, Y. (2005) Photographic surveying of minority carrier diffusion length in polycrystalline silicon solar cells by electroluminescence. *Appl. Phys. Lett.*, **86**, 262108.

6 Ramspeck, K., Bothe, K., Hinken, D., Fischer, B., Schmidt, J., and Brendel, R. (2007) Recombination current and series resistance imaging of solar cells by combined luminescence and lock-in thermography. *Appl. Phys. Lett.*, **90**, 153502.

7 Kasemann, M., Schubert, M.C., The, M., Köber, M., Hermle, M., and Warta, W. (2006) Comparison of luminescence imaging and illuminated lock-in thermography on silicon solar cells. *Appl. Phys. Lett.*, **89**, 224102.

8 Würfel, P., Trupke, T., Puzzer, T., Schäffer, E., Warta, W., and Glunz, W.S. (2007) Diffusion lengths of silicon solar cells from luminescence images. *J. Appl. Phys.*, **101**, 123110.

9 Breitenstein, O., Bauer, J., Trupke, T., and Bardos, R.A. (2008) On the detection of shunts in silicon solar cells by photo- and electroluminescence imaging. *Prog. Photovolt. Res. Appl.*, **16**, 325.

10 Hinken, D., Ramspeck, K., Bothe, K., Fischer, B., and Brendel, R. (2007) Series resistance imaging of solar cells by voltage dependent electroluminescence. *Appl. Phys. Lett.*, **91**, 182104.

11 Bothe, K., Ramspeck, K., Hinken, D., and Brendel, R. (2008) Imaging techniques for the analysis of silicon wafers and solar cells. *ECS Trans.*, **16**, 63.

12 Kasemann, M., Grote, D., Walter, B., Kwapil, W., Trupke, T., Augarten, Y., Bardos, R.A., Pink, E., Abbott, M.D., and Warta, W. (2008) Luminescence imaging for the detection of shunts on silicon solar cells. *Prog. Photovolt. Res. Appl.*, **16**, 297.

13 Fuyuki, T. and Kitiyanan, A. (2009) Photographic diagnosis of crystalline silicon solar cells utilizing electroluminescence. *Appl. Phys. A*, **96**, 189.

14 Kirchartz, T., Rau, U., Kurth, M., Mattheis, J., and Werner, J.H. (2007) Comparative study of electroluminescence from Cu(In,Ga)Se$_2$ and Si solar cells. *Thin Solid Films*, **515**, 6238.

15 Kirchartz, T. and Rau, U. (2007) Electroluminescence analysis of high efficiency Cu(In,Ga)Se$_2$ solar cells. *J. Appl. Phys.*, **102**, 104510.

16 Kirchartz, T., Rau, U., Hermle, M., Bett, A.W., Helbig, A., and Werner, J.H. (2008) Internal voltages in GaInP/GaInAs/Ge multijunction solar cells determined by electroluminescence measurements. *Appl. Phys. Lett.*, **92**, 123502.

17 Kirchartz, T., Helbig, A., Reetz, W., Reuter, M., Werner, J.H., and Rau, U. (2009) Reciprocity between electroluminescence and quantum efficiency used for the characterization of silicon solar cells. *Prog. Photovolt. Res. Appl.*, **17**, 394.

18 Kirchartz, T., Helbig, A., and Rau, U. (2008) Note on the interpretation of electroluminescence images using their spectral information. *Sol. Energy Mater. Sol. Cells*, **92**, 1621.

19 Helbig, A., Kirchartz, T., and Rau, U. (2008) Quantitative information of electroluminescence images, in *Proc. 23rd Europ. Photov. Solar Energy Conf*, WIP Renewable Energies, Munich (eds D. Lincot, H. Ossenbrink, and P. Helm), p. 426.

20 Rau, U. (2007) Reciprocity relation between photovoltaic quantum efficiency and electroluminescent emission of solar cells. *Phys. Rev. B*, **76**, 085303.

21 Koshka, Y., Ostapenko, S., Tarasov, I., McHugo, S., and Kalejs, J.P. (1999) Scanning room-temperature photoluminescence in polycrystalline silicon. *Appl. Phys. Lett.*, **74**, 1555.

22 Ostapenko, S., Tarasov, I., Kalejs, J.P., Haessler, C., and Reisner, E.-U. (2000) Defect monitoring using scanning photoluminescence spectroscopy in

multicrystalline silicon wafers. *Semicond. Sci. Technol.*, **15**, 840.

23 Kittler, M., Seifert, W., Arguirov, T., Tarasov, I., and Ostapenko, S. (2002) Room-temperature luminescence and electron-beam-induced current (EBIC) recombination behaviour of crystal defects in multicrystalline silicon. *Sol. Energy Mater. Sol. Cells*, **72**, 465.

24 Bauknecht, A., Siebentritt, S., Albert, J., and Lux-Steiner, M.Ch. (2001) Radiative recombination via intrinsic defects in $Cu_xGa_ySe_2$. *J. Appl. Phys.*, **89**, 4391.

25 Hönes, K., Eickenberg, M., Siebentritt, S., and Persson, C. (2008) Polarization of defect related optical transitions in chalcopyrites. *Appl. Phys. Lett.*, **93**, 092102.

26 Bauer, G.H., Brüggemann, R., Tardon, S., Vignoli, S., and Kniese, R. (2005) Quasi-Fermi level splitting and identification of recombination losses from room temperature luminescence in Cu$(In_{1-x}Ga_x)Se_2$ thin films versus optical band gap. *Thin Solid Films*, **480–481**, 410.

27 Merdzhanova, T., Carius, R., Klein, S., Finger, F., and Dimova-Malinovska, D. (2004) Photoluminescence energy and open-circuit voltage in microcrystalline silicon solar cells. *Thin Solid Films*, **451–452**, 285.

28 Keevers, M.J. and Green, M.A. (1995) Absorption edge of silicon from solar cell spectral response measurements. *Appl. Phys. Lett.*, **66**, 174.

29 Daub, E. and Würfel, P. (1995) Ultralow values of the absorption coefficient of Si obtained from luminescence. *Phys. Rev. Lett.*, **74**, 1020.

30 Kirchartz, T., Helbig, A., and Rau, U. (2008) Quantification of light trapping using a reciprocity between electroluminescent emission and photovoltaic action in a solar cell. *Mater. Res. Soc. Symp. Proc.*, **1101**, KK08.

31 Basore, P.A. (1993) Extended spectral analysis of internal quantum efficiency. Proc. of the 23rd IEEE Photovoltaic Specialists Conference, IEEE, New York, p. 147.

32 Kirchartz, T., Mattheis, J., and Rau, U. (2008) Detailed balance theory of excitonic and bulk heterojunction solar cells. *Phys. Rev. B*, **78**, 235320.

33 Werner, J.H., Mattheis, J., and Rau, U. (2005) Efficiency limitations of polycrystalline thin film solar cells: Case of $Cu(In,Ga)Se_2$. *Thin Solid Films*, **480**, 399.

34 Vandewal, K., Tvingstedt, K., Gadisa, A., Inganäs, O., and Manca, J.V. (2009) On the origin of the open circuit voltage of polymer–fullerene solar cells. *Nat. Mater.*, **8**, 904.

35 Giesecke, J.A., Kasemann, M., and Warta, W. (2009) Determination of local minority carrier diffusion lengths in crystalline silicon from luminescence images. *J. Appl. Phys.*, **106**, 014907.

36 Hinken, D., Bothe, K., Ramspeck, K., Herlufsen, S., and Brendel, R. (2009) Determination of the effective diffusion length of silicon solar cells from photoluminescence. *J. Appl. Phys.*, **105**, 104516.

37 Glatthaar, M., Giesecke, J., Kasemann, M., Haunschild, J., The, M., Warta, W., and Rein, S. (2009) Spatially resolved determination of the dark saturation current of silicon solar cells from electroluminescence images. *J. Appl. Phys.*, **105**, 113110.

38 Haunschild, J., Glatthaar, M., Kasemann, M., Rein, S., and Weber, E.R. (2009) Fast series resistance imaging for silicon solar cells using electroluminescence. *phys. status solidi (RRL)*, **3**, 227.

39 Werner, J. (1988) Schottky barrier and pn-junction I/V plots – small signal evaluation. *Appl. Phys. A*, **47**, 291.

40 Breitenstein, O., Khanna, A., Augarten, Y., Bauer, J., Wagner, J.-M., and Iwig, K. (2010) Quantitative evaluation of electroluminescence images of solar cells. *phys. status solidi (RRL)*, **4**, 7.

41 Trupke, T., Bardos, R.A., Schubert, M.C., and Warta, W. (2006) Photoluminescence imaging of silicon wafers. *Appl. Phys. Lett.*, **89**, 044107.

42 Abbott, M.D., Cotter, J.E., Chen, F.W., Trupke, T., Bardos, R.A., and Fisher, K.C. (2006) Application of photoluminescence characterization to the development and manufacturing of high-efficiency silicon solar cells. *J. Appl. Phys.*, **100**, 114514.

43 Abbott, M.D., Cotter, J.E., Trupke, T., and Bardos, R.A. (2009) Investigation of edge recombination effects in silicon solar cell

structures using photoluminescence. *Appl. Phys. Lett.*, **88**, 114105.

44 Macdonald, D., Tan, J., and Trupke, T. (2008) Imaging interstitial iron concentrations in boron-doped crystalline silicon using photoluminescence. *J. Appl. Phys.*, **103**, 073710.

45 Kampwerth, H., Trupke, T., Weber, J.W., and Augarten, Y. (2008) Advanced luminescence based effective series resistance imaging of silicon solar cells. *Appl. Phys. Lett.*, **93**, 202102.

46 Herlufsen, S., Schmidt, J., Hinken, D., Bothe, K., and Brendel, R. (2008) Photoconductance-calibrated photoluminescence lifetime imaging of crystalline silicon. *phys. status solidi (RRL)*, **2**, 245.

47 Rosenits, P., Roth, T., Warta, W., Reber, S., and Glunz, S.W. (2009) Determining the excess carrier lifetime in crystalline silicon thin-films by photoluminescence measurements. *J. Appl. Phys.*, **105**, 053714.

48 Glatthaar, M., Haunschild, J., Kasemann, M., Giesecke, J., Warta, W., and Rein, S. (2010) Spatially resolved determination of dark saturation current and series resistance of silicon solar cells. *phys. status solidi (RRL)*, **4**, 13.

49 Pieters, B.E., Kirchartz, T., Merdzhanova, T., and Carius, R. (2010) Modeling of photoluminescence spectra and quasi-Fermi level splitting in μc-Si solar cells. *Sol. Energy Mater. Sol. Cells*, **94**, 1851.

50 Rau, U., Kirchartz, T., Helbig, A., and Pieters, B.E. (2009) Electroluminescence imaging of Cu(In,Ga)Se$_2$ thin film modules. *Mater. Res. Soc. Symp. Proc.*, **1165**, M03.

51 Helbig, A., Kirchartz, T., Schäffler, R., Werner, J.H., and Rau, U. (2010) Quantitative electroluminescence analysis of resistive losses in Cu(In,Ga)Se$_2$ thin-film modules. *Sol. Energy Mater. Sol. Cells*, **94**, 979.

52 Grabitz, P.O., Rau, U., Wille, B., Bilger, G., and Werner, J.H. (2006) Spatial inhomogeneities in Cu(In,Ga)Se$_2$ solar cells analyzed by an electron beam induced voltage technique. *J. Appl. Phys.*, **100**, 124501.

53 Bauer, G.H., Gütay, L., and Kniese, R. (2005) Structural properties and quality of the photoexcited state in Cu(In$_{1-x}$Ga$_x$)Se$_2$ solar cell absorbers with lateral submicron resolution. *Thin Solid Films*, **480**, 259.

54 Romero, M.J., Jiang, C.-S., Noufi, R., and Al-Jassim, M.M. (2005) Photon emission in CuInSe$_2$ thin films observed by scanning tunneling microscopy. *Appl. Phys. Lett.*, **86**, 143115.

55 Romero, M.J., Jiang, C.-S., Abushama, J., Moutinho, H.R., Al-Jassim, M.M., and Noufi, R. (2006) Electroluminescence mapping of CuGaSe$_2$ solar cells by atomic force microscopy. *Appl. Phys. Lett.*, **89**, 143120.

56 Gütay, L. and Bauer, G.H. (2005) Lateral variations of optoelectronic quality of Cu(In$_{1-x}$Ga$_x$)Se$_2$ in the submicron-scale. *Thin Solid Films*, **487**, 8.

57 Gütay, L. and Bauer, G.H. (2007) Spectrally resolved photoluminescence studies on Cu(In,Ga)Se$_2$ solar cells with lateral submicron resolution. *Thin Solid Films*, **515**, 6212.

58 Bothe, K., Bauer, G.H., and Unold, T. (2002) Spatially resolved photoluminescence measurements on Cu(In,Ga)Se$_2$ thin films. *Thin Solid Films*, **403**, 453.

59 Ng-spice http://ngspice.sourceforge.net/index.html (16.02.2010).

60 Rau, U., Grabitz, P.O., and Werner, J.H. (2004) Resistive limitations to spatially inhomogeneous electronic losses in solar cells. *Appl. Phys. Lett.*, **85**, 6010.

4
Capacitance Spectroscopy of Thin-Film Solar Cells
Jennifer Heath and Pawel Zabierowski

4.1
Introduction

Capacitance or, more generally, admittance measurements are particularly suited for probing bulk and interface properties of a buried layer in a working diode-like device, such as a solar cell. The small signal capacitance is by definition sensitive to carrier capture and emission from trap states, being the charge response δQ to a small change of voltage δV, $C = \delta Q/\delta V$. A number of experimental techniques have been developed to exploit this sensitivity and try to map out the sub-bandgap density of states, with the most broadly applied being capacitance–voltage (CV) profiling, admittance spectroscopy (AS), and deep level transient spectroscopy (DLTS). In addition to these, we will also discuss drive level capacitance profiling (DLCP) and photocapacitance techniques. In this chapter, we endeavor to introduce these techniques from introductory concepts, while also commenting on the analysis of data from real photovoltaic materials, which is definitely an advanced topic. Although it is not possible to cover every scenario, we try to give examples, as well as references to a wide range of resources.

The electronic states associated with defects and impurities are typically divided into two broad categories: "shallow" and "deep" states, corresponding, respectively, to centers with extended electronic wavefunctions, and those that are strongly localized [1]. Generally speaking, the techniques discussed here are designed to detect nonradiative transitions involving deep states. Deep states act as traps and recombination centers, reducing the minority carrier mobility, and in some cases pinning the Fermi energy deep in the gap. The electrical activity and even the physical nature of deep states can be complex. Quite energetically shallow states can also be observed in photocapacitance measurements as a valence (or conduction) bandtail; these disorder-induced traps reduce the drift mobility.

Much effort has gone into extending junction capacitance techniques to accurately measure properties of the more continuous densities of sub-bandgap states characteristic of imperfect materials. This includes early work by Losee [2] to treat generally the effect of gap states on admittance in Schottky devices as well as

Advanced Characterization Techniques for Thin Film Solar Cells,
Edited by Daniel Abou-Ras, Thomas Kirchartz and Uwe Rau.
© 2011 Wiley-VCH Verlag GmbH & Co. KGaA. Published 2011 by Wiley-VCH Verlag GmbH & Co. KGaA.

experimental and numerical studies by Cohen and Lang [3, 4] to treat the case of continuous densities of states. In-depth discussions of the field are provided by Refs. [5, 6].

Admittance measurements generally begin with a simple CV evaluation of the diode, including whether significant shunt current exists, whether the film is fully depleted, and a rough estimate of doping density. For suitable devices, AS or DLTS data directly give the thermal trapping time of gap states, while DLTS or other transient measurements indicate whether these are majority or minority carrier traps. We will discuss ways in which additional information can also be gleaned from capacitance data, including the energetic position, density and spatial variation of the defect transition; the carrier capture cross section; and, in some cases, the position of the Fermi energy. The photocapacitance techniques measure optical transitions involving defect states, and are particularly valuable for studying optical properties of the buried absorber layers in working photovoltaic devices. The relationship between optical and thermal transition energies can also lead to a better understanding of lattice relaxation effects.

4.2
Admittance Basics

This chapter makes use of the differential capacitance, $C = \delta Q/\delta V$, as opposed to the dc capacitance definition more commonly used in circuits, $C = Q/V$. These two definitions converge for standard capacitors. The differential capacitance describes the physical quantity measured in capacitance spectroscopy and applies to a more general situation where Q may not vary linearly with V; in diodes we approximate $Q \propto \sqrt{V_{bi} - V}$ where V_{bi} is the built-in potential at the interface. However, in the small signal approximation we can still write $\delta Q \propto \delta V$.

When a small ac voltage is applied to a sample, the linear current response can consist of both a component in phase with the applied voltage, and a component that is 90° out of phase. We can represent the applied voltage V with amplitude V_{ac} and angular frequency ω as a function of time t

$$V = V_{ac}\exp(j\omega t) = V_{ac}[\cos(\omega t) + j\sin(\omega t)]$$

For the ideal resistor, R, and capacitor, C, in parallel, we measure a total current, I

$$I = V(R^{-1} + j\omega C)$$

Note that information about the resistor and capacitor values are clearly separated into the real and imaginary parts of the current, and hence the admittance.

This current can be represented using a phasor diagram on the complex plane, as illustrated in Figure 4.1a and b, and analogous diagrams can be used to represent various types of data, including the voltage difference between two measurement points, and the impedance or admittance of the circuit element. The amplitude and phase of the phasor represent the measurement at time $t = 0$; the entire measure-

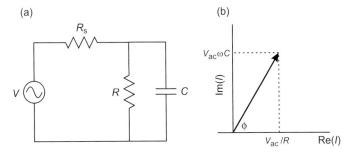

Figure 4.1 (a) Parallel circuit model for the device, with series resistor. (b) Phasor diagram for (a), neglecting R_s.

ment as a function of time corresponds to the real component of the phasor as it rotates counter-clockwise with angular velocity ω.

The small signal response of any sample can be characterized by its complex admittance, $Y = Y' + jY''$, such that $I = YV$, where Y' is the conductance, and Y'' is the susceptance. The complex impedance, $Z = Z' + jZ''$, is the inverse of admittance, such that $V = IZ$. Here, Z' is the resistance, and Z'' is the reactance. Here, we focus on admittance measurements, since the complex admittance normally allows the diode capacitance to be clearly separated from the shunt conductance.

For diodes, including solar cell devices, we frequently assume a simple parallel circuit model, as illustrated in Figure 4.1a and b, such that $C = Y''/\omega$. The most common complication is a too-high series resistance or frequency, such that either the condition $\omega R_s C \ll 1$ or $R_s \ll R$ is violated; in this case the series resistance and capacitance essentially form a low-pass filter, and as ω increases, Y''/ω no longer accurately yields C. This issue can originate either from resistance in the device and contacts, or from spreading resistance in the contact and film. Series resistance can be estimated from the dc current–voltage curve at forward bias, and its influence on the data can additionally be checked by intentionally adding a resistor in series with the device. Problems with spreading resistance can be mitigated by reducing the device area; however, this can in some cases increase the influence of stray capacitance from the edges of the device. Assuming a simple model, as illustrated in Figure 4.1a and b, with a frequency independent series resistance, a peak in the Y'/ω versus ω curve will be observed at $\omega R_s C = 1$, and as ω continues to increase, the apparent capacitance, Y''/ω, will decrease as ω^{-2}. At high frequencies, stray inductance may also be a complicating issue. A more complete discussion of dielectric response than is possible here is found in Ref. [7].

4.3
Sample Requirements

In general, any diode-like devices, including working solar cells, can be studied using the techniques described here. Of course, the diode quality must permit the capacitive response from the space charge region to be clearly distinguished. For simplicity, in this chapter we generally assume a single junction, either a one-sided p^+–n or n^+–p, or Schottky device in which the lightly doped semiconductor is not fully depleted. The treatment discussed here is also consistent for MOS devices as long as they are measured in the depletion regime. Since the oxide is in series with the semiconductor, its effect on the capacitance can easily be subtracted. Interface states also contribute more strongly to the capacitance response in MOS devices, as discussed more below; this can either be a complication or an opportunity. In-depth treatments of the theory appropriate to these more complex device structures are provided elsewhere [5, 8]. Other multilayer devices, including p–i–n structures, can also be studied. In such devices, definitively tying experimental observations to a particular layer or interface of the device becomes more complicated.

In thin-film devices, the reproducibility of sample characteristics and measurement results has been problematic. Sensitivities to light, voltage, temperature, humidity, oxygen, etc., are well documented. In addition, comparisons between samples grown in different labs must be undertaken with caution.

4.4
Instrumentation

To measure admittance, the small ac current response to an applied ac voltage must be measured. This signal consists of two parts: the amplitude and phase. Equivalently, it can be decomposed into components corresponding to the in-phase current and the current that is 90° out of phase. Many LCR meters (impedance analyzers) will directly provide this measurement. Alternatively, a more affordable option is to employ a lock-in amplifier and trans-impedance amplifier (current-to-voltage preamplifier) together to perform the same function, as illustrated in Figure 4.2. The ability to vary the ac frequency is essential. Generally, it is also important to control the ac voltage amplitude and apply a dc voltage offset.

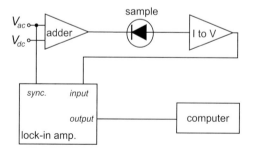

Figure 4.2 Schematic for a lock-in amplifier-based approach to measurement of complex admittance.

In the case where transient signals must be evaluated, for example, after a voltage pulse, several approaches exist. The LCR or lock-in amplifier output can be displayed on an oscilloscope, or collected directly by a data acquisition card and specialized software. In either case, the speed, accuracy, and timing or triggering of the digital data acquisition are essential considerations. The entire transient can be collected and stored for future analysis; collecting the data on a logarithmic time scale allows both short- and long-term contributions to the transient to be analyzed. This may be preferred for measurements of thin-film devices, allowing unusual behavior to be recognized, and allowing the flexibility to attempt different analyses. The analysis of exponential transient signals is discussed by Istratov and Vyvenko [9]. Some purpose-made DLTS equipment uses the Fourier transform approach discussed therein.

Regardless of the admittance measurement approach, careful calibration is required at each measurement frequency. Measurements at higher frequencies, above about 1 MHz, require extra attention to circuit components, connections, and calibration [10].

During the measurement, it is critical to monitor the relationship between conductance and susceptance, characterized by either the quality factor, Q, or dissipation factor D. These factors are defined such that $Q = D^{-1} = Y''/Y = \tan(\phi)$, where ϕ is illustrated in Figure 4.1b. Typically, D should be less than ~ 10. When D is large, such as for large leakage currents or low measurement frequencies, small errors in the measurement of ϕ can lead to large errors in determining Y'' and therefore C. It is straightforward to test the sensitivity and calibration of instrumentation with RC circuits constructed of known components.

In addition to the measurement of capacitance, sample temperature generally must be varied. The required temperature range will depend on the sample, with lower temperatures allowing measurement of transitions from shallower states, but most authors find that an l-N_2 cryostat suffices. The importance of accurately measuring the stabilized sample temperature cannot be overemphasized as it directly influences the accuracy of parameter extraction. It is advisable to carefully consider the temperature sensor and location, calibration, and optimal thermal design of the sensor–sample–cryostat system. For example, thin films grown on glass and placed in a cold-finger-type cryostat in vacuum can lag the cold-finger temperature significantly.

4.5
Capacitance–Voltage Profiling and the Depletion Approximation

The diode capacitance is traditionally analyzed using the depletion approximation. Although this approximation may not be very accurate for thin-film semiconductors, which can have significant densities of deep states, it is still a useful starting point for this discussion of admittance measurements. The depletion approximation assumes that the depletion region is precisely defined, ends abruptly, and is fully depleted of free carriers. In the depletion approximation, the depletion width will vary with applied bias, but the charge density $\rho(x)$ within the depleted region remains constant (where x is measured through the depth of the film, with $x=0$ at the interface), while the bulk region remains neutral. Then, as long as the free carrier relaxation time is short compared with the applied ac frequency, the capacitance response originates from the depletion edge, giving $C = \varepsilon\varepsilon_0 A/W$, where W is the width of the depletion region, A is the area of the device, and ε is the semiconductor dielectric constant.

For an abrupt, one-sided junction or Schottky junction

$$W = \sqrt{\frac{2\varepsilon\varepsilon_0(V_{bi}-V_{dc})}{eN_B}}$$

where N_B is the doping concentration on the lightly doped side of the junction, e is the elementary charge, and V_{dc} is the applied dc bias, where V_{dc} is positive for forward bias [11]. Then

$$C^{-2} = \frac{2(V_{bi}-V_{dc})}{e\varepsilon\varepsilon_0 A^2 N_B}$$

So in a plot of C^{-2} versus V_{dc}, the intercept gives V_{bi}, and the slope yields the CV density, N_{CV}, which in this ideal situation is identical to N_B

$$N_{CV} = -\frac{2}{e\varepsilon\varepsilon_0 A^2}\left[\frac{d(C^{-2})}{dV_{dc}}\right]^{-1} = -\frac{C^3}{q\varepsilon\varepsilon_0 A^2}\left(\frac{dC}{dV_{dc}}\right)^{-1} \quad (4.1)$$

Since, in the depletion approximation, the capacitance response originates solely from the edge of the depletion region, this result also holds true when N_B varies with position, x, through the thickness of the semiconductor [12]. So, $N_B(x)$ can be found using Eq. (4.1), where $x = \varepsilon\varepsilon_0 A/C$ is measured from the junction.

Measurement of $C(V)$ can thus give a one-dimensional profile of the doping density N_B as a function of position through the thickness of the semiconductor, with a spatial resolution characterized by the Debye screening length L_D [13]. The parabolic band bending changes by kT/e over a distance of L_D, where $L_D = \sqrt{\varepsilon\varepsilon_0 kTe^{-2}N_B^{-1}}$. To neglect the transition region at the depletion edge, we require $W \gg L_D$.

In crystalline materials, capacitance measurements as a function of voltage (CV) are considered a straightforward method to profile $N_B(x)$. However, the relatively large density of deep states in thin-film semiconductors can make it impossible to find the free carrier density using CV, as further discussed in Section 4.7.

4.6
Admittance Response of Deep States

When energy levels are present deeper in the band gap, then the depletion approximation no longer necessarily holds true, and carrier capture and emission from these states must also be considered.

The electron capture rate, c_n, into an unoccupied state is

$$c_n = \sigma_n \langle v_n \rangle n$$

where σ_n is the capture cross section for electrons, and n is the density of free electrons moving with rms thermal velocity $\langle v_n \rangle$,

$$n = N_c \exp\left(-\frac{E_C - E_F}{kT}\right)$$

The principle of detailed balance states that in thermal equilibrium, the capture and emission of electrons must be equal, and similarly the capture and emission of holes must balance. For electrons, this gives

$$e_n n_T = c_n (N_T - n_T)$$

where e_n is the emission rate of electrons, N_T is the total density of electron traps, and n_T is the density of occupied traps. Therefore, the fractional occupancy of the trap in thermal equilibrium, which is also given by the Fermi–Dirac distribution function, can be used to relate the capture and emission times to the energetic position of the trap

$$\frac{n_T}{N_T} = \frac{c_n}{c_n + e_n} = \left[1 + \frac{g_0}{g_1}\left(\frac{E_T - E_F}{kT}\right)\right]^{-1}$$

This then gives the emission rate as

$$e_n = \sigma_n \langle v_n \rangle N_c \frac{g_0}{g_1} \exp\left(-\frac{E_c - E_T}{kT}\right)$$

where g_0 and g_1 represent the degeneracy of the initial and final states, respectively.

Since these values, particularly σ_n and E_T, may themselves depend on temperature, they can differ from the measured (apparent) capture cross section σ_{na} and thermal activation energy E_{na} of the trap determined from the thermally activated experimental data. This temperature dependence can have multiple origins. For deep states, a transition corresponds to a change from one charge state to another, so it is common for deep states to be charged, and therefore attractive or repulsive. Deep states may also reside in a larger, local concentration of traps that has an overall net charge and modifies the local free carrier density [14, 15]. Additionally, the change in entropy with the change in trap state occupation may play a role as discussed in Section 4.13.

Electron emission from the defect is thus a thermally activated process with

$$e_n = \gamma \sigma_{na} T^2 \exp\left(-\frac{E_{na}}{kT}\right) \tag{4.2}$$

where, since we can approximate $N_c \propto T^{3/2}$ and $\langle v_n \rangle \propto T^{1/2}$, the temperature dependence of $N_c \langle v_n \rangle$ is isolated by defining $\gamma = N_c \langle v_n \rangle T^{-2}$ (assuming no temperature dependence of σ_{na}). For hole traps, Eq. (4.2) has the identical form, with E_{pa} referenced to the valence instead of the conduction band edge.

This model assumes that the density of deep states obeys the principle of superposition, such that each deep state creates a set of discrete electronic energy transitions that can be added together to give the total density of states within the band gap; interactions between states are neglected. Then, broad densities of states may originate from a single type of defect that is in a variety of different local environments. In real films, deep states can also interact with each other and with the lattice in complex ways that shift and broaden the density of states, and can result in changes in the defect configuration or even diffusion of the trap center upon capture or emission of carriers (see, e.g., [16, 17]).

The differential capacitance of the sample originates from its response to a small voltage perturbation, δV. In the small signal approximation, $\delta V < kT/e$. (Typical values of V_{ac}, around 30 mV, do not strictly satisfy this requirement, which may slightly impact the data [18, 19].) The resulting change in the band bending causes a change in trap state occupation at the location, x_T, where E_T is within kT of E_F, as illustrated in Figure 4.3. Thus, traps contribute to the capacitance in two ways: (i) Traps modify the space charge density and therefore the depletion width W, and (ii) in the vicinity of x_T, the traps may be able to change their charge state dynamically, following the ac voltage, and contributing to $\delta Q/\delta V$.

The characteristic time of the capacitance measurement is defined by the angular frequency of the applied ac voltage, ω, such that states with emission rates $e_n > \omega$

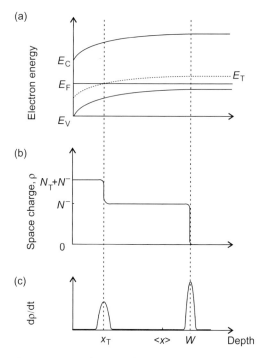

Figure 4.3 (a) Schematic of band bending with one deep trap state. (b) Charge density variation through the depletion region due to the trap state; same scale as (a). (c) In response to a changing bias, dV/dt, changes in space charge density $d\rho/dt$ can occur at both x_T and W, as indicated.

4.6 Admittance Response of Deep States

respond to V_{ac}.[1]) By rewriting Eq. (4.2), we can see that only states with apparent activation energies E_{na} up to the demarcation energy, E_e, will respond dynamically to the ac voltage, where

$$E_e = kT\ln\left(\frac{\gamma\sigma_{na} T^2}{\omega}\right) \quad (4.3)$$

Note that the demarcation energy depends on characteristics of the trap being measured, and so can be difficult to accurately determine; the same ω, T may probe different energies for physically different traps.

A general expression for the capacitance of the diode, which does not assume any specific density of states in the gap, can be derived starting from the identity

$$\frac{d}{dx}\left(x\frac{d\psi}{dx}\right) = \frac{d\psi}{dx} - x\frac{\rho}{\varepsilon\varepsilon_0}$$

where x is measured through the depth of the semiconductor ($x = 0$ at the interface), $\rho(x)$ is the depletion charge density, and $\psi(x)$ is the position-dependent potential in the depletion region, and is defined to be zero far from the interface, such that $\psi(\infty) = 0$ and $d\psi/dx\,(\infty) = 0$. Then

$$\int_0^\infty \frac{d}{dx}\left(x\frac{d\psi}{dx}\right)dx = 0 = \int_0^\infty \frac{d\psi}{dx}dx - \int_0^\infty x\frac{\rho}{\varepsilon\varepsilon_0}dx$$

and the value of ψ can be written as an integral over the charge density

$$\psi(0) = -\int_0^\infty x\frac{\rho}{\varepsilon\varepsilon_0}dx$$

Application of a small voltage, δV, is equivalent to a voltage change at the interface of $-\delta V(0)$ since $\psi(0) = V_{bi} - V_a$, where V_{bi} is the built-in potential and V_a is the voltage applied across the junction. The charge response is $\delta Q = A \int_0^\infty \delta\rho(x)\,dx$, where $\delta\rho$ is the change in charge density due to δV, and A is the area of the contact. Now the capacitance is

$$C = \frac{\varepsilon\varepsilon_0 A \int_0^\infty \delta\rho(x)\,dx}{\int_0^\infty x\delta\rho(x)\,dx} = \frac{\varepsilon\varepsilon_0 A}{\langle x \rangle} \quad (4.4)$$

where $\langle x \rangle$ is the first moment of charge response, also called the center of gravity of the charge response.

---

1) A better approximation in a depletion region dominated by a single deep trap level requires $f_t > \omega$, where $f_t = 2e_n\left(1 + \frac{x_t N_t}{W N_B}\right)$ [5].

Assumptions about the trap occupation and its response to ac and dc bias (and hence, its effect on $\langle x \rangle$) form the foundation of all the techniques discussed here. In both CV and DLTS, it is assumed that the frequency is chosen such that $\omega > e_{n,p}$ for any deep traps; then, they do not respond dynamically to the ac bias, and $\langle x \rangle = W$. In CV it is also assumed that deep traps either do not change their charge state with the dc bias, or are few enough not to have a significant influence on W. In contrast, in DLTS, the dc bias on the sample is assumed to have a strong effect on the trap state occupation. Indeed, the change in occupation of the traps in response to the dc bias yields the DLTS signal, as is further discussed in Section 4.8. In AS and DLCP, the ac frequency and/or the temperature are varied to ramp ω from values below $e_{n,p}$ to above $e_{n,p}$.

4.7
The Influence of Deep States on CV Profiles

Even if trap states do not respond to the ac voltage they may adjust their charge state to the dc bias conditions, which results in artifacts in the CV profiles. To illustrate this behavior we consider a SCAPS1D simulation of the $n^+ - p$ junction containing uniform distributions of shallow and deep ($E_T = E_V + 0.3\,\text{eV}$) acceptors of the concentrations $N_A = 2 \times 10^{15}\,\text{cm}^{-3}$ and $N_T = 2 \times 10^{16}\,\text{cm}^{-3}$, respectively [20, 21]. Figure 4.4a displays $N_{CV}(\langle x \rangle)$, where N_{CV} and $\langle x \rangle$ are determined according to Eqs. (4.1) and (4.4), respectively. These values are simulated at two measurement frequencies: $\omega_L < e_p$ and $\omega_H > e_p$. In this simulation, the deep traps are assumed to be in equilibrium with the dc bias (similar to Figure 4.3). Then, the shallow acceptor density N_A is not reproduced at either frequency. At ω_L, $N_{CV} = N_A + N_T$ is measured, while at ω_H, N_{CV} is a function of V_{dc}, with $N_A \leq N_{CV} \leq N_A + N_T$, since the static charge accumulated at deep states follows the dc voltage sweep and influences the

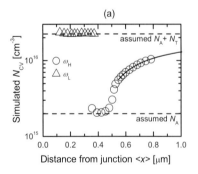
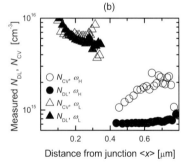
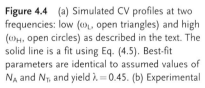

Figure 4.4 (a) Simulated CV profiles at two frequencies: low (ω_L, open triangles) and high (ω_H, open circles) as described in the text. The solid line is a fit using Eq. (4.5). Best-fit parameters are identical to assumed values of N_A and N_T, and yield $\lambda = 0.45$. (b) Experimental CV (open symbols) and DLCP (closed symbols) data for a CuInSe$_2$ device. Device details are described elsewhere [67]. All data were collected at 11 kHz, which corresponds to a low frequency at 280 K (triangles), and a high frequency at 180 K (circles).

capacitance. In this case the charge distribution delivered by CV profiling can be approximated by [22]

$$N_{CV} = N_T(x_T)[1-\lambda/W] + N_A(W) \quad (4.5)$$

where $\lambda = W - x_T$ is assumed to remain constant for applied dc voltages.

Thus, when trap densities are significant, the CV profile does not clearly yield the densities N_A and N_T nor indicate their uniformity. As discussed in Section 4.10 and illustrated in Figure 4.4a and b, DLCP can help sort out these issues. A longer discussion on the influence of deep states, including interface states, on CV and DLCP measurements, can be found in [23].

4.8
DLTS

In the standard approach, DLTS analysis consists of extracting the emission rates of deep levels from a transient capacitance signal, as illustrated in Figure 4.5a–c [23].

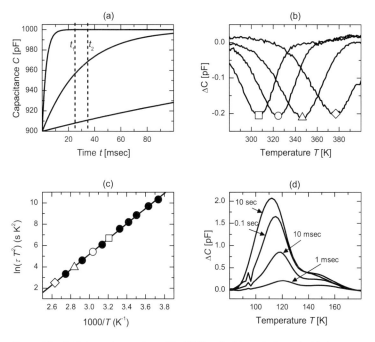

Figure 4.5 (a) Schematic transients. The DLTS signal, $[C(t_1) - C(t_2)]/C_0$, will be maximized when the rate window and emission rate match (Eq. (4.11)). (b) DLTS data for a GaAsN device. Each trace corresponds to a different rate window. (c) Arrhenius plot yielding E_{na} and σ_{na} of the trap. Data points corresponding to peaks in (b) are indicated with the same symbol. (d) Filling pulse dependence of the DLTS signal for the same sample; in this case data for a minority carrier trap is shown. The labels indicate filling pulse length. See [68, 69] for further discussion. Data courtesy of Steve Johnston, National Renewable Energy Laboratory.

These transients result from a voltage or optical pulse applied to the junction. In its simplest application, DLTS allows for determination of deep trap parameters such as the type of the defect (majority or minority carrier trap), activation energy, capture cross sections, and trap concentration.

To understand the DLTS signal, we are interested in calculating how a change of a charge ΔQ at some distance x from the interface ($x = 0$) will influence the junction capacitance. It is assumed that the additional charge is trapped at defect states of the concentration N_T within a layer of the width Δx. Hence, $|\Delta Q| = q\Delta x N_T$. If we do not initially assume a one-sided junction, then this charge will induce a change of the depletion layer width $\Delta W = \Delta W_n + \Delta W_p$, where ΔW_n and ΔW_p are the depletion widths on the n and p side of the junction, respectively. Small capacitance changes ($\Delta C \ll C$) can be approximated as

$$\frac{\Delta C}{C} \approx -\frac{\Delta W}{W} \quad (4.6)$$

By solving the Poisson equation, ΔW can be eliminated from Eq. (4.6) yielding [24]

$$\frac{\Delta C}{C} \approx \mp \frac{N_T \Delta x}{W^2} \left(\frac{x + W_n}{N^-} - \frac{W_p - x}{N^+} \right) \quad (4.7)$$

where N^+ and N^- are the densities of ionized space charge on the n and p sides of the junction, respectively. The negative and positive signs correspond to decrease and increase of negative charge, respectively, or equivalently, an increase and decrease of positive charge. Rewriting Eq. (4.7) for a spatial distribution of trapping states within the depletion layer yields

$$\frac{\Delta C}{C} \approx \mp \int_{-W_n}^{W_p} \frac{N_T(x)}{W^2} \left(\frac{x + W_n}{N^-} - \frac{W_p - x}{N^+} \right) dx \quad (4.8)$$

For highly asymmetrical junctions, (here n^+–p), assuming a uniform defect distribution, and neglecting the transition region, one arrives at

$$\frac{\Delta C}{C} \approx \mp \frac{N_T}{2N^-} \quad (4.9)$$

Thus, by measuring the capacitance transient $\Delta C(t)$, one can easily monitor the occupation of deep levels while the system relaxes after the equilibrium has been disturbed, for example, by application of a voltage or a light pulse.

The above equations show that the sensitivity of the capacitance change depends on the location of the trapped charge within the junction: it is highest at the border of the space charge layer and decreases toward the interface. For $x = 0$, that is, if the charge is accumulated at the very interface, the capacitance does not change at all ($\Delta C = 0$), since by charge neutrality $N^+ W_n = N^- W_p$. For this reason, an MIS-type structure can be an advantage in studying interface states with DLTS. Equations (4.8) and (4.9) also hold true for an MIS diode, or for an n^+–i–p (p^+–i–n) structure, assuming the depletion width W_n (W_p) remains constant [25, 26].

Depending on the nature of the perturbing pulse and the characteristics of the trap states, the transient can be due predominately to either an emission or a capture of majority or minority carriers. In most materials, there may be many different trap states and, assuming as in the SRH model that they are noninteracting and have discrete energy levels, the capacitance transient will consist of a sum over many exponentials.

To observe the emission of majority carriers, the junction is initially maintained in quasiequilibrium at reverse bias $-U_R$. Majority carrier traps of the concentration N_T are ionized in a region of width $W(-U_R) = W_R$ and contribute to the junction capacitance $C(-U_R)$. During the bias pulse, the reverse bias voltage and therefore depletion width are reduced for a time t_p, and these traps capture free carriers that appear in the previously depleted region. When the reverse bias is restored, the excess free carriers will be swept away. However, the trapped majority carriers reduce the space charge density, increasing the depletion width and causing the capacitance to drop below $C(-U_R)$ by ΔC. These trapped carriers are thermally emitted, and the capacitance gradually increases back toward $C(-U_R)$. This will be observed as a capacitance transient.

In the depletion region there are practically no free carriers, and capture processes can be neglected [6]. For an n^+-p junction, assuming $e_p \gg e_n$, which usually holds for acceptor-type defects located in the lower part of the band gap, the capacitance transient $C(t)$ is characterized by a single time constant $\tau = 1/e_p$

$$C(t) = C_\infty \left[1 - \frac{N_T}{2N^-} \exp(-e_p t)\right]$$

where C_∞ is the quasiequilibrium capacitance, in this case equal to $C(-U_R)$. An analogous equation is obtained for a p^+-n junction.

In this kind of experiment, the charge state of minority carrier traps usually does not normally change during the pulse [27]. To investigate such defects it is necessary to inject minority carriers into the depletion region by forward biasing the junction during the pulse or applying an optical pulse. For positively charged donors in p-type material electron capture will prevail ($c_n \gg c_p$) and at the end of the pulse most of the traps will be occupied by electrons. The negative space charge increases, which gives $\Delta C > 0$. After the pulse these electrons will be emitted and the charge relaxation results in a capacitance transient characterized by a single time constant $\tau = 1/e_n$

$$C(t) = C_\infty \left[1 + \frac{N_t}{2N^-} \exp(-e_n t)\right]$$

where N_t is the density of minority carrier (in this case, donor) traps in the material.

The sign of the capacitance transient $C(t)$ is always positive and negative for the emission of minority and majority carriers, respectively, independent of the conductivity type of the semiconductor (p or n). Note that significant series resistance can result in majority carrier traps appearing to have positive (minority carrier) transients. Apparent minority carrier transients should be checked by adding an additional series resistor to the device and observing its influence on the sign of the transient [6].

The normalized DLTS signal is defined as the difference of the capacitance at times t_1 and t_2 ($t_1 < t_2$)

$$S(T) = \frac{C(t_1) - C(t_2)}{\Delta C(0)} \quad (4.10)$$

where $\Delta C(0)$ is the maximal capacitance change [24]. The $S(T)$ curve is called a DLTS spectrum. Since the emission rate[2] is an increasing function of temperature, $S(T) = 0$ for very slow (low T) and very fast (high T) transients. At intermediate temperatures $S(T)$ passes an extremum[3] as the time constant of the transient $\tau_{n,p}$ equals the rate window τ, defined as

$$\tau = \frac{t_1 - t_2}{\ln(t_1/t_2)} \quad (4.11)$$

Activation energy and capture cross section of investigated traps can be calculated from the slope and intercept of Arrhenius plot as illustrated in Figure 4.5c. The trap concentration can be evaluated using Eq. (4.9), provided the shallow net doping level is known.

4.8.1
DLTS of Thin-Film PV Devices

The above considerations were conducted under a number of simplifying assumptions, which very often are not fulfilled in thin-film PV devices. Typical complications that arise if one tries to apply the quoted standard formulae include:

1) In the so-called λ-effect, DLTS peak height is not related to deep trap concentration in a simple equation but depends on the energetic position of the defect level in the band gap. To derive the more accurate formula, $\Delta C/C \approx f_\lambda N_T/2N^-$, where f_λ is a function of trap occupancy, one has to take into account the distance λ between the depletion edge and the point where the trap level and the quasi-Fermi level intersect [28].
2) The approximation $\Delta C/C \approx -\Delta W/W$ breaks down when deep trap concentrations are large, which has severe consequences for the DLTS signal analysis: (i) the capacitance transients become nonexponential, which influences the shape of DLTS peaks, (ii) DLTS peak position, and thus calculated activation energy and capture cross section, start to depend on the N_T/N-ratio, and (iii) the λ-effect becomes more pronounced [29].
3) Since the emission of carriers occurs at large reverse biases, the emission rates can be influenced by the electrical field through Poole–Frenkel and tunneling

---

2) It is assumed that the response of a single level is observed. Ideally, different levels have significantly different emission rates, at a given temperature, and their responses can be clearly separated.
3) A maximum or a minimum occurs for minority and majority carrier traps, respectively.

effects [30, 31]. To avoid this kind of complication, reverse bias DLTS (RDLTS) mode is used [32].
4) If a few closely spaced energy levels contribute to the capacitance transient in the same temperature range, the poor resolution of the standard (R)DLTS method does not allow for a clear separation of the individual components. Application of an inverse Laplace transform to the analysis of capacitance transients, called the Laplace-DLTS method, increases the resolution by an order of magnitude [33, 34]. However, one has to be extremely careful while interpreting emission rate spectra since $s(e_p)$ is always a result of numerical calculations.
5) Nonexponential response, for example, due to defect relaxation, sometimes dominates capacitance transients and can be misinterpreted as the energetic distribution of defects within the band gap. This issue is closely related to metastable phenomena, discussed in Section 4.14.
6) Barriers to carrier capture may require a fairly long filling pulse to fully populate the traps, as illustrated in Figure 4.5d. The dependence of DLTS signal on pulse length can allow the capture barrier energy to be measured [5]. This behavior can also be confused with metastable changes in the device.

4.9
Admittance Spectroscopy

Measurement of the sample admittance as a function of applied ac frequency and temperature is termed AS. This technique can yield the thickness of the film, the position of the Fermi energy in the bulk, the energetic position of dominant defect bands that occur between the Fermi energy and mid-gap, and an estimate of the density of those states. In contrast to both the DLTS and CV techniques, in AS the frequency (or temperature) is ramped so as to cross the transition frequency where the traps just start to respond. Note that AS can only detect traps between the band edge and mid-gap [35].

When the sample is too cold, or the frequency too high, then there is no time for carriers in the bulk, undepleted material to shift in and out of the depletion edge in response to the applied voltage, and a condition called freeze-out occurs. Under these conditions, the capacitance response will be that of the bulk dielectric, $C = \varepsilon\varepsilon_0 A/h$, where h is the distance between the top and back contacts. Increasing the temperature, T, or decreasing the frequency, f, eventually a step will be observed from $C = \varepsilon\varepsilon_0 A/h$ to $C = \varepsilon\varepsilon_0 A/W$.

The dielectric relaxation time, τ_R, is determined by $\tau_R = \rho_s \varepsilon\varepsilon_0$. The resistivity of the semiconductor, ρ_s, depends on the free carrier density, which has a thermal activation energy of E_F, and the mobility. So, this initial step in the capacitance occurs at a characteristic energy of E_F as long as E_F does not vary strongly with T and the mobility is approximately constant. If the charge transfer is instead limited by mobility, usually seen at high frequencies, then these data may give the majority carrier mobility in the film [36].

As T continues to increase or f decrease, trap states can begin to respond. The demarcation energy E_e (Eq. (4.3)) determines the cut-off energy for trap response at a

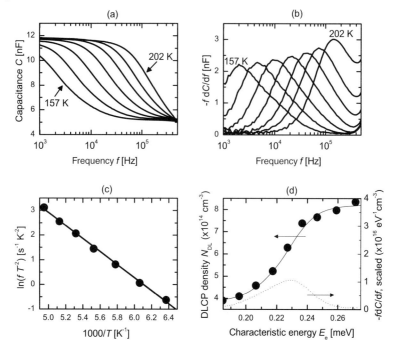

Figure 4.6 (a) Sample AS raw data, for a CIGS device. Data was taken in 8 K increments from 157 to 202 K. (b) The AS data after applying Eq. (4.12), showing the characteristic (peak) frequencies at each temperature. (c) Arrhenius plot of each peak (circles). A linear fit (line) yields $E_{na} = 0.23$ eV and $\gamma\sigma_{na} = 1.0 \times 10^7\, s^{-1}\, K^{-2}$. (d) DLCP data (circles), collected at 45 kHz, at the location $\langle x \rangle = 1.3-1.4\,\mu m$, and at temperatures ranging from 157 to 220 K. The integral over the 181 K AS $(-fdC/df)$ data (solid line) was scaled by a constant factor to show the close agreement of transition energy and magnitude between AS and DLCP. The raw AS $(-fdC/df)$ data at 181 K (dashed line), adjusted by the same scaling factor, thus represents the density of states for the 0.23 eV trap. Data courtesy of Jian V. Li, National Renewable Energy Laboratory.

certain T, f data point. When $E_T = E_e$, the occupation of the state can follow the ac voltage, and its charge state will change at the location x_e. This causes $\langle x \rangle$ to move closer to the interface and C to increase from $\varepsilon\varepsilon_0 A/W$ to $\varepsilon\varepsilon_0 A/\langle x \rangle$ as shown in Figure 4.6a. Successively deeper trap states respond as E_e is further increased.

Steps in C correspond to peaks in G/ω, due to their causal relationship. This correspondence can be calculated using the Kramers–Kronig transformations, which relate the real and imaginary parts of the susceptibility, χ' and χ'', respectively [7]. A slight correction may be necessary to apply these transformations within the p–n junction [37]. After measurement of one component of the data, the Kramers–Kronig transformations can be employed to construct the other, although since frequencies are never truly measured from 0 to ∞ this is done with the assistance of approximations or fits to the data [38, 39]. Particularly in electrochemical measurements, it is common to use this procedure to verify the conductance measurement, but such tests are not necessarily definitive [40].

While the G/ω peaks can be a good way to observe the position of the capacitance step, they can be masked by the leakage conductance in some samples; analogous peaks can also be obtained from the derivative

$$\frac{dC}{dE_e} = -\frac{\omega}{kT}\frac{dC}{d\omega} \qquad (4.12)$$

as illustrated in Figure 4.6b. The (ω, T) data points from each peak can then be plotted on an Arrhenius plot; typically $\ln(\omega/T^2)$ is graphed as a function of $1000/T$ as shown in Figure 4.6c. Using Eq. (4.2), the slope then yields the apparent activation energy E_{na}, while the intercept gives the apparent capture cross section σ_{na}. These quantities may differ from the trap energy and the trap capture cross section, as discussed above in Section 4.6. The graph of $\omega dC/d\omega$ versus ω can be transformed into $N_t(E_e)$ versus E_e by using the measured σ_{na} to rescale the ω axis to E_e, and calculating $N_t(E)$ according to [38]

$$N_t(E_e) \approx -\frac{V_{bi}^2}{W[eV_{bi}-(E_{F\infty}-E_e)]}\frac{\omega}{kT}\frac{dC}{d\omega} \qquad (4.13)$$

where $E_{F\infty}$ indicates the position of the Fermi energy in the bulk. Rescaling of the ω axis to E_e has been illustrated in Figure 4.6d, although in that case the signal magnitude is scaled using DLCP data as described in the caption. For broad densities of states, a clear peak may not be observable in $\omega dC/d\omega$ versus ω, in which case the factor necessary to rescale the ω axis such that all of the $\omega dC/d\omega$ curves overlap can yield an estimate for σ_{na}.

The Arrhenius plot is commonly interpreted to directly yield E_T; however, in many materials the apparently same trap response has been observed with differing E_{na} and σ_{na} values. These are often related by

$$\sigma_{na} = \sigma_{00}\exp\left(\frac{E_{na}}{E_{char}}\right) \qquad (4.14)$$

such that a graph of $\ln\sigma_{na}$ versus E_{na} yields a straight line with slope E_{char}^{-1} and intercept $\ln\sigma_{00}$. The quantities E_{char} and σ_{00} are phenomenologically observed parameters. This relationship is known as the Meyer–Neldel rule, and is further discussed in Section 4.12.

4.10
Drive Level Capacitance Profiling

The technique of DLCP determines the density of states responding dynamically to the ac bias. Like AS, this technique allows different trap response energies to be isolated by adjusting the temperature and frequency of the measurement. At the same time, DLCP yields a density somewhat analogous to CV; however, in DLCP, the *ac* voltage amplitude, V_{ac}, is varied to determine the density.

The value of V_{ac} employed in DLCP no longer satisfies the small signal condition, and the capacitance response, as a function of V_{ac}, is fit to yield higher order corrections

$$C = C_0 + C_1 dV + C_2 dV^2 + \cdots \qquad (4.15)$$

Note that, as V_{ac} is varied, at the same time the dc bias must be adjusted so that the maximum applied voltage remains constant, that is, the voltage waveforms are aligned at their peaks [41]. This gives an accurate value for the small signal capacitance, C_0, as well as an additional experimental parameter C_1. In DLCP, the values of C_0 and C_1 are used to find the gap state density according to [44]

$$N_{DL} \equiv -\frac{C_0^3}{2e\varepsilon\varepsilon_0 A^2 C_1} = p + \int_{E_F^x}^{E_V + E_e} g(E, x_e)\, dE \qquad (4.16)$$

for an n^+–p device, where an analogous equation can be written for a p^+–n device. In Eq. (4.16), N_{DL} is the free carrier density plus the trap density that is able to respond at E_e, and $g(E, x_e)$ is the density of states in the gap at the location x_e.

Since DLCP is a purely ac measurement, the carriers contributing to N_{DL} are exactly those responding in an AS measurement. The same activation of trap response as a function of f, T can be observed, as illustrated in Figure 4.6d. In samples with deep traps, the high-frequency DLCP can give a more accurate estimation of free carrier density than CV profiling data [23]. At low frequencies, DLCP should give the same result as CV profiling, where low frequencies are defined such that all traps that change their charge with the dc bias in CV also respond dynamically to the ac bias. DLCP gives the trap density more directly than AS, not requiring any particular knowledge about the device other than the dielectric constant.

The DLCP measurement can be repeated as a function of V_{dc}, to yield $N_{DL}(\langle x \rangle)$, with the spatial sensitivity roughly determined by the Debye length [42]. The profile $N_{DL}(\langle x \rangle)$ helps to identify spatial variations through the film [23, 43], and responses that are localized or at the interface. The issue of spatial variation is further discussed in Section 4.13.

4.11
Photocapacitance

Photocapacitance experiments come in many variations, all intended to measure the optical energy and cross section for transitions between traps and the conduction or valence band. The change in charge of the junction due to optical excitation of carriers into or out of defects results in a capacitance change that can be measured as a function of photon energy. While there is not space here to delve into the details of these measurements, we give a brief discussion of their benefits and references for further study.

Transient photocapacitance techniques allow a particular thermal transition to be clearly associated with its corresponding optical transition(s), and include both deep level optical spectroscopy (DLOS) [44] and transient photocapacitance spectroscopy (TPC) [45], which are based on the technique of DLTS. Essentially the DLTS experimental setup can be used, with the addition of a monochromatic light source, filters, and a shutter to control optical excitation of the sample. In these measurements, voltage or optical filling pulses are employed to alter trap occupation from its equilibrium value. The ensuing transient is observed under monochromatic optical excitation.

In DLOS, the transient rate is measured at times as close as possible to time = 0 (measured from the end of the filling pulse) to isolate the optically induced excitation of carriers out of the traps from the thermal relaxation. In TPC, transients in the light and in the dark are subtracted to remove the thermal part of the relaxation from the signal. The TPC technique utilizes a very low light intensity, selected such that the signal varies linearly with photon flux.

In both cases, the temperature and filling pulse are chosen carefully from either DLTS or AS data such that the optical response associated with particular thermal transients can be isolated. For example, following a majority carrier filling pulse, at high temperatures a trap may thermally empty very rapidly, more quickly than the response time of the electronics. At a lower temperature, the same trap may have a time constant on the order of seconds, such that it is not significantly thermally emptied during the time scale of the measurement. Only in the latter case could the trap contribute to the optically induced aspect of the transient. By subtracting the high and low temperature spectra, the optical transition(s) associated with a specific defect, which has already been observed thermally, may be identified.

In some cases, dramatically different optical and thermal energies have been observed, allowing a better understanding of the lattice relaxation associated with changes in occupation of a trap; the DX centers in GaAs are a well-known example [46].

These techniques allow what is essentially an optical absorption spectrum of the buried absorber layer in the complete, working device to be measured. Since many characteristics of the film can be altered due to deposition of ensuing layers (such as due to high temperatures, migration of elements, etc.), this can be a valuable tool, for example, to study the role of Ga and Na in CIGS [47, 48].

Finally, a similar measurement can be conducted while measuring the current, rather than capacitance, response of the device. This technique, dubbed transient photocurrent spectroscopy, gives complementary information to TPC. This is because the capacitance response depends on net charge change, p–n, in the device, while the current transient shows the total carriers excited out of the traps, p+n. The difference between these spectra can be used to better understand minority carrier transport in the film [49].

4.12
The Meyer–Neldel Rule

The Meyer–Neldel rule is a very general rule describing thermally activated processes for which the measured prefactors and energies in different samples are interrelated by Eq. (4.14). Meyer and Neldel observed that the activated dc conductivity of several related oxide semiconductors followed this relationship [50]. Since that time, many thermally activated processes have been observed to follow the Meyer–Neldel rule. In a-Si:H, it is observed in such seemingly disparate processes as the thermal annealing of defects and the activation of dc conductivity [51, 52]; AS spectroscopy data for CIGS are discussed in Ref. [53]. The Meyer–Neldel rule is also known as the compensation rule when it is observed in thermally activated rate processes of chemical reactions (see, for example, Ref. [54]).

The Eyring model, essentially the recognition that the measured transition corresponds to an increase in Gibbs free enthalpy, ΔG, for the thermodynamic system upon ionization, is the most generally accepted explanation of this behavior [55]. The trap energy is $E_T - E_V = \Delta G$ (for an acceptor-type trap). But, since ΔG is separated into entropy, ΔS, and enthalpy, ΔH, terms by $\Delta G = \Delta H - T\Delta S$ the measured thermal activation energy corresponds to ΔH rather than ΔG, and the value of ΔS affects the apparent capture cross section.

Then, Eq. (4.2) becomes

$$e_n = \chi_n N_c \sigma_n \langle v_n \rangle \frac{g_0}{g_1} \exp\left(-\frac{\Delta H}{kT}\right)$$

where

$$\chi_n = \exp\left(\frac{\Delta S}{k}\right)$$

For deep states, ΔS is thought to predominantly originate from the change in lattice vibrations occurring with the change in occupancy, and bonding configuration, of the defect. A discussion of ΔS for traps in CIGS is given in [15]. When the electronic transition does not involve phonons, the localized state is not coupled to the local phonon modes, and the electronic degeneracy is neglected, then it has been shown that $\Delta S = 0$ and $\Delta G = \Delta H$ [56]. This includes transitions from the shallow, hydrogenic dopant states.

Explanations attempting to quantify ΔS and relate it to conduction in semiconductors focus on the electron–phonon interaction [57–59]. Unrelated mechanisms can cause apparent Meyer–Neldel behavior, especially when trends are only observed over one or two orders of magnitude. For example, see Refs. [60, 61].

4.13
Spatial Inhomogeneities and Interface States

In thin-film solar cell materials, spatial inhomogeneities and interface states are often present. Here we consider only inhomogeneities through the depth of the film. Inhomogeneity could be observed in N_T, E_{na}, σ_{na}, or all of these. Since changes in band bending or E_F affect x_e, most of the techniques discussed here may probe a range of trap locations over the course of the measurement, and inhomogeneity will lead to anomalous results. One exception is DLCP, in which the trap density can be measured by varying only the frequency, keeping T and V_{dc} constant (even so, the trap response and free carrier response occur at two different locations in the film, and the spatial resolution is limited).

In most cases it is possible to test a hypothesis of inhomogeneity in AS by repeating the measurement at a different value of V_{dc}. For a uniform sample, λ should remain constant with dc bias. So, $\langle x_H \rangle - \langle x_L \rangle = \varepsilon\varepsilon_0 A(C_H^{-1} - C_L^{-1})$ should not vary with V_{dc}, where C_H and C_L correspond to measurements above and below the transition frequency for the trap, respectively. However, this same identity could also hold true for an interface minority carrier trap or for a double-diode device with a nonohmic back contact, as discussed below.

An extremely important issue in thin-film photovoltaic devices are interface states, which represent a special case of inhomogeneity. The characteristic energy at which interface states respond dynamically to the ac bias corresponds to the Fermi energy at the interface. Assuming that E_F is not pinned at the interface, and that the density of interface states $N_I(E)$ is spread fairly uniformly over a wide range of energies, then the apparent activation energy of interface states will shift with V_{dc} (and possibly T); such a measurement also allows $N_I(E)$ to be profiled. These characteristics, particularly a shift of E_{na} with V_{dc}, are highly suggestive of near-interface traps [8] (but not to be confused with freeze-out of the mobility, which can also be bias dependent [39]).

A spatially and energetically localized density of states (including near the interface) could appear as a peak in N_{CV} occurring at a particular value of V_{dc}, because W changes abruptly with V_{dc} as the trap state crosses from above to below E_F; such a peak should generally be suppressed in DLCP except at the exact T and f at which the trap state can respond to the ac voltage [46]. Hence, comparisons of results between different parameter values and different techniques can be extremely useful.

It is also possible for an interface minority carrier trap to be filled and emptied via the highly doped side of the junction. The barrier height for such a response may remain roughly constant with V_{dc}, since there is not a large potential drop on the highly doped side of the interface. This situation would appear like a bulk trap, where $\langle x_H \rangle - \langle x_L \rangle = \varepsilon\varepsilon_0 A(C_H^{-1} - C_L^{-1})$ is constant with V_{dc} [21, 62]. In some cases the two possibilities can be distinguished by the magnitude of $\langle x_H \rangle - \langle x_L \rangle$, which, in the case of an interface trap, will correspond to the relatively small depletion width on the highly doped side of the interface plus the thickness of any intrinsic buffer layer at the interface. Otherwise, variations in the sample preparation, such as replacement of the one-sided junction with a Schottky junction, are necessary to resolve this question.

Such a study has indeed yielded new information on a long-discussed trap in CIGS, showing that the trap response remains even in Schottky devices and indicating instead that the measured response could originate from a nonohmic back contact, which would give a similar $C-V$ result [63]. In this case, the activation energy corresponds to the back contact barrier height. The presence of a nonohmic back contact may be identifiable from $I-V-T$ measurements [64, 65].

In principle, DLTS is much less sensitive for charge changes taking place in the vicinity of the interface. However, both majority and minority carrier traps at the interface can be effectively investigated by DLTS in MIS-like structures, as already indicated. This requires some modifications from the standard DLTS measurement discussed above, and in particular, the pulse amplitude must be carefully chosen. The investigation of interface states in MIS structures by DLTS was discussed extensively by Murray et al. [28].

4.14
Metastability

An additional complication occurs when sample properties can be persistently changed by external factors like voltage stress or illumination, and then recovered, perhaps after heating the sample in a controlled environment and in the dark. In practice, such effects are common, but the details are characteristic to each specific material, and indeed may depend on aspects of device preparation that are closely, but not identically, duplicated in different laboratories. Thus it is critical to carefully control the sample environment until its sensitivities are understood, and to check for repeatability of measurements. Unfortunately, metastable effects may be observed under the conditions necessary for capacitance spectroscopy measurements, which then necessitates careful, and sometimes laborious, experimental procedures. An introduction to these issues for some common solar cell materials can be found in [29, 51, 66–73].

References

1 Feichtinger, H. (1991) Deep centers in semiconductors, in *Electronic Structure and Properties of Semiconductors* (ed. W. Schröter), VCH Publishers, New York, pp. 143–195.

2 Losee, D.L. (1975) Admittance spectroscopy of impurity levels in Schottky barriers. *J. Appl. Phys.*, **46**, 2204.

3 Lang, D.V., Cohen, J.D., and Harbison, J.P. (1982) Measurement of the density of gap states in hydrogenated amorphous silicon by space charge spectroscopy. *Phys. Rev. B*, **25**, 5285.

4 Cohen, J.D. and Lang, D.V. (1982) Calculation of the dynamic response of Schottky barriers with a continuous distribution of gap states. *Phys. Rev. B*, **25**, 5321.

5 Blood, P. and Orton, J.W. (1992) *The Electrical Characterization of Semiconductors: Majority Carriers and Electron States*, Academic Press, San Diego.

6 Schroeder, D.K. (2006) *Semiconductor Material and Device Characterization*, John Wiley and Sons, Inc., Hoboken, NJ.

7 Jonscher, A.K. (1983) *Dielectric Relaxation in Solids*, Chelsea Dielectrics Press, Ltd, Chelsea.

8 Nicollian, E.H. and Brews, J.R. (1982) *MOS (Metal Oxide Semiconductor) Physics and Technology*, Wiley, New York, USA.

9 Istratov, A.A. and Vyvenko, O.F. (1999) Exponential analysis in physical phenomena. *Rev. Sci. Instrum.*, **70**, 1233.

10 Callegaro, L. (2009) The metrology of electrical impedance at high frequency: A review. *Meas. Sci. Technol.*, **20**, 022002.

11 Sze, S.M. (1985) *Semiconductor Devices Physics and Technology*, John Wiley & Sons, Inc., New York.

12 Hilibrand, J. and Gold, R.D. (1960) Determination of the impurity distribution in junction diodes from

capacitance–voltage measurements. *RCA Rev.*, **21**, 245.

13 Johnson, W.C. and Panousis, P.T. (1971) The influence of Debye length on the C–V measurement of doping profiles. *IEEE Trans. Electron. Devices*, **ED-18**, 965.

14 Crandall, R.S. (1980) Trap spectroscopy of a-Si:H diodes using transient current techniques. *J. Electron. Mater.*, **9**, 713.

15 Young, D.L. and Crandall, R.S. (2005) Strongly temperature-dependent free energy barriers measured in a polycrystalline semiconductor. *Appl. Phys. Lett.*, **86**, 262107.

16 Branz, H.M. (1988) Charge-trapping model of metastability in doped hydrogenated amorphous silicon. *Phys. Rev. B*, **38**, 7475.

17 Branz, H.M. and Crandall, R.S. (1989) Ionization entropy and charge-state-controlled metastable defects in semiconductors. *Appl. Phys. Lett.*, **55**, 2634.

18 Los, A.V. and Mazzola, M.S. (2001) Semiconductor impurity parameter determination from Schottky junction thermal admittance spectroscopy. *J. Appl. Phys.*, **89**, 3999.

19 Los, A.V. and Mazzola, M.S. (2002) Model of Schottky junction admittance taking into account incomplete impurity ionization and large signal effects. *Phys. Rev. B*, **65**, 165319.

20 Burgelman, M., Nollet, P., and Degrave, S. (2000) Modelling polycrystalline semiconductor solar cells. *Thin Solid Films*, **361–362**, 527.

21 Ćwil, M., Igalson, M., Zabierowski, P., and Siebentritt, S. (2008) Charge and doping distributions by capacitance profiling in Cu(In,Ga)Se$_2$ solar cells. *J. Appl. Phys.*, **103**, 063701.

22 Kimmerling, L.C. (1974) Influence of deep traps on the measurement of free-carrier distributions in semiconductors by junction capacitance techniques. *J. Appl. Phys.*, **45**, 1839.

23 Heath, J.T., Cohen, J.D., and Shafarman, W.N. (2004) Bulk and metastable defects in CuIn$_{1-x}$Ga$_x$Se$_2$ thin films using drive-level capacitance profiling. *J. Appl. Phys.*, **95**, 1000.

24 Lang, D.V. (1974) Deep-level transient spectroscopy: A new method to characterize traps in semiconductors. *J. Appl. Phys.*, **45**, 3023.

25 Johnston, S.W., Kurtz, S., Friedman, D.J., Ptak, A.J., Ahrenkiel, R.K., and Crandall, R.S. (2005) Observed trapping of minority-carrier electrons in p-type GaAsN during deep-level transient spectroscopy. *Appl. Phys. Lett.*, **86**, 072109.

26 Kurtz, S., Johnston, S., and Branz, H.M. (2005) Capacitance-spectroscopy identification of a key defect in N-degraded GaInNAs solar cells. *Appl. Phys. Lett.*, **86**, 113506.

27 Kukimoto, H., Henry, C.H., and Merritt, F.R. (1973) Photocapacitance studies of the oxygen donor in GaP. I. Optical cross-sections, energy levels, and concentration. *Phys. Rev. B*, **7**, 2486.

28 Murray, F., Carìn, R., and Bogdanski, P. (1986) Determination of high-density interface state parameters in metal-insulator-semiconductor structures by deep-level transient spectroscopy. *J. Appl. Phys.*, **60**, 3592.

29 Zabierowski, P. Electrical characterization of CIGSe-based thin film photovoltaic devices, in *Thin Film Solar Cells: Current Status and Future Trends* (eds A. Romeo and A. Bossio), Nova Science Publishers, Hauppauge, NY.

30 For an example of an exception, see: Grillot, P.N., Ringel, S.A., Fitzgerald, E.A., Watson, G.P., Xie, Y.H. (1995) Minority- and majority-carrier trapping in strain-relaxed Ge$_{0.3}$Si$_{0.7}$/Si heterostructure diodes grown by rapid thermal chemical-vapor deposition. *J. Appl. Phys.*, **77**, 676.

31 Look, D.C. and Sizelove, J.R. (1995) Depletion width and capacitance transient formulas for deep traps of high concentration. *J. Appl. Phys.*, **78**, 2848.

32 Buchwald, W.R., Morath, C.P., and Drevinsky, P.J. (2007) Effects of deep defect concentration on junction space charge capacitance measurements. *J. Appl. Phys.*, **101**, 094503.

33 Frenkel, J. (1938) On pre-breakdown phenomena in insulators and electronic semiconductors. *Phys. Rev.*, **54**, 647.

34 Vincent, G., Chantre, A., and Bois, D. (1979) Electric field effect on the thermal emission of traps in semiconductor junctions. *J. Appl. Phys.*, **50**, 5484.

35 Li, G.P. and andWang, K.L. (1985) Detection sensitivity and spatial resolution of reverse-bias pulsed deep-level transient spectroscopy for studying electric field-enhanced carrier emission. *J. Appl. Phys.*, **57**, 1016.

36 Dobaczewski, L., Peaker, A.R., and Bonde Nielsen, K. (2004) Laplace-transform deep-level spectroscopy: The technique and its applications to the study of point defects in semiconductors. *J. Appl. Phys.*, **96**, 4689.

37 Zabierowski, P. and Edoff, M. (2005) Laplace-DLTS analysis of the minority carrier traps in the Cu(In,Ga)Se$_2$-based solar cells. *Thin Solid Films*, **480–481**, 301.

38 Walter, T., Herberholz, R., Müller, C., and Schock, H.W. (1996) Determination of defect distributions from admittance measurements and application to Cu(In,Ga)Se$_2$ based heterojunctions. *J. Appl. Phys.*, **80**, 4411.

39 Lee, J.W., Cohen, J.D., and Shafarman, W.N. (2005) The determination of carrier mobilities in CIGS photovoltaic devices using high-frequency admittance measurements. *Thin Solid Films*, **480**, 336.

40 Dhariwal, S.R. and Deoraj, B.M. (1992) Contribution of bulk states to the depletion layer admittance. *Solid State Electron.*, **36**, 1165.

41 León, C., Martin, J.M., Santamaría, J., Skarp, J., González-Díaz, G., and Sánchez-Quesada, F. (1996) Use of Kramers–Kronig transforms for the treatment of admittance spectroscopy data of p–n junctions containing traps. *J. Appl. Phys.*, **79**, 7830.

42 Milton, G.W., Eyre, D.J., and Mantese, J.V. (1997) Finite frequency range Kramers–Kronig relations: Bounds on the dispersion. *Phys. Rev. Lett.*, **79**, 3062.

43 van de Leur, R.H.M. (1991) A critical consideration on the interpretation of impedance plots. *J. Phys. D Appl. Phys.*, **24**, 1430.

44 Michelson, C.E., Gelatos, A.V., and Cohen, J.D. (1985) Drive-level capacitance profiling: Its application to determining gap state densities in hydrogenated amorphous silicon films. *Appl. Phys. Lett.*, **47**, 412.

45 Unold, T. and Cohen, J.D. (1991) Enhancement of light-induced degradation in hydrogenated amorphous silicon due to carbon impurities. *Appl. Phys. Lett.*, **58**, 723.

46 Johnson, P.K., Heath, J.T., Cohen, J.D., Ramanathan, K., and Sites, J.R. (2005) A comparative study of defect states in selenized and evaporated CIGS(S) solar cells. *Prog. Photovoltaics*, **13**, 1.

47 Chantre, A., Vincent, G., and Bois, D. (1981) Deep-level optical spectroscopy in GaAs. *Phys. Rev. B*, **23**, 5335.

48 Cohen, J.D. and Gelatos, A.V. (1988) Transient photocapacitance studies of deep defect transitions in hydrogenated amorphous silicon, in *Amorphous Silicon and Related Materials Vol. A* (ed. H. Fritzsche), World Scientific Publishing, Singapore, pp. 475–512.

49 Lang, D. (1992) DX centers in III-V alloys, in *Deep Centers in Semiconductors A State-of-the-Art Approach*, 2nd edn (ed. S.T. Pantelides), Gordon and Breach Science Publishers, Yverdon, Switzerland, pp. 592–665.

50 Heath, J.T., Cohen, J.D., Shafarman, W.N., Liao, D.X., and Rockett, A.A. (2002) Effect of Ga content on defect states in CuIn$_{1-x}$Ga$_x$Se$_2$ photovoltaic devices. *Appl. Phys. Lett.*, **80**, 4540.

51 Erslev, P.T., Lee, J.W., Shafarman, W.N., and Cohen, J.D. (2009) The influence of Na on metastable defect kinetics in CIGS devices. *Thin Solid Films*, **517**, 2277.

52 Cohen, J.D., Unold, T., Gelatos, A.V., and Fortmann, C.M. (1992) Deep defect structure and carrier dynamics in amorphous silicon and silicon–germanium alloys determined by transient photocapacitance methods. *J. Non-Cryst. Solids*, **141**, 142.

53 Meyer, W. and Neldel, H. (1937) Über die Beziehungen zwischen der Energiekonstanten und der Mengenkonstanten *a* in der Leitwerts-Temperaturformel bei oxydischen

Halbleitern. *Zeitschrift für technische Physik*, **12**, 588.

54 Street, R.A. (1991) *Hydrogenated Amorphous Silicon*, Cambridge University Press, Cambridge, p. 224.

55 Crandall, R.S. (1990) Defect relaxation in amorphous silicon: Stretched exponentials, the Meyer–Neldel rule, and the Staebler–Wronski effect. *Phys. Rev. B*, **43**, 4057.

56 Herberholz, R., Walter, T., Müller, C., Friedlmeier, T., Schock, H.W., Saad, M., Lux-Steiner, M.Ch., and Albertz, V. (1996) Meyer–Neldel behavior of deep level parameters in heterojunctions to Cu(In, Ga)(S,Se)$_2$. *Appl. Phys. Lett.*, **69**, 2888.

57 Leffler, J.E. (1955) The enthalpy–entropy relationship and its implications for organic chemistry. *J. Org. Chem.*, **20**, 1202.

58 Glasstone, S., Laidler, K.J., and Eyring, H. (1941) *The Theory of Rate Processes*, McGraw-Hill, New York.

59 Almbladh, C.O. and Reese, G.J. (1982) Statistical mechanics of electronic energy levels in semiconductors. *Solid State Commun.*, **41**, 173.

60 Engström, O. and Alm, A. (1978) Thermodynamical analysis of optimal recombination centers in thyristors. *Solid State Electron.*, **21**, 1571.

61 Yelon, A. and Movaghar, B. (1990) Microscopic explanation of the compensation (Meyer–Neldel) rule. *Phys. Rev. Lett.*, **65**, 618.

62 Yelon, A., Movaghar, B., and Branz, H.M. (1992) Origin and consequences of the compensation (Meyer–Neldel) law. *Phys. Rev. B*, **46**, 12244; Viščor, P. (2002) Comment on "Origin and consequences of the compensation (Meyer–Neldel) law". *Phys. Rev. B*, **65**, 077201; Yelon, A. and Movaghar, B. (2002) Reply to "Comment on 'Origin and consequences of the compensation (Meyer–Neldel) law'". *Phys. Rev. B*, **65**, 077202.

63 Widenhorn, R., Rest, A., and Bodegom, E. (2002) The Meyer–Neldel rule for a property determined by two transport mechanisms. *J. Appl. Phys.*, **91**, 6524.

64 Popescu, C. and Stoica, T. (1992) Meyer–Neldel correlation in semiconductors and Mott's minimum metallic conductivity. *Phys. Rev. B*, **46**, 15063.

65 Niemegers, A., Burgelman, M., Herberholz, R., Rau, U., Hariskos, D., and Schock, H.W. (1998) Model for electronic transport in Cu(In,Ga)Se$_2$ solar cells. *Progr. Photovoltaics: Res. Appl.*, **6**, 407.

66 Eisenbarth, T., Unold, T., Caballero, R., Kaufmann, C.A., and Schock, H.-W. (2010) Interpretation of admittance, capacitance–voltage, and current–voltage signatures in Cu(In,Ga)Se$_2$ thin film solar cells. *J. Appl. Phys.*, **107**, 034509.

67 Niemegeers, A. and Burgelman, M. (1997) Effects of the Au/CdTe back contact on IV and CV characteristics of Au/CdTe/CdS/TCO solar cells. *J. Appl. Phys.*, **81**, 2881.

68 Demtsu, S.H. and Sites, J.R. (2006) Effect of back-contact barrier on thin-film CdTe solar cells. *Thin Solid Films*, **510**, 320.

69 Macdonald, D., Rougieux, F., Cuevas, A., Lim, B., Schmidt, J., Di Sabatino, M., and Geerligs, L.J. (2009) Light-induced boron–oxygen defect generation in compensated p-type Czochralski silicon. *J. Appl. Phys.*, **105**, 093704.

70 Nadazdy, V. and Zeman, M. (2004) Origin of charged gap states in a-Si:H and their evolution during light soaking. *Phys. Rev. B*, **69**, 165213.

71 Shimizu, T. (2004) Staebler–Wronski effect in hydrogenated amorphous silicon and related alloy films. *Jpn. J. Appl. Phys.*, **43**, 3257.

72 Balcioglu, A., Ahrenkiel, R.K., and Hasoon, F. (2000) Deep-level impurities in CdTe/CdS thin-film solar cells. *J. Appl. Phys.*, **88**, 7176.

73 Demtsu, S.H., Albin, D.S., Pankow, J.W., and Davies, A. (2006) Stability study of CdS/CdTe solar cells made with Ag and Ni back-contacts. *Sol. Energy Mater. Sol. C*, **90**, 2934.

Part Three
Materials Characterization

5
Characterizing the Light-Trapping Properties of Textured Surfaces with Scanning Near-Field Optical Microscopy

Karsten Bittkau

5.1
Introduction

In thin-film solar cells based on hydrogenated amorphous silicon (a-Si:H) or hydrogenated microcrystalline silicon (μc-Si:H), the thicknesses of the absorbing layers are too small to absorb the majority of the impinging photons within one single path. The thicknesses are limited owing to the small diffusion lengths of holes in the case of a-Si:H and the demand of cost reductions for both types of solar cells. To increase the effective light path in the absorber layer, textured surfaces are commonly implemented. These textures provide light scattering, which extends the light path in silicon [1]. For larger scattering angles, total internal reflection occurs where the light cannot propagate throughout the absorber layer. This effect is called light trapping.

In common thin-film solar cells, the front contact layer, consisting of ZnO or SnO_2, is textured by wet chemical etching or during the growth. The mechanisms result in a statistical surface morphology with different lateral sizes, heights, and shapes. By engineering the deposition and etching parameters, the distribution of surface features can be varied and optimized for each type of front contact layer and solar cell.

Since the technological effort and the parameter space are large and since each morphology consists of a special distribution of different features, this kind of optimization only leads to a local maximum of the total absorption. To achieve the overall optimum needs two things: a technology to prepare well-defined surface morphologies and a method to analyze the local light-scattering properties of individual surface features. This chapter concentrates on the analysis of local light-trapping properties of textured surfaces with scanning near-field optical microscopy (SNOM).

In Section 5.2, the principle working of SNOM will be discussed. In Section 5.3, the light scattering at statistically textured surfaces will be described in the wave picture of light. The role of evanescent light modes, which only occur in the optical near-field, for the light-trapping properties of thin-film solar cells will be emphasized in Section 5.4. Fast Fourier transformation (FFT) will be discussed in Section 5.5 as an analytical tool to extract local light-trapping properties from SNOM images. In Section 5.6, the link from near-field properties to far-field

Advanced Characterization Techniques for Thin Film Solar Cells,
Edited by Daniel Abou-Ras, Thomas Kirchartz and Uwe Rau.
© 2011 Wiley-VCH Verlag GmbH & Co. KGaA. Published 2011 by Wiley-VCH Verlag GmbH & Co. KGaA.

5.2
How Does a Scanning Near-Field Optical Microscope Work?

SNOM (also found as NSOM) [2] is one kind of scanning-probe microscopy (see Chapter 11) and works similar to an atomic force microscope (AFM) in the noncontact mode.

There are in principle two different SNOM concepts: with and without aperture. The SNOMs with aperture normally make use of a tapered fiber tip which has a metallic coating in the tapered region to achieve a sufficient guidance of light. At the very end of the probe, a small hole is prepared into the coating which serves as an aperture. The fiber modes couple through this aperture to the light modes outside the tip. The optical resolution of such a microscope is defined by the size of the aperture which is much smaller than the wavelength of light. Therefore, spatial resolutions beyond the refraction limit are achieved.

The apertureless SNOMs work without any fiber tip [3, 4]. Instead, an AFM probe is used, which is illuminated with focused laser light. As the tip serves as an antenna with strong field enhancement in the near-field of the tip, light is scattered. The scattering behavior of the probe strongly depends on the interaction of the tip with the surface. This interaction becomes significant only in the near-field. Apertureless SNOMs reach higher resolutions compared to SNOMs using an aperture and the AFM probe is much more robust than a fiber tip. Nevertheless, it only works in the near-field of the surface. The light propagation at larger distances cannot be studied with this kind of near-field microscope. This chapter will focus on SNOMs with aperture, since the study of the light propagation is essential for the light scattering in thin-film solar cells.

The SNOM probe is an Al-coated, tapered fiber tip. The aperture size is about 50–80 nm and defines the optical resolution of the system. The probe is scanning above the surface of the sample by a piezo-controlling system. Distance controlling is achieved by shear-force techniques [5] since the tip is glued to a quartz tuning fork (see Figure 5.1) [6].

By this shear-force technique, the tuning fork is excited by a dither piezo and the tip is oscillating laterally. The damping of this oscillation is measured and the tip-to-sample distance is kept constant by a z-piezo. Therefore, topographic information can be obtained by the controlling voltage of the z-piezo. Additionally, the SNOM tip can collect the local light intensity at the position of the aperture. An incident plane wave illuminates the sample from the back side and the tip collects the transmitted light at the surface [7]. This situation corresponds to the real device where the solar cell is illuminated from the substrate side (superstrate configuration) in p–i–n-junctions.

In addition to this so-called collection mode, an illumination mode is possible, where light is coupled to the fiber and the sample is illuminated by the small SNOM

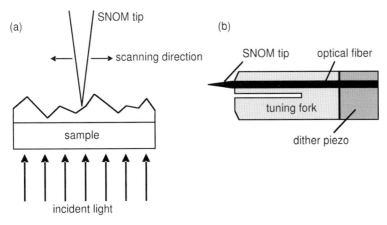

Figure 5.1 (a) Scheme of collection mode SNOM setup in transmission geometry. (b) Shear-force technique for distance controlling.

tip. The transmitted light is collected via a standard microscope objective. These experimental modes are not directly invertible. Also, the tip can both illuminate the sample and collect the light in the so-called luminescence mode, where the re-emitted light is detected with subwavelength resolution. For solar-cell applications, this mode is used to study the local band-gap variations and lateral variations of the splitting of quasi-Fermi levels [8]. By using the sample itself as the detector, local photocurrent and photovoltages can be analyzed with SNOM in illumination mode [9–11]. For the kind of study described in this chapter, the collection mode is most applicable. Therefore, all presented examples are obtained by this technique.

In Figure 5.2, the result of a measurement on a randomly textured ZnO surface is shown. In Figure 5.2a the topography and in Figure 5.2b the local light intensity at a wavelength of 658 nm are given. Both quantities are measured simultaneously by means of SNOM working in collection mode in transmission geometry. The surface consists of laterally distributed craters with different shapes and sizes. A typical

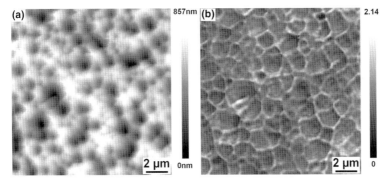

Figure 5.2 (a) Topography of randomly textured ZnO surface. (b) Local light intensity at a wavelength of 658 nm simultaneously measured with SNOM.

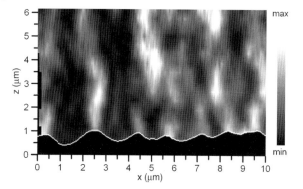

Figure 5.3 Local light intensity above a selected scanning line measured with SNOM at a wavelength of 658 nm.

lateral size of a crater is about 1 μm. By comparing the topography to the local light intensity, it can be found that at the edges of the craters the highest light intensity is found. This behavior is typical for this kind of surface texture. The transmitted light is localized at the ridges between two (or more) craters owing to the strong curvature, which leads to electric-field enhancements.

Since the SNOM tip can be moved in three axes, the local light-intensity distribution in the whole spatial domain above the surface can be obtained. A typical example is shown in Figure 5.3.

For one selected scanning line, all measured intensities are plotted in a cross-sectional image. The spatial distance along the z-direction between two measurements was 350 nm. Missing data points are filled by interpolation [12]. It is found that there are strong localization effects above the surface with a jet-like shape. These jets are found at the highest curved edges in the topography.

5.3
Light Scattering in the Wave Picture

Since interferences occur in the local light intensity distribution at randomly textured surfaces and since geometrical optics is only reasonable for structural sizes larger than the wavelength of light, which is not generally given in thin-film solar cell structures, the description of light propagation in the wave picture is necessary. Therefore, diffuse light scattering at textured surfaces must be understood in the wave behavior.

In this picture, the incident wave is described as a plane wave with a norm of the wave vector given by $k = 2\pi n/\lambda$, where n is the refractive index and λ the wavelength of light. At normal incidence, the planar component of the wave vector is vanishing. Light scattering can be described similarly to diffraction at a grating. Here, a planar wave vector is transferred at the interface by conserving the norm. This leads to well-defined diffraction modes in a periodic grating as illustrated in Figure 5.4. The grating transfers multiples of the reciprocal lattice constant to the planar wave vector

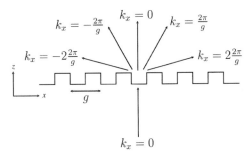

Figure 5.4 Diffraction of light at a periodic grating.

of light. As long as the resulting value for k_x is smaller than the norm k, this mode can propagate in the upper half space. For resulting planar wave vectors larger than the norm, the diffraction mode becomes evanescent. In this case, the amplitude of the electric field decreases exponentially in the upper half space. Therefore, these modes cannot propagate there.

For a thin film or a stack of thin films, for which the refractive indices of the layers are larger than in the upper half space, total internal reflection is possible. The light diffraction can be sufficiently strong to deflect the light in the layer as much that the internal angle becomes larger than the total reflection angle. The planar wave vector of this light is then larger than the norm in the upper half space. Therefore, light, which is totally internally reflected, couples to evanescent light modes in the surrounding media. Inside the absorber layer, this light can be understood as guided waves.

Commonly, in thin-film solar-cell structures, no periodic gratings are applied. Instead, randomly textured surfaces are used. In this case, the situation becomes more complicated. Due to the absence of the periodic grating, no well-defined reciprocal lattice constant exists, and no well-defined diffraction modes are present. A diffuse scattering of light is found. It will be demonstrated in Section 5.6 that the scattering properties known from far-field experiments can be described by light diffraction in the wave picture of light.

In Figure 5.5, the mechanism of light scattering at a randomly textured interface is illustrated. The surface profile can be described by a superposition of periodic gratings with different periodicities. For each individual grating, the normal diffraction process occurs, and the finite result is the superposition of all the diffracted light waves taking into account their different phases. The superposition of the zeroth modes results in the specular transmission, which is hard to explain in a simple geometrical picture. The broad distribution of local periodicities results in a diffuse transmission given by the higher modes.

5.4
The Role of Evanescent Modes for Light Trapping

In Section 5.3, the connection between totally internally reflected light and evanescent diffraction modes was mentioned. But it is still open how the evanescent modes

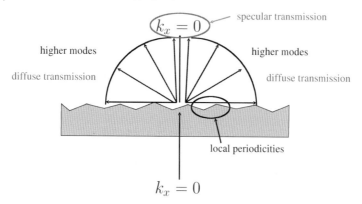

Figure 5.5 Scattering of light at a randomly textured interface.

contribute to the total absorption in a thin silicon film. Since the absorption in the layer is the most important quantity for optical study on thin-film solar cells and since evanescent modes are not detectable in far-field-based optical experiments (e.g., angularly resolved scattering (ARS) haze, reflection/transmission), this is indeed an important question. The answer will have an influence on the choice of experiment and also simulation tools to study the light-trapping properties of surface textures.

Here, a theoretical model was chosen, where Maxwell's equations are solved rigorously for a one-dimensional periodic grating. This model is the so-called C method developed by Chandezon et al. [13, 14]. Within this method, Maxwell's equations are transformed into a curvilinear coordinate system where the surface is flat. By applying periodic boundary conditions, the field components are developed into Fourier series resulting in an algebraic equation system instead of the differential equations. The C method provides two essential advantages: the comparably low demand of computer power and the possibility to separate the evanescent modes. This allows us to study the impact of surface modifications as well as to answer the question how evanescent modes contribute to the total absorption.

In Figure 5.6, the calculated electric field distribution for a single scanning line of an AFM measurement is shown [15]. The thin-film stack consists of a ZnO half space with a surface profile defined by the given scanning line. On top of this surface, an a-Si:H layer with a thickness of 250 nm and a refractive index of $n_{\text{a-Si:H}} = 3.803 + 0.0022i$ is assumed [16]. At the top of this layer, the profile is supposed to be conformal and the upper half space is air. The gray scale is logarithmic for enhanced visualization.

In Figure 5.6a, the distribution of the whole electric field is shown taking into account both propagating and evanescent part of light. In Figure 5.6b, only the intensity distribution of the evanescent part of light is shown. The light focusing effect which forms a jetlike shape (see Figure 5.3) can be seen in Figure 5.6a at, for example, an x-coordinate of 1.5 µm and a z-coordinate of 2 µm. In addition to the information which is collected by SNOM, the local light intensity inside the material can be obtained by these simulations. In the ZnO substrate, an intensity modulation

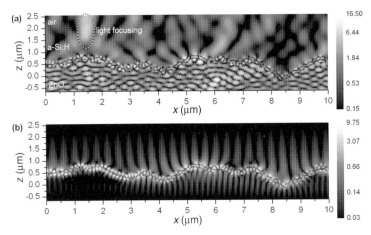

Figure 5.6 (a) Calculated electric field distribution for a single scanning line at a wavelength of 780 nm. (b) Calculated intensity distribution of evanescent part of light for the same scanning line [15]. For both figures, a logarithmic intensity scale was chosen.

is found, which results from the interference effects of the incident plane wave and the diffraction of the reflected waves. Inside the silicon layer, a finer interference pattern is found due to the larger refractive index. From the results of the evanescent part of light, it can be seen that inside the silicon layer, the interference pattern looks very similar to that for the total amount of light. This means that the behavior in the silicon is dominated by guided waves. The exponential decrease in these modes in the surrounding media can be seen in Figure 5.6b by the intensity spikes in air and ZnO.

In Figure 5.7, the calculated absorption enhancement for an a-Si:H thin film on top of a textured ZnO substrate is shown. With the results of the C method, the local absorption can be extracted. This absorption is integrated over the spatial domain of

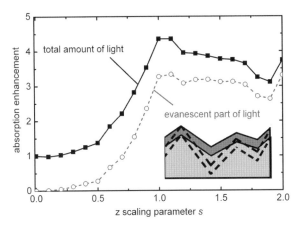

Figure 5.7 Calculated absorption enhancement for a selected scanning line with varying z-scaling parameter for the total amount of light (solid curve) and the evanescent part of light (dashed line) [15]. The inset illustrates the kind of surface variation.

the silicon to get the total absorption. This value is normalized against the corresponding absorption for flat interfaces with an identical layer stack. This gives the enhancement of the absorption due to the texture. The measured surface profile is modified by stretching it along the z-axis. The new profile function is then given by $z(x) = sz_0(x)$. Here, $z_0(x)$ defines the measured profile and s the z scaling parameter. The absorption enhancement is calculated for both the total amount of light (solid curve) and for the evanescent part (dashed curve). Obviously, the total absorption can be increased by a factor of 4 for this kind of profile. The most important result is that the absorption of the evanescent modes nearly follows the absorption enhancement for the total amount of light. This means that the improvement in the quantum efficiency of textured solar cells is strongly dominated by light modes which are guided in the silicon layer. This result is found for all kind of surface profiles and various kinds of modification.

In conclusion, it is demonstrated that light trapping in silicon-based thin-film solar cells is strongly dominated by evanescent light modes. That is, light scattering at rough interfaces must be strong enough to excite these guided modes and that the experimental study must have an access to evanescent light. Since the intensity of these modes decreases exponentially with the distance to the surface, an optical near-field experiment is essential to study the light trapping.

The following sections will focus on the analytical methods for the extraction of the relevant information from SNOM measurements on textured surfaces. The extraction of evanescent modes as well as the ARS in silicon will be discussed.

5.5
Analysis of Scanning Near-Field Optical Microscopy Images by Fast Fourier Transformation

As shown in the previous section, the portion of light which is scattered into evanescent modes is essential for the study of light-trapping efficiency. Since these modes only have finite intensities in the optical near-field regime, where the distance to the surface is less than one wavelength, SNOM is a powerful tool to investigate local light-trapping efficiencies. Nevertheless, the analysis and interpretation of SNOM images is complex as the following questions arise: (i) how does the near-field probe influence the measured signal? (ii) Are the coupling efficiencies of different optical modes comparable? (iii) Since the measured optical modes couple to the same fiber mode, different modes cannot be directly distinguished. Is it possible to distinguish them indirectly? There are certainly more questions concerning the analysis and interpretation of SNOM images. But at this point, it should be enough to concentrate on these three.

The question whether the near-field probe influences the measured signal is reasonable since the tip is metalized and placed in the spatial domain, which is in the focus of investigation. By comparing the measured cross-sectional images (Figure 5.3) to numerical results from finite-difference time-domain simulations, disregarding the existence of the tip, a very good agreement is found for the investigated types of structure [17]. In the direct comparison of the local light intensities at a distance of 20 nm to the surface, deviations are found. A detailed investigation shows that the measured images are better comparable to simulated images at larger distances (about 200 nm for this example). One possible explanation is the influence of the tip on the local light intensity at its position. But it is more reasonable to address this deviation to the second question concerning the coupling efficiencies of different optical modes. A detailed answer to the first question is still under investigation.

The second question deals with the coupling efficiencies of different optical modes to the near-field probe. The optical fiber, with the tip at its end, has a finite angle of acceptance which means that the lateral wave vector of the optical mode inside the fiber is restricted. Optical modes with lateral wave vectors larger than the limit cannot couple directly into the fiber but via scattering at the conical tip which transfers the wave vector into the acceptance cone. Therefore, the coupling into the fiber is in principle possible for all optical modes. Nevertheless, the larger the lateral wave vector is the stronger the scattering at the tip must be in order to detect it. Therefore, it is reasonable that optical modes with large planar wave vectors couple weaker to the fiber as modes which propagate nearly at normal incidence. This limitation affects the evanescent modes in particular. This is the reason why deviations are found by comparing simulated local light intensities at a distance to the surface of 20 nm to the experimental results from SNOM measurements. At the larger distance of 200 nm, the amplitudes of the evanescent modes are decreased since the propagating modes only differ slightly in their intensities. Therefore, the theoretical results at a distance of 200 nm agree better to the measurements at a distance of 20 nm. The absolute value of the discrepancy in the distance is not a universal finding, it rather depends on the individual tip properties.

The third question is the most interesting one. In the spatial domain of investigation, there are different optical modes: specularly scattered light, light with small scattering angles, strongly scattered light with nearly gracing angles, and light which is scattered into evanescent modes. By detecting all these modes by the SNOM probe, they will couple to the same fiber mode and are detected without any information about the lateral wave vector before scattering at the tip. For the study of the scattering behavior of textured surfaces, it is quite important to obtain the information about the transferred wave vectors. Especially, the extraction of evanescent modes is essential.

In Figure 5.8 the formation of periodicities in SNOM images due to the interference of scattered light is illustrated. The wavefronts of the incident plane wave and the scattered light are shown. For the scattered light, only one possible scattering angle θ is depicted. The different wavefronts of the wave which is scattered to the left side interfere with the wavefronts of the light, which is scattered to the right side. This leads to regions of constructive interference, which are oriented perpendicular to the surface. The distance of the interference maxima is given by the scattering angle and, therefore, the transferred planar wave vector.

Regarding all possible scattering angles gives a superposition of periodic light intensity patterns. The individual amplitude for each periodicity can be extracted by

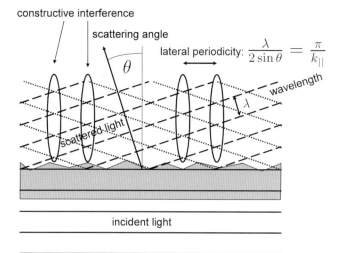

Figure 5.8 Incident and scattered wavefronts. The relation between the periodicity of the interference pattern and the transferred planar wave vector is illustrated for a selected scattering angle.

FFT. This information can be used to generate individual images for the evanescent and propagating light. This is exemplarily shown in Figure 5.9.

In the upper part, the topography (Figure 5.9a) and the local light intensity (Figure 5.9b) at a wavelength of 705 nm is shown for an a-Si:H top cell deposited on a randomly textured ZnO layer. On top of the cell, an intermediate reflector consisting of SiO_x as well as a µc-Si:H cap layer is deposited. This layer stack describes the optical situation of a full tandem solar cell concerning the properties of the a-Si:H top cell [18]. The measured local light intensity is transformed by FFT, and the evanescent (Figure 5.9c) and propagating (Figure 5.9d) part of transmitted light are extracted by high-pass and low-pass filters, where the spatial cut-off frequency was defined by $2/\lambda$. Between the measured image and the propagating part of light, strong similarities can be found. This means that the total transmitted light intensity is dominated by the propagating light modes. The evanescent part shows the light which is trapped inside the solar cell. The image appears as a superposition of spherical waves with the craters as center positions.

In Figure 5.10, the evanescent part of light is shown as the norm of the back-transformed FFT image (see Figure 5.9c for comparison). Both the pure data (Figure 5.10a) and smoothed data (Figure 5.10b) are shown. In the intensity profile of the evanescent modes, nodes are found, which are caused by standing waves. By smoothing the image, this oscillation is damped. Here, a better identification of regions with higher light-trapping efficiencies is possible. A detailed study of these images and the comparison to numerical results are necessary to extract the information relevant for further improvement of the surface texture.

5.5 Analysis of Scanning Near-Field Optical Microscopy Images by Fast Fourier Transformation

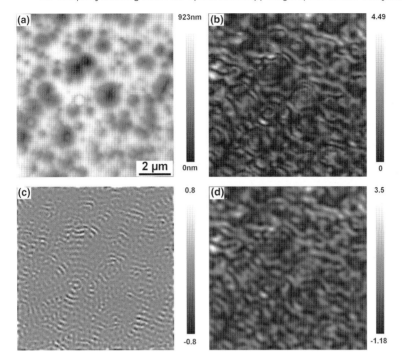

Figure 5.9 (a) Topography of an a-Si:H top cell with intermediate layer and µc-Si:H cap layer deposited on randomly textured ZnO. (b) Local light intensity at a wavelength of 705 nm measured with SNOM. In the lower part, the evanescent (c) and propagating (d) part of the local light intensity is shown.

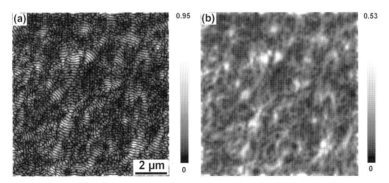

Figure 5.10 Evanescent part of light extracted from the SNOM measurement shown in Figure 5.9. (a) The norm of the FFT-filtered values. (b) A smoothed image to illustrate the envelope of the standing waves.

5.6
How to Extract Far-Field Scattering Properties by Scanning Near-Field Optical Microscopy?

In the previous section, the extraction of evanescent modes from SNOM images was focused. Now, the extraction of far-field scattering properties from SNOM measurements will be discussed. The following questions will be addressed in detail: (i) can the measured local light intensity distribution be used to determine the angularly resolved light scattering? (ii) Is it possible to predict the ARS by the surface morphology? (iii) Can one provide both angularly and spatially resolved light scattering? These questions, in particular the latter one, are essential to study the scattering properties of individual surface elements. This will give a more detailed information as compared to the evanescent part of light concerning the optimization process of the morphology.

As illustrated in Figure 5.8, the periodicity in the measured SNOM images depends on the transferred planar wave vector and, therefore, on the scattering angle. The angularly resolved light scattering [19–21] can easily be determined by FFT analysis of the measured local light intensity distribution. As described in Section 5.3, the scattering mechanism can be understood as diffraction at a superposition of periodic gratings with well-defined diffraction angles.

The FFT generates an image with the intensities and phase information about the individual light modes. The position in the image corresponds to the transferred planar wave vector. Therefore, all light modes with the same norm of the planar wave vector have the same scattering angle [22]. Since the FFT is invertible, it is possible to transform back the image with the intensities and phase information, resulting in an image with the local light intensities. For the transformation of the whole image, this leads to the original SNOM image. After applying a band-pass filter, which sets all intensities outside the given interval of the wave vector to zero, the local light intensities for this interval is determined. This means that the spatial information of light, which is scattered into an angular interval, can be extracted by FFT analysis.

Restricting the wave vector to only one particular value leads to a sinusoidal intensity profile without any spatial information but with the highest angular resolution. Otherwise, taking all values into account leads to the original SNOM measurement with the full spatial resolution but without any angular resolution. Therefore, spatially and angularly resolved light scattering is only possible with a limited resolution for both.

The results for the ARS properties are shown in Figure 5.11. The extracted ARS from the measured local light intensity at a constant distance of 20 nm from the surface is shown as the dotted curve. The prediction from diffraction theory, regarding the topography, is plotted as the dashed curve. With a standard far-field scattering experiment, the real ARS is measured and the result is depicted as the solid curve. Good agreement is found between the various methods. This means that FFT analysis of both the topography and SNOM images is an applicable tool for characterizing the light-scattering properties.

As mentioned above, the Fourier analysis allows one to study the light-scattering mechanism with both angular and spatial information. The reciprocal space is

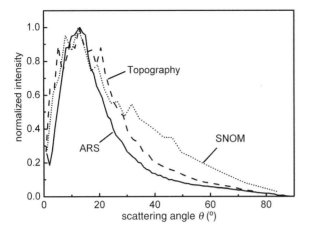

Figure 5.11 Angularly resolved light scattering (ARS) for a randomly textured ZnO surface. The experimentally determined ARS is depicted as a solid curve. The dashed curve shows the predicted ARS from the topography. The results of the FFT analysis of the SNOM measurement are plotted as the dotted curve.

segmented into intervals with $1\,\mu m^{-1}$ in length. For each interval, a band-pass filter was defined and the resulting images were smoothed to get the envelope of the standing waves. The intensity at a particular position as a function of the interval in reciprocal space results in the local scattering intensity.

In Figure 5.12, the spatially and angularly resolved light scattering is shown for four positions on the textured sample studied in Section 5.5 at a wavelength of 705 nm. The angular resolution is chosen to be low to achieve a high spatial resolution. It is easy to see that the scattering properties strongly differ at these

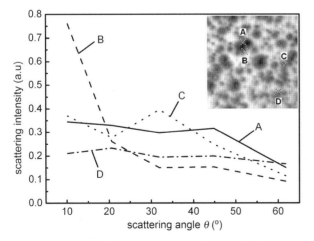

Figure 5.12 Local light-scattering properties at a wavelength of 705 nm evaluated for four different positions of the randomly textured thin-film solar cell.

particular points. Especially points A and B show different behaviors, although the spatial distance between these points is 1 µm, and the local structure looks similar at first sight. While the scattering intensity at point B decreases strongly at lower angles, the scattering at point A shows comparable high intensities for larger angles.

5.7
Conclusion

SNOM is found out to be an excellent tool to study the local light scattering and, therefore, the local light-trapping properties of thin-film solar cells on textured surfaces. The propagation of light, the formation of guided modes, as well as the ARS with high spatial resolution can be obtained by such an experiment. The analytical techniques to characterize the potential of different morphological structures for the light trapping are established. For further improvement of surface texture, these information are essential and can only be obtained in optical near-field experiments. For the interpretation of the collected data, light scattering must be understood in a wave picture.

References

1 Yablonovitch, E. and Cody, G.D. (1982) Intensity enhancement in textured optical sheets for solar cells. *IEEE Trans. Electron Devices*, **ED-29**, 300.

2 Pohl, D.W., Denk, W., and Lanz, M. (1984) Optical stethoscopy: Image recording with resolution $\lambda/20$. *Appl. Phys. Lett.*, **44**, 651.

3 Gleyzes, P., Boccara, A.C., and Bachelot, R. (1995) Near field optical microscopy using a metallic vibrating tip. *Ultramicroscopy*, **57**, 318.

4 Knoll, B. and Keilmann, F. (1998) Scanning microscopy by mid-infrared near-field scattering. *Appl. Phys. A*, **66**, 477.

5 Karrai, K. and Grober, R.D. (1995) Piezoelectric tip–sample distance control for near field optical microscopes. *Appl. Phys. Lett.*, **66**, 1842.

6 Behme, G., Richter, A., Süptitz, M., and Lienau, Ch. (1997) Vacuum near-field scanning optical microscope for variable cryogenic temperatures. *Rev. Sci. Instrum.*, **68**, 3458.

7 Betzig, E., Isaacson, M., and Lewis, A. (1987) Collection mode near-field scanning optical microscopy. *Appl. Phys. Lett.*, **51**, 2088.

8 Gütay, L., Pomraenke, R., Lienau, C., and Bauer, G.H. (2009) Subwavelength inhomogeneities in $Cu(In,Ga)Se_2$ thin films revealed by near-field scanning optical microscopy. *phys. status solidi (a)*, **206**, 1005.

9 Gotoh, T., Yamamoto, Y., Shen, Z., Ogawa, S., Yoshida, N., Itoh, T., and Nonomura, S. (2009) Nanoscale characterization of microcrystalline silicon solar cells by scanning near-field optical microscopy. *Jpn. J. Appl. Phys.*, **48**, 91202.

10 McDaniel, A.A., Hsu, J.W.P., and Gabor, A.M. (1997) Near-field scanning optical microscopy studies of $Cu(In,Ga)Se_2$ solar cells. *Appl. Phys. Lett.*, **70**, 3555.

11 Herndon, M.K., Gupta, A., Kaydanov, V., and Collins, R.T. (1999) Evidence for grain-boundary-assisted diffusion of sulfur in polycrystalline CdS/CdTe heterojunctions. *Appl. Phys. Lett.*, **75**, 3505.

12 Bittkau, K., Beckers, T., Fahr, S., Rockstuhl, C., Lederer, F., and Carius, R. (2008) Nanoscale investigation of light-trapping in a-Si:H solar cell structures

with randomly textured interfaces. *phys. status solidi (a)*, **205**, 2766.
13 Chandezon, J., Maystre, D., and Raoult, G. (1980) A new theoretical method for diffraction gratings and its numerical application. *J. Opt.*, **11**, 235.
14 Chandezon, J., Dupuis, M.T., Cornet, G., and Maystre, D. (1982) Multicoated gratings: a differential formalism applicable in the entire optical region. *J. Opt. Soc. Am.*, **72**, 839.
15 Bittkau, K. and Beckers, T. (2010) Near-field study of light scattering at rough interfaces of a-Si:H/μc-Si:H tandem solar cells. *phys. status solidi (a)*, **207**, 661.
16 Vetterl, O., Finger, F., Carius, R., Hapke, P., Houben, L., Kluth, O., Lambertz, A., Mück, A., Rech, B., and Wagner, H. (2000) Intrinsic microcrystalline silicon: A new material for photovoltaics. *Sol. Energy Mater. Sol. Cells*, **62**, 97.
17 Rockstuhl, C., Lederer, F., Bittkau, K., and Carius, R. (2007) Light localization at randomly textured surfaces for solar-cell applications. *Appl. Phys. Lett.*, **91**, 171104.
18 Rockstuhl, C., Lederer, F., Bittkau, K., Beckers, T., and Carius, R. (2009) The impact of intermediate reflectors on light absorption in tandem solar cells with randomly textured surfaces. *Appl. Phys. Lett.*, **94**, 211101.
19 Schade, H. and Smith, Z.E. (1985) Mie scattering and rough surfaces. *Appl. Opt.*, **24**, 3221.
20 Löffler, J., Groenen, R., Linden, J.L., van de Sanden, M.C.M., and Schropp, R.E.I. (2001) Amorphous silicon solar cells on natively textured ZnO grown by PECVD. *Thin Solid Films*, **392**, 315.
21 Krc, J., Zeman, M., Kluth, O., Smole, F., and Topic, M. (2003) Effect of surface roughness of ZnO:Al films on light scattering in hydrogenated amorphous silicon solar cells. *Thin Solid Films*, **426**, 296.
22 Isabella, O., Moll, F., Krc, J., and Zeman, M. (2010) Modulated surface textures using zinc-oxide films for solar cells applications. *phys. status solidi (a)*, **207**, 642.

6
Spectroscopic Ellipsometry
Sylvain Marsillac, Michelle N. Sestak, Jian Li, and Robert W. Collins

6.1
Introduction

Over the past few decades, spectroscopic ellipsometry has emerged as a nondestructive, noninvasive optical technique for the characterization of thin-film solar cell materials and devices [1–4]. *Ellipsometry* derives its name from the measurement of the output polarization ellipse which is generated after a beam of light with a known input polarization ellipse interacts specularly with a sample [5–7]. The polarization-modifying interaction can occur either when the light beam is transmitted through the sample, as in transmission ellipsometry, or more commonly when it is reflected obliquely from the sample surface, as in reflection ellipsometry. In the absence of depolarization, the pure polarization state of the emerging beam is in general elliptical, and the analysis of this ellipse can provide useful information about the sample such as its optical properties and layer thicknesses.

The instrument that performs such polarization measurements is known as an *ellipsometer* and one of the most popular instrument configurations of an ellipsometer is denoted by PCSA [5–7]. In this configuration, an unpolarized collimated light beam from a source is first polarized elliptically upon transmission through a polarizer (P) and compensator (C), is then specularly reflected from the sample (S), and is finally transmitted through an analyzer (A) before impinging on the detector. The source, polarizer, and compensator (or retarder) work together to generate a state of known polarization before the light beam reaches the sample, whereas the analyzer and detector work together to detect the change in polarization state after the light beam reflects from the sample. Because the change in polarization state depends on wavelength, in a *spectroscopic ellipsometry* experiment either a monochromator must be used in conjunction with the source, or a spectrometer in conjunction with the detector. Overall, a single ellipsometry measurement provides two important interaction parameters, ψ and Δ at a given wavelength, which can be derived from the polarization state parameters of the beam before and after reflection from the sample.

Advanced Characterization Techniques for Thin Film Solar Cells,
Edited by Daniel Abou-Ras, Thomas Kirchartz and Uwe Rau.
© 2011 Wiley-VCH Verlag GmbH & Co. KGaA. Published 2011 by Wiley-VCH Verlag GmbH & Co. KGaA.

These so-called *ellipsometry angles* are defined by: $\tan\psi \exp(i\Delta) = r_p/r_s$, where r_p and r_s are the complex amplitude reflection coefficients of the sample for p and s linear polarization states, for which the electric field vibrates parallel (p) and perpendicular (s) to the plane of incidence, respectively.

Thus, a single ellipsometric measurement performed at one wavelength can provide at most two sample parameters from the ψ and Δ values determined in the experiment [5–7]. When the reflecting sample is isotropic, homogeneous, and uniform, and at the same time presents a single interface to the ambient, meaning that it is atomically smooth and film-free, then the sample parameters of n, the real index of refraction, and k, the extinction coefficient ($k \geq 0$), can be determined from ψ and Δ. Due to the phenomenon of dispersion, (n, k) are functions of wavelength and are characteristic of the material from which the sample is composed (often called "optical functions"). In fact, n and k serve as the real and imaginary parts of the complex index of refraction $N = n - ik$. (The negative sign in this definition arises from the optical electric field phase convention adopted here.) An example of the determination of optical functions from spectra in the ellipsometry angles will be presented in Section 6.2.2.

In the early years of ellipsometry research, before the widespread use of computers, (ψ, Δ) data could only be obtained at one or a few selected wavelength values [8]. The most widespread application of such a measurement was to determine (n, k) and thickness of a thin film on a known substrate, which posed a data analysis challenge due to the limited number of data values. Starting in the 1970s, automatic ellipsometers were developed that could collect (ψ, Δ) point-by-point nearly continuously versus wavelength spanning from the ultraviolet (200–300 nm) to the near-infrared (800 nm) [9]. Acquisition times ranged from several minutes to several hours depending on the number of wavelength points desired. This new type of ellipsometry became known as *spectroscopic ellipsometry*. Using this method, N for one or more components of a sample structure can be extracted as smooth functions of wavelength. Such optical functions can be further analyzed to provide the electronic structure such as the critical point (CP) parameters of a semiconductor material (fundamental and higher band gap energies, their amplitudes and widths) [7, 10–12]. These in turn provide useful information about the sample such as void fraction, crystalline fraction, alloy composition, grain size, strain, and temperature [13–17]. The optical functions of many materials have been studied in-depth and can be found, for example, in Refs. [18] and [19].

The relatively long acquisition times of traditional spectroscopic ellipsometry precluded *in-situ* and real-time measurements during thin-film deposition and processing; however, later developments using high-speed array detectors enabled spectroscopic ellipsometry with acquisition times from tens of milliseconds to tens of seconds [7, 20]. The instrument for such measurements became known as the *multichannel ellipsometer*, and the technique was described as *real-time spectroscopic ellipsometry* (RTSE), which can be very useful in tracking the growth mechanisms of a thin-film material. The advantages of RTSE include measurement of unoxidized surfaces in vacuum environments, the determination of surface roughness from the real-time data set and its correction for highly accurate spectra in N, and the resulting ability to develop a database of optical functions for a given material as a function of deposition parameters and temperature. This optical function database is then useful for analyzing *ex-situ* SE data. RTSE will be the focus of this chapter since it is the most informative type of ellipsometry currently available for photovoltaics research.

6.2 Theory

In order to understand the ellipsometry measurement, it is necessary to start with a basic understanding of polarized light. Once polarized light is well understood, the next step is to use the simplest problem, that of a specularly reflecting single interface, to understand how ellipsometry can be used to determine the optical functions of a material.

6.2.1 Polarized Light

The electric field vector of a monochromatic plane wave of arbitrary polarization state traveling along the z direction can be written as [5]

$$\vec{E}(z,t) = E_{0x}\cos(\omega t - kz + \delta_x)\hat{x} + E_{0y}\cos(\omega t - kz + \delta_y)\hat{y} \quad (6.1)$$

where E_{0x} and E_{0y} are the amplitudes of the electric field along the x and y axes ($E_{0x} \geq 0$; $E_{0y} \geq 0$), respectively, ω is the angular frequency of the wave, k is the magnitude of the wave vector (not to be confused with k the extinction coefficient), and δ_x and δ_y are the phase angles of oscillations along the x and y axes, respectively. In the most general case, the endpoint of this electric field vector traces out an ellipse as a function of time with period $\tau = 2\pi/\omega$ at a fixed position in space; however under special circumstances, the endpoint traces out a circle or a line resulting in circular or linear polarizations, respectively. In particular, when $E_{0y}/E_{0x} = 1$ and $\delta_y - \delta_x = \pm(\pi/2 + 2n\pi)$ ($n = 0, 1, 2 \ldots$) right ($+$) and left ($-$) circular polarization results; when $\delta_y - \delta_x = 2n\pi$ or $\pm(\pi + 2n\pi)$ ($n = 0, 1, 2 \ldots$) linear polarization results. The handedness of circular and, more generally, elliptical polarization describes the sense of rotation of the electric field endpoint. For right- and left-handed polarization states the electric field vector travels clockwise and counterclockwise, respectively, when looking opposite the direction of wave travel. The handedness assignments for $\delta_y - \delta_x$ are valid when using the argument ($\omega t - kz + \delta$) in Eq. (6.1). If an argument of ($kz - \omega t + \delta$) is used, the assignment is opposite; (see, e.g., Refs. [9, 21]). For the latter convention, the sign associated with the imaginary part of N becomes positive.

Thus, the elliptical polarization state can be defined most generally by four parameters: E_{0x}, E_{0y}, δ_x, and δ_y [5]. An alternative set of four parameters can be used, namely, the azimuthal angle Q, the ellipticity angle χ (whose sign defines the handedness), the amplitude A, and the absolute phase δ. The azimuthal angle Q ($-90° < Q \leq 90°$) of the ellipse is the angle between the major axis and a fixed reference direction (such as the p direction). The ellipticity angle χ ($-45° \leq \chi \leq 45°$) is defined as $\chi = \tan^{-1} e$, where $e = b/a$ $(-1 \leq e \leq 1)$ and b and a are the lengths of the semiminor and semimajor axes of the ellipse, respectively. Positive and negative e correspond to right- and left-handed polarization states, respectively. Usually only the shape of the polarization ellipse is of interest in an ellipsometry experiment (an exception being when one is combining ellipsometry with a polarized transmittance and/or reflectance measurement). As a result, the amplitude and absolute phase associated with the polarization ellipse are not of interest and the key polarization state angles (Q, χ) can be derived from the field ratio E_{0y}/E_{0x} and the phase difference $\delta_y - \delta_x$.

6.2.2
Reflection from a Single Interface

Reflection from a single interface between the ambient and an isotropic, homogeneous, and uniform medium ("the sample") provides the simplest example of how ellipsometry can be applied to determine N_s, the complex index of refraction of the given sample [5]. In this case, the complex amplitude reflection coefficients for p and s polarized light are given by the Fresnel equations

$$r_p = \frac{N_s \cos\theta_i - n_a \cos\theta_t}{N_s \cos\theta_i + n_a \cos\theta_t} \qquad r_s = \frac{n_a \cos\theta_i - N_s \cos\theta_t}{n_a \cos\theta_i + N_s \cos\theta_t} \tag{6.2}$$

where n_a is the real index of refraction of the ambient, $N_s = n_s - ik_s$ where n_s is the real index of refraction of the sample and k_s is its extinction coefficient, θ_i is the angle of incidence, and θ_t is the complex angle of transmission. Although these equations appear to have two unknown values, N_s and θ_t, Snell's law can be applied to eliminate θ_t. Then, the ellipsometric parameters ψ and Δ can be expressed in terms of N_s, n_a, and θ_i through

$$\tan\psi e^{i\Delta} = \frac{r_p}{r_s} = \rho \tag{6.3}$$

Upon inversion of the resulting equation, N_s can be determined from the ellipsometry data values ψ and Δ using the known quantities n_a and θ_i. The resulting equation becomes simpler in form when the optical functions n_a and N_s are expressed instead as the corresponding relative electric permittivities, also called the dielectric functions

$$\varepsilon_s = N_s^2 \qquad \varepsilon_a = n_a^2 \tag{6.4}$$

where ε_s is the complex dielectric function of the reflecting sample, and ε_a is the dielectric function of the ambient. The final result, which relates the measured ellipsometry parameters to the dielectric function of the sample, is given as

$$\varepsilon_s = \varepsilon_a \sin^2\theta_i \left[1 + \tan^2\theta_i \left(\frac{1-\rho}{1+\rho}\right)^2\right] \tag{6.5}$$

For a single interface, the requirement on ε_s, that is, that the imaginary part be nonnegative, implies that $0° \leq \psi \leq 45°$ and $0° \leq \Delta \leq 180°$.

6.3
Ellipsometry Instrumentation

Figure 6.1 shows the generic ellipsometer configuration that illustrates the principles of both stepwise wavelength scanning spectroscopic ellipsometry (SE) used in *ex-situ* applications, and multichannel SE used in high-speed real time, mapping, and on-line monitoring applications [5–7, 22]. In wavelength scanning SE, the monochromator in Figure 6.1 is stepped from one wavelength point to the next, with an acquisition time of minutes to hours for a full spectrum in (ψ, Δ). In this case, a single element detector is used. In multichannel ellipsometry, the monochromator is eliminated from the source side of the instrument, and a detection system is used consisting of a spectrograph and linear detector array. Each pixel of the array collects photons over a narrow wavelength band, and array scanning for SE can occur at high speed ($\sim$5 ms) with (ψ, Δ) acquisition times ranging from tens of milliseconds to tens of seconds.

The selected light sources and detectors for both types of instrument depend on the desired spectral range [7, 22–24]. For applications in photovoltaics, a range of $\sim$200–2000 nm (about 0.6–6 eV) is ideal, covering the band-gap onsets and CPs of the key semiconductors used in photovoltaics. This typically requires a tandem source, for example, a quartz–tungsten halogen lamp (400–2000 nm) with a see-through D_2 lamp (200–400 nm). In addition, at least two detectors are required: (i) Si-based detectors for wavelengths below 1000 nm, and (ii) III–V-based detectors above 1000 nm. For wavelength scanning instruments, single element photomultiplier tubes and Pb salt photodetectors can be used, operating below and above 900 nm, respectively.

For both SE instruments, beam sizes are typically 1–5 mm, so that the measurement averages over a macroscopic area on the sample. If the sample is not uniform over this area, depolarization of the reflected beam will occur when the in-plane scale of the thickness or optical property variations is larger than the lateral coherence

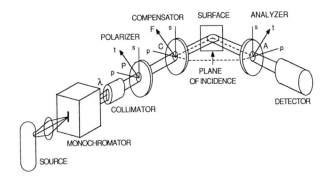

Figure 6.1 Generic spectroscopic ellipsometry (SE) configuration.

length of the beam (typically ~10 μm). For variations smaller than the lateral coherence length (i.e., ~0.05–5 μm), the specularly reflected beam will remain polarized; however in this case, nonspecular scattering is more likely to occur. In all cases of nonuniformity, SE can provide useful information on materials properties and the uniformity of those properties through appropriate modeling [25]. For nonuniformities on a scale much less than the wavelength (<0.05 μm) (i.e., heterogeneities), effective medium theories are used in the modeling of SE data [26].

The most widely used experimental configurations in photovoltaics research are those in which one of the three polarization modifying optical elements rotates at a fixed angular frequency ω versus time t [27, 28]. The elements that can be rotated are: (i) the polarizer with its transmission axis orientation given by $P(t) = \omega_P t - P_S$; (ii) the compensator with its fast axis orientation given by $C(t) = \omega_C t - C_S$; and (iii) the analyzer (which is simply a polarizer on the detection side of the instrument) with its transmission axis orientation given by $A(t) = \omega_A t - A_S$. The zero positions of these angles occur when the transmission or fast axis lies in the plane of incidence established by reflection from a sample. P_S, C_S, and A_S are offsets that account for arbitrary angular positions of the optical elements relative to the plane of incidence at $t = 0$, the onset of data collection. In the following discussion, these offsets are neglected since they can be determined readily in calibration procedures. For the wide spectral range of 200–2000 nm, the polarizer, compensator, and analyzer are fabricated from MgF_2 or fused silica. For the compensator, in fact, monoplates, biplates, or dual biplates have been used, whereby dual biplates act as the combination of a polarization rotator and a compensator (or retarder) and yield more complicated data reduction equations than those provided in the next sections [24, 29].

6.3.1
Rotating Analyzer SE for *Ex-Situ* Applications

The most popular configuration for wavelength scanning ellipsometry incorporates the rotating analyzer principle of polarization state detection [27, 30, 31]. When the analyzer is the rotating element in the configuration of Figure 6.1, the waveform at the detector varies with time according to

$$I_{\exp}(t) = I_0\{1 + \alpha\cos(2\omega_A t) + \beta\sin(2\omega_A t)\} \tag{6.6}$$

where I_0 is the dc Fourier coefficient and α and β are the cosine and sine Fourier coefficients, respectively, normalized to the dc coefficient. For photomultiplier tube or solid-state detectors, the photocurrent is sampled at regular intervals (typically every ~5° of analyzer angle) triggered by an encoder mounted on the motor which drives the analyzer [31]. Assuming the detector is linear, the photocurrent is proportional to the incident irradiance, and Fourier analysis of the sampled current over one-half rotation of the analyzer (one optical cycle) can provide the Fourier coefficients.

From the Fourier coefficients and the settings of the polarizer and compensator, the ellipsometry angles (ψ, Δ) can be determined. The compensator settings include not only the fast axis angle C, but also δ_C, which is the phase shift difference for orthogonal linear polarizations along the fast and slow axes. If the fast axis of the compensator lies in the plane of incidence (so that $C=0°$), then the following equations for (ψ, Δ) hold:

$$\tan\psi = \left|\sqrt{\frac{1+\alpha}{1-\alpha}}\right||\tan P| \quad (6.7)$$

$$\cos(\Delta + \delta_C) = \beta\left|\sqrt{\frac{1}{1-\alpha^2}}\right| \quad (6.8)$$

When source fluctuations limit the achievable precision, then the highest measurement precision occurs for polarizer angle $P = \psi$ and for compensator retardance $\delta_C = \pm 90° - \Delta$. This motivates the use of a polarizer and compensator that can be adjusted automatically to achieve these conditions, wavelength point by point. For the compensator, a Berek monoplate design can be used in which an MgF_2 plate is tilted slightly (~5–10°) about an axis parallel to the fast or slow axis and this tilt angle is varied in order to vary δ_C [29].

6.3.2
Rotating Compensator SE for Real-Time Applications

The rotating analyzer ellipsometer has three limitations in multichannel ellipsometry. In such applications, it may not be possible to perform automatic optimum adjustments in the optical elements to optimize instrument performance or multiple measurements to resolve ambiguities [28]. First, a rotating analyzer cannot be used to detect depolarization by the sample or by optical element imperfections; when such effects are unrecognized and not modeled, they lead to experimental errors. Second, the sign of $\Delta + \delta_C$ is ambiguous, which means that there are two possible solutions for Δ. (When $\delta_C = 0°$, that is, when no compensator is used, this simply means that one cannot detect the handedness of the polarization state entering the rotating analyzer.) Third, there exists a low sensitivity to Δ for nearly linearly polarized light entering the rotating analyzer which occurs in the case of Eq. 6.8 when $\Delta + \delta_C \sim 0°$, 180°, 360° ...

These limitations are overcome by rotating the compensator rather than the analyzer [32, 33]. It is more instructive in this case to consider the situation in which the compensator is placed after the sample, in contrast to the configuration in Figure 6.1. Then one can consider the rotating compensator followed by the fixed analyzer as a polarization state detector for the light beam reflected from the sample, providing the tilt angle Q relative to the plane of incidence ($-90° < Q \leq 90°$), the ellipticity angle χ ($-45° \leq \chi \leq 45°$), and even the degree of polarization p ($0 \leq p \leq 1$), which allows one to quantify possible depolarization. For any rotating compensator ellipsometer, irrespective of the position of the compensator before or after the sample, both $2\omega_C$ and $4\omega_C$ frequencies appear in the irradiance waveform, according to

$$I_{\exp}(t) = I_0\{1 + \alpha_2 \cos(2\omega_C t) + \beta_2 \sin(2\omega_C t) + \alpha_4 \cos(4\omega_C t) + \beta_4 \sin(4\omega_C t)\} \tag{6.9}$$

where $2\omega_C$ can be understood as the fundamental optical frequency.

For a compensator after the sample and an analyzer transmission axis angle A, the polarization state parameters of the beam reflected from the sample are given by [28]

$$Q = \frac{1}{2}\arctan\left(\frac{\beta_4}{\alpha_4}\right) - A \tag{6.10}$$

$$\chi = \frac{1}{2}\arctan\left\{\frac{\sqrt{\alpha_2^2 + \beta_2^2}}{2\sqrt{\alpha_4^2 + \beta_4^2}}\tan\left(\frac{\delta_C}{2}\right)\right\} \tag{6.11}$$

$$p = \frac{\Re_Q}{\cos 2\chi \cos 2(A-Q)\{1 - (1 + \Re_Q)\cos^2(\delta_C/2)\}} \tag{6.12}$$

where

$$\Re_Q = \alpha_4 \cos 4Q + \beta_4 \sin 4Q \tag{6.13}$$

Using more detailed expressions provided elsewhere [28], the sign of the ellipticity angle can be determined, which identifies the handedness of the polarization state. One also notices that for linearly polarized light ($\chi = 0°$) and circularly polarized light ($\chi = \pm 45°$) reflected from the sample, the $2\omega_C$ and $4\omega_C$ coefficients vanish, respectively. As a result, Eqs. (6.12) and (6.13) are optimized for linearly polarized light, which occurs in reflection from a single ambient/semiconductor interface for photon energies below the semiconductor band gap. In general, the ellipsometry angle Δ can be determined from Q and χ using the following simple equation:

$$\tan\Delta = -\csc 2Q \tan 2\chi \tag{6.14}$$

The general expression for ψ is more complicated and can be found elsewhere [28]. If one considers the case of a polarizer set at $P = 45°$, then the equally simple expression

$$\cos 2\psi = -\cos 2Q \cos 2\chi \tag{6.15}$$

is obtained in the absence of depolarization.

A second approach that has been adopted is to place the compensator before the sample as in Figure 6.1 [24]. Such a configuration is less straightforward conceptually, in terms of stepwise data reduction first for reflected beam parameters and then for sample parameters, as in Eqs. (6.9)–(6.15). Operationally however, errors due to system alignment in real time and mapping applications are reduced through this second approach. When the compensator is placed before the sample, one can conclude from time reversal symmetry that Eqs. (6.9)–(6.10) and Eqs. (6.14)–(6.15) hold, but A is replaced by P and vice versa; Δ and δ_C are replaced by $-\Delta$ and $-\delta_C$. In this case, the simple physical interpretation of the degree of polarization is lost.

6.4
Data Analysis

Rarely does a sample assume the simple form of a single interface; the most common sample structure consists of m layers on a substrate and $m + 1$ interfaces. In order to extract one or more complex dielectric functions and the layer thicknesses for such multilayer samples from a pair of ellipsometry spectra (ψ, Δ), a model-based approach must be used [5–7, 22]. Two of the most commonly used approaches include exact numerical inversion and least-squares regression, which have different applications depending on the type of sample being analyzed. Both approaches are based on iterative application of the same three steps: (i) development of a model with starting parameters, (ii) determination of the best-fit parameters within the assumed model, and (iii) evaluation of the reasonableness of the results either through calculation of an error function or through an inspection of the smoothness and consistency of the resulting dielectric function. A third approach, known as virtual interface analysis, is extremely useful for a complex multilayer structure in which case the focus of the analysis is on the near-surface region and not on the underlying multilayer structure.

6.4.1
Exact Numerical Inversion

Exact numerical inversion is useful for analysis of thin-film samples whereby one of the dielectric functions of the sample is unknown, and only approximations to one or more of the thicknesses are available. The simplest inversion method utilizes a Newton–Raphson algorithm; however, more advanced algorithms exist [34, 35]. The first step in this method is to assign the approximate values for the one or more thicknesses in the problem, as well as first guess values of the real and imaginary parts of the one unknown complex dielectric function at one wavelength point (usually the highest value of λ in the case of a semiconductor). Then, (ψ, Δ) values at this wavelength are calculated and are compared to the experimental (ψ, Δ) data. Next, the inversion algorithm is used to bring the first guess values of the complex dielectric function closer to those required for an exact match to the experimental data. These steps are iterated until the calculated and experimental data values agree to within limits smaller than the precision, that is, $<1 \times 10^{-4}$ [31]. The entire process is then repeated for the next spectral point, using the results from the previous spectral point as a first guess for the complex dielectric function. It is important to note that the first guess value must be reasonably close to the actual solution; otherwise the inversion algorithm may diverge. Alternatively, multiple solutions may exist and an incorrect one may be erroneously identified. It should be noted that the final solution is associated with the particular approximate thickness values. There must be an additional iteration loop to identify better thickness values based on the elimination of artifacts in the deduced complex dielectric function. In addition, the result can be checked for Kramers–Kronig consistency (see, e.g., Ref. [7]).

6.4.2
Least-Squares Regression

Least-squares regression is the most commonly used method for extracting the complex dielectric function and thicknesses of a thin-film material from spectroscopic ellipsometry data (ψ, Δ) [36, 37]. The first step in this method is to develop models for the multilayer structure and any unknown complex dielectric functions of the layer components. In this case, in contrast to analysis by inversion, the dielectric functions are expressed as continuous analytical functions of wavelength. Thus, all parameters in the problem are wavelength independent: the parameters in the analytical functions and the thicknesses. Then, (ψ, Δ) spectra are calculated based on the first guess values of the parameters and are compared to the experimental (ψ, Δ) spectra. Next, the least-squares regression algorithm is used to bring the first guess values closer to those that minimize the mean square error (MSE) between the calculated and experimental spectra. As the iteratively adjusted parameter values approach the solution, the MSE will be minimized and the least-squares regression algorithm will be complete. It is important to note that the MSE will never vanish since there are more experimental data values than free parameters in the model, and these experimental data values incorporate both random and systematic errors [37].

6.4.3
Virtual Interface Analysis

Virtual interface analysis is useful for characterizing the near-surface region of a deposited film, while minimizing the complexity of the model by approximating the underlying multilayer structure with a single interface. When applied to RTSE data, this method assumes an initial pseudosubstrate with pseudodielectric function $\langle \varepsilon \rangle$ which is obtained by applying Eq. (6.5) to (ψ, Δ) values collected when $t = 0$ (no outer layer). For all $t > 0$, an optical model is constructed which includes an outer layer of deposited material with dielectric function ε_0 and thickness d_0 on top of the pseudosubstrate described above. Using this model and the (ψ, Δ) values measured at multiple times, it is possible to determine the best dielectric function ε_0 and the deposition rate required to fit the multiple (ψ, Δ) values. In this case, ε_0 is interpreted as the dielectric function of the near-surface region of the deposited film. Repeating this procedure as a function of time will result in a depth profile in the optical functions, which can be further analyzed to extract a depth profile of the structural phase (as in thin Si:H) or the chemical composition (as in $CdTe_{1-x}S_x$ or CIGS solar cell materials) [38, 39].

6.5
RTSE of Thin Film Photovoltaics

In this section, applications of RTSE to the three major thin-film technologies will be described in order of the level to which the capability has been advanced in each case.

For thin film hydrogenated silicon (Si:H), significant progress has been made in understanding the growth mechanisms using RTSE, and guiding principles for the optimization of both amorphous silicon (a-Si:H) and nanocrystalline silicon (nc-Si:H) solar cells have been identified [38]. For thin film CdTe, RTSE studies currently explore the relationships between the nucleation and growth dynamics of as-deposited CdTe and CdS and the resulting solar cell performance [39]. In this case, the complexity of such relationships is enhanced as a result of the postdeposition $CdCl_2$ treatment. Finally for CIGS, RTSE studies are at an early stage, and the initial effort consists of categorizing the optical functions of the relevant phases and compositions of the materials.

6.5.1
Thin Si:H

Thin film Si:H is the most extensively studied of the thin-film solar cell materials since it encompasses the prototypical amorphous and nanocrystalline semiconductors [40]. The best small area solar cells from multijunctions of these materials exhibit a factor of 2 lower efficiency than the corresponding best c-Si solar cells. In spite of this, the thin-film Si:H cells do present certain advantages such as relaxed crystal momentum conservation (hence strong absorption for a much reduced thickness) and a potential for low production costs [40]. The Si:H materials are deposited by plasma enhanced chemical vapor deposition (PECVD), a low temperature process (<300 °C) using silane or disilane and hydrogen gases, making these solar cells excellent candidates for deposition on flexible substrates in roll-to-roll processes. By altering the deposition parameters, most notably the R value which in the examples below is the ratio of hydrogen to disilane gas flows, a thin Si:H film can be made amorphous, mixed-phase, or nanocrystalline. These forms of Si:H thin films have been studied in detail using RTSE as described next [38, 41].

Figure 6.2 compares the deduced dielectric functions of purely amorphous and purely nanocrystalline phases of Si:H (a-Si:H and nc-Si:H, respectively) measured at a calibrated substrate temperature of 200 °C. Although the overall shapes are similar, the nc-Si:H dielectric function shows two well-defined features near 3.4 and 4.2 eV, which are the vestiges of the CPs of single crystal Si [12]. In contrast, the a-Si:H dielectric function is characterized by a single, much broader feature due to complete loss of long range order. This loss leads to a relaxation of crystal momentum (or electron k-vector) conservation and as a result, a larger imaginary part of the dielectric function just above the band gap (1.8 eV) than nc-Si:H. Thus, much stronger absorption is obtained over the photon energy range of the visible spectrum, from 1.8 to 3.1 eV. It is these differences that enable one to distinguish readily between the two phases and to characterize the evolution of mixed-phase materials using RTSE [41].

Figure 6.3 shows the thickness evolution of the volume fraction of the nanocrystalline phase (f_{nc}) in a mixed-phase Si:H intrinsic layer (i-layer), deposited by PECVD under fixed conditions: a nominal substrate temperature of 200 °C and $R = 150$ [38]. In fact, RTSE can be used to calibrate the actual substrate temperature, and the result

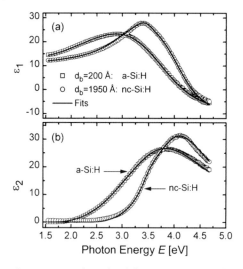

Figure 6.2 (a) The real and (b) imaginary parts of the dielectric functions of intrinsic amorphous and nanocrystalline Si:H (a-Si:H and nc-Si:H, respectively), measured at 200 °C in RTSE studies of 20 and 195 nm thick films [41].

in this case (110 °C) is much lower than the nominal value. The substrate is an a-Si:H n-layer film, which generates the layered configuration used for a standard *amorphous* n–i–p solar cell. The results of Figure 6.3 are obtained by applying the virtual interface analysis technique to RTSE data in order to extract the dielectric function of the top ~1 nm. In this way, a depth profile in the volume fraction of nanocrystallites, f_{nc}, can be obtained by applying the Bruggeman effective medium approximation to the deduced dielectric function assuming a mixture of fully amorphous and nanocrystalline phases. For this deposition, the Si:H i-layer film nucleates on the n-layer as amorphous Si:H, undergoes an amorphous-to-nanocrystalline transition after a thickness of 9 nm, and then becomes fully nanocrystalline after a thickness of 100 nm.

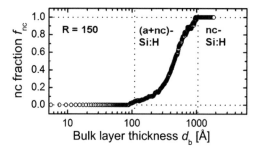

Figure 6.3 Depth profile in the nanocrystalline volume fraction (f_{nc}) for a mixed-phase Si:H i-layer deposited under fixed conditions with $R = 150$ on top of an a-Si:H n-layer [38].

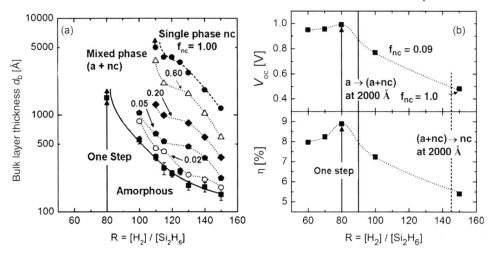

Figure 6.4 (a) Deposition phase diagram for a Si:H i-layers deposited on top of a-Si:H n-layers in the n–i–p configuration used for a-Si:H solar cells; (contours on the diagram indicate f_{nc} values) [38]. (b) V_{oc} and efficiency versus the Si:H i-layer R value for single junction solar cells 200 nm thick. The n-layers used in these solar cells are pure a-Si:H [38].

Similar depth profiles obtained as a function of R from R = 80 to 150 were used to construct a deposition phase diagram for the PECVD films as in Figure 6.4a [38]. In this case, the Si:H i-layer films grown on a-Si:H n-layers remain amorphous throughout the first ~200 nm of bulk layer growth when $R \leq 80$. For higher R ($R \geq 100$), the Si:H i-layer films initially nucleate as a-Si:H, but then undergo an amorphous-to-nanocrystalline transition at a thickness that shifts from above 200 nm to lower thicknesses with increasing R. Performance parameters of single junction solar cells fabricated using a thickness of ~200 nm for the Si:H i-layer, and different R values are shown in Figure 6.4b [38]. The optimum V_{oc} and efficiency for this series of cells occurs when R = 80, which is the maximum R value possible such that the i-layer remains fully amorphous throughout its thickness. For a larger value of R = 100, the fraction of nanocrystallites in the near surface of the 200 nm i-layer is small (~0.09), but these nanocrystallites appear to degrade V_{oc} due to their presence at the i–p interface. This concept of maximum hydrogen dilution developed by RTSE is now applied widely to optimize the efficiencies of a-Si:H solar cells [41]. In fact, RTSE has suggested extensions of this concept using multistep and continuously variable dilution ratios R for improvements over simple single-step processes.

Another deposition phase diagram determined by RTSE is shown in Figure 6.5a, in this case for Si:H i-layers which are deposited on intended nanocrystalline Si:H (nc-Si:H) n-layers at a nominal temperature of 200 °C [38]. This is the layer structure used for a standard *nanocrystalline* n–i–p solar cell, and the resulting diagram exhibits behavior which is significantly different than that of the Si:H i-layers on a-Si:H n-layers shown in Figure 6.4a. Initial stage mixed-phase Si:H growth results from a

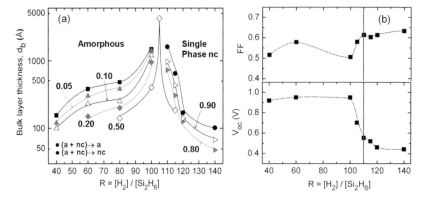

Figure 6.5 (a) Deposition phase diagram for Si:H i-layers on nc-Si:H n-layer substrate films. These i-layers are designed for incorporation into nc-Si:H solar cells [38]; (contours on the diagram indicate f_{nc} values). (b) V_{oc} and FF versus R for intended nc-Si:H solar cells in which case nc-Si:H n-layers are used. The vertical line identifies the optimum nc-Si:H solar cell performance [38].

template effect due to the underlying nc-Si:H n-layer, which is itself mixed-phase, consisting of ~0.5/0.5 vol. fraction ratio of a-Si:H/nc-Si:H. At lower dilution levels ($R < 105$), the mixed-phase i-layers evolve to a-Si:H due to a preference for *amorphous* growth. In this range of R, the mixed-phase nanocrystalline to amorphous [(a + nc) → a] transition shifts to larger thickness with increasing R. At higher dilution levels ($R > 105$), the initial mixed-phase material rapidly evolves to fully nc-Si:H due to the strong preference for *nanocrystalline* growth. In this range, the (a + nc) → nc transition shifts to lower thickness with increasing R. Figure 6.5a reveals a bifurcation value of $R = 105$, which divides the ultimate phase of the film between fully amorphous and fully nanocrystalline – even though the film nucleates as mixed-phase Si:H independent of R.

In the case of 0.4–0.6 μm thick nc-Si:H i-layers used in solar cells with performance shown in Figure 6.5b, the optimum one-step deposition process occurs on the basis of the phase diagram at the minimal value of $R = 110$, that is, the smallest R value possible such that the i-layer evolves to dominantly nc-Si:H during its growth [38]. At $R = 100$ the (a + nc) → a transition occurs at a thickness of 150 nm, and it is clear from the V_{oc} value of 0.95 V in Figure 6.5b that the top of the film is a-Si:H. At $R = 120$ the (a + nc) → nc transition occurs at a thickness of 200 nm, and in this case, it is clear from the V_{oc} value of 0.46 V that the top of the film is nc-Si:H. The highest performance nc-Si:H i-layer in Figure 6.5b is obtained at $R = 110$, just before a more rapid drop in fill factor associated with the bifurcation region. The continued lower fill factor for the low R Si:H films in Figure 6.5b relative to those of Figure 6.4b is attributed to the nc-Si:H phase at the bottom of the i-layer (about 40 nm thick for $R = 60$). Overall these results demonstrate clear correlations between structural evolution from RTSE and device performance that enable single step, multistep, and graded layer optimization.

6.5.2
CdTe

CdTe is used widely in thin-film photovoltaics as a p-type semiconductor serving as the active layer, along with CdS as its n-type heterojunction partner serving as a window layer. Thin films of CdTe can be produced by several different methods including vapor transport deposition (VTD), close-space sublimation (CSS), and rf magnetron sputtering [42]. The work presented below will focus on the use of RTSE to study thin films deposited by magnetron sputtering [43]. Although the sputter deposition rates of CdTe are generally slower, this deposition method is of interest since it allows one to fabricate efficient solar cell devices in a relatively low temperature process (<400 °C). Also, by controlling sputtering parameters such as the temperature, pressure, and power, the growth mechanisms can be controlled, including the surface mobility of the deposited species, the bombardment energy and directionality of incident species, and the rate/kinetics of the deposition, respectively. These mechanisms determine the nano- and microstructure of the sputter deposited film [44, 45].

Figure 6.6a shows the room temperature dielectric functions for bulk single crystal CdTe (c-CdTe) and an as-deposited polycrystalline CdTe film sputtered to 100 nm thickness at 188 °C [39]. CdTe exhibits four CPs over the energy range of $0.75 < E < 6.5$ eV. The CP widths depend on the grain size and measurement temperature, the CP energies depend on the strain in the material and measurement temperature, and the CP amplitudes depend on the density of the material. The first CP energy represents the fundamental band gap of the material, which is slightly blue-shifted for the polycrystalline CdTe film relative to c-CdTe due to strain in the film. The very broad dielectric function features for the thin film are due to its fine

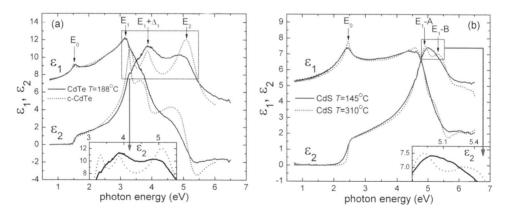

Figure 6.6 (a) Comparison of the room temperature dielectric function of an as-deposited polycrystalline CdTe thin-film magnetron sputtered at 188 °C with that of c-CdTe [39]. (b) Comparison of the room temperature dielectric functions of polycrystalline CdS thin-films magnetron sputtered at 145 °C and 310 °C [39].

grained structure. Figure 6.6b compares the room temperature dielectric functions of polycrystalline CdS films 50 nm thick fabricated at substrate temperatures of 145 and 310 °C [39]. From these results, one can conclude that the grain size of CdS increases with increasing substrate temperature.

In RTSE studies of CdTe thin films, the nucleation and growth dynamics have been characterized as a function of deposition temperature and Ar pressure [46, 47]. These films were deposited onto native oxide covered silicon wafers due to their smoothness and well-known optical properties. Figure 6.7a and b shows the surface roughness thickness evolution for temperature and pressure series, respectively, for the first ~5 nm of growth. In this case, for films sputtered at $T < 267\,°C$ and for films sputtered at $p_{Ar} > 10\,mTorr$, it is clear that significant surface roughness develops before a bulk layer is formed, leading to initial cluster growth followed by layer growth. In contrast, for $T \geq 267\,°C$ and $p_{Ar} \leq 10\,mTorr$, a thin bulk layer is formed before surface roughness starts to develop, leading to initial layer growth followed by cluster growth. These different initial growth modes are more pronounced when viewing plots of the bulk layer thickness at a surface roughness thickness of ~1 monolayer as shown in Figure 6.8a and b for the temperature and pressure series, respectively.

In order to consider these results from RTSE in view of CdTe solar cell performance, several devices were produced at different Ar pressures for the CdTe deposition. Each device underwent $CdCl_2$ postdeposition treatment for times ranging from 15 to 25 min and the resulting efficiency for each device is shown in Figure 6.9. This figure shows that films deposited at low pressure require longer treatment times for higher efficiency, whereas films deposited at high pressure ($p_{Ar} = 18\,mTorr$) require shorter times. The best devices were obtained for $p_{Ar} = 10$ mTorr, where the efficiency is not strongly dependent on treatment time. It is interesting to note that this pressure is consistent with the change of initial growth

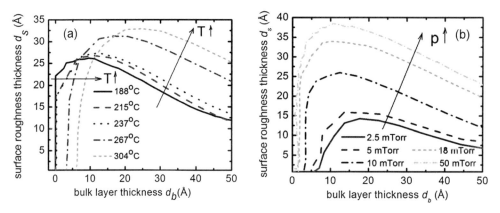

Figure 6.7 (a) Surface roughness evolution for CdTe thin films sputtered at different substrate temperatures [46]. (b) Surface roughness evolution for CdTe thin films sputtered at different Ar pressures [47].

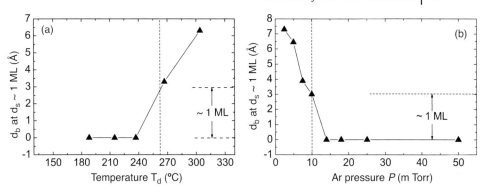

Figure 6.8 (a) Bulk layer thickness d_b at a surface roughness thickness d_s of 1 ML for CdTe thin films sputtered at different temperatures. (b) Bulk layer thickness d_b at a surface roughness thickness d_s of 1 ML for CdTe thin films sputtered at different Ar pressures.

modes as observed by RTSE, suggesting that a microstructure for optimum $CdCl_2$ treatment may be obtained in this transition region.

6.5.3
CuInSe$_2$

The I–III–VI$_2$ chalcopyrite system including CuInSe$_2$ (CIS) and its related alloys is one of the most promising of the second-generation thin-film absorber material technologies for photovoltaics applications [48]. The very rough surfaces of CIS thin films at the thicknesses (1–2 μm) used in photovoltaic devices imply that optical

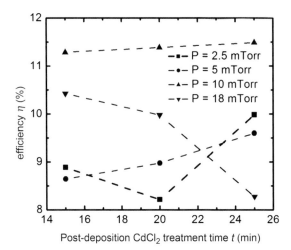

Figure 6.9 Efficiency of CdTe devices prepared at different Ar pressures as a function of the postdeposition $CdCl_2$ treatment time [47].

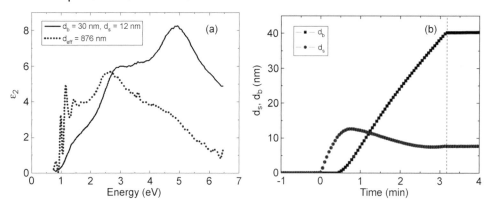

Figure 6.10 (a) Effect of surface roughness on attempts to determine the CIS dielectric function with a simplified optical model. The effective thickness is given by $d_{eff} = d_b + 0.5d_s$. (b) Time evolution of d_s and d_b for a CIS thin film.

function determination can be particularly challenging due to the near-complete loss of specularly transmitted and reflected light waves via scattering. When the roughness layer is very thick, the light beam at high energies does not penetrate to the bulk layer; in addition, the optical model for the roughness as a single layer in such a case is generally oversimplified; (see Figure 6.10a). Because the surface roughness increases in thickness during film growth, one way to avoid this problem is to perform measurements during film growth and to limit the total thickness of the film to ~50 nm. In general, such an approach also enables an accurate surface roughness correction and avoids postdeposition oxidation. The high accuracy dielectric functions deduced versus temperature using this methodology may be applied in future RTSE analyses, not only for the growth dynamics, but also for the structural and electronic properties of the absorber layer.

The CIS thin films studied here were grown by codeposition on Si(100) substrates held at 550 °C and exhibited an average composition of Cu = 22%, In = 23%, and Se = 55%, a total thickness of ~50 nm, a root mean square surface roughness of ~10 nm, and a single α-phase with a slight (112) orientation, as determined by a variety of *ex situ* measurements. Figure 6.10b shows the typical time evolution of both the surface roughness layer thickness (d_s) and the bulk layer thickness (d_b) deduced from RTSE data [49]. The Cu, In, and Se shutters were opened and closed simultaneously at time zero and after 3.2 min of codeposition, respectively. Initial growth of the film is observed in the form of islands which increase to 11 nm in thickness before the onset of bulk layer growth. The islands continue to grow until the onset of coalescence, as indicated by a decrease in d_s with a commensurate more rapid increase in d_b. Thereafter the film continues to smoothen as d_b increases approximately linearly with an average rate of 15 nm/min until shutter closure (dashed vertical line). A weak upturn in d_s visible near 3 min in Figure 6.10b represents the onset of roughening, an effect that would continue until the surface of the film is sufficiently rough that extraction of the dielectric function from RTSE measurements would be more challenging. The final values of surface roughness

($d_s = 7.7$ nm) and effective thickness ($d_b + 0.5d_s = 44$ nm) are in good agreement with those from *ex-situ* measurements.

The RTSE data that provide the evolution of the thicknesses in Figure 6.10b also provide the dielectric function of the bulk layer as shown in Figure 6.11a. Two representative dielectric functions are displayed here; one is obtained from real-time data acquired at the growth temperature of 550 °C, and the other is obtained from data acquired *in-situ* after cooling to 20 °C. For interpretation of these dielectric functions, an optical averaging effect in the polycrystalline films should be taken into account due to the anisotropy of single-crystal CIS [50–54]. Given the uniaxial nature of the tetragonal chalcopyrite crystal structure of CIS and the lack of a strong orientation of the studied films, however, it is expected that the dielectric functions of Figure 6.11a more closely resemble the ordinary component of the single crystal dielectric function [51]. In consideration of this conclusion, the overall amplitude and shape of the room temperature dielectric functions agree well with published literature.

A comparison of the results in Figure 6.11a reveals characteristic red shifts of the CP energies and increases in CP broadening parameters with increasing temperature. These can be quantified by modeling each of the dielectric functions as a sum of direct interband resonances at the band structure CPs using the parabolic band approximation with Lorentzian broadening. Thus, each CP generates a second derivative form for ε given by [11, 12, 55]

$$d^2\varepsilon/dE^2 = Ae^{i\phi}(E - E_0 + i\Gamma)^n \tag{6.16}$$

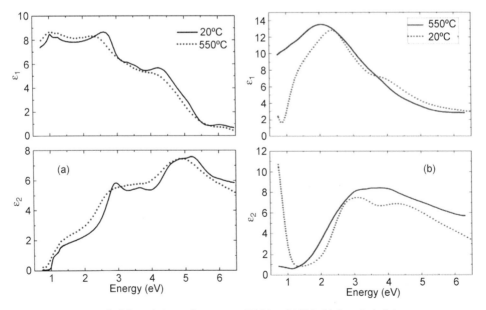

Figure 6.11 (a) CIS bulk layer dielectric functions at 550 °C and 20 °C. (b) $Cu_{2-x}Se$ bulk layer dielectric functions at 550 °C and 20 °C.

where E is the photon energy, E_0 is the CP energy, A is the amplitude, ϕ is the phase, Γ is the broadening energy, and n is the exponent. The exponent was chosen so that the fundamental and higher energy interband transitions were modeled as excitonic and 2D CP line shapes, respectively [50]. The remaining four parameters for each CP were obtained in fits to the second derivatives of the dielectric functions [11, 12]. A sum of up to 12 resonances of the form of Eq. (6.16) was used to simulate the convolved dielectric functions of the polycrystalline CIS thin films along the ordinary and extraordinary rays. The CP energies obtained with such fits agree well with results reported for single crystals [50], whereas three high-energy transitions at 5.11, 5.41, and 5.91 eV were identified for the first time.

After terminating the deposition, the CIS films were cooled in a controlled manner at 50 °C increments while SE spectra were acquired after a period of thermal stabilization at each temperature. Holding d_s and d_b fixed, the CIS dielectric functions were obtained at each temperature and the E_0 transitions were subjected to the analysis of Eq. (6.16). The resulting temperature dependence of the $E_0(A, B)$ critical point energy was in good agreement with previous results reported for lower temperature [56]. As the temperature range studied covers only the linear regime of the empirical Varshni [57] and the Bose–Einstein statistical [58] relations, the data were fit to the linear equation

$$E(T) = E(0) - \alpha T \tag{6.17}$$

The best-fit result leads to $E(0) = 1.05$ eV and $\alpha = 1.03 \times 10^{-4}$ eV/K. This fit may enable the precise measurement of the temperature during CIS deposition over the temperature range studied, and therefore a more precise engineering of CIS solar cells.

Due to the complexity of the CIS material system as well as the solar cell stack itself, many other research activities of interest are in progress. For example, the deposition of the highest efficiency I–III–VI$_2$ solar cells is performed with a so-called three-stage process, based on a variation of the copper content during growth. One of the key components of this process is the $Cu_{2-x}Se$ compound. The analysis of $Cu_{2-x}Se$ in a similar manner as was done for CIS leads to the dielectric functions at 550 and 20 °C displayed in Figure 6.11b. As one can see, these dielectric functions present different features than those of CIS; notably intraband transitions at low energies normally found in metals or semimetals. Such databases should also allow for improved control of the deposition process by controlling more precisely the occurrence of this compound. Other potential future studies include the back contact [59], buffer layer [60] and top window [61], and how their deposition affects the properties of the underlying layers; the effects of alloying group I, III, or VI elements starting from the ternary CIS; or even the effects of the deposition processes (e.g., sputtering, evaporation, or colloidal deposition) or deposition sequences (1-, 2- or 3-stage process) on the overall optical properties and structure of the films, and ultimately the quality of the devices.

6.6
Summary and Future

The optical technique of real-time spectroscopic ellipsometry is becoming increasingly popular as the interest in developing and optimizing thin-film photovoltaic materials and devices grows. RTSE allows one to study the properties of thin film growth ranging from the initial nucleation and coalescence stage to the final film structural depth profile. A sampling of information that can be extracted from RTSE has been provided and includes phase diagrams in the case of Si:H thin films, growth mode determination in the case of CdTe thin films, and optical property development in the case of CIS thin films. These examples provide insights into why solar cells deposited under different conditions result in varying efficiencies, and overall, they help one to study nucleation and growth of thin films deposited by different techniques such as PECVD, magnetron sputtering, and evaporation.

As shown in this chapter, RTSE is well established for the characterization of thin-film photovoltaic materials in a research environment; however, excellent prospects exist for expansion to the pilot and full production environments. For example, RTSE can be applied to the monitoring of solar cell fabrication on flexible substrates such as plastic or stainless steel in a roll-to-roll process. In this case, one can track the initial start-up of the deposition process and the stability at a particular thickness point in the process. Also, using a dielectric function database obtained from RTSE measurements, one can characterize completed solar devices *ex-situ* for relevant information such as component layer thicknesses and compositions. This approach motivates mapping spectroscopic ellipsometry, in which case thickness and compositional maps can provide insights into the uniformity of full scale modules from production lines.

6.7
Definition of Variables

ψ and Δ:	ellipsometry angles of a specularly reflecting sample
r_p and r_s:	complex amplitude reflection coefficients of a sample for orthogonal p (parallel to plane of incidence) and s (perpendicular to plane of incidence) linear polarization states
n and k:	real and imaginary parts of the complex index of refraction N ($N = n - ik$)
E_{0x} and E_{0y}:	amplitudes of the electric field along orthogonal x and y axes
ω:	angular frequency of an optical wave
k:	magnitude of the optical wave vector (not to be confused with k the extinction coefficient)
δ_x and δ_y:	phase angles of electric field oscillations along orthogonal x and y axes
τ:	oscillation period of an optical wave
Q:	tilt angle of a polarization ellipse

χ:	ellipticity angle of a polarization ellipse (whose sign defines the handedness)
A:	amplitude of an optical wave
δ:	phase of an optical wave
θ_i:	angle of incidence
θ_t:	complex angle of transmission
n_s:	real index of refraction of a sample
k_s:	extinction coefficient of a sample
N_s:	complex index of refraction of a sample
ε_s:	complex dielectric function of a sample
n_a:	real index of refraction of an ambient medium
ε_a:	dielectric function of an ambient medium
I_0:	dc Fourier coefficient of the reflected irradiance in an ellipsometry measurement
α and β:	cosine and sine Fourier coefficients normalized to the dc coefficient of the reflected irradiance in an ellipsometry measurement
P_S, C_S, and A_S:	offsets that account for arbitrary angular positions of optical elements (polarizer, compensator, analyzer) relative to the plane of incidence at $t = 0$, the onset of data collection in an ellipsometry measurement
p:	degree of polarization

Acknowledgements

The authors gratefully acknowledge support from DOE Grants No. DE-FG36-08GO18067 and DE-FG36-08GO18073 and from the State of Ohio Third Frontier's Wright Centers of Innovation Program.

References

1 Boccara, A.C., Pickering, C., and Rivory, J. (eds) (1993) *Proceedings of the First International Conference on Spectroscopic Ellipsometry*, Elsevier, Amsterdam.
2 Collins, R.W., Aspnes, D.E., and Irene, E.A. (eds) (1998) *Proceedings of the Second International Conference on Spectroscopic Ellipsometry*, Elsevier, Amsterdam.
3 Fried, M., Humlicek, J., and Hingerl, K. (eds) (2003) *Proceedings of the Third International Conference on Spectroscopic Ellipsometry*, Elsevier, Amsterdam.
4 Arwin, H., Beck, U., and Schubert, M. (eds) (2007) *Proceedings of the Fourth International Conference on Spectroscopic Ellipsometry*, Wiley-VCH, Weinheim, Germany.
5 Azzam, R.M.A. and Bashara, N.M. (1977) *Ellipsometry and Polarized Light*, North-Holland, Amsterdam.
6 Tompkins, H.G. (1992) *A User's Guide to Ellipsometry*, Academic Press, New York.
7 Tompkins, H.G. and Irene, E.A. (eds) (2005) *Handbook of Ellipsometry*, William Andrew Publishing, Norwich, NY.
8 Passaglia, E., Stromberg, R.R., and Kruger, J. (eds) (1963) *Ellipsometry in the Measurement of Surfaces and Thin Films*, National Bureau of Standards Misc. Publ. 256 Washington.
9 Aspnes, D.E. (1976) Spectroscopic ellipsometry, in *Optical Properties of*

     Solids: New Developments (ed. B.O. Seraphin), North-Holland, Amsterdam, p. 799.
10 Wooten, F. (1972) *Optical Properties of Solids*, Academic Press, New York.
11 Lautenschlager, P., Garriga, M., Logothetidis, S., and Cardona, M. (1987) Interband critical points of GaAs and their temperature dependence. *Phys. Rev. B*, **35**, 9174.
12 Lautenschlager, P., Garriga, M., Vina, L., and Cardona, M. (1987) Temperature dependence of the dielectric function and interband critical points in silicon. *Phys. Rev. B*, **36**, 4821.
13 Aspnes, D.E., So, S.S., and Potter, R.F. (1981) *Optical Characterization Techniques for Semiconductor Technology*, vol. 276, SPIE, Bellingham, WA, p. 188.
14 Vedam, K., McMarr, P.J., and Narayan, J. (1985) Non-destructive depth profiling by spectroscopic ellipsometry. *Appl. Phys. Lett.*, **47**, 339.
15 Snyder, P.G., Rost, M.C., Bu-Abbud, G.H., Woollam, J.A., and Alterovitz, S.A. (1986) Variable angle of incidence spectroscopic ellipsometry: Application to GaAs–$Al_xGa_{1-x}As$ multiple heterostructures. *J. Appl. Phys.*, **60**, 3293.
16 Collins, R.W., Yacobi, B.G., Jones, K.M., and Tsuo, Y.S. (1986) Structural studies of hydrogen-bombarded silicon using ellipsometry and transmission electron microscopy. *J. Vac. Sci. Technol. A*, **4**, 153.
17 Irene, E.A. (1993) Applications of spectroscopic ellipsometry to microelectronics. *Thin Solid Films*, **233**, 96.
18 Palik, E.D. (1985) *Handbook of Optical Constants of Solids*, Academic Press, New York.
19 Palik, E.D. (1991) *Handbook of Optical Constants of Solids II*, Academic Press, New York.
20 Collins, R.W. (1990) Automatic rotating element ellipsometers, calibration, operation, and real time applications. *Rev. Sci. Instrum.*, **61**, 2029.
21 Hecht, E. (2002) *Optics: 4th Edition*, Addison-Wesley, San Francisco.
22 Fujiwara, H. (2007) *Spectroscopic Ellipsometry: Principles and Applications*, John Wiley and Sons Ltd, West Sussex, UK.
23 Zapien, J.A., Collins, R.W., and Messier, R. (2000) Multichannel ellipsometer for real time spectroscopy of thin film deposition from 1.5 to 6.5 eV. *Rev. Sci. Instrum.*, **71**, 3451.
24 Woollam, J.A., Johs, B., Herzinger, C.M., Hilfiker, J.N., Synowicki, R., and Bungay, C. (1999) Overview of variable angle spectroscopic ellipsometry (VASE), parts I and II. *Proc. Soc. Photo-Opt. Instrum. Eng. Crit. Rev.*, **72**, 3.
25 Chen, C., Ross, C., Podraza, N.J., Wronski, C.R., and Collins, R.W. (2005) Multichannel Mueller matrix analysis of the evolution of the microscopic roughness and texture during ZnO:Al chemical etching. *Proc. 31st IEEE Photovolt. Spec. Conf.*, IEEE, New York, p. 1524.
26 Fujiwara, H., Koh, J., Rovira, P.I., and Collins, R.W. (2000) Assessment of effective-medium theories in the analysis of nucleation and microscopic surface roughness evolution for semiconductor thin films. *Phys. Rev. B*, **61**, 10832.
27 Collins, R.W., An, I., and Chen, C. (2005) Rotating polarizer and analyzer ellipsometry, in *Handbook of Ellipsometry* (eds H.G. Tompkins and E.A. Irene), William Andrew Publishing, Norwich, NY, p. 329.
28 Collins, R.W., An, I., Lee, J., and Zapien, J.A. (2005) Multichannel ellipsometry, in *Handbook of Ellipsometry* (eds H.G. Tompkins and E.A. Irene), William Andrew Publishing, Norwich, NY, p. 481.
29 Tompkins, H.G. (2005) Optical components and the simple PCSA (Polarizer, Compensator, Sample, Analyzer) ellipsometer, in *Handbook of Ellipsometry* (eds H.G. Tompkins and E.A. Irene), William Andrew Publishing, Norwich, NY, p. 299.
30 Cahan, B.D. and Spanier, R.F. (1969) A high speed precision automatic ellipsometer. *Surf. Sci.*, **16**, 166.
31 Aspnes, D.E. and Studna, A.A. (1975) High precision scanning ellipsometer. *Appl. Opt.*, **14**, 220.
32 Hauge, P.S. and Dill, F.H. (1975) A rotating-compensator Fourier ellipsometer. *Opt. Commun.*, **14**, 431.

33 Aspnes, D.E. (1975) Photometric ellipsometer for measuring partially polarized light. *J. Opt. Soc. Am.*, **65**, 1274.

34 Oldham, W.G. (1969) Numerical techniques for the analysis of lossy films. *Surf. Sci.*, **16**, 97.

35 Comfort, J.C. and Urban, F.K. (1995) Numerical techniques useful in the practice of ellipsometry. *Thin Solid Films*, **270**, 78.

36 Loescherr, D.H., Detry, R.J., and Clauser, M.J. (1971) Least-squares analysis of the film-substrate problem in ellipsometry. *J. Opt. Soc. Am.*, **61**, 1230.

37 Jellison, G.E. Jr. (1998) Spectroscopic ellipsometry data analysis: Measured versus calculated quantities. *Thin Solid Films*, **313–314**, 33.

38 Stoke, J.A., Dahal, L.R., Li, J., Podraza, N.J., Cao, X., Deng, X., and Collins, R.W. (2008) Optimization of Si:H multijunction n–i–p solar cells through the development of deposition phase diagrams. *Proc. 33rd IEEE Photovolt. Spec. Conf.*, IEEE, New York, Art. No. 413.

39 Li, J., Podraza, N.J., and Collins, R.W. (2007) Real time spectroscopic ellipsometry of sputtered CdTe, CdS, and $CdTe_{1-x}S_x$ thin films for photovoltaic applications. *Phys. Status Solidi (a)*, **205**, 91.

40 Deng, X. and Schiff, E.A. (2003) Amorphous silicon-based solar cells, in *Handbook of Photovoltaic Science and Engineering* (eds A. Luque and S. Hegedus), John Wiley and Sons, Inc., Somerset, NJ, p. 505.

41 Collins, R.W., Ferlauto, A.S., Ferreira, G.M., Chen, C., Koh, J., Koval, R.J., Lee, Y., Pearce, J.M., and Wronski, C.R. (2003) Evolution of microstructure and phase in amorphous, protocrystalline, and microcrystalline silicon studied by real-time spectroscopic ellipsometry. *Sol. Energy Mater. Sol. C.*, **78**, 143.

42 McCandless, B.E. and Sites, J.R. (2003) Cadmium telluride solar cells, in *Handbook of Photovoltaic Science and Engineering* (eds A. Luque and S. Hegedus), John Wiley and Sons, Inc., Somerset, NJ, p. 617.

43 Compaan, A.D., Plotnikov, V.V., Vasko, A.C., Liu, X., Wieland, K.A., Zeller, R.M., Li, J., and Collins, R.W. (2009) Magnetron sputtering for II–VI solar cells: Thinning the CdTe. *Mat. Res. Soc. Symp. Proc.*, MRS, Warrendale, PA, vol. 1165, paper M09-01.

44 Messier, R., Giri, A.P., and Roy, R.A. (1984) Revised structure zone model for thin film physical structure. *J. Vac. Sci. Technol. A*, **2**, 500.

45 Mirica, E., Kowach, G., and Du, H. (2004) Modified structure zone model to describe the morphological evolution of ZnO thin films deposited by reactive sputtering. *Cryst. Growth Des.*, **4**, 157.

46 Li, J., Chen, J., Podraza, N.J., and Collins, R.W. (2006) Real-time spectroscopic ellipsometry of sputtered CdTe: Effect of growth temperature on structural and optical properties. *Proc. 4th World Conf. Photovolt. Energy Conv.*, IEEE, New York, p. 392.

47 Sestak, M.N., Li, J., Paudel, N.R., Wieland, K.A., Chen, J., Thornberry, C., Collins, R.W., and Compaan, A.D. (2009) Real-time spectroscopic ellipsometry of sputtered CdTe thin films: Effect of Ar pressure on structural evolution and photovoltaic performance. *Mat. Res. Soc. Symp. Proc.*, MRS, Warrendale PA, vol. 1165, paper M09-02.

48 Repins, I., Contreras, M.A., Egaas, B., DeHart, C., Scharf, J., Perkins, C.L., To, B., and Noufi, R. (2008) 19.9%-efficient ZnO/CdS/CuInGaSe$_2$ solar cell with 81.2% fill factor. *Prog. Photovoltaics*, **16**, 235.

49 An, I., Nguyen, H.V., Nguyen, N.V., and Collins, R.W. (1990) Microstructural evolution of ultrathin amorphous silicon films by real time spectroscopic ellipsometry. *Phys. Rev. Lett.*, **65**, 2274.

50 Alonso, M.I., Wakita, K., Pascual, J., Garriga, M., and Yamamoto, N. (2001) Optical functions and electronic structure of CuInSe$_2$, CuGaSe$_2$, CuInS$_2$, and CuGaS$_2$. *Phys. Rev. B*, **63**, 75203.

51 Alonso, M.I., Garriga, M., Durante Rincon, M.A., and Leon, M. (2000) Optical properties of chalcopyrite $CuAl_xIn_{1-x}Se_2$ alloys. *J. Appl. Phys.*, **88**, 5796.

52 Paulson, P.D., Birkmire, R.W., and Shafarman, W.N. (2003) Optical characterization of $CuIn_{1-x}Ga_xSe_2$ alloy thin films by spectroscopic ellipsometry. *J. Appl. Phys.*, **94**, 879.

53 Kawashima, T., Adachi, S., Miyake, H., and Sugiyama, K. (1998) Optical constants of $CuGaSe_2$ and $CuInSe_2$. *J. Appl. Phys.*, **84**, 5202.

54 Kreuter, A., Wagner, G., Otte, K., Lippold, G., and Schubert, M. (2001) Anisotropic dielectric function spectra from single-crystal $CuInSe_2$ with orientation domains. *Appl. Phys. Lett.*, **78**, 195.

55 Aspnes, D.E. (1980) Modulation spectroscopy/electric field effects on the dielectric function of semiconductors in *Handbook on Semiconductors* (ed. T. Moss), Vol. 2 (ed. M. Balkanski), Chapt. 4A, pp. 109–154.

56 Xu, H.Y., Papadimitriou, D., Zoumpoulakis, L., Simitzis, J., and Lux-Steiner, M.-Ch. (2008) Compositional and temperature dependence of the energy band gap of $Cu_xIn_ySe_2$ epitaxial layers. *J. Phys. D: Appl. Phys.*, **41**, 165102.

57 Varshni, Y.P. (1967) Temperature dependence of the energy gap in semiconductors. *Physica (Utrecht)*, **34**, 149.

58 Vina, L., Logothetidis, S., and Cardona, M. (1984) Temperature dependence of the dielectric function of germanium. *Phys. Rev. B*, **30**, 1979.

59 Walker, J.D., Khatri, H., Ranjan, V., Li, J., Collins, R.W., and Marsillac, S. (2009) Electronic and structural properties of molybdenum thin films as determined by real-time spectroscopic ellipsometry. *Appl. Phys. Lett.*, **94**, 141908.

60 Li, J., Podraza, N.J., Sainju, D., Stoke, J.A., Marsillac, S., and Collins, R.W. (2007) Analysis and optimization of thin film photovoltaic materials and device fabrication by real time spectroscopic ellipsometry. *Proc. Soc. Photo-Opt. Instrum. Eng.*, **6651**, 07.

61 Marsillac, S., Barreau, N., Khatri, H., Li, J., Sainju, D., Parikh, A., Podraza, N.J., and Collins, R.W. (2008) Spectroscopic ellipsometry studies of In_2S_3 top window and back contacts in chalcopyrite photovoltaics technology. *phys. status solidi (c)*, **5**, 1244.

7
Photoluminescence Analysis of Thin-Film Solar Cells
Thomas Unold and Levent Gütay

7.1
Introduction

Luminescence describes the emission of light from a solid arising from deviations from thermal equilibrium, as distinct from black-body radiation which is observed in thermal equilibrium [1, 2]. The thermal equilibrium state can be disturbed by various forms of excitation, such as the application of an external voltage (electroluminescence, see Chapter 3), an incident electron beam (cathodoluminescence, see Section 12.2.5), mechanical stress (mechanoluminescence), a heating ramp releasing carriers from deeply trapped states (thermoluminescence), and also by the absorption of light of sufficient energy, which is commonly called photoluminescence (PL). The emission of PL radiation is caused by the transition of electrons from higher occupied electronic states into lower unoccupied states, under the emission of photons if the transition is dipole-allowed. According to the laws of quantum mechanics, the transition rate can be calculated by first-order perturbation theory using Fermi's golden rule. Using this formalism, optical transitions from an occupied density of initial states to an unoccupied density of final states can be expressed by [2, 3]

$$R_{sp}(E) \propto \int |M_{if}|^2 f(E_i)(1-f(E_i+E))g(E_i)g(E_i+E)dE_i \quad (7.1)$$

where M_{if} represents the matrix element coupling the wavefunctions of the initial and final state, $f(E)$ represents the Fermi–Dirac occupation function, and $g(E)$ denotes the density of electronic states. Usually the optical transition matrix element is calculated in the semiclassical approximation, where the exciting light is treated by classical electrodynamics. Using the fact that the same transition matrix element and density of states govern absorption and emission events, a relationship between the absorption coefficient $\alpha(E)$ and the spontaneous emission rate was derived based on detailed balance arguments in thermal equilibrium [4]

$$R_{sp}^0 = \frac{8\pi n_r^2}{h^3 c^2} \frac{\alpha(E)E^2}{\exp(E/k_B T)-1} \quad (7.2)$$

Advanced Characterization Techniques for Thin Film Solar Cells,
Edited by Daniel Abou-Ras, Thomas Kirchartz and Uwe Rau.
© 2011 Wiley-VCH Verlag GmbH & Co. KGaA. Published 2011 by Wiley-VCH Verlag GmbH & Co. KGaA.

where n_r is the refractive index of the material, h is Planck's constant, c is the speed of light, and k_B is the Boltzmann constant. Equation (7.2) is very useful in the sense that it directly relates the absorption coefficient of a material to its emission spectrum and in principle allows for calculating one quantity if the other quantity is known. Since in PL measurements we are interested in deviations from thermal equilibrium, we write the net spontaneous emission rate as

$$R_{sp} = R_{sp}^0 \frac{np}{n_0 p_0} - R_{sp}^0 = B(np - n_0 p_0) = B n_i^2 (\exp(\Delta\mu/k_B T) - 1) \tag{7.3}$$

where n and p are the total carrier densities, n_0 and p_0 are the equilibrium carrier densities related to the intrinsic carrier density by $n_i^2 = n_0 p_0$, $\Delta\mu = E_{Fn} - E_{Fp}$ represents the quasi-Fermi level splitting between the quasi-Fermi level for electrons E_{Fn} and for holes E_{Fp}, and B defines the material-specific radiative recombination coefficient [1]

$$B = \frac{1}{n_i^2} \frac{8\pi}{h^3 c^2} \int_0^\infty n_r^2 \alpha(E) \exp(-E/k_B T) E^2 dE \tag{7.4}$$

It is useful to distinguish the following experimental conditions: (i) $n \gg n_0$ and $p \gg p_0$. In this case, the PL photon flux is given by $Y_{PL} \propto Bnp$. This situation is referred to as the high-injection condition, where the radiative lifetime $\tau_{rad} = 1/(Bn)$ depends on the injection ratio. (ii) When $p \approx p_0$ for p-type material ($n \approx n_0$ for n-type material), then $Y_{PL} \propto Bnp_0$ and the PL flux depends on the majority carrier density with a radiative lifetime $\tau_{rad} = 1/(Bp_0)$ independent of the injection ratio. This situation is referred to as the low-injection condition. Typical values for the radiative coefficient and radiative lifetime assuming a doping level of 10^{16} cm^{-3} are $B_{Si} = 2 \times 10^{-15}$ cm^{-3} s^{-1} and $\tau_{Si} = 0.025$ s for silicon and $B_{CuInSe_2} \approx 6 \times 10^{-11}$ cm^{-3} s^{-1} and $\tau_{CuInSe_2} \approx 1$ µs for CuInSe$_2$ [1, 5].

In most materials, nonradiative recombination occurs via deep defects in the energy band gap, limiting the recombination lifetime to significantly smaller values than the radiative lifetime since the total recombination rate comprises all individual recombination rates

$$\tau_{tot}^{-1} = \tau_{rad}^{-1} + \tau_{nonrad}^{-1} \tag{7.5}$$

From the ratio of measured lifetime to radiative lifetime, the PL efficiency of a material can be defined as $\eta_{PL} = \tau_{tot}/\tau_{rad}$.

So far, we have been mostly concerned with radiative transitions taking place within the sample. However, the correct calculation of the number of photons emitted from a sample in a PL experiment is much more complicated since details of the absorption profile, diffusion and drift and recombination of carriers with the sample, and the propagation of the emitted photons through the sample surface have to be taken into account. In most PL experiments, the information depth is given by the absorption length $1/\alpha$ of the exciting light or by the diffusion length of minority carriers, which ever of these two quantities is larger [2]. For an exact treatment, closed-form solutions are difficult to obtain and numerical simulation has to be employed. However, assuming homogenous material properties and flat quasi-Fermi levels and taking into account that light is emitted through the sample surface within a narrow emission cone $1/4n_r^2$, the photon flux detected outside the sample can be expressed by [6]

$$Y_{PL}(E) = \frac{1}{4\pi^2 \hbar^3 c^2} \frac{a(E)E^2}{\exp((E-\Delta\mu)/k_B T)-1} \tag{7.6}$$

where the absorptivity is given by $a(E) = (1 - R_f)(1 - \exp(-\alpha(E)d))$ with the front surface reflectivity R_f and the sample thickness d. Equation (7.6) has been referred to as the generalized Kirchhoff's or generalized Planck's law because of its resemblance to these well-known relations valid for thermal equilibrium conditions. Note that Eq. (7.6) is only valid if equilibration between the electronic states involved (e.g., carriers in the conduction band and carriers in the valence band) can occur on the time scale of the radiative recombination time. For most materials, this is true at room temperature for thermal-emission depths smaller than about $E_{gap} - 0.3$ V. However, it is certainly not true for carriers in deep band tail states at temperatures in the range of 10 K. Since the quasi-Fermi level splitting $\Delta\mu$ contained in Eq. (7.6) is related to the maximum open-circuit voltage, V_{oc}, achievable for a photovoltaic device, this relation can be used for predicting this device property using luminescence measurements at room temperature [6–9].

If measurements of the absolute magnitude of the luminescent photon flux emitted from a sample under photoexcitation are available, the quasi-Fermi level splitting can be derived from the magnitude and spectral shape of the luminescence signal. This method in particular allows for the investigation of the influence of subsequent processing steps on the quality of an absorber material. If the optical constants are known for the material, the data may be evaluated using Eq. (7.6) and the correct values for the spectral absorptivity. However, for thin-film compound semiconductor materials, the absorption coefficient very often is not exactly known, and one has to make simplifying assumptions. In this case, the high-energy wing of the PL signal may be evaluated at photon energies sufficiently larger than the band gap, where the absorptivity can be approximated to be constant and $a(E) \approx 1$ is a reasonable assumption. Then, Eq. (7.6) can be rewritten as

$$\ln\left(\frac{Y_{PL}(E)}{10^{23}\ E^2/\text{cm}^2\ \text{eV s}}\right) = -\frac{E-\Delta\mu}{kT} \tag{7.7}$$

allowing for an extraction of the quasi-Fermi level splitting $\Delta\mu$ from a fit to the high-energy wing of the PL spectrum. Note that the equation assumes an absorber layer with homogeneous phase composition and constant quasi-Fermi levels throughout the material. For extracting the local optical threshold energy from a PL spectrum, the determined value $\Delta\mu$ is reinserted in Eq. (7.6) which allows for the extraction of the spectral absorptivity $a(E)$ for the lower energetic region of the luminescence spectrum.

7.2
Experimental Issues

7.2.1
Design of the Optical System

A schematic of a typical PL setup is shown in Figure 7.1. The sample can be mounted in ambient conditions for fast room-temperature measurements or in a cryostat for low-temperature measurements in vacuum or, as is the case in dynamic cryostats, in helium or nitrogen atmosphere. As an excitation source, any light source of suitable luminance and appropriate wavelength range can be used. The most general light source would be a white light from a halogen or xenon lamp filtered by a monochromator, which allows for a wide range of excitation wavelengths, however, at the cost of very low excitation power. Because of the widespread availability and high monochromatic power, in most setups laser excitation sources are used, for example, gas lasers such as helium–neon (633 nm) or argon (514 nm) lasers, or solid-state laser diodes. The excitation source is guided or focused onto the sample by a flat mirror or focusing device (lens or parabolic mirror). The luminescence radiation emitted from the sample is then collected by a light collection device, which again can be given by a lens or also by a parabolic or off-axis parabolic mirror. The collimated light passes an order-sorting filter system which prevents the detection of unwanted higher orders. For most applications, colored glass long-pass filters are chosen with cutoff wavelengths slightly larger than half the measurement wavelength. The luminescence light is then focused into the monochromator system through an entrance slit. As a general comment regarding the focusing assemblies in the whole luminescence setup, we point out that off-axis parabolic mirrors are considered first because they are dispersion free. The luminescence light exits the monochromator through an exit slit and reaching the radiation detector, which can be a photodiode, photomultiplier, or avalanche photodiode, or a one- or two-dimensional detection device such as a photodiode array or CCD array.

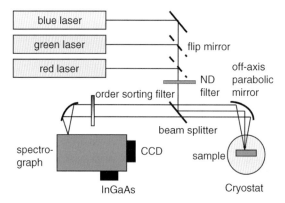

Figure 7.1 Experimental setup for room- and low-temperature PL measurements.

In the case of a single-area detector, the spectral resolution of the system is determined by the dispersion of the grating (nm/mm) times the entrance/exit slit width. Note that the focusing unit directing the light into the monochromator has to be designed to match the aperture of the monochromator specified by the f-number $f/\# = f/D$, where f is the focal length and D collimated beam diameter, to achieve an optimal illumination of the dispersion grating and mirrors inside the monochromator. If the focusing $f/\#$ is too small, then light is scattered inside the monochromator, and if the focusing $f/\#$ is too large then the spectral resolution is decreased. In the case of a diode array, the exit slit is omitted, and the spectral resolution is determined by the dispersion times the diode width, which is typically 10–30 μm. The advantage of using a detector array is the possibility to record a single spectrum in one shot, without the need for scanning the grating. The advantage of using single-area detectors is the higher signal-noise ratio achievable by using low-noise amplification by avalanche diodes or photomultiplier tubes in combination with lock-in-detection. The excitation signal can be modulated by an optical chopper wheel or by direct electrical modulation of a laser diode.

If measurements with high spatial resolution are of interest, a microscope setup can be used [10]. In Figure 7.2, a PL setup utilizing a confocal microscope is outlined. In this configuration, the excitation source is focused onto the sample surface through a microscope lens. The luminescence light is collected through the same lens, separated from the exciting light beam by means of a beam splitter, and focused onto a fiber connected to the detecting system. This system exhibits diffraction-limited resolution of approximately $0.6\lambda/\text{NA}$, where NA is the numerical

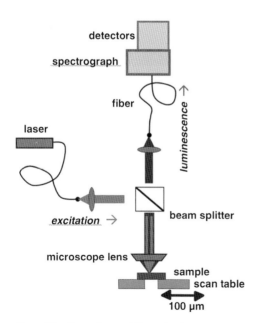

Figure 7.2 Setup for spatially resolved measurements using a confocal microscope.

aperture of the microscope objective and λ is the wavelength of the detected luminescence signal. A spatial resolution of down to about 500 nm can be obtained in an optimized system investigating luminescence light of approximately 1000 nm. The sample is mounted on a *xy*-positioning table, which allows for a two-dimensional scan of the emitted luminescence radiation. As the excitation area in a micro-PL is in the range of $1\,\mu m^2$, very high excitation intensities can be achieved, which may lead to local heating and high-injection effects. This has to be kept in mind when analyzing the data. Efforts should be made to keep the excitation level as low as possible.

7.2.2
Calibration

The two most important calibrations to be performed are wavelength calibration of the monochromator and the determination of the transfer function of the complete optical system. The wavelength calibration is commonly performed using an argon or mercury–argon lamp, which is placed at the sample position. The atomic lines recorded by the detection system are then fitted to the known line spectrum of the lamp type which, for example, may be downloaded from the NIST Website [11]. The transfer function contains the wavelength-dependent attenuation of the propagating luminescence signal after passing all optical components of the collection and detection system, that is, all lenses, mirrors, windows, filters, gratings, and the quantum efficiency of the detector. To measure this function, a radiation source with known emission spectrum is placed at the sample position and its emission spectrum is recorded with all optical components in place. If various filters or gratings are used, each one of these optical elements has to be considered. For absolute calibration of the luminescence photon flux, a black-body calibration source with known emissivity and temperature is used. Calibrated tungsten-band lamps have been useful to perform this task.

In order to obtain the proper luminescence signal, the measured signal has to be multiplied by the transfer function and subsequently converted into energy space by taking into account the change from constant wavelength intervals to constant energy intervals, which requires a multiplication of the luminescence signal by a factor of λ^2.

7.2.3
Cryostat

If low-temperature measurements are of interest, a cryostat has to be used. Commonly used cryostats are either liquid helium or nitrogen flow or closed-cycle helium cryostats. The advantage of a closed-cycle cryostat is that no liquid helium is necessary, at the expense of mechanical vibrations of about $\pm 10\,\mu m$ when compared with continuous-flow cryostats. Helium-flow cryostats allow for lower minimum temperatures of 1.5 K when compared to closed-cycle cryostat with which only 4 K can be reached. When performing low-temperature measurements on thin-film materials, special care has to be taken to correctly determine the sample temperature during the measurement. Because the thin materials are commonly grown on glass substrates which are considerably bad thermal conductors, the temperature of the thin-film sample can differ considerably from the temperature of the cold finger of the cryostat. To properly calibrate the temperature, a second sensor mounted on a glass substrate identical to the sample glass substrate should be employed in the system.

7.3 Basic Transitions

A number of different transitions can occur in a PL measurement depending on the measurement conditions and the material properties. In the following, we give a very brief description of the most important radiative transitions occurring in semiconductor materials. In general, the different transitions can be distinguished by their transition energy and by the change of the transition energy and luminescence yield with varying excitation intensity and temperature. Therefore, it is useful to define the following two dependencies. The luminescence yield of a transition line in general is found to obey the expression [1, 2]

$$Y_{PL}(\phi) = \phi^k \tag{7.8}$$

where ϕ is the excitation intensity and k is a characteristic parameter usually ranging between 0.5 and 2. The luminescence yield of transitions involving localized states generally decreases with increasing temperature owing to the thermal emission of the trapped carriers to the conduction or valence band. This dependence on temperature can be described by

$$Y_{PL}(T) = \frac{1}{1 + C \exp(-E_a/kT)} \tag{7.9}$$

where E_a represents a characteristic activation energy and C is a constant which is proportional to $T^{3/2}$ if the thermal quenching involves thermal emission to the conduction or valence band [12, 13]. Bear in mind that the activation energies can differ significantly if or if not a $T^{3/2}$ dependence is included in the model.

The shift of transition energy with increasing temperature may be influenced by a shift in the optical gap with temperature [1]. Because this applies to all of the transitions discussed in the following section, this will not be discussed individually but only stated here. Typically, a decrease in the band gap with increasing temperature of the order of 10^{-4} eV/K is observed that may compensate positive peak shifts of the order of $k_B T$ in the experiment. Radiative transitions can occur with or without the emission of phonons, depending on the electron–phonon coupling strength and the measurement temperature, thereby often leading to so-called phonon replica of the transitions. Because of limitation in space, these phonon replicas will not be discussed in the present section, and the interested reader is referred to the literature [1, 2].

7.3.1
Excitons

The classical PL experiment is performed at low temperatures in the vicinity of 10 K. At this temperature, the luminescence efficiency can be considerably higher than at room temperatures as the ratio of radiative to nonradiative recombination rate is greatly increased and recombination events arising from electron–hole pairs bound to each other by their Coulomb interaction, so-called free excitons, may be observable. The transition energy of free excitons can be calculated using a simple hydrogenic model yielding

$$E_{FX} = E_g - E_x, \quad E_x = \frac{m_r e^4}{2(4\pi\varepsilon_0 \varepsilon_r \hbar)^2 n^2} \quad (7.10)$$

where E_x is the exciton binding energy and m_r is the reduced electron–hole mass $1/m_r^* = 1/m_e^* + 1/m_p^*$ [1]. The quantum number n specifies the possible excited states of the exciton. It is obvious from Eq. (7.10) that the binding energy of excitons mostly depends on the dielectric constant ε_r and the reduced effective mass. For CuGaSe$_2$ thin films with a dielectric constant of about 13 and effective masses of $m_e^* \approx 0.14$ and $m_p^* \approx 1$, a free exciton-binding energy of 10 meV is expected, which agrees reasonably well with the experimentally determined value $E_{FE} = 13$ meV [14]. Exciton-binding energies increase for large band-gap materials such as ZnO due to their smaller dielectric constants [15]. Optical transitions related to free excitons are only detectable at sufficiently low temperatures when $k_B T < E_x$ and will dissociate at higher temperatures. The thermal quenching of the free exciton luminescence can be described by Eq. (7.9), where the activation energy corresponds to the exciton-binding energy if the quenching process corresponds to the dissociation of the excitons into free carriers. However, if many transitions are present, the activation

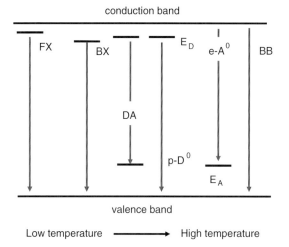

Figure 7.3 Optical transitions observable in luminescence measurements.

energy in Eq. (7.9) may not be unambiguously identified with the free-exciton transition. A more unambiguous determination of the binding energy is possible if excited states of the exciton can be detected [14, 16].

For the change of the PL yield with excitation intensity for free excitons, one expects $k = 1$ for resonant excitation and $k = 2$ for excitation above the band gap [17]. Schmidt et al. showed that if several transitions concur, the exciton k value can take values ranging from 1 to 2 depending on the material and experimental conditions. Excitons do not exhibit a shift in their transition energy, if the excitation intensity is increased.

Free excitons get easily bound to impurities, which leads to a modification of their transition energy by the interaction with the impurity. The transition energy for bound exciton emission is given by

$$E_{BX} = E_g - E_{BE} \tag{7.11}$$

where the term E_{BE} represents the binding energy of the complex [2]. This term is typically a fraction of the ionization energy of the isolated impurity in the case of charged impurities and the sum of the free-exciton binding energy added to a fraction of the impurity binding energy in the case of neutral defects, as described by Hayne's rule [18, 19]

$$E_{BE} = \begin{cases} c_n E_{A/D} + E_{FE} & \text{for neutral defects} \\ c_i E_{A/D} & \text{for charged defects} \end{cases} \tag{7.12}$$

The proportionality factors c_n and c_i depend on the effective mass ratio m_e^*/m_h^* and were estimated for donors in CuGaSe$_2$ to be of the order of $c_n \approx 0.3$ and $c_i \approx 1$ [14]. Considering the binding energy of 13 meV of free excitons in CuGaSe$_2$, binding energies of about 17 meV for excitons bound to shallow donors were obtained [14]. Excitons can also be bound to deeper impurities in which case their binding energy may be much larger than the binding energy of free excitons. The ratio between free and bound exciton emission present in a photoluminescence experiment depends on the number of impurities present in the material and on the measurement temperature. According to Lightowlers [20], for impurity concentrations of $N > 10^{15}$ cm^{-3} in silicon, essentially all free excitons get captured by donors or acceptors and lead to bound-exciton luminescence.

Bound excitons do not have kinetic energy causing their linewidth to be much smaller than the linewidths of order of $k_B T$ found for free excitons. Exciton transitions can also be broadened by inhomogeneities in the material properties such as composition variations and material strain. Since radiative emissions from free and bound excitons can occur with the additional emission of phonons, phonon replica of the exciton emission lines are frequently detected and have to be distinguished from transitions arising from excitons bound at different impurities.

7.3.2
Free-Bound Transitions

In nonideal semiconductor materials, there are always localized states present due to donor or acceptor impurities which can give rise to carrier recombination by transitions between the free carriers in the bands and the localized states in the band gap, so-called free-to-bound (FB) transitions [1]. It is possible to detect both conduction-band-to-acceptor $(e\text{-}A^0)$ and/or valence-band-to-donor state $(p\text{-}D^0)$ transitions. Shallow transitions from the conduction band to donors or from the valence band to acceptors are unlikely, because in these cases, the probability for phonon-related transitions is much higher than the probability of transitions involving the release of a photon [21].

Free-bound transitions can be identified by their spectral signature and specific temperature-dependent und intensity-dependent behavior. The transition lineshape can be described by a modification of Eq. (7.2) in which the square-root dependence of the density of states of parabolic conduction or valence bands in direct-gap semiconductors is considered [2, 22]

$$Y_{PL}(E) \propto E^2 (E-(E_g-E_{A/D}))^{0.5} \exp((E_g-E_{A/D}-E)/kT) \tag{7.13}$$

where $E_{A/D}$ is the donor or acceptor ionization energy. The peak position of the transition as obtained from setting the derivative of Eq. (7.13) equal to zero is related to the optical gap by

$$Y_{PL}(E)_{max} \propto E_g - E_{A/D} + k_B T/2 \tag{7.14}$$

Note that both these expressions strictly only apply to direct-gap semiconductors with ideal parabolic bands. The thermal quenching of the luminescence yield FB transitions is again described by Eq. (7.9), where now the activation energy represents the ionization energy of the impurity state. For the excitation intensity dependence of FB processes $k=1$ is expected in the ideal case since the transition rate depends on one free carrier type, which depends linearly on the excitation intensity, and on the donor or acceptor density, which is independent of the excitation intensity. Note that when impurities are present in large densities, they begin to form impurity bands, which merge with the nearest intrinsic band, making the distinction between free-bound and band–band transitions difficult [1].

7.3.3
Donor–Acceptor Pair Recombination

If both donors and acceptors are present in significant concentrations and the temperature is low enough, it is possible to observe donor–acceptor pair (DAP) recombination processes, which involve transitions between two localized electronic states. These transitions originate from neutral donors and neutral acceptors (having previously captured a free carrier) which recombine to leave two oppositely charged defects [1]. The Coulomb energy between the ionized donor and acceptor is transferred to the emitted photon resulting in an emission energy of

$$E_{DA} = E_{gap} - E_A - E_D + E_{Coul} \tag{7.15a}$$

$$E_{Coul} = \frac{e^2}{4\pi\varepsilon_0 \varepsilon_r r_{DA}} \tag{7.15b}$$

where E_A and E_D are the ionization energies of the donor and acceptor, respectively, and E_{Coul} describes the Coulomb interaction between the ionized donor and acceptor where r_{DA} is the distance between the donor and the acceptor involved in the emission, which depends on the impurity density and the excitation density, and ε_r denotes the static dielectric constant. Since the effective distance between donors and acceptors decreases with the injection level, the effect of the Coulomb interaction is smallest for low excitation intensities and gets largest for high excitation intensities. The dependence of the luminescence peak energy E_{DA} on the excitation intensity, ϕ_{exc}, can be empirically described by [23]

$$E_{DA}(\phi_{exc}) \propto E_{DA}(\phi_0) + \beta \log\left(\frac{\phi_{exc}}{\phi_0}\right) \tag{7.16}$$

where typical values of β are found to be around 1–5 meV per decade of excitation intensity. The maximum possible Coulomb-energy related shift of DAP transitions can be estimated from the binding energy of the spatially more extended defect, usually the donor level, in the hydrogenic impurity model given by [24]

$$E_{Max} = \frac{e^4 m_e^*}{2(4\pi\varepsilon_r\varepsilon_0\hbar)^2} \tag{7.17}$$

For typical values of $\varepsilon_r = 13.6$ and $m_e^* = 0.09$ for CuInSe$_2$, a maximum shift of $E_{max} = 6.6$ meV is estimated [25, 26]. Thermal quenching of DAP-transitions is described by an equation similar to Eq. (7.9), now containing two activation energies

$$I_{PL} = \frac{1}{1 + C_1 \exp(-E_{a1}/kT) + C_2 \exp(-E_{a2}/kT)} \tag{7.18}$$

where again the constants C_1 and C_2 are proportional to $T^{3/2}$ if the quenching occurs by thermal emission of charges to the valence band and conduction band, respectively [12]. Under the assumption that the constants C_1 and C_2 are of similar magnitude, the activation energy observed at lower temperature is related to the ionization energy of the shallower impurity state and the activation energy observed at higher temperatures is related to the ionization energy of the deeper level. However, in some cases it is not possible to get meaningful fits using Eq. (7.18) and the analysis of the data using the single activation energy contained in Eq. (7.9) should be considered [13].

The change of the DAP-lineshape with increasing temperature is complicated, because the recombination process involves a tunneling step which depends on the distance of the recombination pairs. Briefly, for increasing temperatures more distant pairs get thermally re-emitted to the bands before they recombine, increasing the number of close pairs in the radiative recombination. Due to the larger Coulomb term for the close pairs this leads to a blue shift of the order of $k_B T$ of the emission spectrum with increasing temperature [1, 2]. For the luminescence yield as a function of excitation intensity $k = 1$ is expected in the ideal case but as was already discussed above, significantly smaller values than that may be observed in the case of different competing transition types. Schmidt et al. show that for certain experimental conditions the k values of the exciton and DAP transition are related to each other by $k_{exc} = 2\,k_{DAP}$ [17].

7.3.4
Potential Fluctuations

For high concentrations of donors and acceptors when $N_{A/D} a_B^3 > 1$, the Bohr-radii $a_B^* = 4\pi\varepsilon_0 \varepsilon \hbar^2 / m^* e^2$ of the impurity wavefunctions start to overlap, leading to impurity-related band formation [27]. If the donors' and acceptors' concentrations are of similar magnitude, the semiconductor becomes compensated, with a net doping density much lower than the dopant concentration. Since the distribution of dopants in the lattice is statistical, the band structure becomes distorted by potential fluctuations due to local variations in the fixed space charge, which cannot be screened by the low free-carrier density. Such potential fluctuations which are indicated in a schematic band diagram in Figure 7.4 strongly influence the optical and electrical properties as now transitions with significantly smaller transition energy than the standard DAP or FB transition energy become possible [26–30].

The average depth of the fluctuations, γ, can be estimated from the average charge fluctuations occurring within a certain volume defined by the screening radius r_s containing an impurity concentration $N_t = N_A^+ + N_D^-$ screened by a free-carrier density p by [27]

$$\gamma = \frac{e^2}{4\pi\varepsilon_r\varepsilon_0}(N_t r_s)^{0.5} = \frac{e^2}{4\pi\varepsilon_r\varepsilon_0}\frac{N_t^{2/3}}{p^{1/3}} \qquad (7.19)$$

where r_s given by $r_s = N_t^{1/3} p^{-2/3}$ for p-type material. Typical values for CuInSe$_2$ thin films can be estimated assuming $N_A = N_D = 10^{18}\,\text{cm}^{-3}$, $p = 10^{16}\,\text{cm}^{-3}$, and $\varepsilon_r = 13.6$ to yield $\gamma = 78\,\text{meV}$ with a screening length of $r_s \approx 270\,\text{nm}$. Because DAP recombination involves tunneling between the two impurity sites, the transition energy

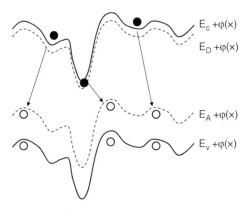

Figure 7.4 Effect of potential fluctuations on the band edges and the radiative transitions between trapped electrons and holes. $\varphi(x)$ denotes the spatial varying electrostatic potential arising from the potential fluctuations and the dashed line represents an acceptor/donor level or band.

can be significantly reduced in the presence of band fluctuations, which experimentally leads to a low-energy tail and broadening of the luminescence transitions. The emission energy of DAP recombination in highly compensated material can be estimated with

$$E_{\text{PF-DA}} = E_{\text{DA}} - 2\gamma \tag{7.20}$$

where E_{DA} is given by Eq. (7.15a) [23]. It is obvious from Eqs. (7.19) and (7.20) that the emission energy for highly compensated DAP transitions depends on the number of impurities, the compensation ratio, and the free-carrier density which in turn depends on the excitation level in the experiment.

Therefore for increasing excitation level, a strong blue shift of the emission line is expected which can greatly exceed the $\beta = 1\text{--}2$ meV expected for the DAP Coulomb term given in Eq. (7.15b). Experimentally peak shifts as large as $\beta = 10\text{--}30$ meV are observed [26, 28, 29].

If the temperature is increased, the emission peak in strongly compensated semiconductors is found to red shift at low temperatures and low excitation intensities. This can be explained with a lack of complete thermalization of the carriers trapped in different potential wells, which leads to an incomplete filling of the lowest energetic wells. If temperature is increased, the carriers become more mobile and also populate the deepest wells, leading to a red shift of the emission line with increasing temperature. If the temperature is further increased, the more distant pairs are increasingly thermally emitted to the bands leaving the closer pairs and thus producing a blue shift, in accordance with the effect observed for DAP transitions. A detailed discussion of the effect of potential fluctuations in compensated semiconductors can be found in the literature [26, 27, 29–31].

7.3.5
Band–Band Transitions

With increasing temperature excitons and impurities become ionized and the conduction and valence bands become increasingly occupied with photoexcited carriers such that band–band transitions become more probable. The lineshape of band–band transitions can be described by Eq. (7.2) using the appropriate absorption coefficient for band–band transitions. For a direct-gap semiconductor with parabolic bands, the following expression is obtained [2]

$$Y_{\text{PL}}(E) \propto E^2 (E - E_g)^{0.5} \exp(-(E - E_g)/kT) \tag{7.21}$$

where E_g ist the band gap of the material. This equation is very similar to the expression obtained for the free-bound emission stated above, which means that in order to distinguish between band–band transitions and free-bound transitions, knowledge of the value of the band gap determined by independent means is very helpful. As for free-bound transitions, the peak position increases with temperature as $k_B T/2$. In band–band transitions, both free-carrier types are involved, both depend on the excitation intensity, and thus k values >1 can be observed. Note that in the case of high injection with band–band transitions dominating recombination $k = 1$ will be observed, because in this case the lifetime is inversely proportional to the injection level as discussed in the introduction.

7.4
Case Studies

In this section, we discuss some typical PL measurement results for chalcopyrite-type $Cu(In,Ga)Se_2$ thin films and also completed solar cells. The defect physics and therefore also the nature of the dominant PL transitions in this material is found to strongly depend on the film composition, most prominently on the [Cu]/[In] + [Ga] ratio. Sharp transitions are observed for material grown under copper-rich conditions, whereas broad, mostly featureless transitions are observed in the case of copper-poor growth conditions [12, 25, 28, 29, 32]. We will also distinguish between low-temperature and room-temperature luminescence analysis. Both techniques have advantages and disadvantages depending on the material and the investigated material properties.

7.4.1
Low-Temperature Photoluminescence Analysis

In Figure 7.5, low-temperature PL spectra of a Cu-rich grown epitaxial $CuGaSe_2$ sample are shown. The sample was grown on GaAs(100) by metal organic vapor phase epitaxy (MOVPE). The Cu/Ga-ratio of the sample was determined as $[Cu]/[Ga] \approx 1.1$ by energy dispersive X-ray analysis (EDX, see Section 12.2.3). The measurement was performed at 10 K and the sample was illuminated by the

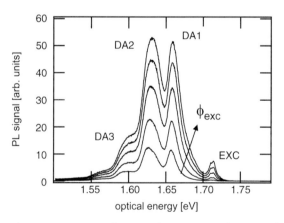

Figure 7.5 PL spectra for a Cu-rich grown epitaxial $CuGaSe_2$ thin film measured at various excitation fluxes and 10 K.

514.5 nm line of an Ar$^+$ laser using different excitation fluxes. The PL spectra exhibit distinct transition peaks at 1.72, 1.66, and 1.63 eV, and a further peak occurring as a shoulder at 1.60 eV with full-widths-at-half-maximum (FWHM) of approximately 10, 19, 27, and 55 meV. These four peaks have been identified in the literature as an exciton-related transition (EXC), and three different donor–acceptor transitions (DA1, DA2, DA3) [14, 29].

This assignment can be justified by evaluating the peak energies and their shift with excitation intensity, and also the corresponding k values describing the dependence of the PL yield on excitation intensity. In Figure 7.6a, the intensity of the DA1, DA2, and EXC transitions are plotted as a function of the excitation intensity using logarithmic scaling of the axes. A fit using Eq. (7.8) yields k values of $k = 0.6$ for the DA1 and DA2 transition and $k = 1.2$ for the EXC transition. Values of $k \leq 1$ are expected for donor–acceptor or free-bound transitions, whereas k values ≥ 1 are

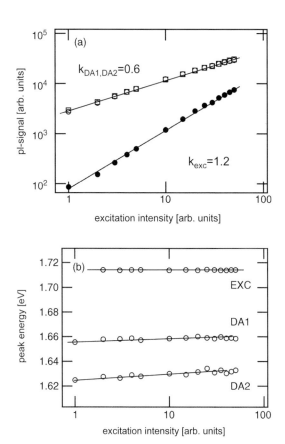

Figure 7.6 (a) Intensity dependence of the PL yield of transitions DA1, DA2, and EXC. (b) Shift of PL peaks as a function of excitation intensity.

expected for exciton or band–band transitions. Considering the known value for the band gap $E_{gap} = 1.73$ eV of CuGaSe$_2$ at 10 K, the EXC transition energy is located 10 meV below the band gap, which agrees well with experimentally determined single-crystal free exciton-binding energies of 13 meV [14].

The shift of the peak energies with excitation intensity shown in Figure 7.6b indicates values of $\beta_{EXC} = 0$ meV, $\beta_{DA1} = 1.5$ meV, and $\beta_{DA2} = 4$ meV. The fact that the EXC transition does not shift with excitation intensity confirms the assignment to an exciton transition. In fact, a more detailed analysis of this transition shows that it can be deconvolved into two closely spaced transitions located at 1.7139 and 1.7105 eV, which have been identified with a free-exciton and a bound-exciton transition. A detailed discussion of this assignment can be found in the literature and is beyond the scope of this article [14]. Peak shifts of $\beta \sim 1$–4 meV are values typically observed for DAP recombination processes, and may arise from the increasing contribution of the Coulomb interaction as the distance between the participating donors and acceptors is decreased when the density of photoexcited carriers increases with excitation intensity. This confirms an assignment of the transitions DA1 and DA2 to a DAP recombination. The fourth transition line (DA3) located at 1.6 eV could be related to a phonon replica of DA2 since its energetic distance between DA2–DA3 ≈ 30 meV is comparable to typical LO-phonon energies $E_{ph} \approx 34$ meV in CuGaSe$_2$ [29]. However, it is also possible that DA3 is a separate DAP transition as has been suggested from cathodoluminescence measurements showing the spatial location of the DA3 transitions to be distinct from the spatial locations of the DA2 transition [33].

We note that although the observed k values are significantly smaller than the values of $k = 1$ and $k = 2$ expected for donor–acceptor and exciton transitions in the ideal case, these values agree well with the theory of Schmidt et al. [17] who predicted $k_{exc} \approx 2k_{DAP}$ for the situation of different concurring radiative and nonradiative transition.

Cu-poor prepared Cu(In,Ga)Se$_2$ films that generally lead to much higher solar conversion efficiencies than Cu-rich prepared films show a very different luminescence behavior at low temperature. In Figure 7.7, the PL spectrum of a coevaporated Cu(In,Ga)Se$_2$ thin-film solar cell with [Cu]/([In + Ga]) = 0.9 and [Ga]/([In + Ga]) = 0.3 in a glass/Mo/Cu(In,Ga)Se$_2$/CdS/ZnO stacking structure is measured at 15 K using a 660-nm laser diode as an excitation source.

Now a single but much broader transition is observed. The peak energy of this transition is 1.11 eV with a FWHM of 60 meV. A plot of the peak energy and intensity versus excitation intensity (not shown here) yields $\beta = 10$ meV and $k = 0.96$ at 15 K measurement temperature. Although this k value is compatible with the values observed for the DAP transitions of the Cu-rich sample, the β value is much larger than the 1–4 meV observed above and also much larger than the total shifts expected for DAP transitions in chalcopyrite materials. Such large β values are a clear indication of potential fluctuations due to strong compensation in the material as already discussed in the Section 7.3.4. This assignment is corroborated by temperature-dependent measurements as shown in Figure 7.8a where the shift of the peak energy of the transition is plotted as a function of temperature.

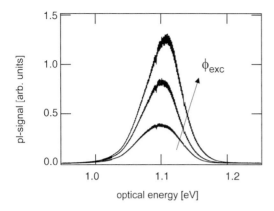

Figure 7.7 PL spectrum of Cu-poor Cu(In,Ga)Se$_2$ with [Cu]/([In] + [Ga]) = 0.9 and [Cu]/([In] + [Ga]) = 0.3 measured at 15 K.

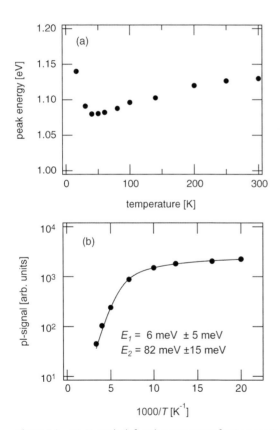

Figure 7.8 (a) PL-peak shift with temperature for Cu-poor Cu(In,Ga)Se2. (b) Dependence of PL yield on temperature.

It can be seen that the peak energy first shifts to smaller energies (redshift) at low temperatures and then to larger energies for temperatures above 50 K (blueshift). This behavior is a characteristic signature of potential fluctuations [25, 26, 29–31] as discussed above. An analysis of the PL yield in Figure 7.8b using Eq. (7.18) yields activation energies of 6 and 82 meV assuming a $T^{3/2}$ dependence of the emission prefactors. Although the errors are relatively large, it can be concluded that a very shallow level at $E_1 = 6$ meV and a deep level at $E_2 = 82$ meV participate in the observed transition. However, we have to caution that the interpretation of transition energies and their temperature dependence is difficult in the presence of potential fluctuations, because thermal redistribution processes within the potential well landscape take place.

7.4.2
Room-Temperature Measurements: Estimation of V_{oc} from PL Yield

When the temperature is raised to room temperature, band–band transitions become very likely, as the bands are now sufficiently populated by photoexcited carriers. A room temperature PL spectrum measured for the sample of Figure 7.7 is shown in Figure 7.9 exhibiting an even broader transition with a peak energy of 1.13 eV and a FWHM ≈ 100 meV (solid line, left scale).

An evaluation of the PL yield with excitation intensity gives a $k = 1.4$ and $\beta = 0$ meV, which allows to identify the recombination process as band–band transitions. Note that the exciton-binding energy is of the order of tens of millivolts in this material leaving excitons fully ionized at 300 K.

Also included in the figure is a PL-spectrum of a coevaporated Cu(In,Ga)Se$_2$ solar cell processed with similar but not identical preparation conditions and a [Ga]/([In + Ga]) ratio of 0.27 [9]. The device structure is glass/Mo/Cu(In,Ga)Se$_2$/CdS/

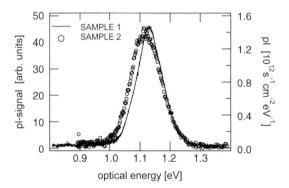

Figure 7.9 Room-temperature PL spectrum of the Cu-poor Cu(In,Ga)Se$_2$ device of Figure 7.7 (left scale, solid line) and a Cu-poor Cu(In,Ga)Se$_2$ device from a different coevaporation process measured in a setup with absolute calibration (right scale, open circles).

ZnO with an active area of 0.5 cm². It can be seen that the spectral shape agrees well for these two samples, with a slight shift in the peak energy by 20 meV. For the measurement, the device was fully illuminated with light from a 780-nm laser diode. In order to allow a prediction of the open circuit voltage corresponding to AM1.5 illumination conditions, the monochromatic excitation photon flux in the experiment was adjusted to correspond to the number of photons that would be absorbed for broad-band AM1.5 excitation, which for a 1.1-eV band-gap material is approximately 2×10^{17} cm²/s or 60 mW/cm². The luminescence radiation was measured by a liquid nitrogen cooled germanium detector in a setup calibrated to yield absolute photon numbers. According to the method outlined in Section 7.1, the absolute quasi-Fermi level splitting is estimated by plotting the quantity ln $(Y_{PL}(E)/10^{23}/E^2)$ versus energy and fitting the high-energy wing using Eq. (7.7) as shown in Figure 7.10a. From the slope, m, the temperature of the photoexcited electron–hole ensemble $T=(1/k_B/m)=320$ K and from the y-axis intercept at $E=0$ eV a quasi-Fermi level splitting of $\Delta\mu=0.58$ eV is obtained. This value agrees

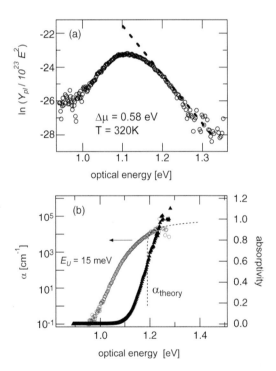

Figure 7.10 (a) Determination of the quasi-Fermi level splitting from calibrated room-temperature PL measurement. (b) Absorptivity (triangles, right scale) and absorption coefficient (open circles, left scale) determined from the calibrated PL measurement. Also included is a direct gap absorption coefficient with square-root dependence on energy (dashed line).

very well with the experimentally determined open-circuit voltage for which $V_{oc} = 0.56$ V was obtained, using exactly the same illumination conditions [9]. The deduced temperature of $T = 320$ K is higher than the measurement temperature $T = 300$ K, most likely due to sample heating from the large-area illumination during the experiment. The good agreement between V_{oc} and $\Delta\mu$ demonstrates the feasibility of the method to judge the electronic quality of the material without need of electrical contacts or electrical measurements. On the other hand, it also shows that for this particular example very little losses occur in the functional layers and at the contacts, leading to a measured V_{oc} almost as high as the bulk value of the quasi-Fermi level splitting.

As described in Section 7.1, the energy-dependent absorptivity function can be determined by plugging the values for the temperature and $\Delta\mu$ back into Eq. (7.6). The resulting absorptivity function $a(E)$ is shown in Figure 7.10b (triangles, right scale). Using the approximation $a(E) \approx 1 - \exp(-\alpha(E)d)$ and assuming a homogeneous sample thickness of 1 µm, the absorption coefficient can then be derived from the absorptivity. The result is shown in Figure 7.10b (open circles, left scale). The absorption coefficient derived from the PL measurement may be compared to an absorption coefficient expected for a direct-gap semiconductor with ideal parabolic bands [1] $\alpha(E) = A(E - E_{gap})^{0.5}$, which is included in Figure 7.10b using a dashed line ($A = 10^5$ cm^{-1} and band-gap value $E_{gap} = 1.19$ eV). It can be seen that the theoretical curve agrees well with the experimental curve in the region above E_{gap}. For energies below E_{gap}, the experimental curve does not vanish rapidly but shows an exponentially decaying band tail with a characteristic energy of $E_0 = 15$ meV. Figure 7.10b shows that a defined value of the absorptivity, for example, $a(E) = 1/e$, may be used to derive an optical threshold from the PL measurement, which is related to the optical band gap of the material. This result will be used in the next section to map spatial inhomogeneities in the optical properties of a sample by PL.

7.4.3
Spatially Resolved Photoluminescence: Absorber Inhomogeneities

Thin semiconductor materials often show significant nonuniformities in their morphology, grain orientation, composition, and their electronic and optoelectronic properties [34–39]. Although the knowledge of the impact of these inhomogeneities on cell performance is still limited, generally, spatial inhomogeneities of the absorber layer are considered to be disadvantageous for the cell efficiency, which was recently shown in several modeling approaches [40–42]. PL performed with microscopic spatial resolution can be used for investigating spatial inhomogeneities of absorber layers and for quantifying the variation of material properties on microscopic length scales [36, 43]. The experimental setup used in a PL-scanning experiment was described in the experimental section.

In the following, micro-PL scans for a Cu(In,Ga)Se$_2$ thin-film sample with [Ga]/([In + Ga]) = 0.3 and [Cu]/([In + Ga]) $\approx$ 0.8 measured at room temperature

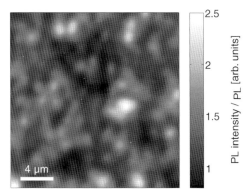

Figure 7.11 Two-dimensional scan of the spectrally integrated PL intensity for a Cu(In,Ga)Se$_2$ thin film at room temperature.

are shown. The sample configuration was glass/Mo/Cu(In,Ga)Se$_2$/CdS. The sample was excited by a 532-nm laser at an excitation flux corresponding to $10^4 \times$ AM 1.5. The detector system of the micro-PL setup was a spectrograph with a photodiode array, which allowed for straightforward measurements of a data set containing a complete PL spectrum for each point of the scan. An image of the spectrally integrated luminescence intensity, which is related to the local recombination lifetime, is shown in Figure 7.11. It can be seen that the integrated luminescence signal varies by a factor of 3 with typical structure sizes on the micrometer scale. To analyze the spatial variation of the quasi-Fermi level splitting and of the optical threshold, the micro-PL spectra in Figure 7.11 are analyzed using the approach outlined above, extracting the quasi-Fermi level splitting from a fit to the high-energy tail of the data. Since in this experiment no absolute calibrated photon counts were recorded, only relative changes of the quasi-Fermi level splitting can be extracted. The results for this quantity for the measured area are shown in Figure 7.12a. It can be seen from the images that the spatial variation closely resembles the variations seen for the integrated PL yield. In Figure 7.12b, a histogram of the distribution of $\Delta\mu$ is displayed showing a maximum variation of 40 meV and a distribution width of FWHM $\approx$ 13 meV. From each recorded PL spectrum, an absorptivity function $a(E)$ and an optical threshold E_{th} related to $a(E_{th}) = 1/e$ is derived by using the method described in the previous section, which we associate with the local optical band gap [43]. The resulting map of optical threshold values and the distribution histogram are shown in Figure 7.13a and b.

It can be seen that the maximum lateral variation of the band gap amounts to approximately 15 meV with a FWHM $\approx$ 6 meV, which is just less than half the value obtained for the distribution width of the quasi-Fermi level splitting. In this example, we have considered the variation of the optical threshold energy to be equal to a local variation of the band gap, which is expected from alloying inhomogeneities in

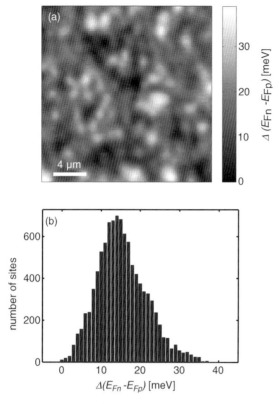

Figure 7.12 (a) Two-dimensional maps of the lateral variation of the quasi-Fermi level splitting. (b) Corresponding statistical distributions of the 2D data.

Cu(In,Ga)Se$_2$, that is, a locally varying Ga content. Generally, the extracted local variation of the optical threshold energy can be caused by further effects than only a shift of the optical gap, such as large local variations of the absorber thickness or change in the characteristic band-tail energy. To strictly exclude thickness variations from the analysis, it would be necessary to monitor the local thickness, for example, by atomic force microscopy and include this information in the evaluation procedure [44]. We note that in general the variation of the quasi-Fermi level splitting tends to decrease with increasing excitation level [45]. Although an excitation flux comparable to AM1.5 conditions would be desirable, it is generally very difficult to achieve for micro-PL measurements at room temperature because of the too small number of photons involved. In summary, the results shown here indicate that the absorber layer contains spatial variations of the quasi-Fermi levels splitting and of the local band gap. As the spatial variation of the optical threshold does not seem to correlate with the spatial variation in the quasi-Fermi level splitting, and the distribution of the optical threshold values is significantly smaller than the distribution of $\Delta\mu$, we

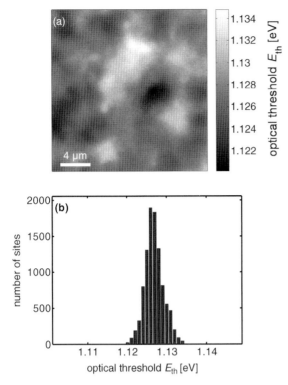

Figure 7.13 (a) Two-dimensional maps of the lateral variations of the optical threshold (local band gap). (b) Corresponding statistical distributions of the 2D data.

conclude that a local variation in recombination lifetime is the main cause of the observed variation in PL yield and $\Delta\mu$.

References

1 Pankove, J.I. (1971) *Optical Processes in Semiconductors*, Dover, New York.
2 Bepp, H.B. and Wiliams, E.W. (1972) Photoluminescence theory, in *Semiconductor and Semimetals*, vol. 8 (eds R.K. Willardson and A.C. Beer) Academic Press, New York.
3 Nelson, J. (2003) *The Physics of Solar Cells*, Imperial College Press, London.
4 van Roosbroeck, W. and Shockley, W. (1954) Photon-radiative recombination of electrons and holes in germanium. *Phys. Rev.*, **94**, 1558.
5 Werner, J.H., Mattheis, J., and Rau, U. (2005) Efficiency limitations of polycrystalline thin film solar cells: Case of Cu(In,Ga)Se$_2$. *Thin Solid Films*, **480**, 399.
6 Würfel, P. (1982) The chemical-potential of radiation. *J. Phys. C – Solid State Phys.*, **15**, 3967.
7 Würfel, P. (2009) *The Physics of Solar Cells*, Wiley-VCH, Weinheim.
8 Schick, K., Daub, E., Finkbeiner, S., and Würfel, P. (1992) Verification of a generalized Planck law for luminescence

radiation from silicon solar cells. *Appl. Phys. A*, **54**, 109.

9 Unold, T., Berkhahn, D., Dimmler, B., and Bauer, G.H. (2000) Open circuit voltage and loss mechanisms in polycrystalline Cu(In,Ga)Se$_2$ heterodiodes from photoluminescence studies, in *Proceedings of the 16th European Photovoltaic Solar Energy Conference*, Glasgow, United Kingdom, May 1–5, 2000 (eds H. Sheer, B. McNelis, W. Paltz, H.A. Ossenbrink, and P. Helm), James & James, London, p. 737.

10 Wilson, T. (1990) *Confocal Microscopy*, Academic Press, New York.

11 http://physics.nist.gov/PhysRefData/ASD/Html/ref.html.

12 Siebentritt, S. (2006) Shallow defects in the wide gap chalcopyrite CuGaSe$_2$, in *Wide-Gap Chalcopyrites, Springer Series in Materials Science*, vol. **86** (eds S. Siebentritt and U. Rau), Springer, Heidelberg, p. 113.

13 Krustok, J., Collan, H., and Hjelt, K. (1996) Does the low-temperature Arrhenius plot of the photoluminescence intensity in CdTe point towards an erroneous activation energy? *J. Appl. Phys.*, **81**, 1442.

14 Bauknecht, A., Siebentritt, S., Alberts, J., Tomm, Y., and Lux-Steiner, M.C. (2000) Excitonic photoluminescence from CuGaSe$_2$ single crystals and epitaxial layers: Temperature dependence of the band gap energy. *Jpn. J. Appl. Phys.*, **39**, 322.

15 Grundmann, M. (2006) *The Physics of Semiconductors*, Springer, Berlin.

16 Yakushev, M., Martin, R.W., Mudryi, A.V., and Ivaniukovich, A.V. (2008) Excited states of the a free exciton in CuInS$_2$. *Appl. Phys. Lett.*, **92**, 111908.

17 Schmidt, T., Lischka, K., and Zulehner, W. (1992) Excitation power dependence of the near-band-edge photoluminescence of semiconductors. *Phys. Rev. B*, **45**, 8989.

18 Haynes, J.R. (1960) Experimental proof of the existence of a new complex in silicon. *Phys. Rev. Lett.*, **4**, 361.

19 Atzmüller, H. and Schröder, U. (1978) Theoretical investigations of Haynes' rule. *phys. stat. sol. (b)*, **89**, 349.

20 Lightowlers, E.C. (1990) Photoluminescence characterisation, in *Growth and Characterization of Semiconductors* (eds R.A. Stradling and P.C. Klipstein), Adam Hilger, Bristol and New York, p. 135.

21 Ascarelli, G. and Rodriguez, S. (1961) Recombination of electrons and donors in n-type germanium. *Phys. Rev.*, **124**, 1321.

22 Eagles, D.M. (1960) optical absorption and recombination radiation in semiconductors due to transitions between hydrogen-like acceptor impurity levels and the conduction band. *J. Chem. Solids*, **16**, 76.

23 Yu, P.W. (1977) Excitation-dependent emission in Mg-, Be-, Cd-, and Zn-implanted GaAs. *J. Appl. Phys.*, **48**, 5043.

24 Zacks, E. and Halperim, A. (1972) Dependence of the peak energy of the pair-photoluminescence band on excitation intensity. *Phys. Rev. B*, **6**, 3072.

25 Wagner, M., Dirnstorfer, I., Hofmann, D.M., Lampert, M.D., Karg, F., and Meyer, B.K. (1998) Characterization of CuIn(Ga)Se$_2$ thin films – I. Cu-rich layers. *phys. status solidi (a)*, **167**, 131.

26 Schumacher, S.A., Botha, J.R., and Alberts, V. (2006) Photoluminescence study of potential fluctuations in thin layers of Cu(In$_{0.75}$Ga$_{0.25}$)(S$_y$Se$_{1(y}$)$_2$. *J. Appl. Phys.*, **99**, 063508.

27 Shklovskii, B.I. and Efros, A.L. (1984) *Electronic Properties of Doped Semiconductors*, Springer, Berlin.

28 Dirnstorfer, I., Wagner, M., Hofmann, D.M., Lampert, M.D., Karg, F., and Meyer, B.K. (1998) Characterization of CuIn(Ga)Se$_2$ thin films – II. In-rich layers. *phys. status solidi (a)*, **168**, 163.

29 Bauknecht, A., Siebentritt, S., Albert, J., and Lux-Steiner, M.C. (2001) Radiative recombination via intrinsic defects in Cu$_x$Ga$_y$Se$_2$. *J. Appl. Phys.*, **89** (8), 4391.

30 Levanyuk, A.P. and Osipov, V.V. (1973) Theory of luminescence of heavily doped semiconductors. *Sov. Phys. Semicond.*, **7**, 721.

31 Krustok, J., Collan, H., Yakushev, M., and Hjelt, K. (1999) The role of spatial potential fluctuations in the shape of the PL bands of multinary semiconductor compounds. *Phys. Scr.*, **T79**, 179.

32 Zott, S., Leo, K., Ruckh, M., and Schock, H.W. (1997) Radiative recombination in CuInSe$_2$ thin films. *J. Appl. Phys.*, **82**, 356.

33 Siebentritt, S., Beckers, I., Riemann, T., Christen, J., Hoffmann, A., and Dworzak, M. (2005) Reconciliation of luminescence and Hall measurements on the ternary semiconductor CuGaSe$_2$. *Appl. Phys. Lett.*, **86** (9), 091909.

34 Sites, J.R., Granata, J.E., and Hiltner, J.F. (1998) Losses due to polycrystallinity in thin-film solar cells. *Sol. Energy Mater. Sol. Cells*, **55**, 43.

35 Eich, D., Herber, U., Groh, U., Stahl, U., Heske, C., Marsi, M., Kiskinova, M., Riedl, W., Fink, R., and Umbach, E. (2000) Lateral inhomogeneities of Cu(In, Ga)Se$_2$ absorber films. *Thin Solid Films*, **361**, 258.

36 Bothe, K., Bauer, G.H., and Unold, T. (2002) Spatially resolved photoluminescence measurements on Cu(In,Ga)Se$_2$ thin films. *Thin Solid Films*, **403**, 453.

37 Grabitz, P.O., Rau, U., Wille, B., Bilger, G., and Werner, J.H. (2006) Spatial inhomogeneities in Cu(In,Ga)Se$_2$ solar cells analyzed by an electron beam induced voltage technique. *J. Appl. Phys.*, **100**, 124501.

38 Yan, Y., Noufi, R., Jones, K.M., Ramanathan, K., Al-Jassim, M.M., and Stanbery, B.J. (2005) Chemical fluctuation-induced nanodomains in Cu(In,Ga)Se$_2$ films. *Appl. Phys. Lett.*, **87** (12), 121904.

39 Abou-Ras, D., Koch, C.T., Küstner, V., van Aken, P.A., Jahn, U., Contreras, M.A., Caballero, R., Kaufmann, C.A., Scheer, R., Unold, T., and Schock, H.W. (2009) Grain-boundary types in chalcopyrite-type thin films and their correlations with film texture and electrical properties. *Thin Solid Films*, **517** (7), 2545.

40 Grabitz, P.O., Rau, U., and Werner, J.H. (2005) Modeling of spatially inhomogeneous solar cells by a multi-diode approach. *phys. status solidi (a)*, **202** (15), 2920.

41 Werner, J.H., Mattheis, J., and Rau, U. (2005) Efficiency limitations of polycrystalline thin film solar cells: Case of Cu(In,Ga)Se$_2$. *Thin Solid Films*, **480–481**, 399.

42 Kanevce, A. and Sites, J.R. (2007) Impact of Nonuniformities on Thin Cu(In,Ga)Se$_2$ Solar Cell Performance. Thin-Film Compound Semiconductor Photovoltaics 2007, Mater. Res. Soc. Symp. Proc. 1012 Warrendale, PA, Matterials Research Society (eds T. Gessert, S. Marsillac, T. Wada, K. Durose, and C. Heske), pp. Y0802.

43 Gütay, L. and Bauer, G.H. (2007) Spectrally resolved photoluminescence studies on Cu(In,Ga)Se$_2$ solar cells with lateral submicron resolution. *Thin Solid Films*, **515**, 6212.

44 Gütay, L., Lienau, C., and Bauer, G.H. (2010) Subgrain size inhomogeneities in the luminescence spectra of thin film chalcopyrites. *Appl. Phys. Lett.*, **97**, 052110.

45 Gütay, L. and Bauer, G.H. (2009) Local fluctuations of absorber properties of Cu(In,Ga)Se$_2$ by sub-micron resolved PL towards "real life" conditions. *Thin Solid Films*, **517** (7), 2222.

8
Steady-State Photocarrier Grating Method
Rudolf Brüggemann

8.1
Introduction

The excess-carrier properties of a photoexcited semiconductor are important indicators of its quality with respect to applications in optoelectronic devices or for studying the recombination physics. In contrast to the majority-carrier properties, which can be determined rather straightforwardly by stationary photocurrent measurements, the minority-carrier properties can only be revealed by more sophisticated methods. In this respect the steady-state photocarrier grating (SSPG) method has had an enormous impact since it was suggested by Ritter, Zeldov, and Weiser, named RZW hereafter, in 1986 [1].

The SSPG method is based on the carrier diffusion under the presence of a spatial sinusoidal modulation in the photogeneration rate G, which induces a so-called photocarrier grating. From photocurrent measurements at different grating periods Λ, the ambipolar diffusion length L can be determined by an analysis that assumes ambipolar transport and charge neutrality.

The proposal by RZW, following papers on the analysis [2, 3] and the simple setup triggered a rapid widespread application [4–8]. In parallel, critical and more in-depth accounts on the underlying theory were given to put the technique on a firm ground or to describe its limits when applied to semiconductors with traps. Ritter *et al.* [9], Balberg *et al.* [10, 11], Li [12], and Shah *et al.* [13] analyzed the transport equations, also with respect to the "lifetime" or "relaxation time" regimes. In the lifetime regime, the carrier lifetimes are longer than the dielectric relaxation time.

The previous publications were later criticized by Hattori *et al.* [14] who performed a second-order perturbation approach and pointed out deficiencies of other earlier analyses. Nevertheless, Hattori *et al.* showed that under the conditions in the "lifetime regime" the analysis and evaluation of the SSPG method are correct. These authors also suggested a correction method to avoid incorrect values of L.

A novel aspect was introduced by Abel and Bauer [15] who numerically solved the transport and Poisson equations and compared the solutions with a generalized theory which enables to study the SSPG results by the variation of Λ and/or electric

Advanced Characterization Techniques for Thin Film Solar Cells,
Edited by Daniel Abou-Ras, Thomas Kirchartz and Uwe Rau.
© 2011 Wiley-VCH Verlag GmbH & Co. KGaA. Published 2011 by Wiley-VCH Verlag GmbH & Co. KGaA.

field E. These authors derive their expressions in terms of the mobility-lifetime product $(\mu\tau)_{min}$ of the minority carriers, which can be determined from the SSPG method and related to L.

More recently, Schmidt and Longeaud [16] developed a generalized derivation of the solution of the SSPG equations at low E. They identified the shortcomings in the previous derivations which differ from the numerical solution. Their approach also allows the SSPG experiment to be used for the density-of-states (DOS) determination.

An important aspect is the above-mentioned association of L with the $(\mu\tau)_{min}$ product of the minority carriers. Together with the majority-carrier mobility-lifetime product $(\mu\tau)_{maj}$ at the same G from the photoconductivity σ_{ph} via $\sigma_{ph} = eG(\mu\tau)_{maj}$, where e equals 1.6×10^{-19} C, the excess-carrier properties can be related to each other with the aim to consistently describe them in relation to models for recombination, the role of the Fermi level E_f, and the relevant DOS in the band gap [17].

The above brief and not complete description of SSPG-related aspects indicates that both the theoretical understanding and the variety of applications have evolved in the last 20 years. While some treatments point to modifications that may be needed for the correct determination of L and the corresponding $(\mu\tau)_{min}$, the original formulation and analysis by RZW is still often used as deviations are considered to be small.

8.2
Basic Analysis of SSPG and Photocurrent Response

8.2.1
Optical Model

Figure 8.1a sketches the arrangement of the SSPG experiment. Two coherent plane waves, which originate from one laser beam that has been split into the beams L_1 and

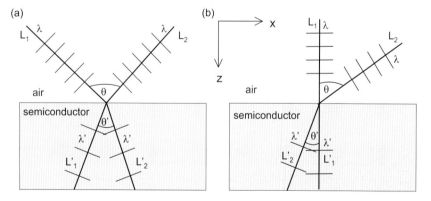

Figure 8.1 Schematics of the interference experiment with two plane waves L_1 and L_2 with wavelength λ. The coordinate system is sketched.

8.2 Basic Analysis of SSPG and Photocurrent Response

L_2 with wavelength λ, impinge symmetrically onto a semiconductor where they suffer refraction according to Snell's law. The angle between the two beams changes from θ in air to θ' in the semiconductor with refractive index n_s. The wavelength changes from λ to λ' where $\lambda' = \lambda\, n_{air}/n_s$. These relations hold accordingly when the surrounding medium is not air but glass. The plane wave of L_1 can be described by the electric field $\vec{E}_1(\vec{r}) = \vec{E}_{10}\exp(i\vec{k}_1 \cdot \vec{r} - i\omega t)$ in air and by $\vec{E}'_1(\vec{r}) = \vec{E}'_{10}\exp\left(i\vec{k}'_1 \cdot \vec{r} - i\omega t\right)$ in the semiconductor, where the variables have their typical meanings. A similar expression holds for $\vec{E}_2(\vec{r})$ of L_2.

In Figure 8.1b, L_1 is perpendicular to the semiconductor surface so that no refraction occurs.

The local photon flux Φ in the semiconductor is related to the square of the local electric field by

$$\Phi(\vec{r}) \propto \left|\vec{E}'_1(\vec{r}) + \vec{E}'_2(\vec{r})\right|^2 = \left|\vec{E}'_{10}\exp\left(i\vec{k}'_1 \cdot \vec{r} - i\omega t\right) + \vec{E}'_{20}\exp\left(i\vec{k}'_2 \cdot \vec{r} - i\omega t\right)\right|^2 \tag{8.1}$$

where $|\vec{k}|$ is given by $(2\pi n/\lambda)$ and we assume low absorption so that there is negligible decay of the photon flux. It is then found from Eq. (8.1) that in the z-direction the value of Φ is constant. In the x-direction the influence of the refractive-index change cancels. The sinusoidal modulation in the x-direction, sketched in Figure 8.2, which forms a so-called grating, has a grating period Λ given by

$$\Lambda = \lambda/[2\sin(\theta/2)] \tag{8.2}$$

Assuming $\vec{E}_{10} = \vec{E}'_{10}$ and for a generation rate $G \approx \alpha\Phi$, independent in z, leads to a variation of G in x according to

$$G(\vec{r}) = G(x) = (G_1 + G_2) + 2\gamma_0(G_1 G_2)^{0.5} \cos(2\pi x/\Lambda) \tag{8.3}$$

where $G_1 = \alpha\Phi_1$, $G_2 = \alpha\Phi_2$. The additional parameter γ_0 has been introduced here by RZW as a grating-quality factor in order to account for a nonideal grating because of optical scattering, nonideal coherence, or mechanical vibrations. The amplitude of the modulation $2\gamma_0(G_1 G_2)^{0.5}$ is typically larger than G_2 so that regions exist in which $G(x)$ with the two beams is less than G_1 only.

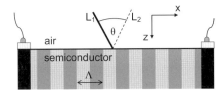

Figure 8.2 Schematic of the interference fringes between coplanar electrodes. The parallel planes of constant photon flux are perpendicular to the air–semiconductor interface.

For the more explicit G profile within the semiconductor with the absorption coefficient α one obtains [18]

$$G(x,z) = G(x) = (G_1 + G_2)\exp(-\alpha z)\left[1 + \frac{2\gamma_0(G_1 G_2)^{0.5}}{G_1 + G_2}\cos(2\pi x/\Lambda)\right] \quad (8.4)$$

An account of the influence of the polarization-dependent reflection on G was given by Nicholson [19].

In the asymmetric case of Figure 8.1b, in which the stationary beam hits the sample at right angle, the analysis yields that the fringes are no longer perpendicular to the surface but that they are tilted. This may cause problems by blurring of the relevant photocarrier grating relative to the electric field that probes its distribution.

8.2.2
Semiconductor Equations

Poisson's equation consists of the local total charge ρ to which the free carriers in the bands, trapped charge, and/or charged defects in addition to any charge from impurities or dopants contribute. In terms of E, it takes the form

$$\frac{1}{e}\frac{dE}{dx} = \frac{\rho}{\varepsilon\varepsilon_0} \quad (8.5)$$

with the dielectric constants ε of the semiconductor and ε_0. For the special case of a homogeneous semiconductor with coplanar electrodes as in Figure 8.2, Poisson's equation reduces to $\rho = 0$ and E is given by the ratio of the applied voltage and the electrode distance.

In addition, with homogeneous free-carrier distributions the current densities are determined by the electric field so that the current equation reduces to

$$j = j_n + j_p = \left(en\mu_n + ep\mu_p\right)E = (\sigma_n + \sigma_p)E \quad (8.6)$$

with the free electron (hole) density n (p) and with the conductivities of the electrons (holes) σ_n (σ_p). The index denotes the respective carrier type of the electrical current density j and the extended-state mobility μ.

The continuity equations with $\operatorname{div} j_n = \operatorname{div} j_p = 0$ become

$$G = R \quad (8.7)$$

with the recombination rate R.

The special case under almost homogeneous G is used for the determination of the mobility-lifetime $(\mu\tau)_{maj}$ product of the majority carriers. The measured photocurrent is assumed to be determined by one carrier type, say electrons. It is then given by the difference of the total to the dark current density. The photoconductivity is determined from Ohm's law $j_{ph} = \sigma_{ph}E$ and it is determined by the excess electron density n_{ph}.

The recombination lifetimes of the free electrons and holes are defined by [20]

$$R = \frac{n_{tot}-n_{th}}{\tau_n} = \frac{n_{ph}}{\tau_n} = \frac{p_{tot}-p_{th}}{\tau_p} = \frac{p_{ph}}{\tau_p} \qquad (8.8)$$

where n_{tot} (p_{tot}) is the total electron (hole) density under illumination and n_{th} and p_{th} are the thermal equilibrium values. With Eq. (8.7) and the substitution of n_{ph} it follows that

$$(\mu\tau)_n = \mu_n \tau_n = \left(\frac{\sigma_{ph}}{eG}\right) \qquad (8.9)$$

which gives experimental access to the majority-carrier $(\mu\tau)_{maj}$ product. Corresponding equations hold if holes are the majority carriers.

8.2.3
Diffusion Length: Ritter–Zeldov–Weiser Analysis

RZW devised the "SSPG technique for diffusion length measurement in photoconductive insulators." Their starting point of analysis are the transport equations of Section 19.2 under the assumption of ambipolar transport, so that only one carrier type is analyzed, and charge neutrality. They reduce to

$$\frac{1}{e}\frac{d}{dx}\left(-e\overline{D}\frac{dn(x)}{dx}\right) = G(x) - R(x) \qquad (8.10)$$

where $\overline{D}$ is an effective diffusion coefficient and where $n(x)$ is the sum of the corresponding homogenous excess density n_0 and the modulated density $n(x) = n_0 + \Delta n(x)$. Splitting off the homogeneous terms results in a differential equation for $\Delta n(x)$ according to

$$\overline{D}\frac{d^2 \Delta n(x)}{dx^2} = g(x) - \frac{\Delta n(x)}{\tau(n)} \qquad (8.11)$$

with $g(x) = 2\gamma_0 (G_1 G_2)^{0.5} \cos(2\pi x/\Lambda)$ and the carrier lifetime τ. The ansatz $\Delta n(x) = \Delta n_0 \cos(2\pi x/\Lambda)$ in phase with $g(x)$ results in

$$\Delta n_0 = \frac{2\gamma_0 \sqrt{G_1 G_2}}{1 + \left(\frac{2\pi L}{\Lambda}\right)^2} \tau \qquad (8.12)$$

where the ambipolar diffusion length L is introduced by

$$L = \sqrt{\overline{D}\tau}. \qquad (8.13)$$

It is necessary to correctly account for a possibly G-dependent τ in Eq. (8.12). Experimentally, the relation between photoconductivity or excess-carrier density and the generation rate is usually given by

$$n \propto G^\gamma \qquad (8.14)$$

with the power-law exponent γ. One may expand Eq. (8.14) to eliminate τ in Eq. (8.12) so that with $\tau \approx \gamma n_0/(G_1 + G_2)$ one determines

$$\Delta n_0 = \frac{2\gamma_0\sqrt{G_1 G_2}}{1 + \left(\frac{2\pi L}{\Lambda}\right)^2} \frac{\gamma n_0}{G_1 + G_2}. \tag{8.15}$$

The local conductivity which is proportional to $n(x) = n_0 (1 + \Delta n(x)/n_0)$ may now be written as

$$\sigma(x) = \sigma(L_1 + L_2)[1 + A \cos(2\pi x/\Lambda)] \tag{8.16}$$

where n_0 is related to $\sigma(L_1 + L_2)$. The photocarrier grating amplitude A is given by

$$A = \frac{2\gamma\gamma_0\sqrt{G_1 G_2}}{G_1 + G_2} \gamma_g \tag{8.17}$$

in which the last term

$$\gamma_g = \frac{1}{1 + \left(\frac{2\pi L}{\Lambda}\right)^2} \tag{8.18}$$

with L and Λ describes the influence of the grating.

So far, the diffusion length L is related to the local conductivity $\sigma(x)$. What is needed for the electrical measurements is a relation between L and the measured current. In a next phenomenological step, RZW thus assumed that the magnitude of the average current density j_{coh} in the direction of the electric field under coherent conditions in the presence of the grating can be modeled by a series resistor model of local resistivities or inverse conductivities, given by

$$j_{coh}(\Lambda, L_1 + L_2)/E = \sigma_{av}(\Lambda) = \frac{\Lambda}{\int_0^\Lambda \frac{dx}{\sigma(x)}} = \sigma(L_1 + L_2)\sqrt{1 - A^2} \tag{8.19}$$

A problem with this approach is that, as j must be constant, $E(x)$ must oscillate if $\sigma(x)$ oscillates, in contradiction to the assumption of no space charge and a constant E.

RZW also suggested that a lock-in technique may be used to determine the parameter β that, as will be shown below, is related to the diffusion length and is defined by RZW by

$$\beta(\Lambda) = \frac{U_{coh}(\Lambda)}{U_{inc}(\Lambda)} \tag{8.20}$$

at a given grating period. Here $U_{coh}(U_{inc})$ denotes the lock-in amplifier (LIA) signal under coherent (incoherent) beam conditions. The LIA detects the difference in the currents with respect to the current density j_1 under the bias level of the beam L_1. The polarization change of $90°$ of L_1 together with the unchanged chopped L_2 results in a comparison of $j_{coh}(\Lambda, L_1 + L_2)$, under coherent condition, with $j_{inc}(\Lambda, L_1 + L_2)$, under incoherent condition. In terms of current densities the parameter β is thus given by

$$\beta(\Lambda) = \frac{U_{coh}}{U_{inc}} = \frac{j_{coh}(\Lambda, L_1 + L_2) - j_1(\Lambda, L_1)}{j_{inc}(\Lambda, L_1 + L_2) - j_1(\Lambda, L_1)} \qquad (8.21)$$

The link between the experimentally determined β and the diffusion length L is adjusted by an approximation of the current densities with the assumed and also typically observed power-law dependence of the photocurrent density $j_{ph} \propto G^\gamma$ and the condition that $G_2/G_1 \ll 1$. The G terms can then be eliminated with, for example, $j_{inc}(\Lambda, L_1 + L_2) - j_{inc}(\Lambda, L_1) \propto (G_1 + G_2)^\gamma - (G_1)^\gamma$ and $j_{coh}(\Lambda, L_1 + L_2)$ from Eq. (8.19) so that $\beta(\Lambda)$ reads

$$\beta(\Lambda) = \frac{\left[(1 + G_2/G_1)^\gamma \sqrt{1-A^2} - 1\right]}{[(1 + G_2/G_1)^\gamma - 1]} \approx \frac{(1 + \gamma G_2/G_1)\left(1 - 2(\gamma\gamma_0\gamma_g)^2 G_2/G_1\right) - 1}{\gamma G_2/G_1}$$

$$\approx 1 - 2\gamma(\gamma_0\gamma_g)^2 \qquad (8.22)$$

where the quadratic terms in G_2/G_1 have been neglected in the last step. Substituting γ_g from Eq. (8.18) yields

$$\beta(\Lambda) = 1 - \frac{2Z}{\left[1 + \left(\frac{2\pi L}{\Lambda}\right)^2\right]^2} \qquad (8.23)$$

with $Z = \gamma\gamma_0^2$.

8.2.3.1 Evaluation Schemes

Equation (8.23) is the central equation of the SSPG analysis: the data sets of the measured β values at different positions Λ can be analyzed with the fit function $\beta(\Lambda)$ and the two fit parameters: the diffusion length L and a fit parameter Z. While L has a physical meaning, the parameter Z is just a fit parameter. However, it can be evaluated for a self-consistency check of the analysis and the measured β data. From the Z value and the separate experimental determination of γ and γ_d, the grating quality factor γ_0 can be calculated. It should be "close to 1."

Balberg et al. [7] suggested rearranging the data according to a linear form as

$$1/\Lambda^2 = [Z^{1/2}/(2\pi L)^2][2/(1-\beta)]^{1/2} - (2\pi L)^{-2} \qquad (8.24)$$

with the ordinate values $1/\Lambda^2$ and the abscissa values $[2/(1-\beta)]^{1/2}$. The diffusion length L can be read from the extrapolation to the abscissa or from the fit parameters for the straight line.

In our laboratory it has been customary to plot the values related to the measured β as $(1-\beta)^{-1/2}$ on the ordinate scale by

$$(1-\beta)^{-1/2} = [(2Z)^{-1/2}(1/L)^2](2\pi/\Lambda)^2 + (2Z)^{-1/2} \qquad (8.25)$$

This kind of linear plot was also used by Nicholson [19].

The advantage of any linear plot is that a disagreement from the theoretical expectation may more easily be identified by deviations from the linear behavior. Ambipolarity is not obeyed by charge separation under high electric fields [21], which can also be identified from a concave instead of a linear shape of the graph according to Eq. (8.24). It is then still possible to evaluate the ambipolar diffusion length from portions of the data set. Any super- or sublinear behavior depends, of course, on the plot specification with respect to abscissa and ordinate values. Balberg [22] analyzed the measurement data for the identification of nonambipolarity under low-field conditions.

For identifying the effect of surface recombination, the rearrangement according to

$$[2/(1-\beta)]^{-1/2} = \left\{[L^2/Z^{1/2}](2\pi/\Lambda)^2 + (Z)^{-1/2}\right\} g_s^2(\Lambda) \tag{8.26}$$

was suggested [18], where $g_s(\Lambda)$ is a function which is 1 for negligible surface recombination and which decreases with increasing Λ. In a linear plot, $g_s(\Lambda)$ thus takes effect at large ordinate values so that L can still be determined from the β data at shorter values of Λ. In the case of surface recombination there is a sublinear increase in the plot.

8.2.4
More Detailed Analyses

Analyses of the SSPG method are available which give a more detailed account by including the effects of the dark conductivity, of traps, E at low and high values and the aspect of space charge. They also relate, in more detail, the measured L with the minority-carrier ($\mu\tau$) product.

8.2.4.1 Influence of the Dark Conductivity
In a following paper [3], RZW also took into account the effect of the underlying dark conductivity σ_d. The introduction of σ_d in Eq. (8.16) yields a slightly modified version of Eq. (8.23).

With the dark-conductivity coefficient γ_d, given by

$$\gamma_d = \frac{\sigma_{ph}}{\sigma_{ph} + \sigma_d}, \tag{8.27}$$

Equation (8.23) reads

$$\beta(\Lambda) = 1 - \frac{2\gamma\gamma_d\gamma_0^2}{[1+(2\pi L/\Lambda)^2]^2} = 1 - \frac{2Z}{[1+(2\pi L/\Lambda)^2]^2} \tag{8.28}$$

where now $Z = \gamma\gamma_d\gamma_0^2$. Formally, the fit functions (8.23) and (8.28) with the two fit parameters L and Z are the same.

8.2.4.2 Influence of Traps
In the original RZW approach, assumptions have been made. Especially with respect to semiconductors with band tails, it will be necessary to introduce the contributions of the trapping states and of the trapped charge. This is often accomplished by the

introduction of effective or drift mobilities in which the reduction in mobility through the trapping and emission processes is expressed. We denote the reference to the total charge by a capital letter of the index so that the drift mobilities read

$$\mu_N = \mu_n \frac{n}{N}$$
$$\mu_P = \mu_p \frac{p}{P} \tag{8.29}$$

with the total density N and P. A ratio θ_n

$$N = n + n_t$$
$$\frac{n}{N} = \theta_n \tag{8.30}$$

can be defined with the trapped-electron density n_t and with a corresponding relation for the holes.

Poisson's equation then takes the form

$$\frac{1}{e}\frac{dE}{dx} = \frac{P-N}{\varepsilon\varepsilon_0} \tag{8.31}$$

where the right-hand side is zero under charge neutrality with $N = P$.

The effective diffusion coefficient D_N may be introduced via the condition

$$\frac{kT}{e}\mu_n \frac{dn}{dx} = D_N \frac{dN}{dx} \tag{8.32}$$

which yields [3]

$$D_N = \frac{kT}{e}\mu_n \left(\theta_n + N \frac{d\theta_n}{dN} \right) \tag{8.33}$$

where k is the Boltzmann constant and T is the temperature.

The ambipolar diffusion equation with $\Delta N = \Delta P$ and a common lifetime τ_R reads

$$\overline{D_{NP}}\frac{d^2 \Delta N}{dx^2} - \frac{\Delta N}{\tau_R} + \Delta G(x) = 0 \tag{8.34}$$

which is similar to Eq. (8.10). The ambipolar diffusion coefficient $\overline{D_{NP}}$ is a rather complicated construct (see Eq. B6 of Ritter et al. [3]). For Boltzmann statistics $\overline{D_{NP}} = 2\frac{kT}{e}\frac{\mu_N \mu_P}{\mu_N + \mu_P}$. It is noted that the parameters $\tau_R = \tau_n/\theta_n = \tau_p/\theta_p$ and $\overline{D_{NP}}$ contain the common response of the total electron and hole contributions and thus pertain to the ensemble of free and trapped carriers.

Upon inspection of the diffusion equation (Eq. (8.34)), the diffusion length L is defined by

$$L = \sqrt{\overline{D_{NP}}\tau_R}. \tag{8.35}$$

The same $\beta(\Lambda,L)$ function as in Eq. (8.28) results from the analysis.

An effective lifetime t_θ is defined by Balberg [10] who obtains

$$L = \sqrt{2 D_\theta t_\theta} \tag{8.36}$$

with the effective diffusion coefficient D_θ. It can be shown that $\overline{D_{NP}} \tau_R = 2 D_\theta t_\theta$.

8.2.4.3 Minority-Carrier and Majority-Carrier Mobility-Lifetime Products

To relate the diffusion length L with the individual mobilities and the individual lifetimes and their products of Eqs. (8.8) and (8.9), we rewrite $\overline{D_{NP}}$ and τ_R according to

$$L = \sqrt{\overline{D_{NP}} \tau_R} = \sqrt{2 \frac{kT}{e} \frac{\mu_n \tau_n \mu_p \tau_p}{\mu_n \tau_n + \mu_p \tau_p}} \tag{8.37}$$

We can then associate either the electrons or the holes with the majority and minority carriers. Writing $(\mu\tau)_{min} = \mu_{min}\tau_{min}$ and $(\mu\tau)_{maj} = \mu_{maj}\tau_{maj}$ with the free-carrier mobilities μ_{min} and μ_{maj} and the free-carrier recombination lifetimes τ_{maj} and τ_{min} embraces the respective mobilities and lifetimes. The corresponding equation for L is

$$L = \sqrt{2 \frac{kT}{e} \frac{(\mu\tau)_{min} (\mu\tau)_{maj}}{(\mu\tau)_{min} + (\mu\tau)_{maj}}} \tag{8.38}$$

For the case that $(\mu\tau)_{min} \ll (\mu\tau)_{maj}$, Eq. (8.38) takes the form

$$(\mu\tau)_{min} \approx \frac{e}{2kT} L^2 \tag{8.39}$$

which allows the separate determination of $(\mu\tau)_{min}$. As a rule of thumb, a room-temperature value of $L = 200$ nm corresponds to $(\mu\tau)_{min} = 8 \times 10^{-9}$ cm^2 V^{-1}.

The factor of 2 stems from the common contribution of the full electron and hole densities in an intrinsic-type semiconductor, with $n \gg n_0$, $p \gg p_0$, to the recombination rate. This relation is also deduced independently without SSPG background for the surface-photovoltage experiment or collection lengths in solar cells [23]. If the above inequality is not met, the respective two unknowns $(\mu\tau)_{min}$ and $(\mu\tau)_{maj}$ can only be determined with the additional photoconductivity measurement and

$$\sigma_{ph} = eG\left[(\mu\tau)_{min} + (\mu\tau)_{maj}\right] \tag{8.40}$$

Equations (8.38) and (8.40) with the two unknowns allow the $(\mu\tau)_{min}$ and $(\mu\tau)_{maj}$ determination but no straightforward correlation whether the electrons or the holes are the majority/minority carriers.

In an analysis based on solutions for the free-carrier densities, Shah et al. [24] deduce the relation

$$L = \sqrt{C \frac{kT}{e} \frac{(\mu\tau)_{min} (\mu\tau)_{maj}}{(\mu\tau)_{min} + (\mu\tau)_{maj}}} \tag{8.41}$$

which is similar to Eq. (8.38). Here, C is a sample-dependent factor between 1 and 2 which depends on the photogeneration-rate dependence of the excess-carrier densities. For n-type a-Si:H samples, Shah et al. [25] estimate $C \approx 1$ so that a discrepancy of a factor of 2 exists in relation with Eq. (8.39). For G-independent lifetimes, C can be shown to be equal to 2.

The factor 2 of Eq. (8.39) also appears in the treatment by Schmidt and Longeaud [16] who suggest to fit Eq. (8.28) and apply Eq. (8.39) for an apparent $(\mu\tau)_{\min}^{\text{app}}$ which can then be corrected.

Abel et al. [15] derive the relation

$$\beta(\Lambda) = 1 - 2\gamma \frac{1 + \left(\frac{2\pi}{\Lambda}\left(2\frac{kT}{e}(\mu\tau)_p \frac{\mu_n \tau_d^{\text{rel}}}{(\mu\tau)_n}\right)^{1/2}\right)^2}{\left[1 + \left(2\pi\left(\frac{kT}{e}\left[2(\mu\tau)_p + \mu_n \tau_d^{\text{rel}}\right]\right)^{1/2}\Big/\Lambda\right)^2 + \left(2\pi\left(\frac{kT}{e}\left[(\mu\tau)_p \mu_n \tau_d^{\text{rel}}\right]^{1/2}\right)^{1/2}\Big/\Lambda\right)^4\right]^2}$$

(8.42)

simplified for the case that electrons are the majority carriers and that the E-related drift terms can be neglected. If the term $\mu_n \tau_d^{\text{rel}}$ with the effective dielectric relaxation time τ_d^{rel} can be neglected, because $\tau_d^{\text{rel}} = (n/N)\tau_d \ll \tau_d$, Eq. (8.42) reduces to the RZW relation (Eq. (8.23)) combined with Eq. (8.39) and $(\mu\tau)_{\min} = (\mu\tau)_p$, including the factor of 2.

Hattori et al. [14] suggest using additional information from frequency-dependent lifetime measurements for a correction of L from the RZW analysis.

8.3
Experimental Setup

Figure 8.3 sketches two of the possible arrangements for the measurements of the β parameter as a function of the grating period Λ. The two beams hit the sample S in the gap between the electrodes. In Figure 8.3a the sample is moved on a linear stage

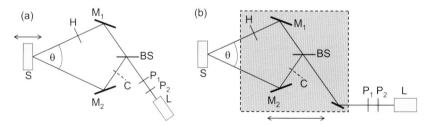

Figure 8.3 Schematics of experimental setups. The double arrows indicate that the sample is moved in (a) and an optical table is moved in (b)[1)] in order to change the angle between the two beams.

1) The set-up of b) has been devised by C. Longeaud and R. Brüggemann at LGEP, Paris and at the University of Oldenburg.

to change the angle between the two beams, split by the beam splitter BS. The mirrors M_1 and M_2 are rotated according to the sample position. The half-wave retardation plate H is positioned in the strong beam, alternatively placed between BS and M_1, in order to rotate the polarization. If H was in the weak beam, chopped by the chopper C, an intensity change by H-rotation would directly enter into the measured signal. In the strong beam, any intensity changes from the rotation of H would only change the light bias slightly. The optional polarization filters define the polarization and can also be used for a variation of Φ by rotation of the polarizer P_2, keeping P_1 fixed.

The linear polarization is typically changed from $0°$ to $+90°$ or $-90°$. A change from $-45°$ to $+45°$ can also be used in order to reduce the polarization-dependent reflection variations of L_1. A change of reflection upon rotating the polarization of L_1 is usually not taken further into consideration as it only slightly changes the bias level.

Figure 8.3b is different through the movement of an optical table on which most of the optical elements are arranged. The sample S is fixed which gives a different option for the cryostat, compared with Figure 8.3a. The advantage here is that the length of the light path is almost the same for every Λ.

Arrangements with one perpendicular beam can also be found in the literature. Niehus and Schwarz [26] achieve long Λ by introducing an additional beam splitter. One beam is transmitted and the second beam is reflected on the beam splitter surface close to the other's transmission position which defines small angles θ.

A glass prism was employed by Nowak [27] while a compact goniometer-like setup is used at Utrecht University where the sample is rotated with no linear movement involved [28].

For *T*-dependent measurements, a suitable cryostat is incorporated into the setups. One should check that the grating quality factor is not reduced by the influence of the cryostat window or additional vacuum-pump-related vibrations. For very thin samples, a cold finger with a hole may be helpful as a beam dump for the transmitted light.

The photocurrent is typically measured by an LIA. The photocurrents under coherent and incoherent illumination conditions for Eq. (8.21) could also be measured in the steady state, for example, by an electrometer but usually the LIA measurements will result in lower experimental errors.

Figure 8.4 sketches a strategy for the alignment of the two beams if their diameters are smaller than the electrode length. After adjusting L_1 and L_2 independently horizontally for maximum photocurrent signal, the vertical position must be optimized for overlap. A slight mismatch in the vertical direction may be difficult to assess by inspection with the eye. However, better alignment can be achieved by adjustment of the vertical position of L_2, as indicated by the double arrow in Figure 8.4a, and by monitoring the photocurrent or the LIA signal under coherent illumination. Best overlap is achieved when the LIA signal shows a minimum for maximum interference, as sketched in Figure 8.4b if β is positive.

For negative β, the phase shift of $180°$ must be taken into consideration. Depending on the experimenter's and the LIA options, the negative values of the real part or the amplitude can be maximized in this case.

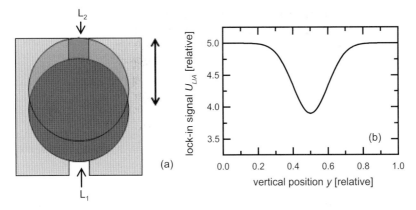

Figure 8.4 Schematic of the adjustment of L_1 and L_2 between the electrodes, (a). L_1 is fixed. The vertical movement of the lightly shaded L_2 results in the typical signal at the LIA, sketched in (b). The position of minimal signal results in optimized overlap, shown here for the Λ range with positive β.

One can always make a consistency check for the fitting results for L and Z. As pointed out in the literature [7, 18], the determined value of the fit parameter Z should be reasonable when comparing its value with $Z = \gamma \gamma_d \gamma_0^2$ of Eq. (8.28) with the experimentally determined γ and γ_d and a grating quality factor γ_0 close to 1. During the measurements, a check of an approximately constant $j_{\text{inc}}(\Lambda)$ or LIA-signal $U_{\text{inc}}(\Lambda)$ at the different grating periods is helpful. For the different measurement positions, the measured β data should decrease with increasing Λ. It may also be helpful initially, after having set-up the experiment, to compare the β values measured with two different setups in order to check the data [29, 30].

8.4
Data Analysis

The experimental results comprise a set of values of the parameter β, measured at a number of grating periods Λ. The (Λ,β) data can be evaluated according to one of the schemes from Eqs. (8.23) to (8.28). A comparison of the evaluation methods is discussed in [31]. Figure 8.5 illustrates graphical representations for long and short L. Through the large value for $L = 200$ nm and $Z = 0.8$, the β values in Figure 8.5a cover a wide range and do not even reach the limit $1 - 2Z = -0.6$ for long Λ. The β values with $Z = 0.1$ cover only a small range between 0.8 and 1; for $L = 200$ and 20 nm. This makes a precise L-determination difficult. There is almost no variation in β for $L = 20$ nm, even for $Z = 0.8$. Ritter et al. [3] consider that $L \approx 20$ nm can just be determined as a lower limit. It must be possible to measure at least some variation in β at the shortest experimental Λ to determine L. It is noted that results for determining β for $\Lambda < 500$ nm are difficult to achieve experimentally.

Figure 8.5b with the linear plots shows that the two cases for $Z = 0.1$ can be more easily distinguished compared to Figure 8.5a but it has been pointed out [31] that any

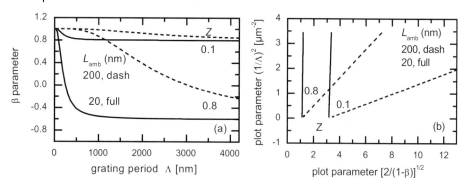

Figure 8.5 Analytical β–Λ plot (a) and linear plot (b) for $L = 200$ and 20 nm and $Z = 0.1$ and 0.8. The β values are almost constant for $L = 20$ nm (a). The almost vertical lines of the data points correspond to $L = 20$ nm (b).

small variation in the large β values would lead to a large error because of the $(1 - \beta)$ term in the denominator of the linear plots.

The two full lines for $L = 20$ nm and the two dashed lines for $L = 200$ nm extrapolate to the same ordinate value, respectively, which represents L. Figure 8.5b has thus very steep straight lines for $L = 20$ nm because the ordinate intercept is a large negative value.

Figure 8.6 illustrates the effect of measurement errors in the determination of β. It is more likely that errors in the beam adjustments yield β values that are rather too large due to incomplete overlap or nonideal grating. The artificial data points with the full circles in Figure 8.6 represent the ideal $\beta(\Lambda)$ variations in Figure 8.6a and b with $L = 140$ nm. The open symbols show the shift by increasing β by 15%.

The dashed curve in Figure 8.6a shows the fit to the open symbols and is only slightly different compared to the full line which is the fit of the full symbols. Because a constraint in a (β, Λ) fit is given by $\beta(\Lambda = 0) = 1$ according to Eq. (8.28), the fitting

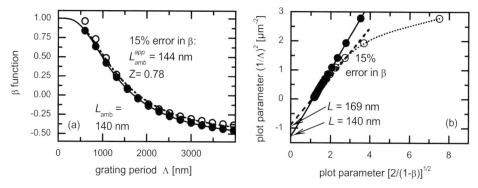

Figure 8.6 The symbols represent artificial measurement data (full symbols are ideal). A 15% error is imposed on β for $\Lambda > 500$ nm (upward shift, $L = 140$ nm, $Z = 0.8$, open symbols). The dashed curves show the fits in (a) and (b).

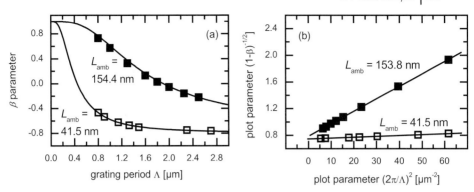

Figure 8.7 Experimental data in different representations in (a) and (b) for two undoped a-Si:H samples. There are data points with both negative and positive values of β.

procedure according to Eq. (8.28) is found to be less affected by too large a value of β at short Λ. The apparent L of 144 nm is close to the original value of 140 nm as is Z with Z = 0.78 instead of 0.8.

The nominally straight line in Figure 8.6b shows some curvature in the range of short Λ, that is, large ordinate values. Restricting the linear fit in the range of abscissa values <4 results in a diffusion length $L = 169$ nm, which is quite a large discrepancy of 29 nm. Only if one restricts the fitting to a more linear region with abscissa values <2, an apparent value of $L < 150$ nm is achieved.

Figure 8.7 shows typical experimental data of a device quality and a high-defect a-Si:H sample in the β(Λ) plot (Figure 8.7a) and in a linear plot (Figure 8.7b). There is very good agreement between the two evaluation schemes with $L = 154$ nm (and a good value of $\gamma_0 = 0.95$ from Z and γ) and $L = 41.5$ nm.

For the evaluation of the experimental data, the following analysis may be instructive. With say m values of (β, Λ), one can estimate the error in L by performing several fits with variations of $(m-1)$ or $(m-2)$ data points to determine the variation in L values. Single measurement data should be skipped, if too large a disagreement between the two evaluation schemes is observed. Especially for large values of β, any of the linear plots with the term $(1-\beta)$ in the denominator suffers from error enhancement and should be handled with caution.

It is noted at the end of this section that the sample alignment in the SSPG experiment is quite robust against small variations in the sample orientation. Experimental tests in our laboratory have shown that such rotations of the sample or the sample holder by a few degrees, do not lead to a significant variation in the measured β values. Unintentional misalignment of the sample is thus not severe. Any larger misalignment is easy to spot by the eye.

Substantial intentional rotation of the sample leads to a decrease of the effect of the modulation of the photocarrier grating. Balberg et al. [7] rotated the sample by 90° so that the photocurrent is parallel to the interference fringes. Meeting the requirements of the photocurrent measurement under incoherent conditions in this way, no half-wave plate is needed.

8.5
Results

This chapter presents and reviews results on a number of thin-film semiconductors. The limited and partially personal account that is given for each semiconductor is restricted to the photoconductive properties in relation to the SSPG method, that is, the minority-carrier properties in the steady state. As these are linked with the majority-carrier properties, the latter will also be discussed in comparison when appropriate. Typically, Eq. (8.39) is used to convert the experimental SSPG-derived L into $(\mu\tau)_p$.

8.5.1
Hydrogenated Amorphous Silicon

Hydrogenated amorphous silicon (a-Si:H) has been in the focus of SSPG applications since the early work by Ritter et al. [3, 32] The main issues in a-Si:H-related research with SSPG deal with the identification of deficiencies of the method, the physics of recombination and its relation to the DOS, the effect of doping on the excess-carrier properties, and the role of the E_f position in general.

8.5.1.1 Temperature and Generation Rate Dependence

The monotonous decrease of $(\mu\tau)_p$ with decreasing T [33] in Figure 8.8 is a signature in a-Si:H. To demonstrate at least one example of numerical-modeling results of the photoresponse and the minority-carrier properties in particular, Figure 8.8b shows

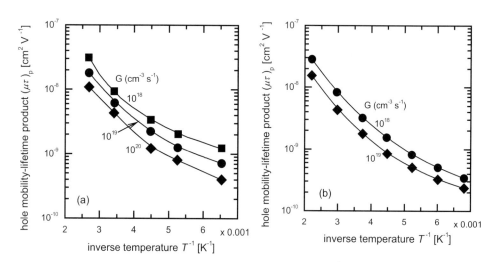

Figure 8.8 T dependence of $(\mu\tau)_p$ of an undoped a-Si:H sample at different G, after 33 (a). Numerical-simulation results with a free-hole mobility of $1\,cm^2\,V^{-1}\,s^{-1}$ which reproduce the decrease of $(\mu\tau)_p$ with decreasing T and increasing G, after [34] (b).

the simulated $(\mu\tau)_p$ that can roughly reproduce the experimental findings of Figure 8.8a [34].

At any T, $(\mu\tau)_p$ drops with increasing G. There has been some controversy in the literature about the consequences of the G dependence with respect to DOS models in the band gap of a-Si:H. Balberg and Lubianiker [35] suggested that one correlated dangling bond (DB) level is not sufficient to explain the experimental data while the Neuchatel group [36] pointed out that a proper balancing of charge in one correlated DB and in the band-tail states can also explain the experimental findings.

8.5.1.2 Surface Recombination

The application of Eq. (8.26) is illustrated in Figure 8.9 with data, measured with a HeNe laser and an Ar laser [18]. From the β versus Λ plot in Figure 8.9a the identification of deviations is not so easy, given that there is always an experimental error. A reduced L will be determined if all the circle-data points are taken for the fit.

In Figure 8.9b, there is a curvature in the data for the Ar illumination especially in contrast to the linear behavior of the squares. It is noted that the β data at the longer Λ coincide from the two measurements. This means, as noted by Haridim et al., that L can be determined in the usual way for longer Λ. For the HeNe measurements with the larger absorption depth all data points are suitable.

Haridim et al. also determined the two surface recombination velocities at the air/a-Si:H and the a-Si:H/substrate by illuminating from the air and substrate sides and by fitting the experimental data with Eq. (8.26) with suitable $g_s(\Lambda)$.

8.5.1.3 Electric-Field Influence

In the early publications on SSPG there has been concern with respect to the electric-field limitations in two ways: E modulation due to space charge and phase shift in the

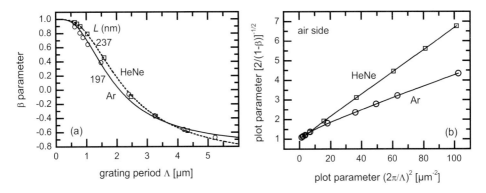

Figure 8.9 Digitized data from Haridim et al. [18] in a β versus Λ plot (a) and in the linear representation (b). The β values were determined under HeNe and under Ar illumination. A force fitting of the data results in a 40 nm different L of the latter. Note that at small abscissa and ordinate values in (b) the data points of the two kinds of experiments fall together.

modulation of $n(x)$ and $p(x)$ because E pulls these apart. Accounts on the "high"-E onset, the value at which such a deviation from the ideal behavior according to Eq. (8.28) occurs, have been summarized by Schmidt and Longeaud [16]. They point out that depending on the theoretical approach, different E values for the low-field limit have been determined. Experimentally, one may study the β dependence on E and take a constant value of β as an indicator of the low-field limit. At higher field values, a drift length has been determined from the field dependence [3, 21].

The analysis by Abel et al. [15], which includes both the effect of space charge and external electric field, has been applied to study the β dependence on Λ and E. Good agreement could be achieved between the experiment and the fit function (Eq. (8.42)) for $\beta(\Lambda, E)$ [37]. Accounts on the effective dielectric relaxation time and space charge could thus be given.

8.5.1.4 Fermi-Level Position

The aim of SSPG measurements with a-Si:H films was also to add complimentary information from the minority-carrier side to the majority-carrier properties from σ_{ph}.

Shortly after the SSPG proposal, the properties of boron-doped films were studied by Yang et al. [6]. Low-concentration boron doping was shown to lead to an increase in L. With higher doping level, L dropped drastically. These results were interpreted as a change in the minority-carrier type. For the higher doping levels and the shift of E_f toward the valence-band edge, the electrons become the minority carriers and the increasing defect density upon doping then leads to the reduction in L. An important observation was a benefit in L at small shifts of E_f toward mid-gap which was thought to be beneficial for the net i-layer carrier collection in a-Si:H based pin solar cells.

In undoped a-Si:H with $E_c - E_f$ in the range of 0.65–0.75 eV, where $E_c - E_f$ is calculated from $E_c - E_f = kT \ln(\sigma_0/\sigma_d)$, with the prefactor $\sigma_0 = 150$–$200\,\Omega^{-1}\,cm^{-1}$, the $(\mu\tau)_{maj}$ values are typically a factor of 50–100 higher than the $(\mu\tau)_{min}$ [17]. There is thus quite a variation in E_f for undoped a-Si:H samples and this variation expresses itself in a variation in the mobility-lifetime products [38]. Typically, $(\mu\tau)_{maj}$ increases with decreasing $E_c - E_f$ [39] accompanied by a decrease in $(\mu\tau)_{min}$. This anticorrelated behavior is also seen when the electrons are the minority carriers in p-type a-Si:H, where $(\mu\tau)_{min}$ decreases with increasing $E_c - E_f > 0.8$ eV [36, 40, 41, 46].

In a field-effect configuration, the Fermi level in the conduction channel can be changed by the gate voltage. This effect was exploited by Balberg and Lubianiker [36] and also Schwarz et al. [42] to study the E_f-dependent variation of σ_{ph} and L without being hampered by a doping-dependent increase in the DB density.

8.5.1.5 Defects and Light-Induced Degradation

The time-dependent decrease of the photocurrent upon illumination is a signature of many a-Si:H films. Any increase in the defect density should also have an influence on the minority-carrier properties so that SSPG may reveal the underlying recombination physics. It is known in the literature that E_f shifts toward mid-gap upon light-

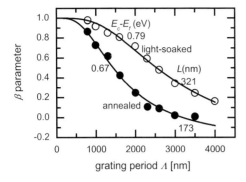

Figure 8.10 Increase of L upon light-soaking of a-Si:H, after [44].

soaking [43]. In view of the Fermi level related discussion above, it is illustrative to point out the possibility that L may increase upon light-soaking, as illustrated in Figure 8.10 [44]. Here, light-soaking with the concomitant increase in $E_c - E_f$ leads to an increase in L while new defects are being created. Defects become less negative and thus less attractive for hole capture, so that the minority-carrier properties improve.

It is usually reported that L does not change in the first hours of light-soaking – with different interpretations. Wang and Schwarz [45] reported that for a-Si:H films with a low DB density, the main recombination channel is via band-tail states. An increase in the DB density upon illumination thus does not lead to a drop in L in the early stage of degradation.

The interpretation by the Neuchatel group [21] was different in that they assumed that the main recombination channel is via DB. Because of conversion of the newly created DB into *neutral* DB they argued that these become then ineffective as the main capture process of holes is capture by *negatively* charged DB.

Sakata et al. [46] argued that errors have been made in the determination of L by Wang and Schwarz [45] because the measured L by SSPG should be corrected. They maintain that there is a drop in the true diffusion length also in the initial stages of degradation.

With respect to the influence of the $(\mu\tau)$ products on E_f and the reported change of E_f upon light-soaking, it appears necessary to always monitor the dark current, too. This has been done by Morgado [47] who determined $(\mu\tau)_p$, $(\mu\tau)_n$, and the E_f variation upon light-soaking and also annealing.

8.5.1.6 Thin-Film Characterization and Deposition Methods

SSPG has also been applied in order to characterize the samples that were deposited by newly developed deposition methods, for example, hot-wire chemical vapor deposition (HWCVD). For example, Mahan et al. [48], Unold et al. [49], Lubianiker et al. [50], and Feenstra [51] presented SSPG results with L in the range of 150–170 nm for typical device quality a-Si:H. These values were correlated with deposition parameters and, for example, with the hydrogen content of the films. Under specific deposition conditions, the so-called polymorphous silicon can be prepared. It has been shown to exhibit favorable minority-carrier properties [52], which enhances the red response in solar cells.

8.5.2
Hydrogenated Amorphous Silicon Alloys

One of the first reports on the application of SSPG was given by Bauer et al. [4] on hydrogenated amorphous silicon germanium (a-$Si_{1-x}Ge_x$:H) alloys. Typically, both σ_{ph} as a signature of the majority carriers and L as a signature of the minority carriers decrease with increasing x but they are affected in a different way by the concomitant shift of E_f toward mid-gap.

Abel et al. [53] report on a decrease in $(\mu\tau)_p$ with increasing x in the range of x up to about 0.43. The values of $(\mu\tau)_p$ stay $>10^{-9}\,cm^2\,V^{-1}$, that is, L of about 100 nm. Fölsch et al. [54] report slightly larger values in this x-range. These authors report a strong decrease in L for x larger than 0.5 at which the combination of increasing σ_d and decreasing σ_{ph} also makes L very difficult to measure.

The data by Gueunier et al. [55] on polymorphous silicon germanium alloys (pm-$Si_{1-x}Ge_x$:H) range between 100 nm and 130 nm, that is, $(\mu\tau)_p$ between about 10^{-9} and $3\times 10^{-9}\,cm^2\,V^{-1}$, for $x\leq 0.35$ at which the optical gap E_{04} (defined as the energy where the absorption coefficient α $(E_{04})=10^4\,cm^{-1}$) was 1.5 eV. Light-soaking was shown to only affect the majority-carrier $(\mu\tau)_n$ product. An L-value of 100 nm was reported by Bhaduri et al. [56] in the range $x=0.4$–0.5.

Yang et al. [57] determined L for hydrogenated amorphous silicon carbide films (a-$Si_{1-x}C_x$:H) of different band gaps and different feed stock gases. These authors find a correlation between the increase of L and better solar cell properties. They make it clear though that, as mentioned above, the sole value of L may not be the only indicator of the defect density of a sample. They deduce from the complementary σ_d- and σ_{ph}-results that E_f-related occupation of defects may have lead to higher values of L of some of their a-$Si_{1-x}C_x$:H samples.

Mohring et al. [33] reported on the T-dependent and G-dependent $(\mu\tau)$ products of majority and minority carriers in optimized a-$Si_{1-x}C_x$:H with an optical band gap up to 1.95 eV. For the alloys, there is a monotonous decrease of $(\mu\tau)_p$ with decreasing T, similar to the findings for a-Si:H. At any T, $(\mu\tau)_p$ decreases with increasing alloy content which has been related by Mohring et al. to the previous observation of increased DB density with increasing x. For a-$Si_{1-x}C_x$:H with a band gap of 1.94 eV, the room-temperature values for $(\mu\tau)_p$ were $(1.2)\times 10^{-9}\,cm^2\,V^{-1}$ and with about a factor of 10 higher for $(\mu\tau)_n$.

8.5.3
Hydrogenated Microcrystalline Silicon

The application of SSPG to hydrogenated microcrystalline silicon (μc-Si:H) is rather straightforward on the one hand as, for example, sample geometries and relevant absorption coefficients are not significantly different. On the other hand, there are some obstacles involved because of the typically higher σ_d and some possible effect of surface roughness, which reduces γ_0.

Goerlitzer et al. [58] and Droz et al. [59] find smaller Z in μc-Si:H because of the small ratio $\sigma_{ph.}/\sigma_d$. With respect to optical scattering, Z was shown to increase after polishing the rough surface of μc-Si:H samples [32].

The T dependence of $(\mu\tau)_p$ of μc-Si:H was studied by Brüggemann and Kunz [60], and also by Balberg [61]. For the analysis, they also performed numerical simulations in order to correlate the experimental and simulation results. Other work is devoted to relating L with the deep defect density [29, 62–64] and the crystalline volume fraction and the deposition technique [29, 65].

There are indications that transport is inhomogeneous in μc-Si:H with respect to the transport path parallel or perpendicular to the film-growth direction [66].

8.5.4
Hydrogenated Microcrystalline Germanium

Badran et al. [67] reported on L in microcrystalline germanium (μc-Ge:H). Because L was quite short, the conclusion was that μc-Ge:H appears suitable for diode applications as a sensor, as was demonstrated previously, under reverse bias but not as a solar cell.

8.5.5
Other Thin-Film Semiconductors

Thin-film chalcopyrite semiconductors were characterized soon after the SSPG method was proposed. Balberg et al. [5] determined the majority and minority-carrier properties of $CuGaSe_2$ thin films. Here, electrons are the minority carriers for which the authors determined L of 115 nm. The reliability of the data was increased by a self-consistency check by comparing the fit parameters Z from SSPG and γ from σ_{ph}. From the intensity dependence of $(\mu\tau)_p$ it was concluded that there is a sharp drop in the DOS close to the band edges – in contrast to a-Si:H.

Menner et al. [68] report SSPG results on $CuGaSe_2$ thin films and determined the highest L for a Cu/Ga ratio at 0.95, that is, close to the stochiometric point. SSPG measurements in the Cu-rich region were impossible because of the high σ_d. Menner et al. also aimed at revealing inhomogeneity in the growth process by application of a laser with a short absorption depth and by illumination from either the air/film and the substrate/film side. They concluded that a better crystal structure exists close to the substrate.

Belevich and Makovetskii [69] give an account of L measurements in polycrystalline p-type $CuInSe_2$. From their analysis they determine G-dependent minority-electron diffusion lengths between 360 and 420 nm, decreasing with increasing G. They relate the values of L to the dimension of the crystallites in the polycrystalline film.

Generally, it is noted that one should not be too optimistic with respect to the application of SSPG to the $Cu(In,Ga)Se_2$ system as it may be hampered by the large σ_d in $Cu(In,Ga)Se_2$, for which Zweigart et al. [70] determined L for some chemical compositions, only. Because the dark-current activation energy is also low, a decrease of T will only help if σ_{ph} does not decrease too much as well.

8.6
Density-of-States Determination

Schmidt and Longeaud [16] developed an analysis of SSPG with emphasis on the DOS determination at the quasi-Fermi energy of the majority carrier. For the energy scale, the quasi-Fermi energy can be shifted by a variation of T or G, so that a scan in energy can be performed. The experimentally determined DOS values are given by a combination of photocurrent and SSPG measurements. The application of the method was demonstrated on a-Si:H [16] and μc-Si:H [71].

8.7
Summary

We have given an account of the SSPG method with some emphasis on the setup, on the necessary and complementary measurement steps, and on consistency checks with the aim to obtain a more complete picture on the excess-carrier properties. The evaluation of the experimental data on the basis of the RZW approach provides the basis of determining L when the experiment is performed under the necessary condition of small modulation depth at low E, short enough dielectric relaxation time and sufficiently high photocurrent values with distinctly different values between $(\mu\tau)_{min}$ and $(\mu\tau)_{maj}$. A simple relationship between L and $(\mu\tau)_{min}$ can then be applied.

Reference has been made to alternative and additional analyses by Ritter *et al.* [3], Balberg *et al.* [10, 11], Li [12], Hattori *et al.* [14], Abel *et al.* [15], Schmidt and Longeaud [16], and shah *et al.* [24] which cover possible shortcomings, limitations, and additional aspects. Some of the relevant literature on the applications of SSPG on thin-film amorphous and microcrystalline semiconductors and on chalcopyrite and other semiconductors was sketched. The examples illustrate the potential of the method for the characterization of the photoelectronic properties of thin-film semiconductors.

References

1 Ritter, D., Zeldov, E., and Weiser, K. (1986) Steady-state photocarrier grating technique for diffusion length measurement in photoconductive insulators. *Appl. Phys. Lett.*, **49**, 791.

2 Ritter, D. and Weiser, K. (1986) Ambipolar drift-length measurement in amorphous hydrogenated silicon using the steady-state photocarrier grating technique. *Phys. Rev. B*, **34**, 9031.

3 Ritter, D., Weiser, K., and Zeldov, E. (1987) Steady-state photocarrier grating technique for diffusion-length measurement in semiconductors – theory

and experimental results for amorphous-silicon and semiinsulating GaAs. *J. Appl. Phys.*, **62**, 4563.

4 Bauer, G.H., Nebel, C.E., and Mohring, H.-D. (1988) Diffusion lengths in a-SiGe:H and a-SiC:H alloys from optical grating technique. *Mater. Res. Soc. Symp. Proc.*, **118**, 679.

5 Balberg, I., Albin, D., and Noufi, R. (1989) Mobility-lifetime products in $CuGaSe_2$. *Appl. Phys. Lett.*, **54**, 1244.

6 Yang, L., Catalano, A., Arya, R.R., and Balberg, I. (1990) Effect of low-level boron doping and its implication to the nature of gap states in hydrogenated amorphous-silicon. *Appl. Phys. Lett.*, **57**, 908.

7 Balberg, I., Delahoy, A.E., and Weakliem, H.A. (1988) Self-consistency and self-sufficiency of the photocarrier grating technique. *Appl. Phys. Lett.*, **53**, 992.

8 Liu, J.Z., Li, X., Roca i Cabarrocas, P., Conde, J.P., Maruyama, A., Park, H., and Wagner, S. (1990) Ambipolar diffusion length in a-Si-H(F) and a-Si,Ge-H,F measured with the steady-state photocarrier grating technique. Conference Record 21st IEEE Photovoltaic Specialists Conference, p. 1606.

9 Ritter, D., Zeldov, E., and Weiser, K. (1988) Ambipolar transport in amorphous-semiconductors in the lifetime and relaxation-time regimes investigated by the steady-state photocarrier grating technique. *Phys. Rev. B*, **38**, 8296.

10 Balberg, I. (1990) The theory of the photoconductance under the presence of a small photocarrier grating. *J. Appl. Phys.*, **67**, 6329.

11 Balberg, I. (1991) Theory of the small photocarrier grating under the application of an electric-field. *Phys. Rev. B*, **44**, 1628.

12 Li, Y.M. (1990) Phototransport under the presence of a small steady-state photocarrier grating. *Phys. Rev. B*, **42**, 9025.

13 Hubin, J., Sauvain, E., and Shah, A.V. (1989) Characteristic lengths for steady-state transport in illuminated, intrinsic a-Si:H. *IEEE Trans. Electron. Devices*, **36**, 2789.

14 Hattori, K., Okamoto, H., and Hamakawa, Y. (1992) Theory of the steady-state-photocarrier-grating technique for obtaining accurate diffusion-length measurements in amorphous-silicon. *Phys. Rev. B*, **45**, 1126.

15 Abel, C.-D., Bauer, G.H., and Bloss, W.H. (1995) Generalized theory for analytical simulation of the steady-state photocarrier grating technique. *Philos. Mag. B*, **72**, 551.

16 Schmidt, J.A. and Longeaud, C. (2005) Analysis of the steady-state photocarrier grating method for the determination of the density of states in semiconductors. *Phys. Rev. B*, **71**, 125208.

17 Brüggemann, R. (1998) Steady-state photoconductivity in a-Si:H and its alloys, in *Properties of Amorphous Silicon and its Alloys* (ed. T. Searle), INSPEC, London, p. 217.

18 Haridim, M., Weiser, K., and Mell, H. (1993) Use of the steady-state photocarrier-grating technique for the study of the surface recombination velocity of photocarriers and the homogeneity of hydrogenated amorphous-silicon films. *Philos. Mag. B*, **67**, 171.

19 Nicholson, J.P. (2000) Fresnel corrections to measurements of ambipolar diffusion length. *J. Appl. Phys.*, **88**, 4693.

20 Blakemore, J.S. (1987) *Semiconductor Statistics*, Dover, New York.

21 Sauvain, E., Shah, A., and Hubin, J. (1990) Measurement of the ambipolar mobility-lifetime product and its significance for amorphous silicon solar cells. Conference Record 21st IEEE Photovoltaics Specialists Conference, Orlando, p. 1560.

22 Balberg, I. and Weisz, S.Z. (1991) Identification of nonambipolar transport in the application of a photocarrier grating to hydrogenated amorphous-silicon. *Appl. Phys. Lett.*, **59**, 1726.

23 Moore, A.R. (1984) *Semiconductors and Semimetals*, vol. 21 (ed. J.I. Pankove), Academic, Orlando, Part C, p. 239.

24 Shah, A., Sauvain, E., Hubin, J., Pipoz, P., and Hof, C. (1997) Free carrier ambipolar diffusion length in amorphous semiconductors. *Philos. Mag. B*, **75**, 925.

25 Shah, A., Hubin, J., Sauvain, E., Pipoz, P., Beck, N., and Wyrsch, N. (1993) Role of dangling bond charge in determining μτ products for a-Si:H. *J. Non-Cryst. Solids*, **164–166**, 485.

26 Niehus, M. and Schwarz, R. (2006) Diffusion lengths in GaN obtained from steady state photocarrier gratings (SSPG). *phys. status solidi (c)*, **3**, 2103.

27 Nowak, M. and Starczewska, A. (2005) Steady-state photocarrier grating method of determining electronic states parameters in amorphous semiconductors. *J. Non-Cryst. Solids*, **351**, 1383.

28 K.F. Feenstra (1998) Hot-wire chemical vapour deposition of amorphous silicon and the application in solar cells, Ph.D. Thesis, Utrecht University.

29 Okur, S., Gunes, M., Goktas, O., Finger, F., and Carius, R. (2004) Electronic transport properties of microcrystalline silicon thin films prepared by VHF-PECVD. *J. Mater. Sci. – Mater. Electron.*, **15**, 187.

30 Balberg, I., Epstein, K.A., and Ritter, D. (1989) Ambipolar diffusion length measurements in hydrogenated amorphous silicon. *Appl. Phys. Lett.*, **54**, 2461.

31 Brüggemann, R. (1998) Improved steady-state photocarrier grating in nanocrystalline thin films after surface-roughness reduction by mechanical polishing. *Appl. Phys. Lett.*, **73**, 499.

32 Ritter, D., Zeldov, E., and Weiser, K. (1987) Diffusion length measurements in a-Si-H using the steady-state photocarrier grating technique. *J. Non-Cryst. Solids*, **97**, 571.

33 Mohring, H.D., Abel, C.D., Brüggemann, R., and Bauer, G.H. (1991) Characterization of high electronic quality a-SiC-H films by mu-tau products for electrons and holes. *J. Non-Cryst. Solids*, **137**, 847; and (1997) Erratum. *J. Non-Cryst. Solids*, **210**, 306.

34 Brüggemann, R. (1993) Ph.D. Thesis, Philipps-Universität Marburg.

35 Balberg, I. and Lubianiker, Y. (1993) Evidence for the defect-pool model from induced recombination level shifts in undoped a-SiH. *Phys. Rev. B*, **48**, 8709.

36 Hubin, J., Shah, A.V., Sauvain, E., and Pipoz, P. (1995) Consistency between experimental-data for ambipolar diffusion length and for photoconductivity when incorporated into the standard defect model for a-Si:H. *J. Appl. Phys.*, **78**, 6050.

37 Brüggemann, R. and Badran, R.I. (2004) Electric-field dependence of photocarrier properties in the steady-state photocarrier grating experiment. *MRS Symp. Proc.*, **808**, 133.

38 Brüggemann, R. (2003) Parameters for photoelectronic characterisation and the Fermi level in amorphous silicon. *Thin Solid Films*, **427**, 355.

39 Anderson, W.A. and Spear, W.E. (1977) Photoconductivity and recombination in doped amorphous silicon. *Philos. Mag.*, **36**, 695.

40 Brüggemann, R., Abel, C.-D., and Bauer, G.H. (1992) Implications of the defect pool model for the simulation of a-Si:H solar cells, in *Proceedings of the 11th E.C. Photovoltaics Solar Energy Conference 12–16 October 1992, Montreux, Switzerland* (ed. L. Guimaraes et al..), Harwood Academic Publishers, Chur, p. 676.

41 Böhme, T., Kluge, G., Kottwitz, A., and Bindemann, R. (1993) Mobility-lifetime products of a-Si:H prepared at high deposition rate with triethylboron. *phys. status solidi (a)*, **136**, 171.

42 Schwarz, R., Wang, F., and Reissner, M. (1993) Fermi-level dependence of the ambipolar diffusion length in amorphous-silicon thin-film transistors. *Appl. Phys. Lett.*, **63**, 1083.

43 Beyer, W. and Mell, H. (1996) Comparative study of light-induced photoconductivity decay in hydrogenated amorphous silicon. *J. Non-Cryst. Solids*, **198–200**, 466.

44 Abel, C.D. (1993) Ph.D. Thesis, Universität Stuttgart, unpublished.

45 Wang, F. and Schwarz, R. (1992) *J. Appl. Phys.*, **71**, 791.

46 Sakata, I., Yamanaka, M., and Sekigawa, T. (1997) Relationship between carrier diffusion lengths and defect density in hydrogenated amorphous silicon. *J. Appl. Phys.*, **81**, 1323.

47 Morgado, E. (2002) Light-soaking and annealing kinetics of majority and

minority carrier mobility-lifetime products in a-Si:H. *J. Non-Cryst. Solids*, **299**, 471.

48 Mahan, A.H., Carapella, J., Nelson, B.P., Crandall, R.S., and Balberg, I. (1991) Deposition of device quality, low H content amorphous silicon. *J. Appl. Phys.*, **69**, 6728.

49 Unold, T., Reedy, R.C., and Mahan, A.H. (1998) Defects in hot-wire deposited amorphous silicon: Results from electron spin resonance. *J. Non-Cryst. Solids*, **227**, 362.

50 Lubianiker, Y., Balberg, I., Fonseca, L., and Weisz, S.Z. (1996) Study of recombination processes in a-Si:H by the temperature dependence of the two carriers phototransport properties. *MRS Symp. Proc.*, **420**, 777.

51 Feenstra, K.F., van der Werf, C.H.M., Molenbroek, E.C., and Schropp, R.E.I. (1997) Deposition of device quality amorphous silicon by hot-wire CVD. *MRS Symp. Proc.*, **467**, 645.

52 Kleider, J.P., Gauthier, M., Longeaud, C., Roy, D., Saadane, O., and Brüggemann, R. (2002) Spectral photoresponses and transport properties of polymorphous silicon thin films. *Thin Solid Films*, **403–404**, 188.

53 Abel, C.D. and Bauer, G.H. (1993) Evaluation of the steady-state photocarrier grating technique with respect to a-Si:H and its application to a-SiGe:H alloys. *Prog. Photovoltaics*, **1**, 269.

54 Fölsch, J., Finger, F., Kulessa, T., Siebke, F., Beyer, W., and Wagner, H. (1995) Improved ambipolar diffusion length in a-$Si_{1-x}Ge_x$:H alloys for multi-junction solar cells. *MRS Symp. Proc.*, **377**, 517.

55 Gueunier, M.E., Kleider, J.P., Brüggemann, R., Lebib, S., Cabarrocas, P.R.I., Meaudre, R., and Canut, B. (2002) Properties of polymorphous silicon-germanium alloys deposited under high hydrogen dilution and at high pressure. *J. Appl. Phys.*, **92**, 4959.

56 Bhaduri, A., Chaudhuri, P., Williamson, D.L., Vignoli, S., Ray, P.P., and Longeaud, C. (2008) Structural and optoelectronic properties of silicon germanium alloy thin films deposited by pulsed radio frequency plasma enhanced chemical vapor deposition. *J. Appl. Phys.*, **104**, 063709.

57 Li, Y.M., Fieselmann, B.F., and Catalano, A. (1991) Conference Record 22nd IEEE Photovoltaic Specialists Conference, Las Vegas, p. 1231.

58 Goerlitzer, M., Beck, N., Torres, P., Meier, J., Wyrsch, N., and Shah, A. (1996) Ambipolar diffusion length and photoconductivity measurements on "midgap" hydrogenated microcrystalline silicon. *J. Appl. Phys.*, **80**, 5111.

59 Droz, C., Goerlitzer, M., Wyrsch, N., and Shah, A. (2000) Electronic transport in hydrogenated microcrystalline silicon: Similarities with amorphous silicon. *J. Non-Cryst. Solids*, **266**, 319.

60 Brüggemann, R. and Kunz, O. (2002) Temperature dependence of the minority-carrier mobility-lifetime product for probing band-tail states in microcrystalline silicon. *phys. status solidi (b)*, **234**, R16.

61 Balberg, I. (2002) A simultaneous experimental determination of the distribution and character of the two band tails in disordered semiconductors. *J. Optoelectron. Adv. Mater.*, **4**, 437.

62 Brüggemann, R. (2005) Mobility-lifetime products in microcrystalline silicon. *J. Optoelectron. Adv. Mater.*, **7**, 495.

63 Brüggemann, R., Brehme, S., Kleider, J.P., Gueunier, M.E., and Bronner, W. Effects (2004) of proton irradiation on the photoelectronic properties of microcrystalline silicon. *J. Non-Cryst. Solids*, **338–340**, 77.

64 Günes, M., Göktas, O., Okur, S., Isik, N., Carius, R., Klomfass, J., and Finger, F. (2005) Sub-bandgap absorption spectroscopy and minority carrier transport of hydrogenated microcrystalline silicon films. *J. Optoelectron. Adv. Mater.*, **7**, 161.

65 Okur, S., Günes, M., Finger, F., and Carius, R. (2006) Diffusion length measurements of microcrystalline silicon thin films prepared by hot-wire/catalytic chemical vapor deposition. *Thin Solid Films*, **501**, 137.

66 Svrcek, V., Pelant, I., Kocka, J., Fojtik, P., Rezek, B., Stuchlikova, H., Fejfar, A., Stuchlik, J., Poruba, A., and Tousek, J.

(2001) Transport anisotropy in microcrystalline silicon studied by measurement of ambipolar diffusion length. *J. Appl. Phys.*, **89**, 1800.

67 Badran, R.I., Brüggemann, R., and Carius, R. (2009) Minority-carrier properties of microcrystalline germanium. *J. Optoelectron. Adv. Mater.*, **11**, 1464.

68 Menner, R., Zweigart, S., Klenk, R., and Schock, H.W. (1993) Ambipolar diffusion length in CuGaSe$_2$ thin-films for solar-cell applications measured by steady-state photocarrier grating technique. *Jpn. J. Appl. Phys. 1*, **32**, 45.

69 Belevich, N.N. and Makovetskii, G.I. (1994) Ambipolar diffusion and ambipolar carrier drift in CuInSe$_2$ films. *Semiconductors*, **28**, 988.

70 Zweigart, S., Menner, R., Klenk, R., and Schock, H.W. (1995) Application of the steady-state photocarrier grating technique for determination of ambipolar diffusion lengths in Cu(In,Ga)(S,Se)$_2$-thin films for solar cells. *Mater. Sci. Forum*, **173**, 337.

71 Souffi, N., Bauer, G.H., and Brüggemann, R. (2006) Density-of-states in microcrystalline silicon from thermally-stimulated conductivity. *J. Non-Cryst. Solids*, **352**, 1109.

9
Time-of-Flight Analysis

Torsten Bronger

9.1
Introduction

For thin-film semiconductors, time-of-flight (TOF) techniques allow for an accurate determination of the mobility of charge carriers, which in turn may lead to insight about the transport mechanisms and the structural properties of the material. Additionally, TOF results include information about the energetic distribution of localized states, which is of particular interest for the disordered structures often found in thin-film semiconductors.

Although it is not limited to them, TOF is especially suitable for low-conductivity materials (early intensively studied materials were highly disordered amorphous semiconductors such as vitreous As_2Se_3). Therefore, it is a good complement to other characterization methods such as conductivity or Hall measurements. However, it is also a quite challenging method because the experimental setup must be able to cover many orders of magnitude.

TOF as described in this chapter has been used since the late 1960s to learn about material properties in various types of semiconductors. Although first attempts to probe the drift of photogenerated carries go back to 1949 [1], it was 20 years later that the first viable setup of the TOF experiment became popular [2].

During the 1970s, thorough theoretical examination by various authors led to a good understanding of the underlying processes of a TOF experiment [25–27]. This development culminated in comprehensive theoretical models such as the multiple-trapping model around 1980, which were in good agreement with the experimental observations [16, 28].

Since then, these models have not been modified significantly. However, they were extended in order to extract even more information out of the different phases of TOF measurements, as well as temperature-dependent results.

Advanced Characterization Techniques for Thin Film Solar Cells,
Edited by Daniel Abou-Ras, Thomas Kirchartz and Uwe Rau.
© 2011 Wiley-VCH Verlag GmbH & Co. KGaA. Published 2011 by Wiley-VCH Verlag GmbH & Co. KGaA.

9.2
Fundamentals of TOF Measurements

The left-hand sketch in Figure 9.1 depicts the basic principle of such an experiment: An intrinsic semiconductor specimen is put between two contact plates, and a voltage is applied. Then, electron–hole pairs are generated as a two-dimensional sheet at one contact in the specimen. By the electric field of the voltage drop V across the sample, they are separated. One type of carriers immediately leaves the sample through the nearby contact. The other type is dragged through the whole sample thickness to the opposite contact.

During the flight through the sample, the excess charge carriers push an electrical potential in front of them. This causes a step in the electrical field. Without the charge, the electrical field in the specimen would be $F_0 = V/L$, with L being the transit length for the charge carriers (for the usual setup, this is identical to the sample thickness). The charge sheet at position $0 \leq x \leq L$ reduces F_0 behind it and increases F_0 in front of it. With F_{back} and F_{front} denoting the resulting electrical field behind and in front of the charge sheet, respectively, this leads to

$$F_{back} = F_0 - \frac{4\pi Ne}{\varepsilon_r \varepsilon_0 A}\left(1 - \frac{x}{L}\right),$$

$$F_{front} = F_0 + \frac{4\pi Ne\, x}{\varepsilon_r \varepsilon_0 A\, L}.$$

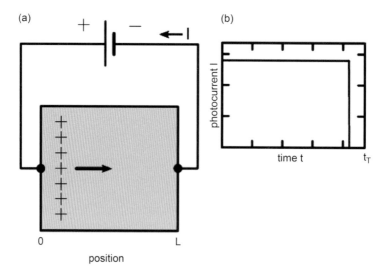

Figure 9.1 (a) Basic procedure of a time-of-flight experiment. (After Figure 1 in Ref. [3].) (b) Ideal photocurrent transient, in the absence of any kind of dispersion.

(ε_r and ε_0 are the relative permittivity of the material and the vacuum permittivity, respectively.)

Additionally, the drifting charge generates a displacement current in the external circuit. If one electron drifts through the whole specimen, exactly one electron is pushed through the external circuit. The same happens if two electrons pass half of the specimen.

This is illustrated in Figure 9.1b. As long as the carriers are drifting in the sample, one measures the current I in the external circuit. When they reach the contact, the current vanishes instantaneously. The integral of the current over time equals the generated excess charge. (Note, however, that we neglect the carriers of the opposite sign here. They would produce a delta peak at $t=0$ and double the integral.)

Thus, the actual measurement quantity in a TOF experiment is the decay of the electrical current with time. This leads to the transit time t_T, which in turn yields the drift mobility μ_d by

$$\mu_d = \frac{v_d}{F} = \frac{L^2}{t_T V}.$$

Here, v_d is the drift velocity.

There are two popular methods for determining the transit time t_T:

- One analyzes the photocurrent decay over time. A drop, or mostly a kink in the curve marks the transit time.
- One integrates the photocurrent and measures the collected charge over time. Then, the moment when half of the total charge is collected is the transit time.

Note that in general, both methods yield slightly different values for t_T. Therefore, it is highly advisable not to mix both within one series of measurements.

9.2.1
Anomalous Dispersion

In general, the photocurrent curve in Figure 9.1 is heavily idealized. Various effects modify its shape, and these modifications contain valuable information about the material. In particular, they may contain information about the energetic distribution of trap levels, as we will see later.

Figure 9.2 shows two more realistic transients. Carrier diffusion leads to "normal dispersion". In this case, the spacial distribution has a Gaussian shape, which broadens with time. The electrical current remains nearly constant during transit, and the transit time t_T can be easily defined as long as the width of the charge sheet is small compared to the sample thickness.

Besides diffusion, other effects also lead to normal dispersion. One example is hopping between localized states of the same energy level if the distance between neighbor states is always the same. Then, the release times of the carriers fluctuate due to the random nature of the hopping process. Another example is trapping of carriers in a single trap level. Here, the re-emission must be assisted by phonons,

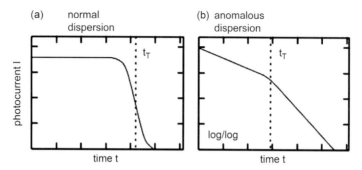

Figure 9.2 Two kinds of dispersive transport [4]. Obviously both cases are very different. While (a) consists of a constant phase followed by a sharp drop, (b) exhibits power-law behavior and has a much less distinct t_T. Note the logarithmic axes of the right-hand plot.

which is a statistical process. And last but not least, the initial width of the photogenerated charge sheet contributes to dispersion.

Typically for thin-film semiconductors, not only diffusion but also trapping in deep traps contribute to the spreading out of the electrical pulse. The resulting dispersion is called *anomalous dispersion*. While Section 9.4.1 will discuss the underlying physical effects in detail, the remainder of this section gives an overview of it.

As you can see in Figure 9.2b, anomalous dispersion is qualitatively different from normal dispersion. The electrical current strongly decays during the whole transit. Additionally, the current exhibits power-law behavior with different exponents. If the transit time t_T is defined as the kink in the curve, the part before t_T is called *pre-transit* or *first branch*, and the part after it is called *post-transit* or *second branch*.

In contrast to normal dispersion, the post-transit decays slowly. This implies that a large fraction of carriers reaches the contact during the post-transit. Thus, while for normal dispersion t_T is the average transit time of the carriers, this is not true for anomalous dispersion. Instead, t_T denotes the transit time for the fastest carriers here (see Ref. [5], Section 2.2).

Figure 9.3 shows a peculiar property of the current transient in the presence of anomalous dispersion: the *relative* dispersion is independent of the applied field and sample thickness. This means, if the time scale is normalized to the transit time t_T, the curves are congruent. This observation is called the "universality" of the current transient. The picture refers to measurements with different voltages; however, this universality can also be seen with different sample thicknesses.

As far as the underlying physical effects are concerned, it is typical of anomalous dispersion that the interaction of the excess carriers with the material does not reach a steady state during the experiment. Instead, as we will see later, the characteristic quantities (first and foremost the so-called demarcation energy) change continuously. This explains why the electrical current does not keep a constant level as in the case of normal dispersion.

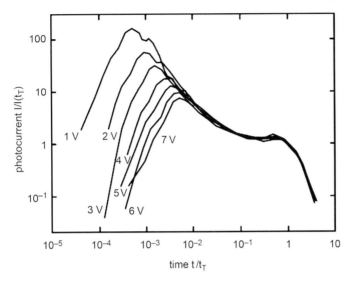

Figure 9.3 Universality of the current transient for different voltages.

9.2.2
Basic Electronic Properties of Thin-Film Semiconductors

Typically, mobilities in thin-film semiconductors are orders of magnitude smaller than in their bulk counterparts since their atomic arrangements are far away from the monocrystalline case. For example, while monocrystalline silicon has an electron mobility beyond $1000\,\text{cm}^2/\text{V}\,\text{s}$, amorphous or microcrystalline silicon only has $0.1\text{--}1\,\text{cm}^2/\text{V}\,\text{s}$.

Numerous kinds of deviations from the ideal crystal may occur. Defects, stacking faults, clusters, or loss of long-range order in an amorphous phase.

This leads to states in the forbidden gap of the semiconductor. Since these states represent local distortions of the atomic arrangement, they are *localized*. Carriers in these states cannot easily drift through the material. It is very important to note the fundamental difference to the *extended* states in the band: these states spread across the whole specimen, and drifting is easy. In amorphous semiconductors and some other thin-film materials, both domains are separated by the *mobility edge*. Below the mobility edge (toward midgap), carrier mobility drops very quickly.

Still, charge carriers in localized states can have a significant influence on the electronic properties of the material. For example, they can hop to neighbor states (close enough so that there is an overlap), especially at low temperatures, where almost no carriers occupy the extended states at higher energy levels anymore. Moreover, localized states can drag the Fermi level closer to midgap in doped semiconductors. And finally, they can trap carriers that are drifting in the extended states.

Therefore, it is interesting to know the density-of-states (DOS) of the material, both within the gap and close to the band edges.

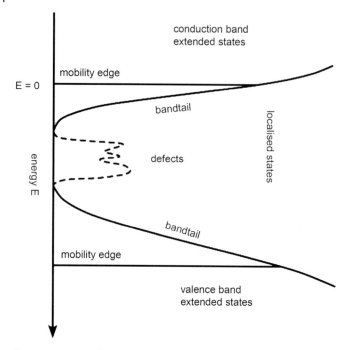

Figure 9.4 Model of the density of states in a disordered semiconductor.

Figure 9.4 shows a popular model of the DOS in a disordered semiconductor. The bands contain the extended states that you can also find in the ideal crystal. The band edges of the ideal crystal are replaced with the mobility edge. From the mobility edge toward midgap, the so-called band tails contain localized states. Their DOS drops exponentially. In some cases, further states close to midgap are assumed due to defects.

9.3
Experimental Details

Today, the most popular setup for TOF experiments is the light-induced generation of electron–hole pairs close to the top layer in a vertically stacked p–i–n diode.

Figure 9.5 shows the basic concept: The sample layer is packed between two doped layers so that the whole stack forms a p–i–n diode. Ideally, the substrate is transparent, allowing for measuring the mobility of holes (illumination through the n-layer) and electrons (illumination through the p-layer) separately. The doped layers should be thin (<30 nm) so that they do not absorb the light pulse significantly.

Shortly before the light pulse, a reverse bias is applied to the sample. The blocking contact prevents electrical carriers entering the sample from the external circuitry. Thus, after the light pulse has generated the free carriers, only their transient through

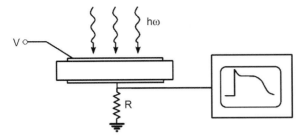

Figure 9.5 The basic principle used in vertical time-of-flight experiments. The resistor R builds a voltage divider together with the sample. This way, the current through the sample can be measured by the voltmeter to the right.

the sample is measured. Other techniques have been tested such as thin isolators between the contact pads and the sample, but diode configurations using p–n or Schottky junctions work more efficiently.

As far as the sample thickness and light pulse duration are concerned, several boundary conditions must be considered. First, the penetration depth of the light pulse – which determines the width of the generated charge sheet – should be much smaller than the sample thickness because otherwise, the features of the current transient are blurred, and they are distorted by the transient of the charge carriers of the opposite sign. Besides, some assumptions in the theoretical models become invalid and interpretation of the experimental data becomes more difficult.

Second, the duration of the light pulse must be smaller than the transient time, that is, the time the charge carriers need to cross the sample, because a too long pulse means that the experiment has no well-defined starting point. The characteristic features of the TOF experiment are transformed to a steady-state experiment with less information.

And third, the sample must be thin enough so that recombination of the photogenerated carriers is insignificant. Otherwise, the results are influenced by an additional unknown quantity and thus more difficult to interpret.

The contacts may be connected to the external circuitry with metallic pads or conductive silver paste. If the substrate is transparent for illumination, a transparent conductive oxide (TCO) like ZnO, SnO, or ITO between the contact and the substrate may help contacting the back contact.

Figure 9.6 shows a full and concrete setup for a TOF experiment. Every cycle starts with the laser emitting a short pulse. The laser is triggered either internally or externally (e.g., by a computer). The trigger frequency is of the order of seconds. Most often the laser is a nitrogen laser pumped dye laser. The wavelength should be chosen so that the penetration depth is small and the quantum efficiency is high. For silicon samples, 488 nm or 500 nm are typical values.

By a beam splitter, a small fraction of the laser pulse is redirected to a photodiode which triggers the function generator. The function generator applies the reverse bias to the sample – usually between 0 and 10 V. It is a rectangular pulse which should be shorter than the dielectrical relaxation time $t_d = \varepsilon_r \varepsilon_0 / \sigma$ (with σ being the

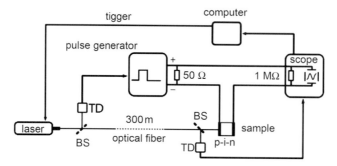

Figure 9.6 Exemplary setup for time-of-flight experiments. TD, trigger diode; BS, beam splitter.

conductivity) and much longer than the expected transit time. For example, for a 2-μm specimen with $\sigma = 10^{-9}$ S/cm, a good value for the length of the electrical pulse is 20 μs.

Then, the light is sent through 300 m of an optical fiber in order to retard the light pulse by 1 ms. This is done so that electrical disturbances due to switching on the voltage are diminished and the field in the sample is stabilized.

The light pulse must pass another beam splitter which reflects the light to a second photodiode, which triggers the oscilloscope. Immediately after that, the light pulse finally reaches the sample and is absorbed in its surface region. The photogenerated excess carriers drift through the external field, which sends the displacement current through the external circuit. This is measured and stored by the oscilloscope.

Finally, the function generator switches off the voltage. Till the next measurement cycle, the sample should have time to get rid of space charge which could affect subsequent transient curves. In the worst case, the transient curve features change gradually from shot to shot. Normally, a few seconds of waiting time is enough for sample relaxation. If this does not help, one can earth both contacts between the cycles while still sending a couple of laser pulses onto the sample so that the trapped charges can recombine with excited charges (Section 5 in [2]).

Typically, this measurement cycle is repeated a couple of thousands of times and linearly averaged in the oscilloscope or in a computer.

9.3.1
Accompanying Measurements

There are a couple of measurements that support TOF experiments. They help with evaluating the TOF results, as well as verifying that the sample and the setup are suitable for a TOF experiment.

9.3.1.1 Capacitance
Capacitance measurements (see also Chapter 4) help to assure that the externally applied field is uniform across the sample. The expected value is $C = \varepsilon_r \varepsilon_0 A / d$, where

ε_r, A, and d are relative permittivity of the material, the area of the contacts, and the sample thickness, respectively. Any deviation from this means that there is an electrical field within the sample, which may perturb the external field.

One can determine the capacitance by measuring the RC time constant of the current decay. For this, a rectangular voltage signal is applied to the sample as a reverse bias by a function generator. The switching off of the signal triggers an oscilloscope, which then stores the current decay through its input resistance R. The curve can be plotted half-logarithmically and RC can be extracted easily.

Another method is to connect the sample with its capacitance C_S and a known capacitor C_g in series. Then, the reverse bias voltage V is applied to both capacitors. Since charge cannot be created or destroyed, and it can only gather on the capacitor plates, both capacitors share the same amount of charge. This amount of charge can be determined by measuring the voltage drop V_g across the known capacitor. This yields

$$C_S = \frac{V_g C_g}{V - V_g}.$$

Figure 9.7 shows the voltage-dependent capacitance for four μc-Si:H samples. As you can see, two samples are very close to the expected result, whereas the two other samples exhibit way too high capacities. Additionally, these two samples have a voltage-dependent capacity.

For TOF experiments, a uniform field is required. For this, the depletion layer of the p–i–n diode should be the whole i-layer. However, this excess capacity means that the electrical field is not uniform. Instead, space charge in the i-layer, probably close to

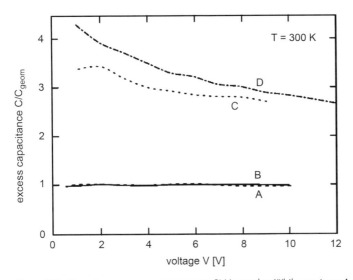

Figure 9.7 Capacitance measurements on μc-Si:H samples. While specimen A and B are close to the values expected from their geometries, C and D show significant, voltage-dependent deviations. After Ref. [6].

the contacts, confines it to a fraction of the sample thickness. Thus, the effective thickness is reduced, which results in a higher capacity. Higher voltages reduce the relative effect of the space charge but it does not vanish.

9.3.1.2 Collection

An important and simple test for a TOF setup is the total collection of charge carriers Q_0. It is the time integral of the photocurrent. Theoretically, the number of electrons drifting through the sample, and being eventually collected in one contact, must equal the number of photons absorbed in the sample, which can be estimated by measuring the energy of one laser shot. Of course, there are several effects that cause deviations from this ideal picture, most notably the quantum efficiency. Nevertheless, it is very helpful to check whether the total collection has the expected order of magnitude.

9.3.1.3 Built-in Field

The sample has a built-in voltage V_{bi} due to its p–i–n structure. This voltage should be determined because it enlarges the externally applied reverse-bias voltage. Thus, the actual drift mobility for a p–i–n diode is

$$\mu_d = \frac{d^2}{t_T(V + V_{bi})}.$$

Therefore, knowing V_{bi} is significant for accurate interpretation of the results, especially at low voltages. In particular, it allows for a transient measurement even without an external field. Conversely, such a transient measurement can be used to determine the built-in field itself [7].

Figure 9.8 visualizes how the determination of V_{bi} is realized. The inverse transit time over the applied voltages exhibits a linear dependency in the small-voltage regime. With a linear regression, one can extrapolate the voltage for which no transit takes place anymore. This is the built-in voltage.

Figure 9.9 shows another method for determining the built-in field, the so-called Hecht plot. It depicts the dependence of carrier collection on applied voltage. For small voltages, there is more or less linear behavior, which can be used to determine the built-in voltage, which is the voltage no carriers are collected at.

9.3.2
Current Decay

The best way to depict the photogenerated current decay is a log–log plot: the current is on the logarithmic y-axis, and the time is on the logarithmic x-axis. Most experimental and theoretical work has revealed power-law behavior of the photocurrent in disordered semiconductors, including the multiple-trapping model that will be explained later, so these axes help to visualize the dependency properly.

Figure 9.10 is such a plot for a 4.3-μm sample of μc-Si:H. The current decay is clearly visible, as are the linear sections of the curves, namely the pre- and the post-

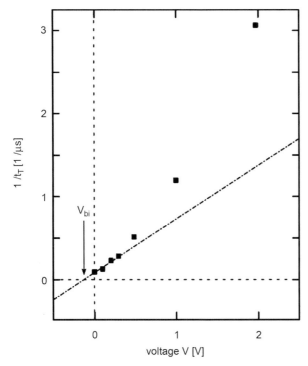

Figure 9.8 Dependence of inverse transit time on applied voltage. For small voltages, the dependence is linear. The extrapolation to infinite transit time yields the build-in voltage. After Ref. [8].

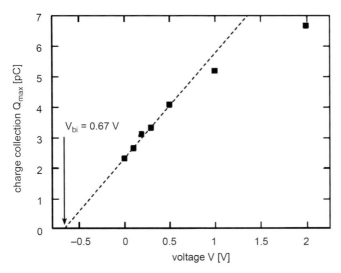

Figure 9.9 Dependence of total collection on applied voltage (Hecht plot). For small voltages, the dependence is linear. The extrapolation to vanishing charge collection yields the built-in voltage. After Ref. [8].

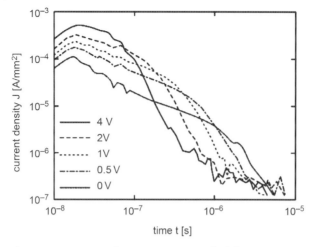

Figure 9.10 Transient photocurrent in µc-Si:H for different voltages. For higher voltages, the kink moves to the left to shorter transit times. After Ref. [9].

transit, also known as the first and the second branch. The kink between both branches denotes the transit time t_T. The transients become shorter for higher voltages, because the mobility is field-independent.

Figure 9.11 shows the current transient for different input resistances of the oscilloscope. One can achieve different input resistances by applying a parallel resistor to the input of the oscilloscope. Apparently, while higher resistances are good for the post-transit regime where they lead to less noise due to higher voltages

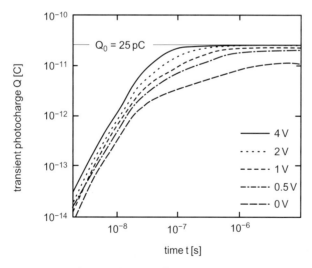

Figure 9.11 Schematic depiction of a sectioned transient photocurrent with different load resistors. The common right-hand part of the curves illustrates that one can increase the dynamic range of the measurement by using different resistors.

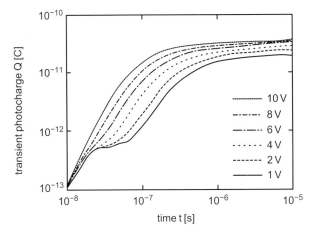

Figure 9.12 Normalized current transients for different voltages. The normalization leads to a common pretransit curve, which shows that all measurements share the same value of α. After Ref. [10].

for the same current, they suppress the first part of the curve. By stepping together the sections, one can construct a transient with a high dynamic range.

As we will see later, the pretransit consists almost exclusively of carriers that drift through the sample according to $v_d = \mu F$. Consequently, one expects ohmic behavior of the pretransit. Therefore, Figure 9.12 plots the current decay $I(t)$ for different applied voltages, normalized by that voltage to get $\tilde{I}(t)$:

$$\tilde{I}(t) = I(t) \frac{d^2}{Q_0(V + V_{bi})}$$

As you can see, there exists an envelope line which corresponds to a power law. In the multiple-trapping model that will be discussed below, this envelope is $\mu_d(t)$ (note that the y-axis has the unit of a mobility). By using this plot, the exponent of the pretransit can be determined rather easily. Additionally, [24] uses the crosspoint with the $0.8 \cdot \mu_d(t))$ line, in order to have a quite well-defined transit time.

9.3.3
Charge Transient

Figure 9.13 shows the transient photocharge measurement, i.e. the integrated photocurrent, on a sample at different applied voltages. As already explained, the upper limit of all curves, that is, the saturation value for $V \to \infty$, is called Q_0. Every single branch has its own saturation value $Q(\infty)$. The transit time t_T is reached when $Q_0/2$ is collected. Sometimes for small voltages, $Q_0/2$ is never reached. In this case, t_T cannot be determined by this method. Generally, high voltages lead to more accurate results for the charge collection.

As far as to choosing a method is concerned, whether the "kink" in the photocurrent or the point of half-charge, it depends on whether the kink is reliably

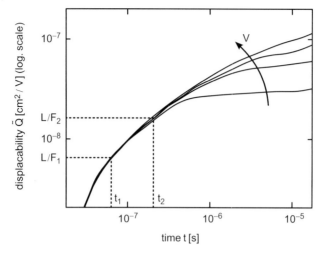

Figure 9.13 Charge collection of a μc-Si:H sample at various voltages. For higher voltages, the intersection with $Q_0/2$ moves to the left to shorter transit times. For 0 V, there is no intersection anymore. After Ref. [9].

observable. If it is, the photocurrent measurement is more accurate, especially for voltage-dependent examinations. If the kink cannot be seen well, use the half-charge method.

Figure 9.14 demonstrates how a nonuniform field within the sample distorts the photocharge measurement. The screening of the electrical field by the leading charge carriers causes a retarded transient, which can be seen in the dent of the curves. Such results do not allow for the determination of a transit time, which can be correlated with any of the available theoretical models.

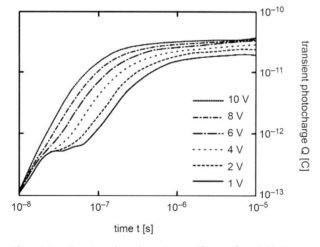

Figure 9.14 Transient photocharge in case of nonuniform field. The current breaks down already at very short times. Here, no accurate values for the transit time t_T can be extracted anymore. After Ref. [11].

9.3 Experimental Details

Figure 9.15 illustrates an alternative approach to determine the transit time. By normalizing the photocharge according to

$$\tilde{Q}(t) = Q(t) \frac{d^2}{Q_0(V + V_{bi})} = \frac{L(t)}{F}$$

and reducing the photocharge curves to the pretransit $t \ll t_T$, one can eliminate the voltage dependence. Thus, different voltages can be plotted together and form a common continuous curve. The resulting quantity L/F is called *displacability*, with L being the distance covered by the carriers (the *displacement*) and F being the electrical field in the specimen. By identifying L with the sample thickness, the transit time t_T can be extracted directly from the plots for various values of L/F [29].

Section 9.4.3.1 will explain how information about the band tails can be obtained from these plots, too.

9.3.4
Possible Problems

There are some caveats with TOF measurements. In the following sections, some of them are discussed.

9.3.4.1 Dielectric Relaxation

The dielectric relaxation is a reaction of the sample material to the applied voltage. The free carriers drift in the external field, building up their own field which weakens

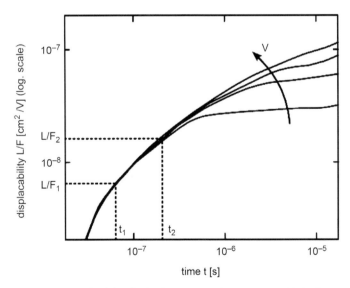

Figure 9.15 Displacability plot for the same sample as in Figure 9.13. By choosing the proper value of the displacability, that is, the ratio of sample thickness and electric field in the sample L/F, one can determine the transit time t_T for that case. After Ref. [12].

the external one. The more conductive the material is, the quicker this screening field is created:

$$t_d = \frac{\varepsilon_r \varepsilon_0}{\sigma_d}.$$

Here, t_d is the relaxation time, ε_r the relative permittivity, and σ_d the *dark* conductivity of the material. The latter is very important: We do not consider the photogenerated excess carriers here (they may distort the applied field by their own which is discussed below). Instead, we consider only the free carriers in the material which are in equilibrium. Therefore, we must use the dark conductivity.

The dielectric relaxation is a significant problem for the TOF experiment because by the partial screening of the external field, it is not well-defined anymore. This may render the results uninterpretable. Besides, the transient then takes extremely long and is blurred by carrier diffusion [13].

The only way to cope with it is to keep the measurement time short enough so that the relaxation cannot occur in the first place. Additionally, the time without the externally applied voltage must be much longer than the time with it. However, this is very simple to achieve.

By the way, concerning this aspect, the very small conductivity of most thin-film materials is an advantage. That is one of the reasons why early TOF experiments were mostly done with very low conductivity material: It was much simpler to ensure a homogeneous electrical field throughout the sample.

9.3.5
Inhomogeneous Field

In the previous section, we have already discussed a potential cause for an inhomogeneous electrical field in the sample, namely the dielectric relaxation. We also discussed why field homogeneity is important for being able to sensibly interpret the results of the TOF experiment.

There are also other reasons why the field may not be uniform. One of them are the photogenerated carriers themselves. If too many of them are generated (i.e., their area density is larger than that of the charge carriers in the contacts), those that are close to the opposite contact screen out the external field for those following them, which means that the drift saturates for high light intensities.

Figure 9.16 shows the current transient in this case, compared with an ordinary transient. The first surprising feature is the strong initial peak. It is possibly due to carrier diffusion, which is not affected by the reduced field, and which is especially strong in the beginning of the experiment, when the concentration gradient is large.

After the initial peak, the current goes through a minimum which is a typical feature of space-charge-limited in this case, the experiment must be repeated with attenuated laser light.

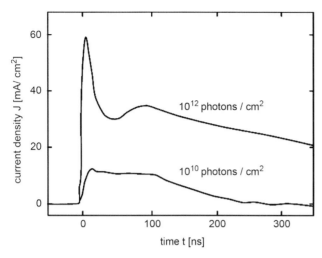

Figure 9.16 Effect of too many photoexcited carriers on the current transient. After Ref. [13].

9.4
Analysis of TOF Results

9.4.1
Multiple Trapping

The "multiple-trapping" model is a popular approach to explain the anomalous dispersive transport in disordered semiconductors. The transport properties are determined by three basic assumptions:

1) Actual transport takes place only in the extended states at the mobility edge.
2) Charge carriers in extended states may be trapped in localized states below the mobility edge.
3) By thermal excitation, trapped carriers may be re-emitted into the extended states (where retrapping is possible).

Then, the pronounced broadening of the packet of carriers in anomalous dispersive transport is mainly due to the exponential nature of the thermal excitation process out of the traps.

9.4.1.1 Overview of the Processes
The TOF experiment starts with the photoexcitation of carriers. At first, these excess carriers occupy states above the mobility edge in the conduction band. However, it takes only picoseconds for the carriers to thermalize down to the mobility edge (for amorphous silicon, see Ref. [14]).

Now, drift (and diffusion) at the mobility edge begins. However, every carrier has a certain probability to be trapped in a localized state. When this happens, it becomes totally immobile, and the only way to get out of the trap is by thermal excitation into the extended states. There, the carrier drifts again, but it may be trapped and re-emitted multiple times before it reaches the collecting contact.

Obviously, the time needed to be thermally excited out of a trap depends exponentially on trap depth. This makes a distinction between those carriers that have been trapped very many times, and those that have been trapped only once because it had been a deep trap.

As we will see later, this distinction is a pretty sharp energy level called demarcation energy. It separates the deep traps from the shallow traps. Carriers in shallow traps have continually switched between the shallow traps and the mobile extended states. (Unless of course, they were unlucky enough to fall into a deep trap.) By contrast, carriers in deep traps just stayed there. During the experiment, the demarcation energy moves downward, releasing carriers from the deep trap levels.

Throughout the model of multiple-trapping, the so-called dispersion parameter α determines the behavior. For exponential band tails, it is

$$\alpha = \frac{k_B}{\Delta E} T,$$

with ΔE being the width of the band tail. As we will see later, $\alpha = 1$ is an important special case, which is not a theoretical one: for typical values of the bandwidth, it lies in the middle of the experimentally accessible temperature range (approx. 100 °C).

9.4.1.2 Energetic Distribution of Carriers

The initial distribution of the photogenerated excess carriers is illustrated in Figure 9.17. Zero energy is at the mobility edge, increasing toward midgap. The solid line is the DOS. We assume here exponential band tails below the mobility gap, and the crystalline square root behavior above it.

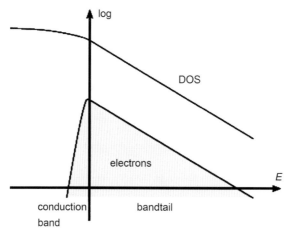

Figure 9.17 Distribution of carriers shortly after the very first trapping events (i.e., $t \approx \omega_0^{-1}$). In the tail, the density of electrons is proportional to the DOS.

The filled shape denotes the states occupied by excess carriers. In its basic form, the multiple-trapping model assumes the same capture cross section for all traps. This means that the fraction f of occupied states of a certain energy level is the same for all energies. Thus, the distribution of occupied states is parallel to the DOS on the logarithmic scale.

Note, however, that f is not constant in *time*. Instead, it obeys to a power law according to (see [15])

$$f(t) = \frac{N}{\Delta E} \frac{\sin(\alpha\pi)}{\alpha\pi} (\omega_0 t)^\alpha.$$

ω_0 is the attempt-to-escape frequency, which in first order equals the phonon frequency.

Let us make a rough estimation of the initial f for amorphous silicon: $\omega_0 t = 1$ (initial condition), $\alpha = 0.5$, $g_0 = 10^{22}\,\mathrm{cm}^{-3}\,\mathrm{eV}^{-1}$, $\Delta E = 50\,\mathrm{meV}$, and $N = \frac{10\,\mathrm{pC}}{e}/10^{-7}\,\mathrm{cm}^3 = 6.2 \times 10^{14}\,\mathrm{cm}^{-3}$ leads to $f = 8 \times 10^{-7}$. Furthermore, one can estimate that even after 1 ms, f is still only at $0.03 \ll 1$.

The mobility edge separates the mobile from the immobile carriers. At this early stage, a distinction between shallow and deep traps does not make sense.

Note that the theory of multiple trapping is valid only for $t \geq \omega_0^{-1}$. Actually, not before $t \gg \omega_0^{-1}$ the predictions are really reliable. This is due to the fact that some aspects of the theory average over many trapping events, and there must be a sufficiently large number of them for meaningful statistics.

Let us have a look at the distribution of carriers after some arbitrary time t as shown in Figure 9.18.

f has increased because more and more carriers were trapped in deep traps and stayed there. This is indicated by the arrow pointing upward.

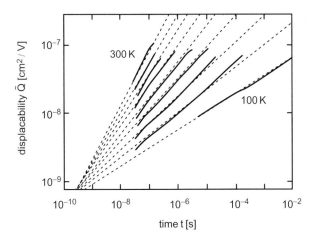

Figure 9.18 Distribution of carriers at the time $t \approx \omega_0^{-1}$. The demarcation energy E^* separates shallow traps to the left from deep traps to the right. The arrows illustrate the development of E^* and the density of electrons with time.

As already explained, carriers are constantly trapped and re-emitted. The emission probability strongly depends on the trap depth. If you look at the number of emission events by trap depth, it decreases exponentially. At some point, the number becomes smaller than one. Since a fraction of an emission event is impossible, this means a very sharp drop. The energy where this happens is the demarcation energy

$$E^* = k_B T \ln \omega_0 t. \tag{9.1}$$

The demarcation energy moves to deeper levels with time, as indicated by the arrow pointing to the right. Below that energy ($E > E^*$), there are *deep* traps with no emissions so far. Because we assumed the same capture cross section for all localized states, the density of carriers is still parallel to the DOS for the deep traps.

Above E^*, there are *shallow* traps with frequent trapping and emitting. This trapping and thermal emitting is much faster than all other actions in this model (e.g., carriers leaving the sample, shift of the distribution toward deeper energy levels). Therefore, the shallow traps simply are thermalized according to a Boltzmann distribution:

$$f_{\text{shallow}} = f \exp\left(\frac{E - E^*}{k_B T}\right).$$

This means that the distribution of carriers has a sharp maximum at E^* as can be seen in Figure 9.18 (mind the logarithmic scale). [16] concludes that this effectively simplifies the trap distribution to one single trap level. Thus, the mobility can be expressed in terms of the concentration of excess carriers in extended states n_{free} and excess carriers in traps n_{trap}:

$$\mu_{\text{TOF}} = \mu_0 \frac{n_{\text{free}}}{n_{\text{free}} + n_{\text{trap}}} \approx \mu_0 \frac{n_{\text{free}}}{n_{\text{trap}}}.$$

As a further simplification, we assume that the density of traps at $E = 0$ equals the effective DOS in the band. Then, the TOF mobility is

$$\mu_{\text{TOF}} = \mu_0 \alpha (1 - \alpha)(\omega_0 t)^{\alpha - 1}. \tag{9.2}$$

Note that due to $\alpha < 1$, μ_{TOF} is decreasing in time. This is due to the shift of E^* to deeper traps.

The resulting drift time for $\alpha < 1$ is

$$t_T = \omega_0^{-1} \left(\frac{1}{1-\alpha} \frac{L}{\mu_0 F}\right)^{1/\alpha}.$$

While these equations for μ_0 and t_T can be used for fitting experimental data (Section 9.4.3.1 will say more on this), the results do not differ significantly from those obtained by examining the current decay curves as described in Section 9.3.3. In practice, the latter method is much easier and comparably accurate.

9.4.1.3 Time Dependence of Electrical Current

The direct measurement quantity is the electrical photocurrent. Therefore, its time-dependence is of particular interest. From Eq. (9.2), it follows directly the pre-transit photocurrent

$$I(t < t_T) = \mu_0 \frac{eFg}{L} \alpha(1-\alpha)(\omega_0 t)^{\alpha-1},$$

where g is the total number of electrons injected at $t = 0$. However, we must also explain the so-called post-transit (or "second branch") of the transient curve, that is, the photocurrent decay after the transit time t_T. The qualitative change of the transport between pre- and post-transit is very important for understanding why one can define a transit time t_T at all.

After the transit, the single trap level at $E^*(t)$ has moved down so much that the DOS at this level is very small. It is small enough so that the trapping probability is much smaller than the emitting probability. In other words, emitted carriers almost always leave the sample before they are retrapped at $E^*(t)$.

This is in contrast to the pretransit when carriers could thermalize down to $E^*(t)$ due to very many trapping–emitting events. But during post-transit, they would not get that far. They rather leave the sample than being trapped in the region below $E^*(t_T) < E^*(t)$.

This makes the photocurrent during post-transit quite easy to describe: It simply consists of all thermally excited particles from the single-trap level (which still is the maximum of the carrier distribution). This is

$$I(t > t_T) = \mu_0 \frac{eFg}{L} \alpha(1-\alpha)(\omega_0 t_T)^{2\alpha}(\omega_0 t)^{-\alpha-1}. \tag{9.3}$$

9.4.2
Spatial Charge Distribution

The spatial distribution of charge carriers in the presence of anomalous dispersion surely is not the simple charge sheet drifting through the sample while widening symmetrically due to diffusion. In fact, the problem can only be solved numerically for the general case.

However, in [17], the special case $\alpha = 0.5$ is presented in detail because it can be calculated easily. Figure 9.19 shows the temporal development for the free excess carriers, assumed that $\alpha = 0.5$.

9.4.2.1 Temperature Dependence

Since temperature is the primary parameter for thermal activation, which in turn is responsible for the emission of excess carriers into the mobile states, the dependence of the TOF experiment on temperature is significant.

In the multiple-trapping model, different temperatures lead to different dispersion parameters α. As already said, for exponential band tails, both quantities are even proportional. Figure 9.20 shows the photocurrent in a log–log plot for a series

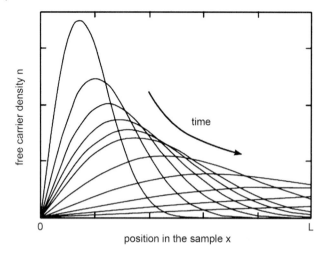

Figure 9.19 Spacial distribution of carriers at different times of the experiment.

of temperatures. Obviously, for $T \to 0$, the dispersion becomes more and more pronounced, whereas for $T \to \Delta E/k_B \Leftrightarrow \alpha \to 1$, the dispersion vanishes.

Although the equations presented so far assumed $0 < \alpha < 1$, the theory of multiple trapping can easily be extended to $\alpha \geq 1$, or in other words, to high temperatures.

Figure 9.21 depicts the energetic distribution of the excess carriers for exponential band tails if $T > \Delta E/k_B/ \Leftrightarrow /\alpha > 1$. Apparently, most of the carriers are now at the mobility edge. This is a qualitative difference to the low-temperature case. In particular, it means that there is no dispersive behavior anymore.

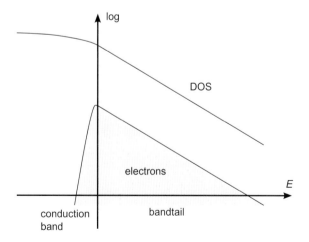

Figure 9.20 Transient curves for different temperatures. Obviously, the transit time t_T, if defined by the location of the kink, becomes larger for lower temperatures. After Ref. [18].

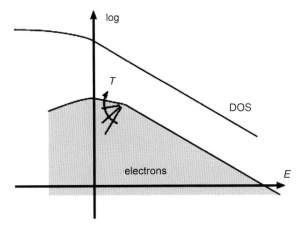

Figure 9.21 Distribution of carriers for high temperature $k_B T > \Delta E$. The maximum of the carrier distribution no longer is in the tail but at the mobility edge.

Consequently, the TOF mobility does not exhibit the time-dependence of Eq. (9.2), but instead [16]

$$\mu_{TOF} = \mu_0 \left(1 - \frac{\Delta E}{k_B T}\right),$$

and the transit time t_T does not depend on the electrical field or sample thickness. In order to determine μ_{TOF} this way, one must know the band tail width, though.

9.4.3
Density of States

Beyond the deep understanding of the processes in the material and the sensible definition of a carrier mobility, the multiple-trapping model also greatly helps in determining the DOS in the material.

Note, however, that you must first assure that multiple trapping is the dominating process. Especially the time and temperature dependence of the photocurrent should follow the expected behavior.

9.4.3.1 Widths of Band Tails
For calculating the band tail width, we only need to know α, which yields the tail width by $\Delta E = k_B T / \alpha$. The approach used here for determining α is the displacability because it is an easily accessible measurement quantity. At the same time, it can be calculated within the multiple-trapping model as follows.

Accoring to [15], the case of exponential band tails leads to

$$\mu_{TOF} = \mu_0 \frac{\sin(\alpha\pi)}{\alpha\pi} (\omega_0 t)^{\alpha-1}.$$

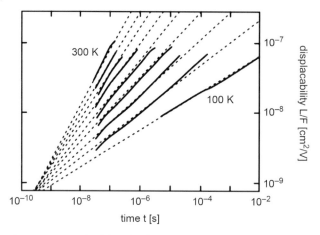

Figure 9.22 Displacability plots for different temperatures of a μc-Si:H sample with $d = 4.3$ μm and $V = 1$ V. The dashed lines are the fit curves. The fit yields μ_0, α, and ω_0 as fit parameters. After Ref. [20].

(Note that the prefactor in Schiff's calculation slightly differs from that in Eq. (9.2) because Schiff uses slightly different approximations.) By integration over time, this yields for the displacability [19]

$$\tilde{Q}(t) = \frac{\sin(\alpha\pi)}{\alpha\pi(1-\alpha)} \left(\frac{\mu_0}{\omega_0}\right)(\omega_0 t)^\alpha. \tag{9.4}$$

Figure 9.22 revisits the displacability plots of Section 9.3.3. This time, the temperature is varied. Eq. (9.4) can now be used to fit these curves. There is enough raw data so that one can obtain the fit parameters μ_0, ω_0, and α. Finally, the band tail width is obtained from α.

9.4.3.2 Probing of Deep States

The post-transit curve contains information to deduce the DOS deep in the mobility gap from it. The more orders of magnitude the measurement covers, the deeper it will be able to probe the gap. Ideally, the time axis should span at least six decades. Therefore, this method is not feasible for high-mobility samples because their post-transit cannot be determined over a sufficient time scale. As an illustration of this, it can be noted that the determination of the valence band tail of amorphous silicon succeeded earlier than that of the conduction band tail, due to the much lower hole mobility in this material [21].

The main property of the post-transit that is exploited here is that it is dominated by thermal excitation of carriers at the demarcation energy as described in Section 9.4.1.3. As this energy moves deeper into the gap with time, the electrical current generated by the collected carriers samples the DOS.

After deconvoluting the current signal with an inverse Laplace transform [21], the DOS is

$$g(E) = \frac{2g(0)I(t)t}{Q_0 t_0 \omega_0}, \quad t \equiv \frac{1}{\omega_0} \exp\left(\frac{E}{k_B T}\right). \tag{9.5}$$

Here, $g(0)$ is the DOS at the mobility edge, Q_0 the total collected charge as in Section 9.3.3, ω_0 the attempt-to-escape frequency, and t_0 the free transit time, that is, the transit time for carriers that have not been trapped. It can be estimated by $t_0 = d^2/\mu_0(V + V_{bi})$, with V being the applied voltage, V_{bi} the built-in voltage, and d the sample thickness.

An estimate for ω_0 for the material of interest can be taken from the fit of the previous section, or it can be calculated by applying Eq. (9.1) to the temperature activation of t_T as done in section C of [21]. A well-chosen ω_0 lets the DOS curves of different temperatures form a continuous line.

The conversion of time into energy of the right-hand side of Eq. (9.5) identifies E with the demarcation energy E^* from Eq. (9.1). The assumption is derived from the fact that E^* marks a strong peak in carrier distribution.

Since Eq. (9.5) is equivalent to $g(e) \propto I(t) \cdot t$, a photocurrent decay of $I(t) \propto t^{-1}$ means a constant DOS (on the energy scale). Similarly, any deviation above t^{-1} stands for a raise of the DOS, and below it for a decay.

Note that this matches well with Eq. (9.3), which suggests $t^{-\alpha-1}$ for the expected time dependency of the current. From $g(E) \propto t^{-\alpha-1}t = t^{-\alpha}$ and $\alpha = k_B T/\Delta E$ follows

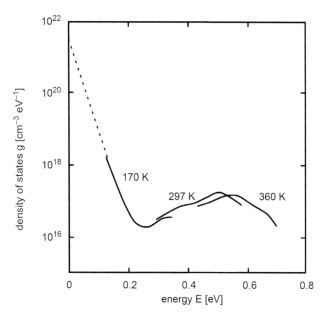

Figure 9.23 The DOS at the conduction band of μc-Si:H. The three TOF measurements at different temperatures can be used as patches for a continuous DOS curve. The dashed line is the extrapolated slope of the band tail [23].

$$g(E) \propto \left[\exp\left(\frac{E}{k_B T}\right)\right]^{-k_B T/\Delta E} = \exp(-E/\Delta E).$$

which is exactly the assumed slope of the band tail.

Due to approximations made in its derivation, Eq. (9.5) smoothes all sharp features in the DOS curve. More precisely, the curve appears as if a mollifier with a Lorentzian kernel of width $k_B T$ was applied to it. While this diminishes its usefulness for the deep-states region only slightly, it may distort band tails, except at low temperatures. More sophisticated techniques exist which do not have this limitation [22] but Eq. (9.5) is sufficient for most practical purposes.

The temperature-dependent mapping of the time axis to the energy axis by the right-hand side of Eq. (9.5) means that by varying temperature, different energy intervals of the DOS can be sampled. At low temperatures, it may even be possible to probe the rim of the band tail.

Figure 9.23 shows the DOS of a μc-Si:H sample, determined with the method as outlined in this section. The bump at 0.5 eV are defect states, and the drop on the left is the rim of the conduction band tail. As you can see, the three TOF measurements at three different temperatures combine to a quite continuous and plausible DOS curve.

References

1 Haynes, J.R. and Shockley, W. (1949) Investigation of hole injection in transistor action. *Phys. Rev.*, **75**, 691–691.

2 Spear, W. (1969) Drift mobility techniques for the study of electrical transport properties in insulating solids. *J. Non-Cryst. Solids*, **1**, 197–214.

3 Tiedje, T. (1984) Information about Band-Tail States from Time-of-Flight Experiments. *Semiconductors and Semimetals*, part C, Academic Press, vol. 21, pp. 207–238.

4 Dylla, T. (2004) Electron Spin Resonance and Transient Photocurrent Measurements on Microcrystalline Silicon, Ph.D. Thesis at Freie Universität Berlin, p. 16.

5 Marshall, J. (1983) Carrier diffusion in amorphous semiconductors. *Rep. Prog. Phys.*, **46**, 1235–1282.

6 Dylla, T. (2004) Electron Spin Resonance and Transient Photocurrent Measurements on Microcrystalline Silicon, Ph.D. Thesis at Freie Universität Berlin, p. 86.

7 Wyrsch, N., Beck, N., Meier, J. et al. (1998) Electric field profile in μc-Si:H pin devices. *Mater. Res. Soc. Proc.*, **420**, 801–806.

8 Lehnen, S. (2008) Elektronen- und Löcherbeweglichkeit am Übergang zwischen amorpher und mikrokristalliner Phase von Silizium. Diploma thesis Universität Köln, p. 52.

9 Dylla, T. (2004) Electron Spin Resonance and Transient Photocurrent Measurements on Microcrystalline Silicon, Ph.D. Thesis at Freie Universität Berlin, p. 91.

10 Dylla, T. (2004) Electron Spin Resonance and Transient Photocurrent Measurements on Microcrystalline Silicon, Ph.D. Thesis at Freie Universität Berlin, p. 92.

11 Dylla, T. (2004) Electron Spin Resonance and Transient Photocurrent Measurements on Microcrystalline Silicon, Ph.D. Thesis at Freie Universität Berlin, p. 88.

12 Dylla, T. (2004) Electron Spin Resonance and Transient Photocurrent Measurements on Microcrystalline

Silicon, Ph.D. Thesis at Freie Universität Berlin, p. 28.

13 Tiedje, T., Wronski, C.R., Abeles, B., and Cebulka, J.M. (1980) Electron transport in hydrogenated amorphous silicon: Drift mobility and junction capacitance. *Sol. Cells*, **2**, 301–318.

14 White, J.O., Cuzeau, S., Hulin, D., and Vanderhaghen, R. (1998) Subpicosecond hot carrier cooling in amorphous silicon. *J. Appl. Phys.*, **84**, 4984–4991.

15 Schiff, E.A. (1981) Trap-controlled dispersive transport and exponential band tails in amorphous silicon. *Phys. Rev. B*, **24**, 6189–6192.

16 Tiedje, T. and Rose, A. (1981) A physical interpretation of dispersive transport in disordered semiconductors. *Solid State Commun.*, **37**, 49–52.

17 Tiedje, T. (1984) *Time-Resolved Charge Transport in Hydrogenated Amorphous Silicon. The Physics of Hydrogenated Amorphous Silicon II: Electronic and Vibrational Properties*, vol. 56. Springer, Berlin, Heidelberg, pp. 261–300.

18 Dylla, T. (2004) Electron Spin Resonance and Transient Photocurrent Measurements on Microcrystalline Silicon, Ph.D. Thesis at Freie Universität Berlin, p. 94.

19 Dylla, T. (2004) Electron Spin Resonance and Transient Photocurrent Measurements on Microcrystalline Silicon, Ph.D. Thesis at Freie Universität Berlin, p. 95.

20 Dylla, T. (2004) Electron Spin Resonance and Transient Photocurrent Measurements on Microcrystalline Silicon, Ph.D. Thesis at Freie Universität Berlin, p. 96.

21 Seynhaeve, G.F., Barclay, E.P., Adriaenssens, G.J., and Marshall, J.M. (1989) Post-transit time-of-flight currents as a probe of the density of states in hydrogenated amorphous silicon. *Phys. Rev. B*, **39**, 10196–10205.

22 Main, C. (2002) Interpretation of photocurrent transients in amorphous semiconductors. *J. Non-Cryst. Solids*, **299–302**, 525–530.

23 Reynolds, S., Smirnov, V., Main, C. et al. (2003) Localized states in microcrystalline silicon photovoltaic structures studied by post-transit time-of-flight spectroscopy. *Maert. Res. Soc. Proc.*, **762**, 327–332.

24 Marshall, J.M., Street, R.A., and Thompson, M.J. (1986) Electron drift mobility in amorphous Si:H, *Philos. Mag. B*, **54** (1).

25 Scharfe, M.E. (1970) Transient photoconductivity in vitreous As_2Se_3, *Phys. Rev. B*, **2**.

26 Scher, H. and Montroll, E.W. (1975) Anomalous transit-time dispersion in amorphous solids, *Phys. Rev. B*, **12**.

27 Marshall, J.M. and Owen, A.E. (1971) Drift mobility studies in vitreous arsenic triselenide, Philosophical Magazine 24.

28 Orenstein, J. and Kastner, M. (1981) Photocurrent Transient spectroscopy: measurement of the density of localized States in a-As_2Se_3, *Phys. Rev. Lett.*, **46**.

29 Wang, Q., Antoniadis, H., and Schiff, E.A., Guha, S. (1993) Electron-drift-mobility measurements and exponential conduction-band tails in hydrogenated amorphous silicon-germanium alloys, *Phys. Rev. B*, **47**.

10
Electron-Spin Resonance (ESR) in Hydrogenated Amorphous Silicon (a-Si:H)

Klaus Lips, Matthias Fehr, and Jan Behrends

10.1
Introduction

Hydrogenated amorphous silicon (a-Si:H) is frequently used as absorber layers in thin-film solar cells [1]. Its electrical properties are controlled by defect states which, among other factors, limit the efficiency of the solar cells [2]. The most prominent defects in a-Si:H are threefold coordinated silicon atoms (dangling bonds, db) and strained (weak) bonds that induce localized states in the band gap and act as traps for charge carriers. Latter states are known to limit the electronic mobility. Electron spin resonance (ESR), also often named electron paramagnetic resonance (EPR), is one of the few experiments which give structural information about such defect states. The defect spin acts as a local probe which very accurately senses the local magnetic field distribution in its vicinity. This is monitored by the line position and line shape of the corresponding ESR spectrum or its dynamics. The purpose of this chapter is to introduce the reader without ESR experience to the basic principles of ESR and its application to a-Si:H.

A necessary prerequisite for ESR is that the electronic state to be investigated is paramagnetic. A typical example is the neutral state of a broken silicon bond. If this state is not paramagnetic, for example, the state is not occupied, one has to change the occupancy of the state by shifting the (quasi-) Fermi level. This can be accomplished by charge injection, illumination, changing measurement temperature, or bombardment of the sample with high-energy electrons or ions.

ESR is also capable of quantifying the number of defects in a given sample volume with a sensitivity of about 10^{10-11} spins at 9 GHz. This mode is particularly interesting to material researchers that want to optimize the material quality. The main advantage of ESR is that almost any kind of sample geometry can be studied nondestructively, provided that the sample fits into the resonator of the ESR spectrometer. Typical sample volumes that can be studied are $4 \times 4 \times 10\,\text{mm}^3$ and the sample's state can be gaseous, liquid or frozen solution, powder, crystal, or thin film. The conductivity of the sample has to be small enough in order not to affect the properties of the ESR resonator. When studying thin films or interfaces, the sensitivity limit of

Advanced Characterization Techniques for Thin Film Solar Cells,
Edited by Daniel Abou-Ras, Thomas Kirchartz and Uwe Rau.
© 2011 Wiley-VCH Verlag GmbH & Co. KGaA. Published 2011 by Wiley-VCH Verlag GmbH & Co. KGaA.

ESR is easily reached. Here special sample preparation techniques or alternative ESR methods such as electrically detected magnetic resonance (EDMR) are mandatory.

The following paragraphs will introduce the reader to the theoretical and experimental principles of continuous wave (cw) ESR focusing on those issues that are relevant for a-Si:H. In Section 10.2, the reader is introduced to the basic spectroscopic principle of ESR. In Section 10.3, the experimental details of ESR spectroscopy are presented with particular emphasis on the requirements that are necessary for studying thin-film samples. In Section 10.4, the influence of magnetic interactions and how they affect the ESR are discussed, whereas in Section 10.5, the main experimental results from the a-Si:H literature are briefly presented and discussed. For those readers that are interested in increasing the absolute sensitivity of the setup through electrical detection, they will receive in chapter 10.6 an easy to understand. For those who are interested in a more detailed treatise, we refer to the ESR textbooks of Atherton [3] or Poole [4].

10.2
Basics of ESR

A non-interacting electron in free space, often referred to as the free electron, is an ideal paramagnetic state with total spin, $S = 1/2$, to explain the basics of ESR spectroscopy. The free electron has a magnetic moment

$$\vec{\mu}_e = \gamma \vec{S} = -\frac{g_e}{\hbar} \frac{e\hbar}{2m_e} \vec{S} = -\frac{g_e}{\hbar} \mu_B \vec{S} \tag{10.1}$$

with γ the gyromagnetic ratio, $g_e = 2.002319$ the Landé factor of the free electron, $\mu_B = 5.788 \times 10^{-5}$ eV/T the Bohr magneton, which is determined by the mass of the electron, m_e. If the electron is captured in an atomic orbital, its g value changes dramatically due to interaction with the magnetic fields induced by the orbital motion [3]. If the electron, however, is captured in a db state in a-Si:H ($g_{db} = 2.0055$), only a 0.16% shift is induced since the orbital momentum of the db is quenched by the crystal field (cf. Section 10.4.1).

Without an external magnetic field, the energy level with respect to the orientation of the electron spin expressed by the magnetic quantum number, which is either $m_S = 1/2$ or $-1/2$, is degenerated but will be split by an external magnetic field,[1] $B_0 = \mu_0 H_0$, into two Zeeman levels with energy $E = m_S \mu_B g_e B_0$ as shown by the solid lines in Figure 10.1. Between these two Zeeman levels, magnetic dipole transition can be induced through electromagnetic radiation which is in resonance with the energy splitting

$$\Delta E = h\nu = \mu_B g_e B_0 \tag{10.2}$$

---

[1] Note that magnetic fields in ESR are expressed in units of the magnetic flux density B, which is in Tesla.

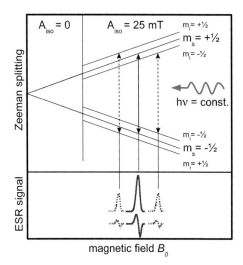

Figure 10.1 Energy levels of an electron ($S = 1/2$) as a function of the magnetic field for two cases: solid lines, without nuclear spin interaction ($m_I = 0$, $A_{iso} = 0$); dashed lines, with nuclear spin interaction ($m_I = \pm 1/2$, $A_{iso} = 25$ mT). Details of the hyperfine interactions are explained in Section 10.4.2. The predicted ESR resonances are shown in absorption (pESR) and derivative mode (cw ESR).

In ESR spectroscopy, the frequency of the electromagnetic radiation, typically microwaves (mw), is kept fixed while B_0 is swept, thereby bringing the spins into spin resonance. Due to historical reasons, the frequencies are classified in different mw bands with the following nomenclatures; S band: 2–4 GHz, X band: 8–12 GHz, Q band: 30–50 GHz, W band: 75–110 GHz. The selection rule for the ESR transition is $\Delta m_S = \pm 1$ and is observed in ESR as absorption (and sometimes as emission, not shown) of the mw at the resonant magnetic field B_0 (solid lines in Figure 10.1). The transition then is identified through the characteristic ESR spectrum, which directly appears as an absorption line when pulsed ESR (pESR) techniques are used or as the derivative in case of standard cw ESR. In X band, the required resonant magnetic field for free electrons is $B_0 \approx 0.35$ T inducing an energy splitting ΔE between the two spin levels of about 30 µeV, which corresponds to a temperature $T \approx 0.36$ K. The absorption of the mw is proportional to the occupation difference between the energetically upper and lower level in Figure 10.1. At thermal equilibrium, the fractional population of the upper and lower spin level can be easily calculated using Boltzmann's law as

$$\frac{N_{low} - N_{up}}{N_{low} + N_{up}} = \frac{\Delta N}{N} = \frac{1 - e^{-\Delta E/kT}}{1 + e^{-\Delta E/kT}} \tag{10.3}$$

with N the total number of spins in the measured volume, N_{low} and N_{up} the population of the lower and upper spin state, respectively, and k the Boltzmann factor. It becomes obvious from Eq. (10.3) that for $g_e = 2$, $T = 300$ K, and $B_0 = 0.35$ T the fractional population (spin polarization) $\Delta N/N \approx 0.0008$ whereas at $T = 4$ K this value increases to $\Delta N/N = 0.058$. This shows that even for X band and cryogenic conditions, the population of $S = 1/2$ states are still almost equally populated ($\Delta N/N \ll 1$). This implies that both spin orientations, up and down, are almost equally likely to exist and hence the net magnetization, M, being the sum of all individual spin moments that can be polarized in a magnetic field, is rather small. To achieve a spin polarization of 93%, it requires $B_0 = 10$ T and $T = 4$ K. Nevertheless, ESR can measure M with high accuracy even under environmental conditions by determining the mw absorption at spin resonance thereby fulfilling Eq. (10.2).

Since the mw absorption, I_{ESR}, is proportional to the population difference ΔN of both spin states, it can be easily shown from Eqs. (10.2) and (10.3) that [3]

$$I_{ESR} \propto \Delta N = N_{low} - N_{up} \approx N \frac{g_e \mu_B B_0}{2kT}, \text{ when } \frac{\Delta E}{kT} \ll 1 \quad (10.4)$$

According to Eq. (10.4), the mw absorption I_{ESR} (ESR response) is directly proportional to the total number of spins in the measured sample volume. Note that through the absorption process, the thermal equilibrium of the spin system is disturbed. The time that the spins need to relax back to thermal equilibrium is referred to as the spin-lattice relaxation time, T_1. This relaxation is accompanied by the emission of an mw photon and is a resonant process.

Since quite a number of experimental parameters such as mw power, mw-field distribution within an ESR cavity or sample morphology, and sample shape determine the ESR response, reference measurements with a sample of comparable shape and morphology with known spin concentration is required for calibration purposes. Details on the procedure on how to determine the db concentration of thin-film silicon can be found in Section 10.3. Since I_{ESR} increases linearly with resonance field B_0 (which is equivalent to increasing mw frequency) and reciprocal with temperature, the signal-to-noise (S/N) ratio of an ESR experiment can be increased by measuring at higher mw frequency and/or lower T. Note that Eqs. (10.3) and (10.4) are only valid for isolated spins such as db states which are not coupled (low spin concentration). Equation (10.4) which is also known as Curie's law is definitely not valid for spins that are occupying conduction- or valence-band states (such as free electrons).

For the understanding of a-Si:H, it is also very beneficial to study the magnetic properties of their nuclear spins (cf. Section 10.4.2). Nuclear spins introduce internal magnetic fields through the hyperfine interaction (HFI, dashed lines in Figure 10.1), which can be very sensitively measured. Nuclear spins in a-Si:H are ^{1}H, ^{29}Si, and ^{31}P in typical concentrations of 10 at.%, 4.7 at.%, and below 0.1 at.%, respectively. The principle concept discussed so far also holds for nuclear spins. In the above equations one only has to be replaced μ_B by μ_N (nuclear magneton) and ge by g_N, the nuclear g value. Since the Bohr magneton is proportional to the reciprocal of the mass of the particle (see Eq. (10.1)), μ_N is a factor of 1836 smaller than μ_B. Hence, the nuclear-spin polarization is approximately three orders of magnitude smaller compared to electron spins. At the magnetic field used in ESR, the nuclear spins can be considered to be completely unpolarized (equal population of the spin states). The spectroscopy of nuclear spins (nuclear magnetic resonance (NMR)) is therefore orders of magnitude less sensitive than ESR. On the other hand, the necessary frequency to induce a nuclear-spin transition (Eq. (10.2)) is also lower by three orders of magnitude and hence technically much easier to access and control. This is one of the reasons why time-domain NMR experiments were introduced decades before time-domain ESR.

10.3
How to Measure ESR

10.3.1
ESR Setup and Measurement Procedure

Figure 10.2 schematically shows the basic components of an ESR spectrometer. In contrast to most optical experiments, ESR is usually carried out as a reflection measurement in order to maximize the sensitivity. The sample is placed in an mw resonator, often referred to as cavity, which serves two purposes: first, the magnetic field, B_1, of the mw is well separated from the electric field, E_1, so that the detrimental interaction between the E_1 and the sample is minimized. Second, the cavity dramatically enhances B_1 – and concomitantly the sensitivity – at the sample position. The amplification of B_1 is quantified by the cavity quality factor Q (specifies how much energy can be stored in the resonator) which amounts to a few 1000 for a typical X-band ESR cavity. Q is strongly determined by the design of the cavity. The mw

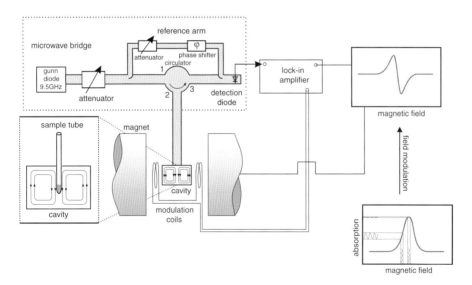

Figure 10.2 Simplified scheme of an ESR spectrometer. On the right side is shown how field modulation results in the derivative form of the ESR absorption signal. The B_1 distribution is indicated for a rectangular cavity. E_1 vanishes at the position of the sample in the center of the cavity at the maximum of B_1.

frequency is tuned to the resonance frequency of the cavity, and the coupling between the mw waveguide and the resonator is adjusted in such a way that no mw are reflected but the energy is completely stored in the cavity. This adjustment procedure, which is referred to as *matching* and *coupling*, must be carried out whenever the sample is changed or when the experimental conditions that have an influence on resonance frequency or resonator coupling are varied. When continuous wave (cw) mw radiation is absorbed by the sample at resonance, Q decreases. This leads to a change in the coupling and results in enhanced mw reflection from the resonator which is detected by an mw-sensitive detector (typically a diode or mixer). To ensure that only the reflected part of the mw reaches the diode, the signal is decoupled in a mw bridge configuration from the incoming mw radiation through a circulator (for details see Ref. [4]). The output current of the diode is recorded as a function of B_0. The current generated by the detector diode is proportional to the square root of the mw power, P_{MW}, only in a limited range (the *linear* regime). To ensure operation of the detector in this regime, a part of the mw power from the source bypasses the cavity (cf. Figure 10.2) and is directly guided to the detector via the *reference arm*. This mw bias is adjusted with respect to amplitude and phase in order to operate the detection diode in the optimum range and to match the phase at the detector.

Phase-sensitive detection can drastically improve S/N ratio because it minimizes the influence of noise or interfering signals. In analogy to optical experiments where the exciting light is chopped at a fixed frequency and lock-in detection is performed at the same frequency, it is likewise possible to modulate the mw amplitude in ESR. However, it turned out to be more favorable for most ESR experiments to modulate B_0 at frequency ν_{mod}. To this end a magnetic field with periodically changing amplitude (typically less than 1 mT) is superimposed with B_0. In consequence, the signal at the detection diode is modulated at the same frequency as well, and lock-in detection can be employed. Due to the magnetic-field modulation, the signal measured is the derivative of the ESR absorption signal with respect to the magnetic field, that is, dI_{ESR}/dB_0. Consequently, the resonance line shape is the derivative of the associated absorption line. In order to achieve optimum S/N ratio, the modulation frequency is often set to the maximum value which is 100 kHz for most spectrometers.

In cw ESR, the line width is usually determined as the peak-to-peak width of the derivative of the absorption line, ΔB_{PP}. It has to be kept in mind that field modulation can dramatically distort the measured resonance line shape. To obtain the true resonance signal, $1/\nu_{mod}$ has to be large compared to the spin-relaxation times in the sample. Obeying this rule is particularly important when performing cw ESR measurements on disordered silicon at cryogenic temperatures. Here, the standard value $\nu_{mod} = 100$ kHz is usually too high and hence experimentally obtained line shapes are likely to be affected. This is called *passage effect* [5], that is, the spins do not have sufficient time to relax between two successive modulation cycles. Further, the modulation amplitude, B_{mod}, should be smaller than the narrowest line in the spectrum. Otherwise overmodulation will distort and increase the observed line width. $B_{mod} < 0.3 \cdot \Delta B_{PP}$ is sufficient in case that line shapes have to be determined accurately. As a rule of thumb, a good compromise between spectral resolution and S/N is achieved when $B_{mod} \approx \Delta B_{PP}$.

In an ESR spectrometer, P_{mw} can be set by an attenuator (see Figure 10.2). At low P_{mw}, $I_{ESR} \propto \sqrt{P_{MW}}$, and ΔB_{PP} is independent of P_{mw}. However, upon increasing P_{mw} beyond a value, which strongly depends on the type of sample and the experimental conditions, the increase in the ESR signal with mw power becomes weaker and finally even decreases. In addition, the experimentally obtained line is broadened. Both effects are called *saturation*. It means that at high P_{mw}, the population between both energy levels is equalized and the transition is thus saturated (see Eq. (10.4)). The mw power level where saturation sets in is determined by the interplay between P_{mw} and the spin-relaxation times T_1 and T_2 (the latter will be defined in Section 10.4.3). It critically depends on the underlying line-broadening mechanisms (cf. Section 10.4.3). A detailed theoretical treatment of saturation is quite complex and beyond the scope of this chapter; a review can be found in Ref. [4]. The absence of saturation is a necessary prerequisite for determining line shapes and defect density. P_{mw} where saturation starts to influence the signal can easily be determined by measuring spectra for different power levels and analyze I_{ESR} as well as the line width as a function of $(P_{mw})^{1/2}$.

One of the main advantages of ESR is that the spin concentration can be determined, provided that the necessary experimental conditions are well established, that is, the sample volume contains enough spins. Increasing the sample volume generally leads to a gain in S/N until the sample starts to interact with E_1 which would lead to a reduction in Q. In consequence, obtaining optimum sensitivity on such "lossy" samples always requires a compromise between the number of paramagnetic centers and the sample volume that can be tolerated. Two categories of losses can be distinguished [6]: (i) dielectric losses in nonconducting samples which originate from mw absorption due to the imaginary part of the dielectric constant of the material, and (ii) losses in conducting samples due to surface currents induced by the mw. Both effects are relevant when studying semiconductor materials and result in a decrease in Q. This is of fundamental importance when performing quantitative ESR measurements and comparing a (lossy) sample with an unknown spin concentration to a reference sample containing a known number of spins. In this case, both samples either have to be measured under identical conditions (e.g., in a double cavity, for details see Ref. [4]) or Q has to be explicitly included in the calculation. When the sample contains only a thin lossy layer, for example, a transparent conducting oxide on top of a semiconducting solar-cell absorber layer, the reduction in Q can be minimized by exact positioning of the sample so that the conductive layer is situated in the minimum of the electric field (usually in the center of the cavity).

States which are not paramagnetic can often be made paramagnetic by illuminating the sample with light. This technique is commonly referred to as *light-induced ESR* (LESR). For LESR, one has to choose the appropriate light intensity and frequency. Note that often LESR signals, in particular at low T, persist even after switching off the light. To restore the dark situation, the sample has to be heated back to room temperature or illuminated with infrared light (IR) to depopulate the states. In cases when a white light source like a halogen lamp is used, it is important to cut off the IR part since this may otherwise lead to a simultaneous depopulation of the spin states.

To further increase the S/N ratio or to reduce effects due to spin-lattice relaxation, the samples can be cooled thereby enhancing $\Delta N/N$. Cooling is performed by placing the ESR sample in the flow of cooled N_2 or He gas. To decouple the sample from the environment, a glass Dewar is inserted into the center of the cavity. A heater in close proximity to the gas nozzle heats the cooling gas. Stabilization of sample temperature is achieved through a feedback loop with a temperature controller. It is worthwhile noting that water, which may condense in the cavity during low-T operation, leads to strong dielectric losses. To prevent the presence of water, the cavity should be purged with N_2 gas.

10.3.2
Pulse ESR

Although cw ESR is the appropriate technique for the extraction of relevant interaction parameters or spin densities, the vast potential of ESR spectroscopy for the experimental investigation of structure and dynamics, in particular in the presence of disorder in the material, cannot fully be explored with traditional cw methods. While the technical demands are higher for pulse (p) ESR, the experiments are often more easily interpreted and less prior assumptions are needed as compared to cw ESR. pESR relies on the coherent manipulation of the electron spin by short (ns) intense mw pulses, instead of a steady perturbation as in cw ESR. Since it is beyond the scope of this chapter to cover the details of pESR, we instead refer to the excellent textbook by Schweiger and Jeschke [7].

Field-swept pESR spectra can be recorded by measuring the transient signal created by a sequence of mw pulses. The ESR spectrum is obtained by integrating the so-called *Hahn-echo* [8] as a function of the external magnetic field. Since a field-swept echo experiment directly measures the absorption spectrum (referred to as echo-detected ESR) instead of the derivative spectrum resulting from magnetic field modulation in cw ESR (cf. Section 10.3.1), it is often superior to cw ESR for the detection of broad resonance lines due to better baseline stability. Another advantage is the absence of passage effects at low temperatures, suppression of broad fast-relaxing background signals. In addition, signal saturation – often a severe problem in solid-state cw ESR – is easily avoided by increasing the time span between repeating mw pulse sequences.

Since pESR is a time-domain technique, it provides direct experimental access to spin-relaxation times and the dynamics of spin-related processes (e.g., charge capture, separation, or emission). Small hyperfine couplings (cf. Section 10.4.2), which are often not resolved in the ESR spectrum, can be studied using double-resonance techniques such as electron-nuclear double resonance (ENDOR), which is a combination of an ESR and NMR experiment, and electron-spin echo envelope modulation (ESEEM) [7].

10.3.3
Sample Preparation

To ensure that material parameters obtained from ESR measurements apply to the material incorporated in thin-film solar cells, deposition conditions and sample structures have to be identical for both ESR samples and solar cells. However, when studying device-grade material, the number of spins in a thin film of less than 1 µm is usually well below the detection sensitivity of conventional ESR, and the above requirement cannot be fulfilled. There are several ways to increase the ESR sample volume and accordingly the number of detectable paramagnetic defects. Three commonly applied approaches to prepare ESR samples are briefly described in Table 10.1 along with their advantages and disadvantages.

When the material has a high defect concentration, it can be deposited on a substrate which is likewise used as solar-cell substrate (e.g., glass). In this way, one can assure similar boundary conditions as for the layer growth during solar-cell preparation. Depending on the substrate thickness and resonator dimensions, several samples may be stacked in order to increase S/N. The maximum active sample volume of a typical X-band resonator is $10 \times 5 \times 5 \, mm^3$. This volume is determined by the distribution of B_1 and B_{mod}. Note that in many cases thin-film silicon samples are grown in reactors where the substrate is constantly exposed to a plasma. This produces ions that impinge on the surface and can lead to the creation of paramagnetic defects. In other cases the substrate may contain defects even before deposition, for example, through a TCO or contact layers. Such ESR signals may superimpose the ESR spectrum of the silicon film and thus complicate the extraction of line shapes and defect densities. If such effects are not taken into account, large errors can easily be introduced for quantitative ESR.

When studying device-grade Si, the material is often deposited on Al foil (typical size $10 \times 10 \, cm^2$). Afterward, the film is removed from the substrate by diluted hydrochloric acid, and the Si flakes are collected in an ESR tube. This technique facilitates measurements on diluted spin systems; however, it is accompanied by the

Table 10.1 Properties of thin-film sample preparation techniques for ESR.

Sample structure	Advantage	Disadvantage
Thin film on substrate	Growth conditions like during solar-cell preparation	Low number of spins, parasitic signals from substrate, or TCO
Powder sample	Large sample volume and hence number of spins	Growth conditions differ from those used for solar-cell preparation, removal of sacrificial substrate (or layer) may induce additional defects
Film on scotch tape	Growth conditions like during solar-cell preparation, large number of spins	Removal of sacrificial layer (TCO) may induce additional defects

problem of different film growth conditions on Al foil compared to deposition on glass substrates. Moreover, the removal of the Al foil can introduce additional defects. Often it is also feasible to deposit Si on molybdenum foil. The samples can be peeled off by bending the metal substrate. In both cases, the Si powder has to be collected, weighed, and filled into sample tubes, which are then sealed under He atmosphere to ensure good thermal contact with the cooling gas. In addition, the He atmosphere prevents the samples from oxidizing.

Alternatively, Si may be deposited on ZnO-coated substrates which are also used as substrates for solar-cell preparation. After deposition, the film is covered with adhesive tape, and the ZnO sacrificial layer is subsequently etched away. This procedure maintains the initial structure of the film and at the same time allows a stack of many samples to be put into the resonator. It is worthwhile noting that the ZnO layer can be replaced by other materials, for example, SiN_X, provided that the acid for the etching process is changed accordingly.

10.4
The g Tensor and Hyperfine Interaction in Disordered Solids

In Section 10.2, we already noted that electron spins confined to solids or molecules exhibit a different g value than free electrons. This shift in g value is a direct consequence of the interaction of the electron spin with its environment and is therefore a significant source of information for the microstructure of the paramagnetic center. Moreover, the resonant magnetic field in ESR of the electron spin is influenced by superimposed internal magnetic fields of magnetic nuclei, acting as magnetic dipoles. This effect, known as HFI, contributes significantly to the spin's resonance frequency. Both Zeeman and HFI can be summarized in a quantum mechanical description with the following spin Hamiltonian:

$$H = \mu_B \vec{B}_0 \cdot \overleftrightarrow{g} \cdot \frac{\vec{S}}{\hbar} + \sum_i \vec{I}_i \cdot \overleftrightarrow{A}_i \cdot \vec{S}, \quad \overleftrightarrow{A}_i = A_{iso} 1_{3\times 3} + \overleftrightarrow{A}_{dip} \tag{10.5}$$

where the first term describes the Zeeman energy, the second term the HFI, $\vec{S}$ and $\vec{I}_i$ denote the electron and nuclear spin operators, while $\overleftrightarrow{g}$ and $\overleftrightarrow{A}_i$ are the g and HFI tensor, respectively, and $1_{3\times 3}$ denotes the identity matrix. Note that in the above Hamiltonian, spin–spin interactions such as exchange coupling, dipolar coupling, as well as the nuclear Zeeman and quadrupolar interactions have been neglected. For a good overview of such effects, we refer to Ref. [3]. In the following, we will describe the two terms of the spin Hamiltonian in more detail, highlighting their physical origin and their potential for structural identification.

10.4.1
Zeeman Energy and g Tensor

The g value of an electron in a defect state is shifted with respect to the free electron value (g_e) due to the fact that the magnetic moment created by the electron spin interacts with the orbital angular momentum of the charge carrier. This spin–orbit coupling induces an additional term in the spin Hamiltonian that depends linearly on the spin–orbit coupling constant, λ (for details see Ref. [3]). Although the spin–orbit interaction in general is quite suitable to obtain microscopic information, it vanishes in first-order approximation, since the orbital momentum is quenched by the crystal field. Treating the spin–orbit interaction as a perturbation, the original ground state is mixed with other states and a certain degree of orbital momentum L_{eff} is restored.

In the presence of B_0, an additional term in the Hamiltonian arises, which couples the partially restored orbital momentum to B_0

$$H_{SO} = \lambda \vec{L}_{eff} \cdot \vec{S} = \mu_0 \vec{B}_0 \cdot [\Delta g] \cdot \vec{S}; [\overleftrightarrow{\Delta g}]_{ij} = -2\lambda \sum_{k \neq 0} \frac{\langle 0|L_i|k\rangle \langle k|L_j|0\rangle}{E_k - E_0} \quad (10.6)$$

Here L_{eff} is the effective orbital angular momentum which depends on the wavefunctions of the defects ground and excited states, E_0 and E_k, respectively. The g tensor in Eq. (10.6) is a function of the local defect symmetry and the chemical nature of the center, since it depends on λ which strongly depends on the nuclear charge of the center. Hence with increasing λ, strong shifts of g are observed in a-Si:H [9].

The g tensor of most paramagnetic centers is anisotropic due to asymmetries in their wavefunction and the crystal field. The Zeeman term in the spin Hamiltonian in Eq. (10.5) is therefore influenced by the relative orientation of the external magnetic field and the principle axes of the g tensor. This is shown in Figure 10.3 for a threefold coordinated atom of a Si crystal (c-Si), a so-called dangling-bond defect. Such an ideal defect has axial symmetry with two principle components of the g tensor labeled $g_{\parallel}$ and $g_{\perp}$ as indicated in Figure 10.3. When the whole crystal is rotated with respect to the magnetic field B_0 in the plane of Figure 10.3, a variation in the g value (resonance position) as indicated will be observed, reflecting the symmetry of the defect state. Through such rotation experiments, a complete mapping of the g tensor and its

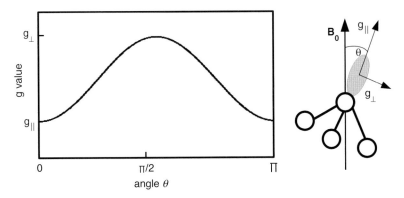

Figure 10.3 Schematic structure of a silicon db oriented in an angle θ with respect to B_0. The plot shows the variation of the observed g value, when the db is rotated in the plane. The extreme of g at 0 and 90° denote the principle values $g_{\parallel}$ and $g_{\perp}$, respectively.

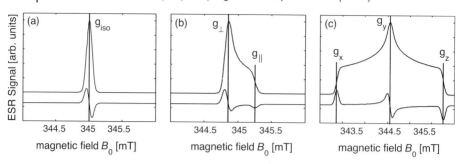

Figure 10.4 Numerical simulation of ESR powder patterns for different g tensor symmetries. The ESR spectra are plotted as absorption lines (upper curves) and derivative spectra (lower curves) for (a) isotropic, (b) axial (using data obtained for dbs in a-Si:H [10, 11]), and (c) rhombic symmetry. The principle values of the g tensor for orientations of the magnetic field along the canonical directions are indicated. Simulations are carried out with the numerical simulation package *easyspin* [12] using the following parameters $\nu = 9.7$ GHz, $g_{iso} = 2.0055$, $g_\perp = 2.0065$, $g_\parallel = 2.0040$, $g_x = 2.0150$, $g_y = 2.0081$, $g_z = 2.0011$, Gaussian line broadening $\Delta B_{PP} = 0.1$ mT.

symmetry is possible. For true powder samples, all crystal orientations appear with equal probability. In such orientationally disordered systems, all relative orientations between the B_0 field vector and principal axes of the g tensor occur with equal probability. Although a certain broadening of the ESR line shape by g tensor anisotropy is introduced, the resulting spectra, usually referred to as powder patterns, still reflect the principle values and the symmetry of the g tensor. Solely its principle axes in the molecular frame cannot be analyzed in a powder pattern. In Figure 10.4, three exemplary spectra for different g tensor symmetries are computed.

In case of an isotropic g tensor, only a single symmetric line is observed as shown in Figure 10.4a. Paramagnetic centers with axial symmetry, like dbs in a-Si:H with a sp-hybrid orbital, exhibit a typical powder pattern shown in Figure 10.4b. If B_0 is oriented along the symmetry axis of such a center, the effective g value is given by $g_\parallel$. If it is oriented perpendicular to the symmetry axis, the effective g value is given by $g_\perp$ (see also Figure 10.3). Since more orientations contribute to the latter situation, the spectral intensity at $g_\perp$ is larger than that at $g_\parallel$. In many cases, the microscopic structure of paramagnetic centers is more complex and the axial symmetry is lifted. In that case, all three principle values of the g tensor, g_x, g_y, and g_z, are nondegenerate and a rhombic powder pattern arises (Figure 10.4c).

It becomes obvious from Figure 10.4 that when line broadening is smaller than the g anisotropy, the principal g values and hence the symmetry of the spin-carrying center can be easily determined even for random distribution of the crystals (multi- or poly-crystalline Si) or in an amorphous material like a-Si:H. If the orientation distribution in the thin film is not random (e.g., preferential orientation in crystal growth), deviation from the depicted powder pattern may occur. In case line broadening is stronger than the g anisotropy, the powder pattern features may be completely hidden.

10.4.2
Hyperfine Interaction

The magnetic interaction with nuclear spins present in the vicinity of an electron spin will introduce, in addition to the external magnetic field B_0, an internal magnetic field B_{int}. The interaction of the electron spin with B_{int} is referred to as HFI, which is determined by the sum of the magnetic interactions with the magnetic moments of all nuclei, I_i. The magnitude and orientation dependence of HFI are determined by the electron-spin density[2] at the position of the respective nuclei and by the distance of the electron spin to the nuclear spins in its vicinity. Hence, to measure HFI is a very important source of information to identify the microscopic structure of a spin state.

The interaction with the nuclei is generally expressed by the HFI tensor A, which can be written as a sum of the isotropic Fermi contact interaction, A_{iso}, and the electron-nuclear dipole–dipole coupling, A_{dip} (see Eq. (10.5)). The Fermi contact interaction is given by

$$A_{iso} = \frac{2}{3}\frac{\mu_0}{\hbar} g_e \mu_e g_n \mu_n |\psi_0(0)|^2 \tag{10.7}$$

where $|\psi_0(0)|^2$ is the electron-spin density at the nucleus at position $r=0$. If the electron spin is distributed over several nuclei, determination of the individual HFI allows a complete mapping of the spin density.

In Figure 10.5, the influence of $A_{iso} = 4.2$ mT on the Zeeman splitting of an electron spin is depicted assuming that it is interacting with only one nuclear spin $I = 1/2$ as is the case for an electron trapped at a phosphorous donor atom (^{31}P) in crystalline silicon at low doping concentration ($N_D < 10^{17}$ cm^{-3}). Both Zeeman levels of the electron spin ($m_S = \pm 1/2$, dashed line in Figure 10.1) are split by the same amount in two levels belonging to the two nuclear-spin states with $m_I = \pm 1/2$. Since in an ESR experiment only the electron spin is in resonance and not the nuclear spin, the ESR selection rule is $\Delta m_S = \pm 1$ and $\Delta m_I = 0$. From this, it becomes evident that the ESR line is split into two resonances which are separated by $A_{iso} = 4.2$ mT symmetrically about the resonance position of the non-interacting spin (cf. Figures 10.1 and 10.5a). Note that the intensity of the lines is reduced by a factor of two since the total number of spins is not reduced but spread over two lines. According to Eq. (10.7), the magnitude of the HFI is determined by the probability finding the electron at the site of the nucleus (position $r=0$). Since for a p-type wavefunction $|\psi_0(0)|^2 = 0$, the isotropic HFI vanishes if spin-polarization effects due to spin–spin interaction are not important. This simple case already proves the strength of ESR spectroscopy in determining the local structure of paramagnetic states. Recent advances of numerical quantum-chemical calculations, such as *ab-initio* density-functional theory (DFT), improved the reliability of structure determination. They allow a quantitative evaluation of ESR parameters,

---

[2] In the a-Si:H ESR literature the number of paramagnetic centers per volume is also often referred to as the "spin density." To avoid confusion, the latter will be referred to as N_S or N_D (donor) through the article.

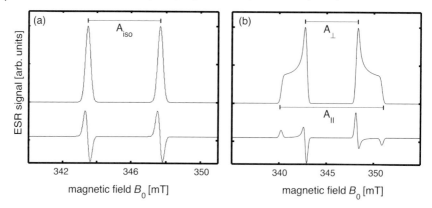

Figure 10.5 ESR powder pattern for a single-electron spin $S = 1/2$ and a single-nuclear spin $I = 1/2$, with an isotropic g tensor ($g_{iso} = 2.0055$); (a) $A_{iso} = 4.2$ mT (117 MHz[3]); (b) anisotropic HFI tensor ($A_\perp = 140$ MHz, $A_\parallel = 320$ MHz or $A_{iso} = 200$ MHz, $A_{dip} = 60$ MHz).

instead of qualitative interpretation mainly based on symmetry analysis of the HFI tensor [13].

In contrast to isotropic Fermi-contact HFI, anisotropic HFI induced by dipolar interaction depends on the relative orientation between B_0 and the vector connecting electron spin to nuclear spin. This anisotropic contribution to the HFI can be expressed by a tensor $\overleftrightarrow{A}_{ij}^{dip}$ [7]:

$$\overleftrightarrow{A}_{ij}^{dip} = \frac{\mu_0}{4\pi h} g_e \mu_e g_n \mu_n \left\langle \psi_0 \left| \frac{3 r_i r_j - \delta_{ij} r^2}{r^5} \right| \psi_0 \right\rangle = \begin{pmatrix} A_x & 0 & 0 \\ 0 & A_y & 0 \\ 0 & 0 & A_z \end{pmatrix} \quad (10.8)$$

where $\vec{r}$ is the vector connecting the electron spin to the nuclear spin. $\overleftrightarrow{A}_{ij}^{dip}$ is a traceless, symmetric tensor which can be diagonalized to obtain the three principle hyperfine values $A_{x,y,z}$. Anisotropic HFI vanishes for electrons in s-orbital and is usually only important for electrons in orbitals with partial p-character. The wavefunction of dbs in a-Si:H can be described by a sp-hybrid orbital (cf. Section 10.5.1) with axial symmetry which is localized at an Si atom. If the Si isotope carries a nuclear spin $I = 1/2$, as is the case for ^{29}Si, strong HFI arises. For the principle axis perpendicular to the db symmetry axis, A_x and A_y are degenerated because of symmetry and will be denoted by $-A_{dip}$. Since the dipolar HFI tensor is traceless, the principle value parallel to the symmetry axis A_z is given by $2A_{dip}$. In addition to anisotropic HFI, the silicon db also exhibits isotropic HFI to its ^{29}Si host atom, due to its sp-orbital. The principle values of the full axially symmetric HFI tensor are then in analogy to the notation of g values given by

$$A_\parallel = A_{iso} + 2 A_{dip}$$
$$A_\perp = A_{iso} - A_{dip} \quad (10.9)$$

3) HFI parameters are often given in frequency units. For $g = 2$, 28 MHz is equivalent to 1 mT.

The resulting ESR spectra, using typical isotropic and anisotropic HFI values for states in c-Si and a-Si:H, are shown in Figure 10.5b. Note that with the presence of anisotropic HFI the spectrum becomes rather complex if the hyperfine contributions are not well separated or are masked by other ESR signals. To entangle such complex spectra, ESR measurements at various frequencies are mandatory.

If the electron spin is fairly well localized – a situation often encountered in disordered solids – we can approximate the spin as a magnetic point dipole, with all its spin density concentrated at one point. In this case, we can simplify $\overleftrightarrow{A}^{dip}$ for this two-particle case as

$$\overleftrightarrow{A}^{dip} = \frac{\mu_0}{4\pi\hbar} \frac{g_e\mu_e g_n\mu_n}{r^3} \begin{pmatrix} -1 & & \\ & -1 & \\ & & 2 \end{pmatrix} = \begin{pmatrix} -A_{dip} & & \\ & -A_{dip} & \\ & & 2A_{dip} \end{pmatrix} \quad (10.10)$$

This point-dipole approximation is often applied to extract distances of the electron spin to weakly coupled nuclei, which allows a deep insight into the microscopic structure surrounding the paramagnetic center. Nevertheless, the applicability of this approximation in disordered solids has to be elaborated before conclusions about a microstructure can be made.

10.4.3
Line-Broadening Mechanisms

In the above analysis, we assumed a certain intrinsic line width for the ESR resonance, which we now discuss in more detail. ESR resonances can be either homogeneously or inhomogeneously broadened. If all electron spins possess the same resonance frequency, which implies that their microscopic environment is identical (e.g., defect structure Figure 10.6a), the ESR line shape is homogeneous and its line width $\Delta\omega$ and shape are determined by spin-relaxation properties, that is, the lifetime determined spin coherence T_2 (often call spin–spin relaxation time) via the Heisenberg uncertainty principle: $\Delta\omega \approx 1/\hbar T_2$. The resulting line shape is a Lorentzian (Figure 10.6) and its width is defined by $2/T_2$. As shown in Figure 10.6, the EPR spectrum is the sum of a large number of identical spin packets producing identical lines with the same Larmor frequency (g value) and line width.

If the resonance frequencies are not identical for all electron spins, the ESR resonance is composed of different spin packets, each with its own distinct resonance frequency. Each of the individual spin packets is homogeneously broadened, with line width $2/T_2$ as in the case discussed before. The proliferation of resonance frequencies is induced by a variation of the spin interaction with its environment either by dipolar spin coupling, HFI, or a distribution of g values as in Figure 10.6 and giving rise to Gaussian line shape.

We will see that although the idealized analysis of the g-tensor symmetry allows the precise determination of the full g tensor, its application in materials with small g tensor anisotropies, such as silicon, is usually hampered by inhomogeneous

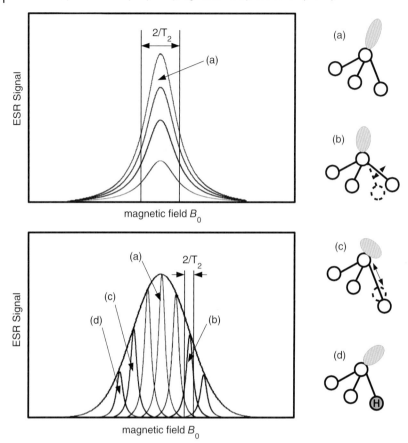

Figure 10.6 Upper plot: homogeneously broadened ESR line giving rise to a Lorentzian. The line is the superposition of many identical lines due to db structure (a). Lower plot: inhomogeneously broadened ESR line induced by variation in bond angle and bond length and due to unresolved HFI as indicated by defect structures (a)–(d), respectively. Each db structure produces its respective line as indicated in the plot leading to Gaussian broadening.

broadening. To illustrate this effect, we again consider the case of paramagnetic dbs in a-Si:H as indicated in Figure 10.6.

Inhomogeneous broadening of the db resonance due to HFI is induced by highly abundant magnetic nuclei, such as ^{1}H and ^{29}Si. Their random distribution in the matrix of localized dbs causes a variation in the dipolar HFI field experienced by each db. The matrix nuclei are located at a rather large distance from the electron spin (>4 Å) [14]; hence the HFI is small and not resolved in the ESR spectrum. The inhomogeneous broadening due to unresolved HFI is frequency independent (see Eqs. (10.7) and (10.8)) if the external magnetic field exceeds the HFI (high-field limit), a condition fulfilled in most ESR experiments. This nuclear-spin-induced disorder effect introduces a Gaussian line broadening, ΔA.

In amorphous materials like a-Si:H, the g tensor analysis is further complicated by inhomogeneous broadening due to a distribution of g tensors. dbs are not well defined in terms of their microscopic structure due to disorder, which results in variation of bond lengths and bond angles as is shown on the right side of Figure 10.6. This induces a change in the wavefuction and tends to result in a distribution of g tensors. In the ESR literature, this is referred to as g strain, Δg, and leads to a significant field-dependent broadening (see Eq. (10.2)) of the ESR lines and ultimately limits the resolution of the full g tensor in the spectrum.

Both discussed inhomogeneous broadening mechanisms, ΔA and Δg, in X band are illustrated in Figure 10.7 for the line with axial g tensor shown in Figure 10.4b. Obviously, the powder pattern that is visible if no inhomogeneous broadening is present is completely masked in the X-band spectrum. In case of unresolved HFI, the spectrum is almost symmetric and, without detailed knowledge, one is tempted to associate this with an isotropic g value. Note that although the lines in Figure 10.7a and b have the same g tensor, their peak position is different. Again one might associate this effect with a different g value and hence assign this to different defect

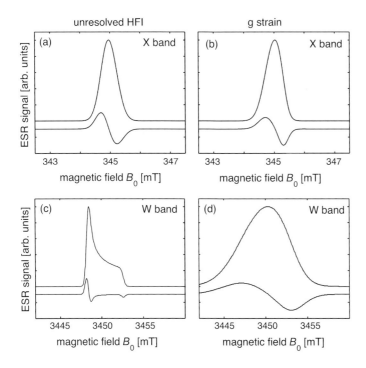

Figure 10.7 X-band ESR spectra (a) and (b) simulated assuming a defect with axial g tensor (see Figure 10.4b) but now assuming line-broadening effects due to (a) unresolved HFI ($\Delta A_{iso} = 15$ MHz), (b) g strain ($\Delta g_\perp = 0.0037$, $\Delta g_\parallel = 0.0019$). Note that in both cases, the typical powder pattern can no longer be resolved. At W band (c, d), resolution enhancement can only be gained for the case of unresolved hyperfine (c) since the line-broadening mechanism is field (and hence frequency) independent.

states. In fact one might discover that both lines have an obvious asymmetry and associate this with the fact that two lines are superimposed. There is only one way out of this dilemma and that is to measure ESR spectra at higher mw frequency. This is performed in the simulation (Figure 10.7c and d) for W band and due to the different frequency dependence of the broadening mechanism, the g tensor in case of unresolved HFI can now easily be recovered, whereas the line exhibiting g strain is only broadened at high frequency. This example shows how difficult sometimes ESR interpretations are if only X-band data are used for the evaluation of the line parameters. The concept of multifrequency ESR is now widely used in the community and commercial spectrometers from 1 GHz up to 263 GHz are now available.

10.5
Discussion of Selected Results

In the previous sections, we have discussed the essence of ESR spectroscopy focusing on thin-film materials. We have shown how microscopic parameters such as g value or HFI influence the ESR spectra and what the challenge is in the interpretation of the data in the presence of disorder. In this section, we will review how ESR has succeeded to unravel some very important aspects about the microstructure of a-Si:H. Nevertheless, it will become evident that many open questions remain, in particular, on light-induced degradation and on the interpretation of the ESR parameters. At the end of this section, we will highlight some of those challenges and address future research strategies to finally unravel the microstructure of defects in a-Si:H.

10.5.1
ESR on Undoped a-Si:H

The first report of ESR measurements in a-Si was by Brodsky and Title in 1969 [16]. They observed a Lorentzian-shaped line with $g = 2.0055$ with a paramagnetic defect density $N_S = 10^{20}\,\text{cm}^{-3}$ (about one defects for every 500 Si atoms), which concomitantly resulted in material with very poor electronic quality. The ESR line width is determined by strong spin–spin interaction effects. Due to the similarity of the a-Si ESR line with that of dangling bond defects observed at disordered c-Si surfaces [17], Brodsky and Title speculated that the ESR signal originates from ruptured Si–Si bonds, although they assumed that the structure of the material was microcrystalline. After the discovery that the introduction of hydrogen into the a-Si network dramatically reduces the defect density and improves the electronic quality [18], progress has been made over the years and today in device-grade a-Si:H $N_S < 10^{16}\,\text{cm}^{-3}$ is easily reached [19]. The defect density of the $g = 2.0055$ line was used from early on as a measure of the electronic quality of a-Si:H although it is now well established that details of the deposition conditions also strongly influence the g value [19]. Currently no thorough understanding of this behavior exists. For ESR measurements on device-grade a-Si:H, powder samples with a

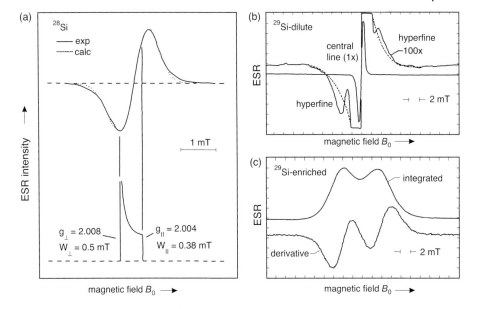

Figure 10.8 X-band ESR spectrum of undoped a-Si:H powder samples, which were (a,b) naturally abundant a-Si:H, and (c) ^{29}Si-enriched a-Si:H. Note that the sign reversal of the spectra results from a 180° phase shift of the lock-in amplifier setting with respect to the field modulation and is not due to an inversion of the spin population, which would lead to an emission line. Due to the strong amplification, the central line in (b) is clipped. Figures (a, b) reprinted with permission from Ref. [10] by Stutzmann et al. Figure (c) reprinted with permission from Ref. [15] by Biegelsen et al. Copyright (1989, 1986) by the American Physical Society.

volume between 0.01 and 0.1 cm^{-3} (5–50 mg weight) have to be prepared or many thin-film samples have to be stacked. In device-grade a-Si:H, the ESR line has Gaussian shape but is slightly asymmetric as shown in Figure 10.8a with an average g value (zero crossing of the field axis) $g = 2.005$–2.0055. The ESR line obeys almost a perfect Curie law behavior [20, 21] indicating that the defect configuration is rather robust and that its occupation is little affected by temperature. Stutzmann and Biegelsen analyzed the g tensor of the line according to the procedures as described before and found that the line shape at X band could be described with an axial symmetric g tensor with $g_{||} = 2.004$ and $g_\perp = 2.008$ [10]. Due to structural disorder, the ESR spectrum of the db is broadened leading to an orientation-dependent Gaussian line shape due to strain with $\Delta H_{||} = 0.38$ mT and $\Delta H_\perp = 0.5$ mT at X band. Since the g strain in Ref. [10] is of the same order of magnitude as the g anisotropy, extending the range of ESR experiments to higher frequency did not introduce substantial improvement of data evaluation ($g_{||} = 2.0039$, $g_\perp = 2.0063$, $\Delta g_{||} = 0.0019$, $\Delta g_\perp = 0.0048$) [11]. When deuterium (D) is incorporated in the film replacing hydrogen during the deposition process, the line width of the 2.0055 line remains unchanged [15, 22]. This was an unexpected result, since the magnetic moment of D is smaller than that of H (H: $I = 1/2$ $g_N^H = 5.585$; D: $I = 1$, $g_N^D = 0.857$).

From this experiment it was concluded that hydrogen (or D) is not located in the direct vicinity of the defect, for example, at a backbonded position. Since the g tensor in first-order approximation was found to have axial symmetry with the tensor's principal components being comparable to P_b-like defects at the c-Si/SiO$_2$ interface [23, 24], the ESR signal was associated with db states, supporting the early assignment of Brodsky and Title. Since the assignment of the 2.0055 line purely from g tensor arguments is not possible in a-Si:H due to the small g shifts, the early speculation about its identification with dbs generated controversial discussions [10, 25–28].

One additional piece of information that is required is HFI of the db wavefunction with the ^{29}Si nuclei [10, 11, 15]. Naturally abundant silicon consists of 92.2 at.% ^{28}Si ($I=0$) and 4.7 at.% ^{29}Si ($I=^1/_2$) plus some other isotopes. When the magnetic-field range and the amplification of the lock-in amplifier are extended, two small satellites can be observed in the ESR spectrum as shown in Figure 10.8b with the satellites on the low field side contributing to about 2.0(5)% of the total integrated signal intensity. When we assume that the spin density is located primarily at one atom and is not extended over many silicon atoms, only this nuclear spin can contribute to the total HFI. In that case, due to the abundance of ^{29}Si, each of the two HFI satellites is expected to contain 2.35% of the ESR signal. This is in good agreement with the experimental observation. When the a-Si:H was produced from 93 at.% ^{29}Si-enriched SiH$_4$, the $g=2.0055$ line completely vanishes and only two hyperfine satellites are observed as shown in Figure 10.8c.

The HFI observed in the ESR spectrum was clearly identified to originate from ^{29}Si. Here different situations have to be taken into account to understand the spectrum. The spin of the db can interact with nuclear spins that are in the vicinity of the db. Every Si atom including the threefold coordinated Si atom of the db defect itself can carry a nuclear spin with a certain probability depending on the isotope enrichment of ^{29}Si. In first-order approximation, from the isotropic HFI the distribution of the spin density, and hence the amount of s character of the wavefunction at the db state can be determined and from the anisotropic HFI its p contribution. Unfortunately, disorder of the amorphous network induces considerable g strain, which makes it extremely difficult to extract the above HFI features from X-band data. Except for Q band [11], no further multifrequency ESR data were available. Nevertheless, a rough estimate of the HFI parameters $A_{iso}=7$ mT and $A_{dip}=2$ mT [10] assuming axial symmetry of the HFI could be deduced from spectra as shown in Figure 10.8 [10, 11, 15].

Watkins and Corbett introduced a local-hybrid model [29], approximating the wavefunction $|\Psi\rangle$ of the unpaired db electron ($g=2.0055$) by a linear combination of atomic orbitals (LCAOs)

$$|\Psi\rangle = \sum_i \alpha_i(\sigma_i|s\rangle) + (\pi_i|p\rangle) \quad \text{with} \quad \sum_i \alpha_i^2 = 1, \sigma_i^2 + \pi_i^2 = 1 \qquad (10.11)$$

where i indexes all atoms within the extend of $|\Psi\rangle$ ($i=0$ being the db Si atom) and the projection coefficients α_i, σ_i, π_i of s-type and p-type contributions to the wavefunction. In case the db would be a pure sp^3 hybrid, one would expect $\alpha_0=1$, $\sigma_0=\sqrt{1/4}=0.5$, and $\pi_0=\sqrt{3/4}=0.87$. Assuming that the wavefunction of the

2.005 line is primarily localized at one atom, one can experimentally determine the projection coefficients by $\alpha_0^2 \sigma_0^2 = A_{iso}/A(s)$, $\alpha_0^2 \pi_0^2 = A_{dip}/A(p)$, and $\sigma_0^2 + \pi_0^2 = 1$, where $A(s)$ and $A(p)$ are the HFI constants of pure $|s\rangle$ and $|p\rangle$ orbitals as they were determined from Hartree–Fock calculations, respectively (for details see Ref. [29]). Assuming typical literature values for $A(s)$ and $A(p)$, it was found that $\alpha_0^2 = 0.45-0.69$, $\sigma_0^2 = 0.06-0.11$, and $\pi_0^2 = 0.89-0.94$ [10, 11]. Therefore, the sp hybridization has ≈10% s- and ≈90% p-character and the unpaired db electron is localized with more than 50% on the threefold-coordinated Si atom itself. With the above assumptions, it was shown that the a-Si:H db is similar to the P_b-like dangling-bonds defects [30]; however, the P_b center is more localized as compared to the a-Si:H db. This picture of db defects in a-Si:H is up to now broadly accepted in the research community. However, it was pointed out that the local-hybrid model has serious deficiencies in reliably predicting spin-density distributions [31]. The model assumes that the atomic wavefunctions of Si are not perturbed in the solid state and completely neglects the influence of spin polarization due to exchange interaction on the HFI. It has been shown by DFT calculations that such effects are extremely important [32], especially for predicting isotropic HFI.

Despite the enormous efforts of the early years in determining the g tensor and HFI parameters, it was recently shown within the German research cluster *EPR-Solar* [33] that with multifrequency ESR and pulsed ENDOR experiments a much more complex picture of magnetic interaction parameters were determined showing that simple axial symmetry is not sufficient in describing experiments [34]. In particular, it is shown that the g tensor has rhombic symmetry. This was already suspected from earlier calculations of the g tensor and HFI parameters [22, 35].

In conjunction with other experimental data (for an overview see Ref. [2]), it was found that the db state is localized in the band gap and is amphoteric; hence the state can be occupied with 0, 1, or 2 electrons thereby changing the charge of the state from positive, neutral to negative, respectively (Figure 10.9). Since a closed silicon bond is occupied by two electrons, a broken bond contains one electron and is a neutral state which is paramagnetic ($S = 1/2$). With the number of electrons residing on the localized state, the transition energy to populate or depopulate the state, $D^{0/+}$ or $D^{0/-}$, will shift with respect to the band edges by the correlation

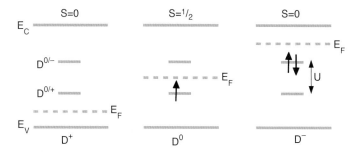

Figure 10.9 Energy transition levels of dangling bonds in their different charge and spin states. Both charged states, D^- and D^+, are not visible in ESR measurements.

energy, U (Figure 10.9). U is determined by Coulomb repulsion between the two electrons and the energy shift induced by the change of the local bonding configuration. Since the temperature dependence of the db signal was shown to have an almost ideal Curie law behavior, $U = 0.2$–0.3 eV was estimated from the deviations of the Curie law [21]. Note that if db states would have negative U, they would not be observable in ESR. Negative-U centers are well known in chalcogenide glasses and are characterized by having an unstable singly occupied state [36].

The occupation of the db can be changed either by doping or charge injection (light, applied bias) giving rise to D^+, D^0, and D^- as shown in Figure 10.9. The positively charged D^+ state (no electron, but two holes) has a total spin $S = 0$. To occupy this state with one electron, the Fermi-level E_F has to move toward the conduction band. Once E_F passes the transition energy $D^{0/+}$, the db is turned into a neutral D^0 state. The db will stay paramagnetic until E_F passes the second transition energy $D^{0/-}$. Now it is energetically more favorable to have two electrons residing on one defect and turning it into the negatively charged D^- state. Due to the Pauli principle, both electrons must have an opposite spin orientation resulting in a vanishing total spin, $S = 0$. From this picture it becomes obvious that D^0 states must obey Curie's law in first-order approximation since the position of E_F is pinned by the db state itself when no doping is assumed. Deviations from this behavior are due to the fact that disorder in a-Si:H broadens the energy transition levels of the db states to Gaussian energy distributions and unintentional doping effects producing moderate n-type behavior [2]. Hence, when interpreting ESR paramagnetic defect-density results one has to carefully consider E_F shifts.

10.5.2
LESR on Undoped a-Si:H

Despite the complexity of the microscopic structure of db states, many of the electronic properties of a-Si:H are determined through their interaction with excess charge carriers generated by light or through injection [2]. Such charge carriers may by trapped in localized states of the band and may change the ESR spectrum.

Using X-band LESR with above band-gap light, two additional ESR signals could be observed in undoped a-Si:H, a narrow line ($\Delta B_{pp} = 0.6$ mT) at $g = 2.004$ and a broad one ($\Delta B_{pp} = 2.0$ mT) at $g \approx 2.01$–2.013 [37, 38]. These signals were assigned to electrons ($g = 2.004$) and holes ($g = 2.01$–2.013) trapped in band-tail states of the conduction and valence band, respectively [38] (see Figure 10.10). Band-tail states are believed to originate from weak bonding and antibonding Si–Si bonds [39], which are induced by the variation in the disorder potential [40]. This leads to localized states that may act as trapping centers and thereby strongly reduces charge carrier mobility as compared to extended states [2]. In an alternative explanation, the LESR states were assigned to positively and negatively charged dbs that are made paramagnetic through trapping of light-generated charge carriers forming the $g = 2.004$ and 2.01 line, respectively [41–43]. These interpretations all assumed the presence of a high density of charged defects as it was introduced through the charged db models [41, 44].

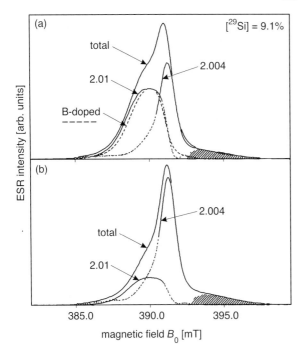

Figure 10.10 Echo-detected LESR spectra of ^{29}Si-enriched undoped a-Si:H measured at $T = 40$ K by (a) a two-pulse Hahn echo and (b) three-pulse stimulated echo. Figure reprinted with permission from Ref. [45] by Umeda et al. Copyright (1996) by the American Physical Society.

In a more detailed pulsed LESR measurement by Umeda et al. [45] on undoped a-Si:H with varying ^{29}Si concentration, the HFI of the $g = 2.004$ line with ^{29}Si could be well resolved, partially due to the fact that the superimposed db and 2.01 signal in the LESR spectrum could be suppressed due to different relaxation behavior. As shown in Figure 10.10, the line at $g = 2.004$ contains on the high-field side clearly resolved shoulders (gray area) which stem from HFI with ^{29}Si. The intensity of the shoulder increased corresponding to the ^{29}Si concentration. However, the relative contribution to the central LESR line ($g = 2.004$) is about twice as large as what was reported for the db line (cf. Section 10.5.1). From this, Umeda et al. concluded that the wavefunction of the 2.004 center spreads equally over two Si atoms, different from the ^{29}Si of the db line. This result is in good agreement with the assignment of the $g = 2.004$ line to an electron being localized on a weak Si—Si bond with its s-like wavefunction centered in between two Si atoms instead of on a threefold-coordinated Si atom as is expected for charged dbs.

10.5.3
ESR on Doped a-Si:H

The assignment of the two LESR signals is nicely supported by the dependence of the ESR signal on substitutional doping by boron (B) and phosphorous (P). Doping is accomplished by introducing small amounts of PH_3 or B_2H_6 to the SiH_4 into the PECVD growth chamber. The ratio of the respective concentration is in the following considered as the doping concentration, for example, 1% P doping indicates $[PH_3] = 0.01 \cdot [SiH_4]$.

By B doping, E_F is shifted in the direction of the mobility edge of the valence band, by P doping in the opposite direction. With increasing B concentration, an ESR line at $g = 2.01$–2.013 appears which has similar line parameters as the broad LESR line shown in Figure 10.10a [20, 38, 45]. This line was assigned to shallow band-tail states of the valence band. Through the shift of E_F, shallow doubly occupied band-tail states are depopulated into singly occupied paramagnetic states. With P doping, on the other hand, the intensity of the 2.004 signal is increased which is naturally explained by the fact that E_F is shifted in the conduction band tail [20]. In Figure 10.11, the measured spin concentrations of the respective ESR lines discussed so far are plotted as a function of E_F. The steep increase in N_S of the 2.01 and 2.004 lines toward the mobility edges is associated with the exponentially increasing density of states (DOSs) of the respective band tails. The fact that the db spin concentration, N_S^{db}, is reduced with increasing doping level, however, does not imply that the total db density (including charged dbs) is reduced when shifting E_F. On the contrary, with increasing [P] or [B] the db density increases dramatically due to doping-induced defect formation (often referred to as autocompensation) [2, 46]. Most P or B states form defect complexes which leads to nondoping configurations, for example, P_3^0 or B_3^0. The excess electrons and holes, which are introduced by the few active fourfold-coordinated dopants, are captured at defect states rather than band-tail states, thereby positively or negatively charging the defects as becomes obvious from Figure 10.9. At high doping levels, such doping-induced defects are not visible in ESR but they can be easily detected and analyzed through sub-band gap optical and electrical spectroscopy [2].

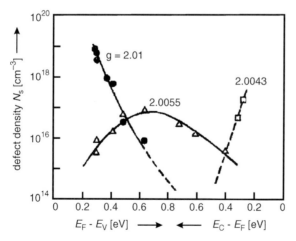

Figure 10.11 Dependence of spin densities on the distance of E_F from the respective mobility edges for the three discussed ESR resonances. Figure reproduced with permission from Ref. [48] by Dersch et al.

Since it is believed that the neutral fourfold-coordinated boron acceptor state in a-Si:H (B_4^0) lies energetically very close to the valence-band mobility edge, it can easily capture an electron from an energetically higher lying band-tail state thereby charging the boron (B_4^-) and making it diamagnetic and therefore invisible for ESR [47]. Due to the low doping efficiency and the high density of db and band-tail states, boron acceptor states cannot be made paramagnetic even for very high illumination levels. In compensated a-Si:H – this refers to the case when $[PH_3] = [B_2H_6]$ – the situation is different. Due to the compensation of both charges introduced into the a-Si:H network, E_F remains in the middle of the band gap. It was found that in this case, the db density is similar to that of undoped a-Si:H and that the doping efficiency is two orders of magnitude higher [47]. From this important result, it was concluded that defect formation in a-Si:H is related to the position of the Fermi energy in the material.

While the B-acceptor state is not observable in a-Si:H, the P donor is easily identified in the ESR spectrum through the HFI with the 100% abundant ^{31}P nuclei ($I = 1/2$). In c-Si, the P-donor is observed at low temperatures when the electrons of the conduction band are trapped at P forming a paramagnetic state, P_4^0. This state reveals the typical P signature as shown in Figure 10.5a, namely a signal split by a HFI with $A_{iso} = 4.2$ mT [49]. As shown in Figure 10.12 in 1% P-doped a-Si:H also two satellites with equal intensity were observed [50]. Note that the splitting between the satellites is about six times larger than what is found for P in c-Si. The satellites are centered almost symmetrically around a strong narrow line with $g = 2.004$ (line is clipped in Figure 10.12). This 2.004 line has been discussed before with respect to LESR experiments and is associated with band-tail states of the conduction band. These states are populated by the doping-induced shift of E_F in the direction of the mobility edge of the conduction band. The correlation energy of the conduction band-tail states is about 20–30 meV as derived from LESR experiments [51]. The intensity of the satellites increased with the square root of [PH_3], whereas the increase in the 2.004 line is less pronounced. To rule out other sources of HFI due to nuclear spins with $I = 1/2$ such as ^{1}H, ^{13}C, ^{15}N, or even $I = 1$ such as ^{14}N, Stutzmann and coworkers deposited 1% P-doped a-Si:D replacing H by D ($I = 1$) as well a similarly doped a-Si:H sample under ultrahigh vacuum (UHV) conditions, thereby minimizing the impurity concentration [52]. As shown in Figure 10.12, all three samples exhibit the same satellites lending support to the identification of the HFI with P_4^0.

For a quantitative evaluation of the HFI, second-order effects have to be taken into account [3]. The first-order HFI approximation is valid for $A_{iso} < 10$ mT and produces $(2I + 1)$ satellites that are distributed equally spaced around the resonance position of the non-interacting spin. The second-order approximation introduces an additional quadratic term in B_0, which shifts all lines to lower fields and the spacing between the lines increases from low field to high field. Assuming that the HFI in P-doped a-Si:H is isotropic, Stutzmann and Street determined $A_{iso} = 24.5$ mT taking into account a large distribution of HFI-coupling strengths induced by disorder. This is described by A strain (see Section 10.4.3) which for the case of P is found to be $\Delta A = 16$ mT. From second-order HFI, the g value of the P states can be calculated to $g_P^{a\text{-}Si} = 2.003(2)$ [52], which is similar to that of band-tail states

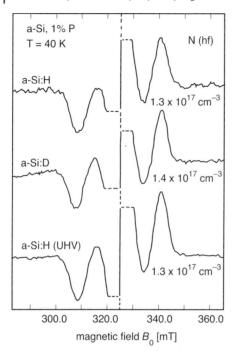

Figure 10.12 ESR spectra of 1% P-doped a-Si:H, a-Si:D, and a-Si:H deposited at UHV conditions. The central line of the ESR spectrum at $g = 2.004$ is clipped due to the large amplification. Figure reprinted with permission from Ref. [52] by Stutzmann et al. Copyright (1987) by the American Physical Society.

identified through LESR and supports the assumption that the P states are distributed over a wide energy range among band-tail states of the conduction band. $g_P^{a\text{-Si}}$ is considerably larger than what is found in c-Si ($g_P^{c\text{-Si}} = 1.9985$), which indicates that the P state in a-Si:H can no longer be approximated by a shallow, effective-mass state as in case of c-Si [52], since otherwise $g_P^{c\text{-Si}} < 2.0023$. The six-times larger HFI interaction that is observed in a-Si:H is explained by the fact that the wavefunction of the neutral P atom in the a-Si matrix is much stronger localized as compared to c-Si. Stutzmann and Street estimate a donor radius from the HFI of about 9 Å [50] instead of 16.7 Å as found in c-Si [53]. As stated before, with P doping the db density dramatically increases thereby reducing the doping efficiency. Stutzmann and coworkers estimated that from 0.1% [PH$_3$] in the gas phase about 10^{20} cm^{-3} P atoms are incorporated in the a-Si:H film. Of these, only about 10^{18} cm^{-3} are fourfold coordinated and only 10% of them (10^{17} cm^{-3}) are P_4^0 states and hence observable in ESR [52].

Figure 10.13 summarizes the main achievements of ESR on doped and undoped a-Si:H. The sketch depicts the currently accepted picture of the DOS for nominally undoped a-Si:H as it is derived from many different experimental results [2]. From the mobility edges band-tail states with exponentially decreasing DOS with a slope of about 25 and 45 meV for the conduction and valence band tail with ESR signatures

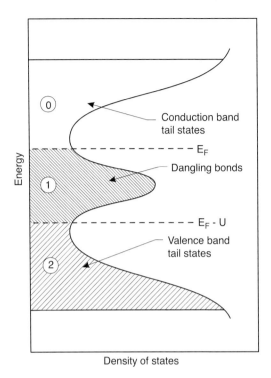

Figure 10.13 Density of states of a-Si:H showing the occupation of localized states for a certain position of the Fermi energy. States that lie within the correlation energy U energetically beneath E_F will be paramagnetic (region 1). Lower lying states will be doubly occupied (region 2 and states above E_F are unoccupied. Figure reprinted with permission from Ref. [54] by Stutzmann et al. Copyright (1983) by the American Physical Society.

at $g = 2.01$ and 2.004, respectively, are observed. Only those states in region 1 are paramagnetic, all other states are diamagnetic and will not be observed in ESR. By shifting the Fermi level in the respective tails either through illumination or doping, the ESR signatures of P_4^0 donor states or conduction and valence band-tail states can be observed. For strongly P- and B-doped materials, the db density strongly increases in regions 0 and 2, respectively, through the shift of E_F. These defects will no longer be observed in ESR since they are diamagnetic (see Figure 10.9).

10.5.4
Light-Induced Degradation in a-Si:H

In 1977, Staebler and Wronski observed that intense light illumination of undoped electronic quality a-Si:H decreases both the dark- and photoconductivity (Staebler–Wronski effect – SWE) [55, 56]. They found that annealing to temperatures of about 150 °C completely reverses the effect. Subsequent investigations discovered that the observed conductivity decrease is due to a shift in the Fermi level toward

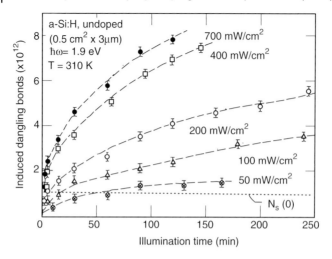

Figure 10.14 Number of light-induced dbs versus illumination time measured by ESR at various light intensities. Sample and illumination details are displayed in the upper left corner. The horizontal dotted line indicates the number of dbs $N_s(0)$ prior to illumination. Figure reprinted with permission from Ref. [39] by Stutzmann et al. Copyright (1985) by the American Physical Society.

midgap and a severe decrease in the excess charge carrier lifetime, while the mobility is essentially unchanged. It was not until 1980 that Dersch et al. [57] investigated a-Si:H by ESR methods realizing that the observed degradation is caused by a reversible increase in the db absorption signal. Today it is clear that the light-induced increase in the db concentration is the most important, but not the only degradation mechanism of a-Si:H [58]. dbs contribute most to the decrease in the excess charge carrier lifetime, which is strongly correlated to the defect-density N_s [39].

It was found experimentally that the number of metastable dbs increases with $N_s \propto t^{1/3} I^{0.6}$ under cw illumination for time t with light of intensity I and is independent of low impurity concentration in the material (cf. Figure 10.14). Several authors proposed intrinsic microscopic models for the SWE to explain these observations. Since a complete review of all published microscopic models for the SWE is beyond the scope of this publication, we discuss only the three following models.

10.5.4.1 Excess Charge-Carrier Recombination and Weak Si–Si Bond Breaking

In a series of experiments, Staebler and Wronski showed that a metastable increase in the db concentration can also be generated by injecting excess charge carriers in a-Si:H pin diodes and can be prevented when light illumination is combined with a strong reverse bias to remove light-generated charge carriers from the i-layer. It is therefore clear that the degradation is induced by excess charge-carrier trapping or recombination and not by the initial light-absorption event itself [59]. From these experiments, Stutzmann et al. [39] concluded that bimolecular recombination of excess charge carriers is the cause of db creation and derived a differential equation for the db concentration N_s

$$\frac{dN_s}{dt} = c_{sw} n p \tag{10.12}$$

where n, p are the electron and hole concentrations, respectively, and c_{sw} denotes an empirical proportionality factor depending on microscopic details of the db creation mechanism. If there is already a certain amount of dbs available, Eq. (10.12) can be easily solved by assuming that most of the recombination is taking place at db sites. In this monomolecular recombination limit, the charge carrier densities n and p are proportional to G/N_s, where G is the photocarrier generation rate. Together with Eq. (10.12), we then obtain the experimentally observed db creation dynamics $N_s \propto t^{1/3} G^{2/3}$. This result also reproduces the intensity dependence, if we assume that G is proportional to the light intensity I. As for the microscopic process leading to db creation, Stutzmann et al. proposed a breaking of weak Si–Si bonds in the valence band tail after trapping and bimolecular recombination of electrons with holes at the *same* weak bond site. In their model, the weak bond enters an excited state after trapping an electron–hole pair, which leads to the formation of two adjacent dbs (cf. Figure 10.15a). Since such a configuration is unlikely to be stable with an annealing energy barrier of 1.1 eV and is not compatible with the absence of strong exchange coupling in the ESR properties, the dbs must be separated on a short timescale after their creation.

Several microscopic models involving H were proposed in the literature to explain the separation of dbs. Among them are hydrogen bond switching [39, 60], dissociation of H_2^*-complexes [61], and hydrogen migration on internal hydrogenated

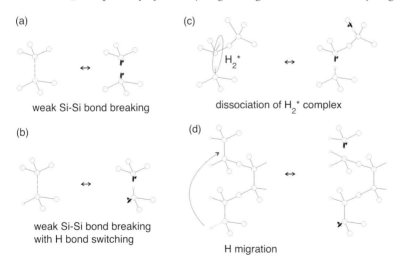

Figure 10.15 Different microscopic defect reactions proposed for the creation of metastable dbs in the SWE. (a) breaking of weak Si-Si bonds, (b) hydrogen bond switching, (c) dissociation of H_2^*-complexes, and (d) hydrogen migration on internal hydrogenated surfaces. Gray objects denote Si atoms, open objects denote H atoms, and dbs are indicated by a schematic wavefunction and an arrow indicating the electron spins.

surfaces [62] (Figure 10.15b–d). All these models predict dbs which are spatially correlated to H atoms.

10.5.4.2 Si–H Bond Dissociation and Hydrogen Collision Model

The fact that ESR experiments do not show dbs spatially correlated to H atoms stimulated a variety of alternative microscopic models for the SWE. Direct emission of H atoms from Si–H bond, leading to the formation of a db, and subsequent diffusion of H away from the broken bond was excluded as a source of the SWE, since this process is irreversible [63]. Branz extended the Si–H bond dissociation model by proposing a collision step between mobile H atoms, originally emitted from Si–H bonds [64]. The two mobile H atoms form a metastable H-complex denoted by $M(Si-H)_2$ in a strongly exothermic reaction as shown in Figure 10.16. By annealing to 150 °C for several hours, the process is reversed and the metastable complexes dissociate and saturate the dbs. The predicted time evolution of the db concentration is again $N_s \propto t^{1/3} G^{2/3}$, as in the case of the weak Si–Si bond breaking model, but derived from different physical processes.

10.5.4.3 Transformation of Existing Nonparamagnetic Charged Dangling-Bond Defects

Another model accounting for the light-induced increase in paramagnetic dbs was introduced by Adler and proposes a charge transfer between originally charged dbs

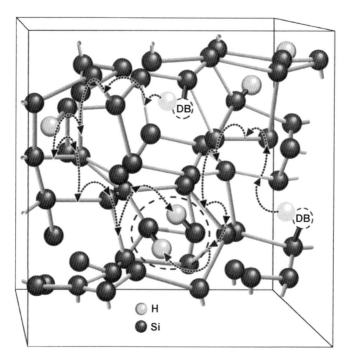

Figure 10.16 Schematic representation of the microscopic processes within the framework of the H collision model. Figure reprinted with permission from Ref. [64] by Branz et al. Copyright (1999) by the American Physical Society.

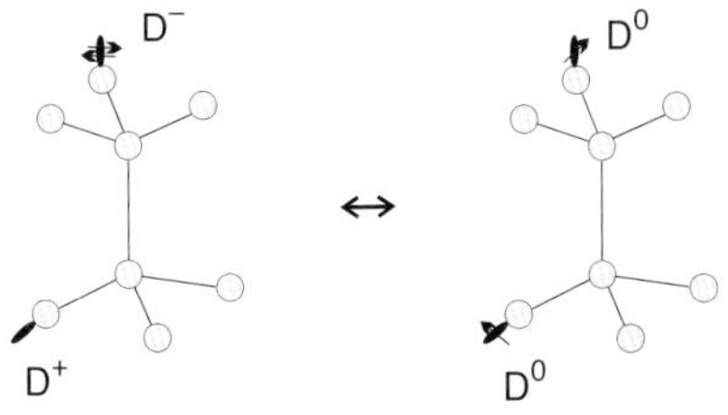

Figure 10.17 Schematic representation of the charge-transfer model proposed by Adler [41]. Gray objects denote Si atoms and dbs are indicated by a schematic wavefunction and an arrow indicating the electron spin.

(D^+, D^-) with an effective negative correlation energy [41] (cf. Figure 10.17). This process does not create new dbs, but transforms nonparamagnetic centers into paramagnetic dbs, which are then observable by ESR. However, charged db densities appear to be too low in undoped a-Si:H to support this model and the experimentally determined time evolution of the db density is not compatible with the one-carrier capture mechanism (see [64] and references therein).

In the preceding discussion we have seen that different models, based on completely different microscopic processes, do actually result in the same unusual sublinear time dependence. It is therefore most difficult to distinguish between the available theories by merely considering the time evolution of the db concentration. Another approach to deduce the microscopic origin and dynamics of the SWE is a microscopic investigation of the final metastable state by ESR. Here different models do actually predict quite different microscopic centers for the metastable state and its environment, for example, the hydrogen distribution around the defect, offering a possibility to test the available theoretical models.

In contrast to the H collision model (cf. Section 10.5.4.2), the weak Si–Si bond-breaking model (cf. Section 10.5.4.1) predicts a spatially correlated H distribution for light-induced dbs (light-soaked a-Si:H – state B) but not for native dbs (as-grown or annealed a-Si:H – state A) as noted earlier. We have seen in Section 10.4.2 that close magnetic nuclei like H induce a characteristic splitting of the db resonance by HFI. Since native dbs, present in as-deposited a-Si:H or in the annealed state A, do not exhibit the predicted defect-H correlation, db ESR spectra of degraded samples should differ significantly. Several studies were performed to address this question. Brandt et al. [65] carried out EDMR experiments, while Isoya et al. [66] conducted ESEEM experiments to measure the HFI of H atoms directly.

ESEEM measurements clearly showed a contribution of distant matrix H atoms with a small HFI to the db, but no larger HFI, which would indicate the presence of close H atoms. Relying on these observations, Isoya et al. concluded that the immediate vicinity ($r < 4$ Å) is actually depleted from H, an interpretation which disagrees with the predictions of the model described in Section 10.5.4.1. In a recent contribution [14], pulsed ENDOR measurements showed that this interpretation is not correct and the authors of Ref. [66] were mislead by a typical

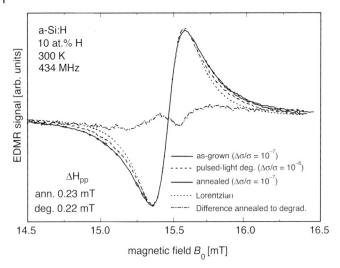

Figure 10.18 EDMR spectra of undoped a-Si:H at an excitation frequency of 434 MHz at room temperature for different sample conditions indicated in the legend. Samples are light-soaked by pulse light degradation. Figure reprinted from Ref. [65] by Brandt et al. with permission from Elsevier.

artifact of two-pulse ESEEM experiments, observed in a wide range of disordered materials [67].

Brandt et al. [65] performed a more reliable experiment to determine the HFI of H atoms in the vicinity of dbs. By applying EDMR at low excitation frequencies ($\nu = 434$ MHz), contributions due to g anisotropy were avoided and the line shape is only determined by HFI. Figure 10.18 compares EDMR spectra for native dbs (state A) and light-induced dbs (state B). The obtained spectra are nearly identical, hence no evidence for the presence of spatially correlated db–H pairs is found.

The fundamental problem inhibiting a complete resolution of the microscopic defect structure and its environment is the disorder-induced inhomogeneous broadening. Once spectral lines are subject to such excessive broadening, they become undetectable due to a finite experimental resolution and sensitivity. However, there is still room for improvement by the application of advanced ESR techniques at higher magnetic fields. We recently managed to increase resolution and sensitivity by the application of a Q-Band pulse ENDOR technique at a magnetic field of 1.2 T, using a laboratory-built ESR resonator [14]. Investigation of HFI between native dbs in state A and H atoms shows that HFIs of up to 25 MHz are observed (see Figure 10.19a), indicating that H atoms can be very close to the db. By comparing the ENDOR spectrum with typical spectra for free SiH_3, we can conclude that a small amount of H atoms are bonded to db silicon atoms at a back-bonded site. However, the results show that there is no observable short-range order and the H distribution is well described by a homogeneous distribution (see Figure 10.19b).

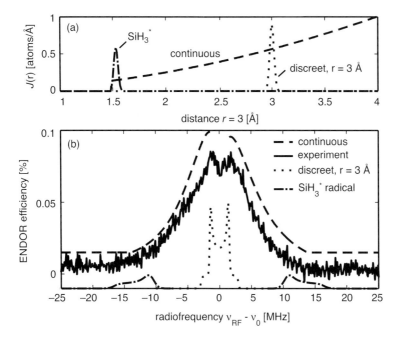

Figure 10.19 ENDOR spectroscopy of native db in undoped a-Si:H. (a) Radial distribution function $J(r)$ for a continuous (dashed line) and discreet (dotted line) distribution of H atoms around dbs. The position of H atoms in free SiH_3 radicals is shown by the dash-dotted line. (b) Corresponding ENDOR spectra for distributions in (a) and comparison with the experimental spectrum. The dashed, dotted, and dash-dotted spectra are offset vertically with respect to the experimental spectrum for better overview. For further details please refer to Ref. [14].

10.6
Alternative ESR Detection

The application of ESR to state-of-the-art a-Si:H thin films or even completely processed a-Si:H solar cells remains a challenge because the sensitivity of conventional ESR is usually not sufficient to reliably detect the paramagnetic defects. Moreover, it is often difficult to discriminate between signals from the cell structure, contact layers, or background signals from the substrate [68]. Although a correlation between the defect density as determined by ESR on powder samples and the efficiency of resulting solar cells is well established [69], it still remains questionable if the defect properties of thin films in the device and in powder samples are identical, because the boundary conditions for the layer growth are different. The sensitivity limit of ESR, which prevents its successful application to thin-film devices, can be overcome by using indirect detection techniques. These methods do not rely on the detection of reflected (or absorbed) mw radiation, but monitor the influence of ESR on sensitively measurable macroscopic observables such as (photo-) conductivity, capacitance, optical absorption, photoluminescence, or electroluminescence. To elucidate the impact of paramagnetic defects on charge transport in solar cells, the (photo-) conductivity is an appropriate observable. For this reason, we restrict this paragraph to the description of EDMR. Due to the fact that we detect resonant changes in the (photo-) current, only electrically active defect states contribute to the spectrum. In consequence, paramagnetic states which do not influence the conductivity and are thus not relevant for solar-cell operation (e.g., defects in the substrate) do not affect the EDMR signal.

10.6.1
History of EDMR

EDMR, which combines the microscopic selectivity of ESR with the sensitivity of a current measurement, has a long-standing history in the field semiconductor defect analysis. It was first demonstrated in 1966 by Maxwell and Honig [70, 71] as well as Schmidt and Solomon [72] who measured the influence of ESR on the photoconductivity in c-Si. The underlying mechanism was spin-dependent scattering of electrons at impurities. In 1972, Lepine utilized EDMR for the detection of charge carrier recombination via defect states at a silicon surface [73]. Since then, EDMR was applied to large variety of organic and inorganic semiconductors. In particular, this technique provided insight into transport and recombination pathways in a-Si:H thin films and devices (for a review see [74]). Until recently, EDMR experiments were exclusively carried out like cw ESR experiments, that is, the sample was continuously subjected to mw radiation while sweeping an external magnetic field and detecting the conductivity. Thus, spectroscopic information (e.g., g values, line widths, and line shapes) about the current-influencing paramagnetic states could be extracted. Towards the beginning of this century, the field of EDMR got a new impetus when the first pulse (p) EDMR measurements employing short mw pulses were demonstrated [75]. By the use of pEDMR, it is not only possible to study the kinetics of spin-dependent transport and recombination processes, but also to harness coherent spin effects [76] in a similar way like in pulsed ESR. An analysis of the time evolution in two-dimensional pEDMR measurements (ΔI vs. B_0 and time) allows the deconvolution of spectrally overlapping signals [77]. In this way, pEDMR signals may be assigned to defect states in the individual layers of a pin solar cell. Moreover, for spin-dependent processes involving two different types of paramagnetic centers (e.g., a hopping process between conduction band-tail states and phosphorous donor states in n-doped a-Si:H) the contributing pairs of paramagnetic centers can be identified based on the fact that the associated pEDMR signals exhibit the same time behavior.

The high sensitivity of EDMR originates from two effects: first, a current can be detected with high accuracy. Second, the EDMR signal intensity is not necessarily limited by the thermal polarization of the contributing paramagnetic centers which may be very small under standard ESR conditions. In fact, two spin-dependent mechanisms frequently encountered in disordered thin-film materials – namely recombination and hopping transport via localized states – rely on a spin-pair mechanism

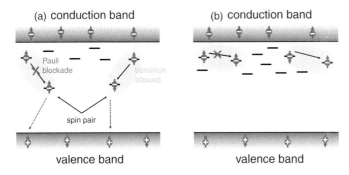

Figure 10.20 Spin-dependent processes involving localized states in the band gap as observed in thin-film silicon. For simplicity, only localized states in the upper half of the band gap are depicted. (a) Spin-dependent recombination. (b) Spin-dependent hopping transport. The arrows indicate the orientation of the electron and hole spins.

in which the relative orientation of two spins determines the probability for a current-influencing transition. This model was developed by Kaplan et al. [78] and is referred to as the *KSM model*. It is equally applicable to spin-dependent recombination as well as hopping transport. Both processes are shown schematically in Figure 10.20.

In the case of a recombination process as shown in Figure 10.20a, the spin pair consists of two localized states with different energy. Although it would be energetically favorable if both electrons occupied the lower lying state, owing to the Pauli principle this transition is only allowed when both spins are aligned antiparallel. Thus, invoking that the spin state is conserved during the process, the transition is blocked in the case of parallel spins. The application of mw radiation, which is resonant with either of the two spin-pair constituents, alters the respective spin state and the initially forbidden transition becomes allowed. The recombination process is finally completed by the capture of a hole from the valence band. In addition, an electron from the conduction band can be trapped at the unoccupied defect state. All in all, one electron from the conduction band and one hole from the valence band are annihilated, resulting in a decrease in the sample conductivity. In the case of hopping transport as shown in Figure 10.20b, the spin pair consists of two states having similar energy close to E_F. Analogous to the recombination process, the transition between adjacent singly occupied states can be blocked by the relative spin orientation. Again, resonant mw radiation can flip one spin and enable an initially forbidden transition. The resonant enhancement of the hopping rate can be considered to enhance the mobility in the hopping transport path. For a detailed review on spin-dependent hopping see Ref. [79].

10.6.2
EDMR on a-Si:H Solar Cells

The application of EDMR to fully processed semiconductor devices dates back to the early days of EDMR. In 1976, Solomon performed spin-dependent conductivity measurements on a commercial diode based on crystalline silicon [80]. He observed an EDMR signal associated with recombination of excess charge carriers that were injected by applying a bias voltage and found a strong dependence of the signal intensity on the voltage. This voltage dependence could be modeled when taking into account the different recombination mechanisms (i.e., diffusion- and recombination-limited processes) that occur in p–n diodes. It was concluded that the spin-dependent signal is sensitive only to recombination in the space-charged regions, which renders EDMR suitable to differentiate between diffusion- and drift-limited effects. Several studies on c-Si p–n diodes provided a quite detailed picture of spin-dependent recombination via intrinsic defects as well as impurity states and defect complexes [81–84]. Although the mechanisms of EDMR in devices can be quite complex in detail, an identification of the contributing defect states is often possible based on the line parameters which in many cases agree well with those found in ESR spectra of the respective material. In addition, extensive literature is available on EDMR in a-Si:H thin films [74].

After the demonstration that EDMR can successfully be applied to a-Si:H pin solar cells [86], spin-dependent recombination studies revealed a strong variation of the EDMR signal at room temperature with bias voltage and illumination conditions [85, 87]. Figure 10.21 shows EDMR spectra of an a-Si:H pin solar cell measured at different bias voltages without illumination. When the db states are brought into spin resonance, the recombination in the space charge region of the pin solar cell is enhanced. Since the dark currents in this bias regime are limited by recombination, an increase in the current is observed. The turnover between an enhancing and a quenching EDMR signal upon varying the bias voltage was shown to be related to the

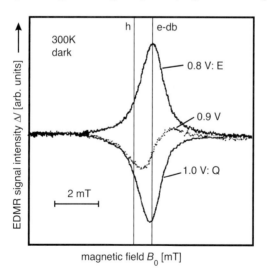

Figure 10.21 Dark EDMR spectrum obtained at $T = 300$ K on a-Si:H pin solar cell at different forward bias. Note that the signal changes sign at about 0.9 V. The peak of the resonance is at $g \approx 2.005$. Figure reprinted with permission from Ref. [85] by Lips et al. Copyright (1993), American Institute of Physics.

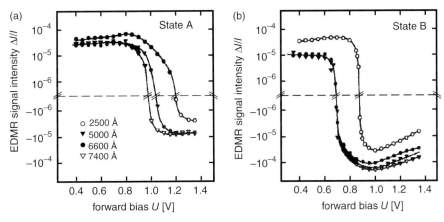

Figure 10.22 EDMR signal intensity ($\Delta I/I$) as a function of the bias voltage applied to a-Si:H pin solar cells for several thicknesses of the i-layer as given in the legend. (a) annealed (State A), (b) degraded (State B). Figure reprinted from Ref. [88] by Lips et al. with permission from Elsevier.

transition from diffusion- to drift-dominated behavior – similar to crystalline p–n diodes. Moreover, it was demonstrated by means of EDMR that charge transport and carrier collection are controlled by recombination in the bulk of the i-layer [87]. The finding that light-induced defect creation in this layer (cf. Section 10.5.4) is detrimental for the solar cell performance together with the fact that EDMR probes recombination via db states make this technique well suited to investigate degradation mechanisms in a-Si:H-based solar cells at ambient conditions [88]. Figure 10.22 illustrates this fact by comparing the bias-voltage dependence of the EDMR signal intensity recorded under dark forward bias injection conditions ($\Delta I/I$, with ΔI is the current change that is induced by the resonant mw) of annealed (Figure 10.22a) and degraded (Figure 10.22b) a-Si:H pin solar cells with varying i-layer thicknesses. The degradation was carried out by subjecting the cells to a high forward current to ensure homogeneous degradation throughout the i-layer (150 mA cm^{-2} for 3 h). Figure 10.22 indicates a decrease in the turnover voltage (change of sign from positive to negative EDMR signal) by approx. 0.3 V upon degradation. Further, the quenching signal amplitude increases by more than one order of magnitude. The situation with respect to recombination in the solar cell is very complex and EDMR results cannot be interpreted right away since recombination may both enhance and decrease the dark current. Nevertheless it is possible to directly simulate EDMR with solar-cell simulation tools only assuming that through ESR the capture cross section of electrons in neutral db states is slightly increased by the mw [89]. From comparison to simulations, it became clear that degradation is predominantly a bulk effect in the i-layer of the pin a-Si:H solar cell. These observations demonstrate the sensitivity of EDMR with respect to defects created by degradation of a-Si:H pin solar cells.

Phosphorus-doped a-Si:H layers are frequently used in pin solar cells with microcrystalline silicon (μc-Si:H) absorber. Charge transport in these n-doped layers with typical thicknesses of a few 10 nm is relevant for solar-cell operation because

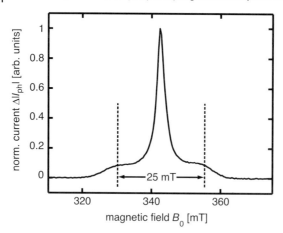

Figure 10.23 X-Band pEDMR spectrum (resonant change in photoconductivity, $\Delta|I_{ph}|$, normalized to 1 as a function of the magnetic field) of a µc-Si:H pin solar cell with amorphous n-layer measured at $T = 10$ K. The experiment was performed under reverse bias ($U = -1.0$ V), and the sample was illuminated by a halogen lamp. The dashed vertical lines denote the positions of two ^{31}P hyperfine lines with their center-of-gravity at $g = 2.003$.

electrons generated in the absorber have to transverse them in order to be collected at the metal electrode. Especially at low temperatures, transport in n-doped a-Si:H occurs via hopping among localized band tail and phosphorus donor states and can thus be observed by EDMR. Figure 10.23 shows a pEDMR spectrum of a µc-Si:H pin solar cell with amorphous n-layer which clearly reveals the ^{31}P hyperfine lines (with a splitting of approx. 25 mT) known from ESR measurements on n-doped a-Si:H powder samples (cf. Section 10.5.3). The central line originates from spin-dependent hopping of electrons via conduction band-tail states in the n-doped a-Si:H layer as well as in the intrinsic µc-Si:H absorber [90]. Similar effects can be observed in high-efficiency heterojunction solar cells consisting of c-Si as absorber and n-doped a-Si:H as emitter layer [91].

These examples demonstrate that EDMR is capable of sensitively detecting electrically active paramagnetic states in fully processed solar cells and can identify the underlying transport and recombination mechanisms.

10.7
Concluding Remarks

ESR has contributed significantly to the understanding of the microscopic structure of defect states in a-Si:H. In particular, db states have been shown to be the dominant midgap defects in a-Si:H and are responsible for the electronic degradation of a-Si:H after light soaking. ESR was able to unravel the structure and influence of dopant states. However, most ESR results rely on experiments with only limited resolution and quantum chemical methods, which did not allow to link the magnetic interaction parameters with the defect structure and distribution. This generally insufficient resolution and detection sensitivity of conventional EPR has dramatically changed with the advent of novel multifrequency EPR spectrometers and detection schemes as well sophisticated DFT simulation tools, which allow determining information about spin-coupling parameters inaccessible until then. Recently, a German network called *EPR-Solar* was formed with the aim to transfer these advances in EPR instrumentation and theory to solve some of the remaining open questions related to defect states in a-Si:H materials, in particular the SWE [33]. Latest ESR results [34] and DFT calculations [92] show that db states in a-Si:H have a complex structure and cannot be described by simple sp-hybridized states. In order to transfer the obtained ESR results into the microscopic structure of defect states with their respective spin-density distribution, DFT calculations of the magnetic interaction parameters (e.g., HFI with ^{29}Si, g tensor) of many different db models are required. Currently such tools have become available [93–95] and are currently used by *EPR-Solar* to shed light on the microscopic structure of a-Si:H defects and their involvement in the SWE.

References

1 Shah, A., Torres, P., Tscharner, R., Wyrsch, N., and Keppner, H. (1999) Photovoltaic technology: The case for thin-film solar cells. *Science*, **285**, 692.

2 Street, R.A. (1991) *Hydrogenated Amorphous Silicon*, Cambridge University Press, Cambridge, New York.

3 Atherton, N.M. (1993) *Principles of Electron Spin Resonance*, Ellis Horwood, PTR Prentice-Hall, New York.

4 Poole, C.P. (1983) *Electron Spin Resonance: A Comprehensive Treatise on Experimental Techniques*, Wiley, New York.

5 Weger, M. (1960) Passage effects in paramagnetic resonance experiments. *Bell Syst. Tech. J.*, **39**, 1013.

6 Feher, G. (1957) Sensitivity considerations in microwave paramagnetic resonance absorption techniques. *Bell Syst. Tech. J.*, **36**, 449.

7 Schweiger, A. and Jeschke, G. (2001) *Principles of Pulse Electron Paramagnetic Resonance*, Oxford University Press, Oxford, UK, New York.

8 Hahn, E.L. (1950) Spin echoes. *Phys. Rev.*, **80**, 580.

9 Stutzmann, M. and Stuke, J. (1983) Paramagnetic states in doped amorphous-silicon and germanium. *Solid State Commun.*, **47**, 635.

10 Stutzmann, M. and Biegelsen, D.K. (1989) Microscopic nature of coordination defects in amorphous-silicon. *Phys. Rev. B*, **40**, 9834.

11 Umeda, T., Yamasaki, S., Isoya, J., and Tanaka, K. (1999) Electron-spin-resonance center of dangling bonds in undoped a-Si: H. *Phys. Rev. B*, **59**, 4849.

12 Stoll, S. and Schweiger, A. (2006) Easy spin, a comprehensive software package for spectral simulation and analysis in EPR. *J. Magn. Reson.*, **178**, 42.

13 Vandewalle, C.G. and Blochl, P.E. (1993) 1st-principles calculations of hyperfine parameters. *Phys. Rev. B*, **47**, 4244.

14 Fehr, M., Schnegg, A., Teutloff, C., Bittl, R., Astakhov, O., Finger, F., Rech, B., and Lips, K. (2010) Hydrogen distribution in the vicinity of dangling bonds in hydrogenated amorphous silicon (a-Si:H). *phys. status solidi (a)*, **207**, 552.

15 Biegelsen, D.K. and Stutzmann, M. (1986) Hyperfine studies of dangling bonds in amorphous-silicon. *Phys. Rev. B*, **33**, 3006.

16 Brodsky, M.H. and Title, R.S. (1969) Electron spin resonance in amorphous silicon, germanium, and silicon carbide. *Phys. Rev. Lett.*, **23**, 581.

17 Walters, G.K. and Estle, T.L. (1961) Paramagnetic resonance of defects introduced near surface of solids by mechanical damage. *J. Appl. Phys.*, **32**, 1854.

18 Chittick, R.C., Alexande, Jh., and Sterling, H.F. (1969) Preparation and properties of amorphous silicon. *J. Electrochem. Soc.*, **116**, 77.

19 Astakhov, O., Carius, R., Finger, F., Petrusenko, Y., Borysenko, V., and Barankov, D. (2009) Relationship between defect density and charge carrier transport in amorphous and microcrystalline silicon. *Phys. Rev. B*, **79**, 104205.

20 Dersch, H., Stuke, J., and Beichler, J. (1981) Temperature-dependence of electron-spin-resonance spectra of doped a-Si:H. *phys. status solidi (b)*, **107**, 307.

21 Lee, J.K. and Schiff, E.A. (1992) Modulated electron-spin-resonance measurements and defect correlation energies in amorphous-silicon. *Phys. Rev. Lett.*, **68**, 2972.

22 Ishii, N., Kumeda, M., and Shimizu, T., (1982) The effects of H and F on the electron-spin-resonance signals in a-Si. *Jpn. J. Appl. Phys. 2*, **21**, L92.

23 Caplan, P.J., Poindexter, E.H., Deal, B.E., and Razouk, R.R. (1979) ESR centers, interface states, and oxide fixed charge in thermally oxidized silicon wafers. *J. Appl. Phys.*, **50**, 5847.

24 Poindexter, E.H., Caplan, P.J., Deal, B.E., and Razouk, R.R. (1981) Interface states and electron-spin resonance centers in thermally oxidized (111) and (100) silicon-wafers. *J. Appl. Phys.*, **52**, 879.

25 Pantelides, S.T. (1986) Defects in amorphous-silicon – a new perspective. *Phys. Rev. Lett.*, **57**, 2979.

26 Stutzmann, M. and Biegelsen, D.K. (1988) Dangling or floating bonds in amorphous-silicon. *Phys. Rev. Lett.*, **60**, 1682.

27 Pantelides, S.T. (1987) Defects in amorphous-silicon – a new perspective – reply. *Phys. Rev. Lett.*, **58**, 2825.

28 Fedders, P.A. and Carlsson, A.E. (1987) Energetics of single dangling and floating bonds in amorphous Si. *Phys. Rev. Lett.*, **58**, 1156.

29 Watkins, G.D. and Corbett, J.W. (1964) Defects in irradiated silicon – electron paramagnetic resonance + electron-nuclear double resonance of Si-E center. *Phys. Rev. A*, **134**, 1359.

30 Brower, K.L. (1983) 29Si hyperfine-structure of unpaired spins at the Si/SiO$_2$ interface. *Appl. Phys. Lett.*, **43**, 1111.

31 Cook, M. and White, C.T. (1988) Hyperfine interactions in cluster-models of the Pb defect center. *Phys. Rev. B*, **38**, 9674.

32 Neese, F. and Munzarova, M. (2004) Historical aspects of EPR parameter calculations, in *Calculation of NMR and EPR Parameters: Theory and Applications* (eds M. Kaupp, M. Bühl, and V.G. Malkin), Wiley-VCH.

33 http://www.helmholtz-berlin.de/forschung/enma/si-pv/projekte/bmbf/index_de.html (12 July 2010).

34 Teutloff, C., Fehr, M., Schnegg, A., Astakhov, O., Rech, B., Finger, F., Lips, K., and Bittl, R. (2010) Multifrequency EPR on defects in hydrogenated amorphous silicon. *to be published*.

35 Ishii, N. and Shimizu, T. (1997) Cluster-model calculations of hyperfine coupling constants of dangling bond and weak bond in a-Si:H. *Solid State Commun.*, **102**, 647.

36 Shimakawa, K., Kolobov, A., and Elliott, S.R. (1995) Photoinduced effects and metastability in amorphous semiconductors and insulators. *Adv. Phys.*, **44**, 475.

37 Knights, J.C., Biegelsen, D.K., and Solomon, I. (1977) Optically induced electron-spin resonance in doped amorphous silicon. *Solid State Commun.*, **22**, 133.

38 Street, R.A. and Biegelsen, D.K. (1980) Luminescence and electron-spin-resonance studies of defects in hydrogenated amorphous-silicon. *Solid State Commun.*, **33**, 1159.

39 Stutzmann, M., Jackson, W.B., and Tsai, C.C. (1985) Light-induced metastable defects in hydrogenated amorphous-silicon – a systematic study. *Phys. Rev. B*, **32**, 23.

40 Anderson, P.W. (1958) Absence of diffusion in certain random lattices. *Phys. Rev.*, **109**, 1492.

41 Adler, D. (1983) Origin of the photoinduced changes in hydrogenated amorphous-silicon. *Solar Cells*, **9**, 133.
42 Hautala, J. and Cohen, J.D. (1993) Esr studies on a-Si-H – evidence for charged defects and safe hole traps. *J. Non-Cryst. Solids*, **166**, 371.
43 Morigaki, K. (1981) Spin-dependent radiative and nonradiative recombinations in hydrogenated amorphous-silicon – optically detected magnetic-resonance. *J. Phys. Soc. Jpn.*, **50**, 2279.
44 Adler, D. (1981) Defects in amorphous chalcogenides and silicon. *J. Phys.-Paris*, **42**, C43.
45 Umeda, T., Yamasaki, S., Isoya, J., Matsuda, A., and Tanaka, K. (1996) Electronic structure of band-tail electrons in a Si:H. *Phys. Rev. Lett.*, **77**, 4600.
46 Street, R.A., Biegelsen, D.K., and Knights, J.C. (1981) Defect states in doped and compensated a-Si-H. *Phys. Rev. B*, **24**, 969.
47 Stutzmann, M. (1987) Electron-spin resonance of shallow defect states in amorphous-silicon and germanium. *J. Non-Cryst. Solids*, **97–8**, 105.
48 Dersch, H., Stuke, J., and Beichler, J. (1981) Electron-spin resonance of doped glow-discharge amorphous-silicon. *phys. status solidi (b)*, **105**, 265.
49 Fletcher, R.C., Yager, W.A., Pearson, G.L., Holden, A.N., Read, W.T., and Merritt, F.R. (1954) Spin resonance of donors in silicon. *Phys. Rev.*, **94**, 1392.
50 Stutzmann, M. and Street, R.A. (1985) Donor states in hydrogenated amorphous-silicon and germanium. *Phys. Rev. Lett.*, **54**, 1836.
51 Schumm, G., Jackson, W.B., and Street, R.A. (1993) Nonequilibrium occupancy of tail states and defects in alpha-SiH – implications for defect structure. *Phys. Rev. B*, **48**, 14198.
52 Stutzmann, M., Biegelsen, D.K., and Street, R.A. (1987) Detailed investigation of doping in hydrogenated amorphous-silicon and germanium. *Phys. Rev. B*, **35**, 5666.
53 Feher, G. (1959) Electron spin resonance experiments on donors in silicon.1. Electronic structure of donors by the electron nuclear double resonance technique. *Phys. Rev.*, **114**, 1219.
54 Stutzmann, M. and Biegelsen, D.K. (1983) Electron-spin-lattice relaxation in amorphous-silicon and germanium. *Phys. Rev. B*, **28**, 6256.
55 Staebler, D.L. and Wronski, C.R. (1977) Reversible conductivity changes in discharge-produced amorphous Si. *Appl. Phys. Lett.*, **31**, 292.
56 Staebler, D.L. and Wronski, C.R. (1980) Optically induced conductivity changes in discharge-produced hydrogenated amorphous-silicon. *J. Appl. Phys.*, **51**, 3262.
57 Dersch, H., Stuke, J., and Beichler, J. (1981) Light-induced dangling bonds in hydrogenated amorphous-silicon. *Appl. Phys. Lett.*, **38**, 456.
58 Fritzsche, H. (2001) Development in understanding and controlling the Staebler–Wronski effect in a-Si: H. *Ann. Rev. Mater. Res.*, **31**, 47.
59 Stutzmann, M. (1997) Microscopic aspects of the Staebler-Wronski effect, in *Amorphous and Microcrystalline Silicon Technology – 1997* (eds S. Wagner, M. Hack, E.A. Schiff, R. Schropp, and I. Shimizu), Materials Research Society, Warrendale.
60 Morigaki, K. (1988) Microscopic mechanism for the photo-creation of dangling bonds in a-Si-H. *Jpn. J. Appl. Phys. Part 1*, **27**, 163.
61 Zhang, S.B., Jackson, W.B., and Chadi, D.J. (1990) Diatomic-hydrogen-complex dissociation: A microscopic model for metastable defect generation in Si. *Phys. Rev. Lett.*, **65**, 2575.
62 Carlson, D.E. (1986) Hydrogenated microvoids and light-induced degradation of amorphous-silicon solar cells. *Appl. Phys. A – Mater.*, **41**, 305.
63 Stutzmann, M., Jackson, W.B., Smith, A.J., and Thompson, R. (1986) Light-induced metastable defects in amorphous-silicon – the role of hydrogen. *Appl. Phys. Lett.*, **48**, 62.
64 Branz, H.M. (1999) Hydrogen collision model: Quantitative description of metastability in amorphous silicon. *Phys. Rev. B*, **59**, 5498.
65 Brandt, M.S., Bayerl, M.W., Stutzmann, M., and Graeff, C.F.O. (1998) Electrically detected magnetic resonance of a-Si: H at

low magnetic fields: The influence of hydrogen on the dangling bond resonance. *J. Non-Cryst. Solids*, **230**, 343.

66 Isoya, J., Yamasaki, S., Okushi, H., Matsuda, A., and Tanaka, K. (1993) Electron-spin-echo envelope-modulation study of the distance between dangling bonds and hydrogen-atoms in hydrogenated amorphous-silicon. *Phys. Rev. B*, **47**, 7013.

67 Höfer, P. (1994) Distortion-free electron-spin-echo envelope-modulation spectra of disordered solids obtained from 2-dimensional and 3-dimensional hyscore experiments. *J. Magn. Reson., Ser. A*, **111**, 77.

68 Lee, C., Ohlsen, W.D., Taylor, P.C., and Engstrom, O. (1987) ESR in pin structures based on a-Si:H, in *18th International Conference on the Physics of Semiconductors* (ed. O. Engstrom) World Scientific, Singapore, Stockholm, Sweden.

69 Neto, A.L.B., Lambertz, A., Carius, R., and Finger, F. (2002) Relationships between structure, spin density and electronic transport in 'solar-grade' microcrystalline silicon films. *J. Non-Cryst. Solids*, **299**, 274.

70 Honig, A. (1966) Neutral-impurity scattering and impurity zeeman spectroscopy in semiconductors using highly spin-polarized carriers. *Phys. Rev. Lett.*, **17**, 186.

71 Maxwell, R. and Honig, A. (1966) Neutral-impurity scattering experiments in silicon with highly spin-polarized electrons. *Phys. Rev. Lett.*, **17**, 188.

72 Schmidt, J. and Solomon, I. (1966) Modulation De La photoconductivite dans le silicium a basse temperature par resonance magnetique electronique des impuretes peu profondes. *Compt. Rend. Acad. Sci. B*, **263**, 169.

73 Lepine, D.J. (1972) Spin-dependent recombination on silicon surface. *Phys. Rev. B*, **6**, 436.

74 Stutzmann, M., Brandt, M.S., and Bayerl, M.W. (2000) Spin-dependent processes in amorphous and microcrystalline silicon: A survey. *J. Non-Cryst. Solids*, **266–269**, 1.

75 Boehme, C. and Lips, K. (2001) Time domain measurement of spin-dependent recombination. *Appl. Phys. Lett.*, **79**, 4363.

76 Boehme, C. and Lips, K. (2003) Electrical detection of spin coherence in silicon. *Phys. Rev. Lett.*, **91**, 246603.

77 Behrends, J., Schnegg, A., Fehr, M., Lambertz, A., Haas, S., Finger, F., Rech, B., and Lips, K. (2009) Electrical detection of electron spin resonance in microcrystalline silicon pin solar cells. *Philos. Mag.*, **89**, 2655.

78 Kaplan, D., Solomon, I., and Mott, N.F. (1978) Explanation of large spin-dependent recombination effect in semiconductors. *Journ. de Phys. Lettr.*, **39**, L51.

79 Boehme, C. and Lips, K. (2006) The investigation of charge carrier recombination and hopping transport with pulsed electrically detected magnetic resonance techniques, in *Charge Transport in Disordered Solids with Applications in Electronics* (ed. S. Baranovski), Wiley, Chichester, England, Hoboken, NJ.

80 Solomon, I. (1976) Spin-dependent recombination in a silicon para-normal junction. *Solid State Commun.*, **20**, 215.

81 Rong, F., Poindexter, E.H., Harmatz, M., Buchwald, W.R., and Gerardi, G.J. (1990) Electrically detected magnetic-resonance in P–N-junction diodes. *Solid State Commun.*, **76**, 1083.

82 Christmann, P., Wetzel, C., Meyer, B.K., Asenov, A., and Endros, A. (1992) Spin dependent recombination in Pt-doped silicon P–N-junctions. *Appl. Phys. Lett.*, **60**, 1857.

83 Xiong, Z. and Miller, D.J. (1993) General expression for the electrically detected magnetic-resonance signal from semiconductors. *Appl. Phys. Lett.*, **63**, 352.

84 Müller, R., Kanschat, P., von Aichberger, S., Lips, K., and Fuhs, W. (2000) Identification of transport and recombination paths in homo and heterojunction silicon solar cells by electrically detected magnetic resonance. *J. Non-Cryst. Solids*, **266–269**, 1124.

85 Lips, K. and Fuhs, W. (1993) Transport and recombination in amorphous P–I–N-type solar-cells studied by electrically detected magnetic-resonance. *J. Appl. Phys.*, **74**, 3993.

86 Homewood, K.P., Cavenett, B.C., Spear, W.E., and Lecomber, P.G. (1983) Spin effects in P+-I-N+ a-Si-H cells – photo-voltaic detected magnetic-resonance (PDMR). *J. Phys. C*, **16**, L427.

87 Fuhs, W. and Lips, K. (1993) Recombination in a-Si-H films and pin-structures studied by electrically detected magnetic-resonance (EDMR). *J. Non-Cryst. Solids*, **166**, 541.

88 Lips, K., Block, M., Fuhs, W., and Lerner, C. (1993) Degradation of a-Si-H P-I-N solar-cells studied by electrically detected magnetic-resonance. *J. Non-Cryst. Solids*, **166**, 697.

89 Lips, K., Boehme, C., and Fuhs, W. (2003) Recombination in silicon thin-film solar cells: A study of electrically detected magnetic resonance. *IEE Proc. – Circ. Dev. Syst.*, **150**, 309.

90 Behrends, J., Schnegg, A., Boehme, C., Haas, S., Stiebig, H., Finger, F., Rech, B., and Lips, K. (2008) Recombination and transport in microcrystalline pin solar cells studied with pulsed electrically detected magnetic resonance. *J. Non-Cryst. Solids*, **354**, 2411.

91 Boehme, C., Behrends, J., von Maydell, K., Schmidt, M., and Lips, K. (2006) Investigation of hopping transport in n-a-Si:H/c-Si solar cells with pulsed electrically detected magnetic resonance. *J. Non-Cryst. Solids*, **352**, 1113.

92 Jarolimek, K., de Groot, R.A., de Wijs, G.A., and Zeman, M. (2009) First-principles study of hydrogenated amorphous silicon. *Phys. Rev. B*, **79**, 155206.

93 Pickard, C.J. and Mauri, F. (2002) First-principles theory of the EPR g tensor in solids: Defects in quartz. *Phys. Rev. B*, **88**, 086403.

94 Gerstmann, U., Seitsonen, A.P., and Mauri, F. (2008) Ga self-interstitials in GaN investigated by *ab-initio* calculations of the electronic g-tensor. *physica status solidi (b)*, **245**, 924.

95 Ceresoli, D., Gerstmann, U., Seitsonen, A.P., and Mauri, F. (2010) First-principles theory of orbital magnetization. *Phys. Rev. B*, **81**, 060409.

11
Scanning Probe Microscopy on Inorganic Thin Films for Solar Cells

Sascha Sadewasser and Iris Visoly-Fisher

11.1
Introduction

Photovoltaic thin films are highly nonhomogenous at the micron- and submicron scale, involving layers of different composition, polycrystallinity, point- and extended-defects. Spatially resolved characterization at high resolution is therefore adequate to resolve the properties of these different parts and their effect on the device properties, without averaging over the entire material. Scanning probe microscopy allows such characterization, as will be demonstrated in this chapter.

The invention of the scanning tunneling microscope (STM) by Binnig and Rohrer in 1982 revolutionized the field of surface science [1]. The STM provided the means to obtain the first real space images showing atomic resolution on a Si(111) 7×7 surface. The STM uses the quantum mechanical tunneling current through a gap between a sharp metallic tip and the sample surface as a control parameter; therefore, it is limited to the study of conducting surfaces. A solution to this restriction was provided by the atomic force microscope (AFM), which employs a sharp tip supported by a cantilever beam as the measuring probe [2]. In AFM, the tip is in contact with the surface and the deflection of the cantilever is monitored by optical beam deflection detection. Thus, the AFM also provided access to insulating surfaces. A different technique in AFM was developed with the noncontact (or dynamic) mode [3], in which the cantilever is vibrated close to its resonance frequency. Tip–sample interaction changes the vibration amplitude or frequency, which serves as a feedback to maintain a constant distance to the sample surface while scanning across the sample. Forces exerted by the tip on the sample are minimal in noncontact mode; therefore, it is the method of choice for soft samples, as for example biological or polymer samples.

In the subsequent years, the AFM was developed to a very versatile technique by the combination of the AFM with other measurement methods. This provided access to additional sample properties on a lateral scale in the nanometer range. Combination with the macroscopic Kelvin probe technique led to the development of

Advanced Characterization Techniques for Thin Film Solar Cells,
Edited by Daniel Abou-Ras, Thomas Kirchartz and Uwe Rau.
© 2011 Wiley-VCH Verlag GmbH & Co. KGaA. Published 2011 by Wiley-VCH Verlag GmbH & Co. KGaA.

the Kelvin probe force microscope (KPFM). Application of an ac bias and detection of the capacitance resulted in scanning capacitance microscopy (SCM). The measurement of currents in the contact mode of the AFM with an applied bias between tip and sample is the core of the conductive AFM (C-AFM).

These and other techniques will be introduced in this chapter, providing the reader with the most important basics about each technique, and allowing understanding experiments performed by these techniques. Additionally, this chapter will also present some selected important results that have been obtained applying the various techniques to characterize thin-film solar cells. While only selected examples are discussed in this chapter, the interested reader is referred to the extensive list of references for further reading and exploration of the application of AFM- and STM-based techniques for the characterization of nanoscale properties of thin-film solar cell materials and devices.

11.2
Experimental Background

11.2.1
Atomic Force Microscopy

An AFM [2] consists of a probe in the shape of a cantilever with a small tip at its free end, a laser, a four-quadrant photodiode, and a scanner unit. The latter is usually constructed from piezoelectric elements. The laser beam is focused onto the back of the free end of the cantilever and from there reflected to the four-quadrant photodiode; this allows detecting the bending of the cantilever with high precision. While scanning the tip across a sample surface, the force interaction between tip and sample leads to changes either in the static bending of the cantilever (when the tip is in contact) or in the resonance frequency of the cantilever oscillation (when it is vibrated and the tip is at a small distance from the surface). The first operating mode is called contact mode and the second is called dynamic or noncontact mode. Both modes will be presented in detail in the subsequent subsections. Figure 11.1 schematically shows this basic AFM experimental setup.

The forces interacting between the tip and the sample consist of various contributions, including the chemical binding force F_{chem}, the van der Waals force F_{vdW}, electrostatic and magnetic forces F_{el} and F_{mag}, respectively. The short-range chemical forces originate from a quantum mechanical overlap of the electron wave functions of tip and sample. They can be described by an exponential distance dependence, and the interaction range is of the order of ~5 Å [4]. The long-range van der Waals force originates from electromagnetic field fluctuations. Typically, in AFM it is described by the interaction between a sphere of radius R, representing the tip, and an infinite plane, representing the sample surface [5]

$$F_{vdW} = -\frac{H R}{6 z^2} \qquad (11.1)$$

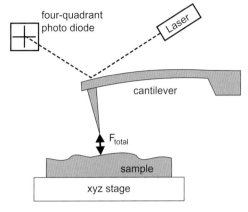

Figure 11.1 The working principle of an AFM showing the detection system, consisting of a laser and a position-sensitive photodiode, and the sample on a xyz-stage.

where H represents the Hamaker constant for the tip and sample material, and z the distance between the sphere and the plane. The electrostatic force F_{el} can be expressed by the capacitive force between the tip and the sample

$$F_{el} = -\frac{1}{2}\frac{\partial C}{\partial z}V^2 \qquad (11.2)$$

where C is the capacitance and V is the total voltage. In the present case of AFM, only the force perpendicular to the surface along the z-direction has to be considered. More details on F_{el} will be given in Section 11.2.4, where Kelvin probe force microscopy is discussed. The magnetic force is not relevant, as only nonmagnetic tips and samples are considered here.

Chemical forces are responsible for atomic scale contrast in AFM imaging. When experiments on a larger scale are performed, the tip–sample distance in noncontact-AFM is typically larger than 5 Å, and chemical forces do not have to be considered. In this case, topography imaging is governed by the long-range van der Waals and possibly by electrostatic forces.

11.2.1.1 Contact Mode

The point of contact between the tip and the sample surface is defined as the tip–surface distance at which tip–surface attractive forces (described above) are replaced by repulsive ones, originating from short-range Coulomb and Pauli repulsion between ion cores and electron clouds of atoms at the tip and on the sample surface. Further reducing the tip–surface distance will result in a normal deflection of the cantilever, which is proportional to the repulsive force acting on it according to Hooke's law

$$F = k_N z \qquad (11.3)$$

where k_N is the normal spring constant of the cantilever, determined by its material mechanical properties, shape, and dimensions. Typical spring constants of commercially available cantilevers range from 0.01 to 75 N/m, enabling detection of forces down to 10^{-9} N. For topography imaging, a feedback loop is used to keep the cantilever deflection constant by changing the probe height z while scanning in x and y. Thus, a nearly constant force is maintained between the tip and surface, and a map is created by recording the changes in z-position as a function of the x and y position, which can be interpreted as a topographical map. Lateral forces due to friction cause the cantilever to torque, and can also be followed by the four-quadrant photodiode. The lateral resolution of measurements is strongly dependent, among other factors, on the tip–surface contact area, which is increased as the tip and sample are driven toward each other. Several models can be used to calculate this contact area using the probe and surface mechanical properties [5]. Typically, this contact area will be circular with a diameter of 10–50 nm.

The advantages of the method include high lateral resolution and the ability to map the topography as well as adhesion forces, and measure surface mechanical properties at high resolution. This scan mode is also used for conductive-AFM mapping. The disadvantages include potential harm to the tip and/or surface and relatively low force resolution, issues addressed by the noncontact imaging mode. Applications of contact-mode AFM mapping in thin-film solar cells include the characterization of surface morphology, grain size and shape and its distribution [6], grain orientation (by measuring the angles between grain facets) [7], surface defects, and detecting the presence of impurity particles by following the surface mechanical properties.

11.2.1.2 Noncontact Mode

In noncontact AFM (nc-AFM) [8], the cantilever is mechanically oscillated at or close to its resonance frequency using a piezoelectric element. The four-quadrant photodiode signal is used to monitor the oscillation, which will be influenced by the tip–sample interaction. The oscillation of the cantilever is described by the equation of motion, where the tip is considered as a point-mass spring (mass m) and only the motion in z-direction, perpendicular to the sample surface, is considered [5]

$$m\ddot{z} + \frac{m\omega_0}{Q}\dot{z} + kz = F_0 \cos(\omega t) + F_{\text{total}} \tag{11.4}$$

where Q and k describe the quality factor and spring constant of the cantilever, respectively, and F_0 and ω represent the amplitude and angular frequency of the driving force, respectively. The tip–sample interaction is given by F_{total}.

For the case of no tip–sample interaction, that is, when the tip is far away from the surface, the free resonance frequency ω_0 is given by

$$\omega_0 = \sqrt{\frac{k}{m}} \tag{11.5}$$

where m is considered as the effective mass, which accounts for the specific geometry of the cantilever. When approaching the tip to the surface, forces between tip and surface become relevant; this influences the oscillation of the cantilever. As a result, the resonance frequency is changed according to [5]

$$k_{\text{eff}} = k - \frac{\partial F_{\text{total}}}{\partial z} \tag{11.6}$$

Essentially, the effective resonance frequency is modified according to the force gradient between tip and sample. For small force gradients and attractive forces, the resonance curve is shifted to lower frequencies and vice versa. The approximation in Eq. (11.6) is only valid for small oscillation amplitudes; the exact solution for larger amplitude can be found in Ref. [5].

Two different detection modes can be used in nc-AFM. In the amplitude modulation mode (AM-mode), the tip–sample distance is controlled such that the oscillation amplitude remains constant [3]. For this purpose, the cantilever oscillation is excited at a constant frequency slightly off resonance. The AM-mode is typically applied in measurements in air. When the AFM is introduced into a vacuum system, the quality factor Q of the cantilever increases significantly (typically above 10^5) as a result of the reduced damping. In this case, a very slow response time of the system results because of the reduced bandwidth [8]. Therefore, in vacuum, typically the frequency modulation mode (FM-mode) is used, in which the cantilever is always excited on resonance and the shift of the resonance curve due to the tip–sample interaction is directly measured [8]. The resonance frequency is determined by a phase-locked loop (PLL) or a FM demodulator. The tip–sample distance is controlled by maintaining a constant frequency shift Δf with respect to the free resonance frequency f_0 of the cantilever [5]

$$\Delta f = -\frac{f_0}{2k} \frac{\partial F_{\text{total}}}{\partial z} \tag{11.7}$$

Both the AM- and the FM-modes measure the topography corresponding to a surface of constant force gradient.

11.2.2
Conductive Atomic Force Microscopy

One of the simplest ways of characterizing electrical properties at high resolution is by applying voltage between the sample and a C-AFM probe and measuring the resulting current flowing between them as a function of lateral tip location while scanning the sample surface. A current map is obtained in parallel with a topography map measured at constant force. Typically measured currents are in the range of pA–μA. Conductive probes can be obtained by coating standard Si probes by a conductive layer or by making the probe entirely of conductive materials such as metals, doped diamond, or highly doped Si [9]. The coating may increase the size of the tip apex and is prone to damage by friction during scanning, whereas all-conductive probes are rare due to manufacturing difficulties. The choice of a specific conductive material also depends on work function matching between the tip and surface materials to reduce a possible current barrier upon contact [10]. The resolution of the current mapping is directly proportional to the tip–surface contact area, determined by the tip and surface mechanical properties and the applied force (see Section 11.2.1.1). Screening of the electrical potential limits the resolution in semiconductors to the relevant Debye length(s).

Historically, the method was developed by combining AFM with scanning tunneling microscopy (STM, see Section 11.2.5) to measure currents tunneling through an oxide layer on metal surfaces [11], which critically depend on the local oxide layer thickness, defects, and charge traps. Such measurements require a more sensitive current amplifier, with a lower noise level, and are sometimes referred to as tunneling AFM (TUNA). Other than the characterization of dielectric films, this method can also be used to map samples of low conductivity, and typically measured currents are in the range of tens of fA–100 pA [12].

In cases where there is no significant surface current barrier (ohmic contact), and the sample can be considered a semi-infinite body of uniform resistivity, the current passes through the small probe-surface contact area and spreads laterally into a cone-shaped current path toward the large counter-electrode. The cone's dimensions and shape vary with the resistivity and homogeneity of the sample. This "spreading resistance" of the sample can be mapped by C-AFM and is referred to as scanning spreading resistance microscopy (SSRM). If an oxide layer is present on the surface, the tip must penetrate through it for SSRM. For a circular tip–sample contact area of radius r, the electrical resistance R is related to the SSRM measured resistivity ρ through [12]

$$R = \frac{\rho}{4r} \tag{11.8}$$

SSRM is commonly used to map variations in charge carrier density [13], and it measures a wide range of currents (10 pA–100 µA) often using a logarithmic current amplifier.

Examples of applications of C-AFM in inorganic thin-film solar cells include studies of current routes in these highly nonhomogenous materials, and studying the effect of deposition parameters on the morphology–electrical properties relations, in the dark and under illumination [14–17].

11.2.3
Scanning Capacitance Microscopy

In SCM a conductive probe forms a capacitor with the sample surface, in a configuration where no or negligible currents flow between them under bias. Small changes in the tip–sample capacitance are measured, which can be generated by changes in the tip–surface separation or by changes in the local dielectric properties of the sample. The first demonstration of the method used a profilometry-like apparatus [18], and was soon adopted in STM- and nc-AFM-based systems [19, 20]. In these initial demonstrations the tip height was modulated and variations in the capacitance, followed by a lock-in amplifier at the same frequency, were used as the input to the feedback loop, which acted to maintain a constant capacitance change by changing the tip–surface distance. The resulting map corresponded both to height variations and to changes in the dielectric properties. Nowadays, a normal force feedback in contact mode AFM is commonly used to control the tip–surface separation and capacitance changes are measured, independently and simultaneously, by applying an ac bias of fixed amplitude and using lock-in detection [19, 21]. To avoid low-frequency noise of the capacitance sensor, the derivative of the capacitance with respect to bias dC/dV (phase signal) is recorded at high ac frequency, typically several kHz. As the capacitance with a contact area radius of a few nanometer is in the range of attofarads, the dynamic change in capacitance, that is, the derivative, is easier to measure using a special sensor circuit [19]. A circuit for direct ultralow capacitance measurements was also recently developed [22].

The charge carrier concentration in a semiconductor can be mapped using SCM and is interpreted through metal-oxide-semiconductor (MOS) capacitor physics, formed between the metal-coated probe, the semiconductor, and an intermediate dielectric surface layer such as SiO_2 (Figure 11.2). The MOS capacitance is voltage dependent, and the slope of the capacitance–voltage (C–V) curve depends on the type and concentration of the charge carriers in the semiconductor. Due to the high ac frequency used for SCM detection, minority carriers cannot follow the change in voltage fast enough; hence only majority carriers affect the measured dC/dV signal. When positive bias is applied to the metal tip, electrons in an n-type semiconductor will accumulate at the semiconductor-oxide interface, and the measured capacitance will be that of the oxide layer only. When negative bias is applied to the metal tip, a depletion layer will develop in the semiconductor, acting as an additional dielectric layer in series with the oxide layer, and the capacitance will decrease. As the negative bias is increased the depletion layer width will grow, and the capacitance will further decrease, until a breakdown field is reached. The velocity of the capacitance decrease with bias, that is, the slope of the C–V curve, depends on the carrier concentration. The C–V curve of a p-type semiconductor is a mirror image of that of an n-type semiconductor [10, 19]. Hence, the sign of the measured dC/dV signal points to the type of majority carriers in the device, and its magnitude can be interpreted in terms of the carrier concentration [23]. Quantification and analysis of the measured signal requires detailed modeling [19], and is sometimes done by calibrating the SCM signal with similar samples of known carrier concentration [24]. Metals and insulators

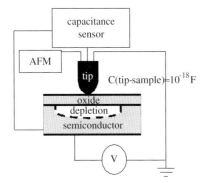

Figure 11.2 Schematic illustration of the MOS capacitor configuration in SCM.

display zero SCM signals, since their capacitance is not voltage dependent, and are thus difficult to distinguish by SCM.

The SCM resolution is determined not only by the tip–surface contact area but also by other factors such as the carrier concentration, with a lower carrier concentration resulting in wider depletion regions around the tip apex, hence lower resolution. Precise modeling can supply data at better resolution than the tip contact area [19]. A modification of the commonly used measurement method includes "closed loop" mapping, in which a second feedback loop is used to adjust the ac bias amplitude to preserve a constant dC/dV signal. The resulting ac bias amplitude is the measured quantity in this method. The advantage is that the depletion region width is constant, hence the resolution is not modified between areas of different carrier concentrations in the sample. SCM was rarely used to characterize inorganic thin-film solar cells due to the difficulty in modeling and quantification in nonhomogenous materials. It was used for imaging qualitative carrier concentration profiles near grain boundaries and for detection of the electrical junction location in cross-sectional studies (see Section 11.3) [25–27].

11.2.4
Kelvin Probe Force Microscopy

The KPFM uses a similar principle to the macroscopic Kelvin probe technique [28]. The electrostatic force between the AFM tip and the sample is compensated by applying a dc bias (V_{dc}) to the sample, which corresponds to the contact potential difference (CPD). The CPD is given by the work function difference between the tip and sample material: $e \cdot V_{CPD} = \Phi_{sample} - \Phi_{tip}$ (e is the elementary charge). To detect the CPD, an additional ac bias (amplitude V_{ac}, frequency ω_{ac}) is applied (see Figure 11.3). This bias induces an oscillation of the cantilever, which results from the electrostatic force between tip and sample

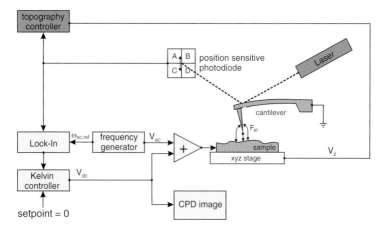

Figure 11.3 Schematic drawing of a typical KPFM setup for AM-mode detection of the CPD.

$$F_{el} = -\frac{1}{2}\frac{\partial C}{\partial z}(V_{dc} - V_{CPD} + V_{ac}\sin(\omega_{ac}t))^2 \qquad (11.9)$$

Computing the square of the voltages, a spectral component at the frequency of the ac bias can be calculated as

$$F_{\omega_{ac}} = -\frac{\partial C}{\partial z}(V_{dc} - V_{CPD})V_{ac}\sin(\omega_{ac}t) \qquad (11.10)$$

Additionally, a dc term F_{dc} results, which gives a constant contribution to the topography signal, and a term at the frequency $2\omega_{ac}$, which can be used to image the capacitance gradient [29]. Equation (11.10) shows that the induced oscillation of the cantilever at the frequency of the ac bias is reduced to zero, when the dc bias is controlled to match the CPD. In many studies, the work function of the tip is calibrated on a reference sample with a known work function (e.g., highly oriented pyrolytic graphite (HOPG)), and therefore the work function of the sample can be determined [7, 30, 31]. However, this is only reasonable for measurements under ultrahigh vacuum (UHV) conditions since the work function of a sample is considerably modified by adsorbates and oxidation, when experiments are performed in air [32].

As in nc-AFM, KPFM measurements can also be taken in AM- and FM-modes. In this case, in AM-mode, the dc bias is controlled by minimizing the amplitude of oscillation at ω_{ac}. For this purpose, the signal of the position-sensitive photodiode is analyzed by a lock-in amplifier tuned to the ac frequency. Frequently, ac frequencies of several kilohertz to several tens of kilohertz are used, and to get sufficient sensitivity, ac biases of 1–3 V have to be applied [33, 34]. An improvement in sensitivity can be gained by tuning the ac frequency to a resonance mode of the cantilever. This resonance-enhanced detection allows using lower ac voltages, which is preferable for semiconductor samples to avoid tip-induced band bending effects [35]. One possible implementation uses a two-pass method, where in a first scan the topography is determined, and in a second scan the mechanical oscillation is switched off and instead an ac bias at the fundamental resonance is applied, while the tip retraces the same scan line [36]. A different realization uses a higher oscillation mode for the ac frequency, that is, the second oscillation mode [37, 38]; this provides the advantage of measuring simultaneously the topography and the CPD signal in one scan line. Due to a limitation of the bandwidth of the position-sensitive photodiode in many commercial AFM systems, the fundamental resonance frequency is limited to 70–80 kHz, which sets the second resonance mode at 400–470 kHz (the ratio between second and fundamental resonance frequency is ∼6.3 for rectangular shaped cantilevers) [39].

In FM-mode KPFM, the ac bias-induced oscillation in the frequency shift signal Δf is used to detect the dc bias necessary to compensate the CPD. In this case, the ac bias is limited at the lower end by the bandwidth of the z-controller (if the ac bias is too low, the z-feedback will oscillate with the ac frequency), and at the higher end by the bandwidth of the PLL or the FM demodulator (at high frequencies the Δf modulation cannot be resolved) [40]. Typically, for FM-KPFM ac frequencies between 1 and 3 kHz and ac voltages of 1–3 V are used.

The main difference between AM- and FM-KPFM is that AM-KPFM is sensitive to the electrostatic force, while FM-KPFM is sensitive to the electrostatic force gradient. As a consequence, the spatial resolution in AM-KPFM is limited due to the long-range character of the electrostatic forces [40, 41]. In this case, the electrostatic interaction between tip and sample is not confined to the tip apex, but also part of the tip–cone and possibly the cantilever contribute to the interaction and therefore decrease the spatial resolution [40, 42]. The FM-mode promises to provide a better spatial resolution due to the more confined character of the electrostatic force gradient [40–42]. However, using resonance-enhanced AM-KPFM, recently even atomic scale contrast in the CPD was demonstrated [43].

11.2.5
Scanning Tunneling Microscopy

The STM was the first development of a series of scanning probe microscopy methods and gained the inventors [1] the Nobel Prize in 1986. The setup is similar to the one of an AFM (see Figure 11.1), however the probe consisting of a cantilever and tip is replaced by a sharp conducting tip. Typically, PtIr or W is used as tip material. The tip is then approached to the sample until a tunneling current is detected. This quantum mechanical tunneling current shows an exponential distance dependence

$$I \propto \rho_s\, V\, e^{-2\kappa d} \tag{11.11}$$

where ρ_s is the density of states of the sample close to the Fermi energy, V the applied voltage, d the distance between the tip and the sample and κ is given by

$$\kappa = \frac{\sqrt{2m(U-E)}}{\hbar} \tag{11.12}$$

where U is the vacuum barrier. From Eq. (11.11) it becomes clear that the tunneling current is very sensitive to the distance between the tip and the sample surface. Thus, a distance controller can be used to adjust the tip–sample distance to maintain a constant tunneling current. It is important to recognize that STM does not image the surface morphology directly but always represents a convolution of the electronic surface properties with the morphology. The surface obtained by STM corresponds to a surface of constant density of states. This is especially important when atomic scale images are obtained.

An additional experimental possibility of STM is to perform spectroscopy measurements by maintaining the tip–sample distance constant (by switching off the feedback circuit) and performing a bias-sweep of the applied voltage. Such a voltage spectroscopy measurement ramps the voltage from low to high values and provides information about the electronic structure of the sample surface. For negative voltages, electrons tunnel from occupied states of the sample into empty states of the tip. When the voltage increases, beyond the valence band maximum, the tunneling current vanishes, as the Fermi level of the tip aligns within the band gap of the sample. Only when the bias voltage is large (positive) enough, that is, when the

Fermi level of the tip aligns with the conduction band minimum, electrons can tunnel from the tip into empty states in the conduction band of the sample. Thus, under ideal conditions, the band gap of the sample can be locally determined. However, such measurements are easily impeded by surface defect states, tip-induced band bending, and surface contamination.

11.2.6
Issues of Sample Preparation

Thin-film solar cells are typically characterized by a rough top surface, which is coated by an electrical contact electrode in the completed device. This surface, lacking the contact, is used for scanning probe mapping using the methods described above. Surface roughness can cause various artifacts in scanning probe microscopy. For example, changes in the local contact area at rough areas may locally change the signal in contact mode AFM, C-AFM, and SCM although no changes in the sample properties are present. Another effect is changes in the electrical field distribution at sharp features. A possible solution is to mechanically or chemically polish the sample surface. However, the resulting flat surface may not represent the true thin-film properties in the solar cell device, in which the film did not undergo polishing. Alternatively, rough surfaces can be characterized, provided variations in the mapped property are shown to be topography independent. This can be done, for example, by comparison to other samples with similar topography but different electrical properties, or by showing that the property of interest varies with the measurement parameters.

Special care has to be taken when mapping cross-section samples of cleaved thin-film inorganic solar cells. Fracturing a multilayered, polycrystalline sample is likely to result in large height differences between the different layers. Such differences are hard to accommodate in a single AFM scan and should be avoided as much as possible. A way to achieve a relatively flat cross section is by fracturing the sample while the film is under tensile stress and the substrate is under compressive stress, as described in Ref. [44]. Polishing of such a cross section can be problematic since the different material layers may polish at different rates, and soft layers may spread over other layers' cross sections [9].

SCM requires careful surface preparation of the semiconductor, such that the oxide layer thickness is thick enough to block significant currents but also thin enough to maximize the dC/dV signal. The oxide should effectively passivate semiconductor surface states and be free of charge traps, which significantly affect the measurement in SCM [19, 45].

It should be noted that many artifacts may affect mapping of electronic properties using AFM-based methods, which are different in different methods, and should be taken into account in the analysis of the obtained results. For example, C-AFM mapping of semiconductor surfaces at high bias may induce oxide formation on the scanned surface, attributed to chemical reactions induced by the electric field involving the water layer adsorbed on the sample surface due to ambient humidity [17, 46]. This may affect subsequent mapping and can be avoided by using lower bias and/or scanning in a controlled-humidity or vacuum environment.

11.3
Selected Applications

This chapter serves to give a few examples of the application of the various microscopy techniques and thereby allows the reader to see what kind of information is accessible by the various methods.

11.3.1
Surface Homogeneity

Typical thin-film solar cells employ a stack of various semiconductor and metal layers forming the device structure. After generation of the electron–hole pair by light absorption in the absorber material, it is separated in the built-in electric field of the pn-junction. In thin-film solar cells, the electronic pn-junction frequently coincides with a material junction, or at least is very close to it. Thus, the main electronic transport is perpendicular to the material junctions and charge carriers have to cross the interface from one material to the subsequent one. From an electronic point of view, this implies the transition of electric fields (represented by curvature of the band in a band diagram) and band offsets due to interface dipoles (represented by vertical offsets in the band diagram). It is therefore highly important to study the spatial homogeneity of a material interface to obtain information of local variations in band bending or band offsets. One possible access to such information is to study the surface of the absorber material prior to the deposition of the subsequent layer. An example of this approach for a $CuGaSe_2$-absorber was presented by Sadewasser et al. [7] using KPFM. For this purpose a comparative study of $CuGaSe_2$ deposited on a Mo/glass substrate and on a ZnSe(110) single crystalline substrate was performed [7]. Figure 11.4a and b show the topography and work function of the $CuGaSe_2$/Mo/glass thin film, respectively. Comparison of the two images shows that areas of distinct and constant work function can be associated with specifically oriented facets of the individual grains seen in the topography image. A better understanding of this observation is gained by the experiments on the oriented $CuGaSe_2$ thin film on the ZnSe(110) substrate. By X-ray diffraction, the [220] direction of $CuGaSe_2$ was determined to be perpendicular to the substrate surface. Also here, the work function image (Figure 11.4d) shows areas of distinct but constant work function, where these areas coincide with specific facets as observed by comparison to the topography image in Figure 11.4c. The known orientation of the film in this case was used to perform a detailed analysis of the geometric angles between various facets of single grains and the substrate orientation. By this geometric analysis, the orientation of various facets could be determined and is indicated in the work function image in Figure 11.4d. It is apparent that a majority of the facets show a {112} or {$\bar{1}\bar{1}\bar{2}$} orientation; this orientation is known to preferentially develop during crystal growth [47], therefore it is not surprising to also frequently find this facet orientation in the present experiments.

The presence of distinct work function values for differently oriented facets can be explained by a surface dipole. This will sensitively depend on the specific surface

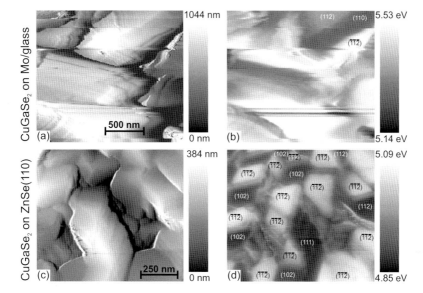

Figure 11.4 KPFM measurements on CuGaSe$_2$ thin films showing the different work function of differently oriented facets of individual grains [7]. (a) Topography of a CuGaSe$_2$ thin film on Mo/glass with height variations (gray scale) of 1044 nm and (b) the corresponding work function image ($\Delta\Phi = 5.14$–5.53 eV) with indicated facet orientations. (c) Topography of CuGaSe$_2$ on ZnSe(110) with height variations of 384 nm and (d) the corresponding work function image ($\Delta\Phi = 4.85$–5.09 eV) with indicated facet orientations.

structure, as the atomic arrangement, surface relaxation, and reconstruction [48]. Specifically, the {112}-facet is metal-terminated, whereas the {$\bar{1}\,\bar{1}\,\bar{2}$}-plane is Se-terminated, which explains the assignment of the lower work function facets to the metal-terminated {112} facet.

Characterization of film morphology as a function of its deposition parameters is the most common use of scanning probe microscopy in thin-film inorganic solar cells [49]. Detection of currents routes in microcrystalline Si via C-AFM and STM mapping enabled Azulay *et al.* to develop a comprehensive model of conduction as a function of the crystalline content, resulting from the manufacturing conditions [50–52]. The optimized crystalline content for solar cell operation is suggested to be associated with a specific connectivity of a conductive structural network [51].

When characterizing photovoltaic devices it is also useful to study the surface homogeneity under illumination, which can point to photoactive defects and to subsurface inhomogeneities. Examples of such characterization were demonstrated using photovoltage mapping in SNOM (scanning near field microscopy, see chapter 5) of microcrystalline Si and Cu(In,Ga)Se$_2$/CdS/ZnO solar cells [53, 54]. In both cases, areas of lower photo response that are not related to clear topographical features, with dimensions of 100–250 nm, were noted. Such phenomena may arise not only from surface defects but also from defects in the internal, sub-surface electrical junction, and reduce the effective active area of the device.

11.3.2
Grain Boundaries

Grain boundaries present the significant difference between poly- and single crystalline material and likely influence material properties and consequently device performance. Crystal defects and impurities at grain boundaries can induce localized energy states within the band gap, leading to trapped, localized charges. These form an electrostatic potential barrier (band bending) for majority-carrier transport across grain boundaries [55], as well as enhanced recombination of photogenerated carriers [56]. On the other hand, segregation of defects and impurities to the grain boundaries may be beneficial for improving the grain's crystalline quality. Characterizing a *single* GB using scanning probe microscopy obviates the need to average over different grain boundary lengths and directions, and over the entire inhomogenous material, as is the case of macroscopic measurements.

In the case of chalcopyrite-based solar cells, different models have been used to understand the physics of grain boundaries and their effect on device performance [55, 57–61]. KPFM has been used to study the electronic properties of grain boundaries, providing an excellent tool to access the properties of individual grain boundaries even in polycrystalline films with a typical grain size of the order of $\sim 1\,\mu m$ [62–67]. The main result of all these studies is the observation of a lower work function at the grain boundaries, with a drop of the order of about 100 mV. As an example for these studies, we show results on a $CuGaSe_2$ absorber in Figure 11.5a–d. The sample was studied by UHV-KPFM, where the measurements were taken on the backside of the absorber, obtained by peel-off of the $CuGaSe_2$ thin film from the Mo/glass substrate inside the UHV [65, 68]. This ensures a clean and nonoxidized sample surface. The topography and work function of this surface are shown in Figure 11.5a and b, respectively. Clearly, a dip in the work function is observed along grain boundaries. Under illumination, the contrast of the work function image looks very similar (see Figure 11.5c), however, the values are elevated with respect to the dark measurement. This is also illustrated by the line profiles shown in Figure 11.5d, showing two grain boundaries with a differently deep work function dip. Also the change of the work function dip under illumination is different for the two grain boundaries. This indicates that differently oriented grain boundaries have different electronic properties.

The shape of the work function decrease at grain boundaries, observed in the above presented experiments as well as many others, is in agreement with a space charge region, providing evidence for the presence of charged defects with a concentration of the order of 10^{11}–$10^{12}\,cm^{-2}$ [62]. However, as was shown by simulations of the spatial resolution of KPFM, care has to be taken with this interpretation, as also steplike band-offsets could result in a similar shape in KPFM imaging [69].

More recently, a correlation between structural and electronic properties has also been achieved by investigation of single grain boundaries grown on epitaxial substrates [70, 71]. These samples provided the possibility to comparatively investigate the grain boundary properties by electron backscatter diffraction (EBSD), Hall effect, and KPFM. On a $CuGaSe_2$ bicrystal containing a twin grain boundary [70, 71]

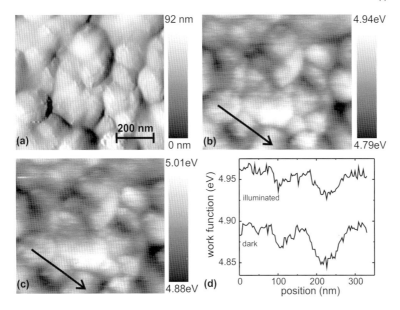

Figure 11.5 KPFM measurement on the back side of a CuGaSe$_2$ absorber film peeled off in UHV [65]. (a) Topography (height range = 92 nm), (b) work function in the dark ($\Delta\Phi = 4.79$–4.94 eV), (c) work function under illumination ($\Delta\Phi = 4.88$–5.01 eV). (d) Line profiles of the work function in the dark and under illumination along the line shown in (b) and (c).

no change of the work function was observed by KPFM, a result contrary to those found on polycrystalline samples. However, Hall-effect measurements did show a small barrier for majority carrier transport, which led to the conclusion that a neutral barrier (i.e., by means of a valence band offset) is present at the $\Sigma 3$ twin grain boundary. These findings are in agreement with theoretical predictions [57, 72].

C-AFM was used to identify current routes in photovoltaic thin film, and such current routes were found along grain boundaries in CuInSe$_2$ and Cu(In,Ga)Se$_2$ polycrystalline films deposited on glass/Mo substrate, when the top surface (before CdS deposition) was mapped [15]. Using bias-dependent C-AFM mapping in the dark and under illumination, Azulay et al. showed evidence for grain boundary band bending in the order of 100 mV, in agreement with KPFM measurements, by a change in polarity of the measured photocurrent at that bias. However, higher currents observed along grain boundaries in the dark may imply an inversion of the dominant carrier type at the boundaries, requiring a much larger band bending there, of the order of 400 mV. The authors propose a spatially narrow band offset of 300 mV, which adds to the 100 mV band bending but is narrower than the KPFM spatial resolution hence cannot be detected. Current flow through the grains was mapped only at significantly higher bias (above 1 V), supporting this band offset [15].

Hole depletion near CdTe grain boundaries was previously deduced from measurements of a single grain boundary in a bicrystal and from macroscopic lateral transport measurements in CdTe polycrystalline films, with grain boundary barrier heights ranging from 0.1 to 0.8 eV [73–76]. Visoly-Fisher *et al.* used a combination of several AFM-based methods for the electrical characterization of CdTe grain boundaries in CdTe/CdS solar cells [25, 77, 78]. Direct evidence for grain boundary depletion was obtained from SCM, showing brighter SCM signals (lower hole concentration) at the grain boundaries compared with that at the grain surface (Figure 11.6b), and from KPFM showing lower work function values at grain boundaries than the outer CdTe grain surface. The hole depletion width changes from one grain boundary to another, and extends for 100–300 nm on each side of the grain boundary, which is of the order of the Debye length (150 nm) in CdTe with a bulk charge carrier concentration of $7 \times 10^{14}\,\text{cm}^{-3}$ [25]. Further mapping with high-resolution probes using C-AFM and SCM under illumination showed evidence for carrier-type inversion at the grain boundary core: surprisingly high photocurrents were observed at the cores of most grain boundaries, as well as SCM signals approaching zero at the core (Figure 11.6b), typical of insulating materials

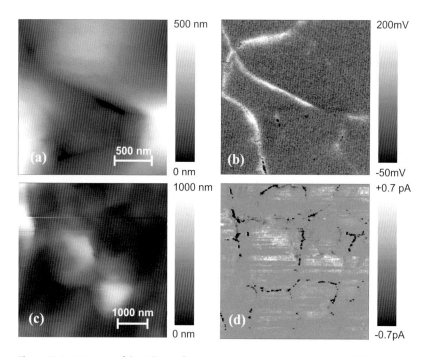

Figure 11.6 Mapping of the CdTe surface in CdTe/CdS cells without the back contact under illumination [77, 78] (Copyright Wiley-VCH Verlag GmbH & Co. KGaA. Reproduced with permission). Top: Simultaneously collected (a) AFM topography and (b) SCM images. Note the GB at the bottom left corner, showing a lower SCM signal at its core. Scan size $2 \times 2\,\mu\text{m}$, height range = 500 nm. Bottom: Simultaneously collected (c) AFM topography and (d) C-AFM images. Scan size $5\,\mu\text{m} \times 5\,\mu\text{m}$, height range = 1000 nm, V_{dc} (CdTe) = 0.5 V (slightly lower than the cell's V_{oc}), current range = −0.7–0.7 pA. The current at GB cores is at opposite polarity to that at grain surfaces.

(with negligible concentration of free carriers, as in the case of inversion of the type of majority carriers) [78]. These unique grain boundary electronic properties develop during the $CdCl_2$ treatment used in the manufacture of CdTe/CdS cells [17, 77].

Variations in the mapped photocurrent as a function of applied bias generally followed those observed in cells, that is, the current reversed direction near the open-circuit voltage, V_{oc}. At grain boundary cores, however, a smaller bias was needed to reverse the current direction (Figure 11.6d), indicating a lower barrier in the CdTe/CdS junction at the points of intersection of grain boundaries with the junction compared with that away from these intersections. Thus, V_{oc}(grain boundary) < V_{oc}(grain surface), and the cell junction can be viewed as a series of junctions of varying barrier heights, connected in parallel. Decreased V_{oc} at the grain boundaries is additional evidence for depletion at grain boundaries [77]. The authors suggest that the unique doping profile that forms near grain boundaries is beneficial for cell performance, by assisting in separation of photo-generated electron–hole pairs, followed by improved electron transport via the grain boundary core toward the junction. In this way, CdTe grain boundaries decrease the recombination rate (by improving the crystalline quality of the bulk because of gettering of defects, accompanied by a low recombination rate at the grain boundary core), and also increase the electron diffusion length in the absorber [77, 78].

11.3.3
Cross-Sectional Studies

The electronic working principle of a solar cell device can be illustrated and analyzed along the band diagram of the structure. However, experimental determination of the band diagram of a device is very difficult if not impossible. With techniques such as photoemission spectroscopy (see Chapter 15), this problem is approached from the interface point of view. KPFM can be used to follow a different approach. The measurement of the work function along the cross section of a complete device provides valuable information about the electronic properties of the different layers [30], the presence of impurity phases [79] and built-in electric fields [80].

As one example of cross-sectional KPFM, we illustrate here the investigation of the Ga-distribution in a $Cu(In_{1-x},Ga_x)S_2$ solar cell device [81]. A study by energy dispersive X-ray (EDX) diffraction in a scanning electron microscope (SEM) showed that the absorber exhibits two distinct layers, where the $Cu(In_{1-x},Ga_x)S_2$ shows a significantly higher Ga-content toward the Mo back contact and a significantly higher In-content toward the CdS buffer layer. KPFM imaging was performed on the very same position of the cross section, thus allowing to compare exactly the obtained electronic information to the compositional one. The KPFM image of the CPD and a line profile across the various layers are shown in Figure 11.7a and b, respectively. Clearly a higher work function for the back part of the absorber is observed. The fairly sharp transition between the two work function regions coincides with the transition from the In- to the Ga-rich part of the absorber layer. This increase of the electric

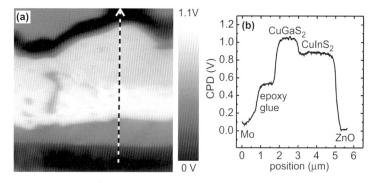

Figure 11.7 KPFM measurement on the cross section of a complete Mo/Cu(In$_{1-x}$,Ga$_x$)S$_2$/CdS/ZnO solar cell device [81]. (a) Contact potential difference (CPD) image showing the variation with the different layers (color scale = 1.1 V). (b) Line profile of the CPD along the line in a, clearly showing the different CPD for the In-rich front and the Ga-rich back part of the absorber.

potential indicates the presence of a built-in electric field, which is oriented in such a way that it accelerates electrons toward the In-rich region of the absorber layer and therefore keeps them away from the back contact of the solar cell. This proves the existence of a back surface field in the device, which reduces recombination at the back contact. Comparison to the quantum efficiency (QE) of the device reveals an increased QE for photons with energies near the band gap, which are absorbed deep inside the absorber. For the generated electron–hole pairs, collection is improved by the presence of this back surface field [81].

The metallurgical junction in heterojunction thin-film solar cells, defined by a change in overall chemical composition, does not necessarily coincide with the electronic junction, defined by the change in electrical properties. In the CdTe/CdS cell, Te from CdTe and Cu from the cell's back contact may, at sufficiently high concentrations, type-convert CdS [82]. S diffusion into the CdTe may type-convert some of the p-CdTe. Both effects would lead to a buried homo- rather than a heterojunction. Visoly-Fisher et al. have shown, by a combination of SCM and KPFM cross-sectional mapping, that the cell is a heterojunction, within the experimental uncertainty (50 nm), with no evidence for CdS-type conversion [26]. Figure 11.8 shows the layer sequence in the cell: the insulating glass substrate is coated with a low-resistance (LR) SnO$_2$:F layer 300–500 nm thick. This layer shows unstable, noisy SCM signal probably related to unwanted current flow in the sensor circuit, due to high conductivity and lack of a surface dielectric layer, resulting in erroneous SCM results. The adjacent n-type layer (dark under any dc bias between −2 and +2 V) consists of both a high-resistance (HR) SnO$_2$ layer and the CdS layer, which are electronically indistinguishable. A structure lacking the CdS layer showed a layer sequence similar to that of a conventional cell, but with a 50–80 nm thinner n-type layer, adjacent to the CdTe. A layer of CdTe grains with very weak (almost zero) SCM signal is seen adjacent to the HR-SnO$_2$/CdS layer, 200–350 nm

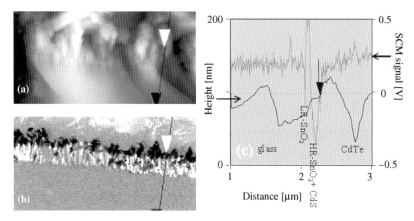

Figure 11.8 Glass/SnO$_2$:F/CdS/CdTe cell cross-sectional mapping. (a) Topography image and (b) SCM image, taken simultaneously. Reprinted with permission from ref. [26]. Copyright [2003], American Institute of Physics. The gray color of the glass substrate denotes a "zero" SCM signal of an insulator. Brighter colors denote p-type semiconductors; darker colors denote n-type semiconductors. Scan size 2 µm × 4 µm, $V_{ac} = 1$ V, $V_{dc} = 0$ V. (c) SCM and topography line sections of the cross section. The lines in the images (a, b) indicate the locations of the line sections. The black arrow in the line sections shows the location of the white triangular markers in the images. At this point the metallurgical junction and the electronic junction are shown to coincide. The metallurgical junction is identified by the transition point between the layers in the topography image, and a sharp change in slope in the topography line section, and the electronic junction is identified as the transition point between bright positive, p-type signal to dark negative, n-type signal in the SCM image, and zero SCM signal in the SCM line section.

thick. These are interpreted to be recrystallized and regrown CdTe grains due to cell processing [83]. The SCM signal shows clear p-type behavior further into the CdTe layer.

A KPFM profile without external bias (not shown) indicates the change in CPD between the different layers, expected from their different work functions [26]. The junction's built-in electric field is shown by a drop in the KPFM signal across certain layers, otherwise expected to show a constant signal related to their work function. Such voltage drop is noted across the HR-SnO$_2$ layer, indicating that this layer supports the junction's built-in electric field/open circuit voltage. This indicates that to support the high open-circuit voltage, the n-type layer must have some minimal thickness (around 300 nm in the cells studied here). The role of the HR-SnO$_2$ that replaces part of the CdS is to improve the cell's blue response, due to its larger band gap, and make good electrical contact to CdS, due to alignment of the conduction band minima. A thin CdS layer is still needed to provide a photovoltaic junction with low defect concentration and high open-circuit voltage [84]. This work demonstrates how combined SCM and KPFM of CdTe/CdS cells show the location of the internal junctions and the roles of different layers in the structure.

11.4
Summary

Thin-film solar cells are made of complex materials; hence the prediction of device properties cannot rely on the properties of model systems and simple junction physics. Physical characterization of the polycrystalline films in use requires the understanding of their spatially resolved properties on the nanoscale. Such characterization, provided by scanning probe microscopy in its numerous variations, in combination with macroscopic analysis, can link the material properties and device performance, and allow proper optimization of its energy conversion efficiency.

References

1 Binnig, G., Rohrer, H., Gerber, C., and Weibel, E. (1982) Surface studies by scanning tunneling microscopy. *Phys. Rev. Lett.*, **49**, 57.

2 Binnig, G., Quate, C.F., and Gerber, C. (1986) Atomic force microscopy. *Phys. Rev. Lett.*, **56**, 930.

3 Martin, Y., Williams, C.C., and Wickramasinghe, H.K. (1987) Atomic force microscope-force mapping and profiling on a sub 100-A scale. *J. Appl. Phys.*, **61**, 4723.

4 Pérez, P., Payne, M.C., Stich, I., and Terukura, K. (1997) Role of covalent tip-surface interactions in noncontact atomic force microscopy on reactive surfaces. *Phys. Rev. Lett.*, **78**, 678.

5 García, R. and Pérez, R. (2002) Dynamic atomic force microscopy methods. *Surf. Sci. Rep.*, **47**, 197.

6 Al-Jassim, M.M., Yan, Y., Moutinho, H.R., Romero, M.J., Dhere, R.D., and Jones, K.M. (2001) TEM, AFM, and cathodoluminescence characterization of CdTe thin films. *Thin Solid Films*, **387**, 246.

7 Sadewasser, S., Th, G., Rusu, M., Jager-Waldau, A., and Lux-Steiner, M.C. (2002) High-resolution work function imaging of single grains of semiconductor surfaces. *Appl. Phys. Lett.*, **80**, 2979.

8 Albrecht, T.A., Grütter, P., Horne, D., and Rugar, D. (1991) Frequency modulation detection using high-Q cantilevers for enhanced force microscope sensitivity. *J. Appl. Phys.*, **69**, 668.

9 Oliver, R.A. (2008) Advances in AFM for the electrical characterization of semiconductors. *Rep. Prog. Phys.*, **71**, 076501.

10 Sze, S.M. (1981) *Physics of Semiconductor Devices*, 2nd edn, Wiley-Interscience, New York.

11 Morita, S., Ishizaka1, T., Sugawara, Y., Okada, T., Mishima, S., Imai, S., and Mikoshiba, N. (1989) Surface conductance of metal surfaces in air studied with a force microscope. *Jpn. J. Appl. Phys.*, **28**, L1634.

12 De Wolf, P., Brazel, E., and Erickson, A. (2001) Electrical characterization of semiconductor materials and devices using scanning probe microscopy. *Mater. Sci. Semicond. Process.*, **4**, 71.

13 De Wolf, P., Clarysse, T., Vandervorst, W., Snauwaert, J., and Hellemans, L. (1996) One- and two-dimensional carrier profiling in semiconductors by nanospreading resistance profiling. *J. Vac. Sci. Technol. B*, **14**, 380.

14 Shen, Z.H., Gotoh, T., Eguchi, M., Yoshida, N., Itoh, T., and Nonomura, S. (2007) Study of nano-scale electrical properties of hydrogenated microcrystalline silicon solar cells by conductive atomic force microscope. *Jpn. J. Appl. Phys. Part 1*, **46**, 2858.

15 Azulay, D., Millo, O., Balberg, I., Schock, H.W., Visoly-Fisher, I., and Cahen, D. (2007) Current routes in polycrystalline $CuInSe_2$ and $Cu(In,Ga)Se_2$ films. *Sol. Energy Mater. Sol. Cells*, **91**, 85.

16 Cavallini, A., Cavalcoli, D., Rossi, M., Tomasi, A., Pizzini, S., Chrastina, D., and Isella, G. (2007) Defect analysis of hydrogenated nanocrystalline Si thin films. *Phys. B-Cond. Mat.*, **401**, 519.

17 Moutinho, H.R., Dhere, R.G., Jiang, C.S., Al-Jassim, M.M., and Kazmerski, L.L. (2006) Electrical properties of CdTe/CdS solar cells investigated with conductive atomic force microscopy. *Thin Solid Films*, **514**, 150.

18 Matey, J.R. and Blanc, J. (1985) Scanning capacitance microscopy. *J. Appl. Phys.*, **57**, 1437.

19 Williams, C.C. (1999) Two-dimensional dopant profiling by scanning capacitance microscopy *Ann. Rev. Mater. Sci.*, **29**, 471.

20 Williams, C.C., Hough, W.P., and Rishton, S.A. (1989) Scanning capacitance microscopy on a 25nm scale. *Appl. Phys. Lett.*, **55**, 203.

21 Barrett, R.C. and Quate, C.F. (1991) Charge storage in a nitride-oxide-silicon medium by scanning capacitance microscopy. *J. Appl. Phys.*, **70**, 2725.

22 Lee, D.T., Pelz, J.P., and Bhushan, B. (2002) Instrumentation for direct, low frequency scanning capacitance microscopy, and analysis of position dependent stray capacitance. *Rev. Sci. Instrum.*, **73**, 3525.

23 Duhayon, N., Clarysse, T., Eyben, P., Vandervorst, W., and Hellemans, L. (2002) Detailed study of scanning capacitance microscopy on cross-sectional and beveled junctions. *J. Vac. Sci. Technol. B*, **20**, 741.

24 Smoliner, J., Basnar, B., Golka, S., Gornik, E., Loffler, B., Schatzmayr, M., and Enichlmair, H. (2001) Mechanism of bias-dependent contrast in scanning-capacitance-microscopy images. *Appl. Phys. Lett.*, **79**, 3182.

25 Visoly-Fisher, I., Cohen, S.R., and Cahen, D. (2003) Direct evidence for grain-boundary depletion in polycrystalline CdTe from nanoscale-resolved measurements *Appl. Phys. Lett.*, **82**, 556.

26 Visoly-Fisher, I., Cohen, S.R., Cahen, D., and Ferekides, C.S. (2003) Electronically active layers and interfaces in polycrystalline devices: Cross-section mapping of CdS/CdTe solar cells. *Appl. Phys. Lett.*, **83**, 4924.

27 Maknys, K., Ulyashin, A.G., Stiebig, H., Kuznetsov, A.Y., and Svensson, B.G. (2006) Analysis of ITO thin layers and interfaces in heterojunction solar cells structures by AFM, SCM and SSRM methods. *Thin Solid Films*, **511**, 98.

28 Kelvin, L. (1898) Contact electricity of metals. *Phil. Mag.*, **46**, 82.

29 Hochwitz, T., Henning, A.K., Levey, C., Daghlian, C., Slinkman, J., Never, J., Kaszuba, P., Gluck, R., Wells, R., Pekarik, J., and Finch, R. (1996) Imaging integrated circuit dopant profiles with the force-based scanning Kelvin probe microscope. *J. Vac. Sci. Technol. B*, **14**, 440.

30 Glatzel, T., Fuertes Marrón, D., Schedel-Niedrig, T., Sadewasser, S., and Lux-Steiner, M.C. (2002) CuGaSe$_2$ solar cell cross section studied by Kelvin probe force microscopy in ultra high vacuum. *Appl. Phys. Lett.*, **81**, 2017.

31 Sadewasser, S. (2006) Surface potential of chalcopyrite films measured by KPFM. *phys. status solidi (a)*, **203**, 2571.

32 Nonnenmacher, M., O'Boyle, M.P., and Wickramasinghe, H.K. (1991) Kelvin probe force microscopy. *Appl. Phys. Lett.*, **58**, 2921.

33 Shikler, R., Meoded, T., Fried, N., Mishori, B., and Rosenwaks, Y. (1999) Two-dimensional surface band structure of operating light emitting devices. *J. Appl. Phys.*, **86**, 107.

34 Shikler, R., Meoded, T., Fried, N., and Rosenwaks, Y. (1999) Potential imaging of operating light-emitting devices by Kelvin force microscopy. *Appl. Phys. Lett.*, **74**, 2972.

35 McEllistrem, M., Haase, G., Chen, D., and Hamers, R.J. (1993) Electrostatic sample–tip interactions in the scanning tunneling microscope. *Phys. Rev. Lett.*, **70**, 2471.

36 Rosenthal, P.A., Yu, E.T., Pierson, R.L., and Zampardi, P.J. (1999) Characterization of Al$_x$Ga$_{1-x}$As/GaAs heterojunction bipolar transistor structures using cross-sectional scanning force microscopy. *J. Appl. Phys.*, **87**, 1937.

37 Sommerhalter, C., Matthes, T.W., Glatzel, T., Jäger-Waldau, A., and Lux-Steiner, M.C.

(1999) High-sensitivity quantitative Kelvin probe microscopy by noncontact ultra-high-vacuum atomic force microscopy. *Appl. Phys. Lett.*, **75**, 286.

38 Kikukawa, A., Hosaka, S., and Imura, R. (1996) Vacuum compatible high-sensitive Kelvin probe force microscopy. *Rev. Sci. Instrum.*, **67**, 1463.

39 Butt, H.-J. and Jaschke, M. (1995) Calculation of thermal noise in atomic force microscopy. *Nanotechnol.*, **6**, 1.

40 Glatzel, T., Sadewasser, S., and Lux-Steiner, M.C. (2003) Amplitude of frequency modulation-detection in Kelvin probe force microscopy. *Appl. Surf. Sci*, **210**, 84.

41 Colchero, J., Gil, A., and Baró, A.M. (2001) Resolution enhancement and improved data interpretation in electrostatic force microscopy. *Phys. Rev. B*, **64**, 245403.

42 Zerweck, U., Loppacher, C., Otto, T., Grafström, S., and Eng, L.M. (2005) Accuracy and resolution limits of Kelvin probe force microscopy. *Phys. Rev. B*, **71**, 125424.

43 Enevoldsen, G.H., Glatzel, T., Christensen, M.C., Lauritsen, J.V., and Besenbacher, F. (2008) Atomic scale Kelvin probe force microscopy studies of the surface potential variations on the $TiO_2(110)$ surface. *Phys. Rev. Lett.*, **100**, 236104.

44 Ballif, C., Moutinho, H.R., Hasoon, F.S., Dhere, R.G., and Al-Jassim, M.M. (2000) Cross-sectional atomic force microscopy imaging of polycrystalline thin films. *Ultramicroscopy*, **85**, 61.

45 Bowallius, O. and Anand, S. (2001) Evaluation of different oxidation methods for silicon for scanning capacitance microscopy. *Mater. Sci. Semicond. Process.*, **4**, 81.

46 Avouris, P., Hertel, T., and Martel, R. (1997) Atomic force microscope tip-induced local oxidation of silicon: Kinetics, mechanism, and nanofabrication. *Appl. Phys. Lett.*, **71**, 285.

47 Jaffe, J.E. and Zunger, A. (2001) Defect-induced nonpolar-to-polar transition at the surface of chalcopyrite semiconductors. *Phys. Rev. B*, **64**, 241304.

48 Mönch, W. (1993) Semiconductor surfaces and interfaces, in *Springer Series in Surface Science*, vol. 26, Springer, Berlin.

49 Durose, K., Asher, S.E., Jaegermann, W., Levi, D., McCandless, B.E., Metzger, W., Moutinho, H., Paulson, P.D., Perkins, C.L., Sites, J.R., Teeter, G., and Terheggen, M. (2004) Physical characterization of thin-film solar cells. *Prog. Photovoltaics: Res. Appl.*, **12**, 177.

50 Rezek, B., Stuchlik, J., Fejfar, A., and Kocka, J. (2002) Microcrystalline silicon thin films studied by atomic force microscopy with electrical current detection. *J. Appl. Phys.*, **92**, 587.

51 Azulay, D., Balberg, I., Chu, V., Conde, J.P., and Millo, O.O. (2005) Current routes in hydrogenated microcrystalline silicon. *Phys. Rev. B*, **71**, 113304.

52 Azulay, D., Millo, O., Savir, E., Conde, J.P., and Balberg, I. (2009) Microscopic and macroscopic manifestations of percolation transitions in a semiconductor composite. *Phys. Rev. B*, **80**, 245312.

53 McDaniel, A.A., Hsu, J.W.P., and Gabor, A.M. (1997) Near-field scanning optical microscopy studies of $Cu(In,Ga)Se_2$ solar cells. *Appl. Phys. Lett.*, **70**, 3555.

54 Gotoh, T., Yamamoto, Y., Shen, Z., Ogawa, S., Yoshida, N., Itoh, T., and Nonomura, S. (2009) Nanoscale characterization of microcrystalline silicon solar cells by scanning near-field optical microscopy. *Jpn. J. Appl. Phys.*, **48**, 091202.

55 Seto, J.Y.W. (1975) The electrical properties of polycrystalline silicon films. *J. Appl. Phys.*, **46**, 5247.

56 Fahrenbruch, A.L. and Bube, R.H. (1983) *Fundamentals of Solar Cells: Photovoltaic Solar Energy Conversion*, Academic Press, New York.

57 Persson, C. and Zunger, A. (2003) Anomalous grain boundary physics in polycrystalline $CuInSe_2$: The existence of a hole barrier. *Phys. Rev. Lett.*, **91**, 266401.

58 Yan, Y., Jiang, C.-S., Noufi, R., Wei, S.-H., Moutinho, H.R., and Al-Jassim, M.M. (2007) Electrically benign behavior of grain boundaries in polycrystalline $CuInSe_2$ films. *Phys. Rev. Lett.*, **99**, 235504.

59 Metzger, W.K. and Gloeckler, M. (2005) The impact of charged grain boundaries on thin-film solar cells and characterization. *J. Appl. Phys.*, **98**, 063701.

60 Gloeckler, M., Sites, J.R., and Metzger, W.K. (2005) Grain-boundary recombination in Cu(In,Ga)Se$_2$ solar cells. *J. Appl. Phys.*, **98**, 113704.

61 Taretto, K., Rau, U., and Werner, J.H. (2005) Numerical simulation of grain boundary effects in Cu(In,Ga)Se$_2$ thin-film solar cells. *Thin Solid Films*, **480–481**, 8.

62 Sadewasser, S., Glatzel, T., Schuler, S., Nishiwaki, S., Kaigawa, R., and Lux-Steiner, M.C. (2003) Kelvin probe force microscopy for the nano scale characterization of chalcopyrite solar cell materials and devices. *Thin Solid Films*, **431–432**, 257.

63 Jiang, C.-S., Noufi, R., AbuShama, J.A., Ramanathan, K., Moutinho, H.R., Pankow, J., and Al-Jassim, M.M. (2004) Local built-in potential on grain boundary of Cu(In,Ga)Se$_2$ thin films. *Appl. Phys. Lett.*, **84**, 3477.

64 Jiang, C.-S., Noufi, R., Ramanathan, K., AbuShama, J.A., Moutinho, H.R., and Al-Jassim, M.M. (2004) Does the local built-in potential on grain boundaries of Cu(In,Ga)Se$_2$ thin films benefit photovoltaic performance of the device? *Appl. Phys. Lett.*, **85**, 2625.

65 Fuertes Marrón, D., Sadewasser, S., Glatzel, T., Meeder, A., and Lux-Steiner, M.C. (2005) Electrical activity at grain boundaries of Cu(In,Ga)Se$_2$ thin films. *Phys. Rev. B*, **71**, 033306.

66 Jiang, C.-S., Noufi, R., Ramanathan, K., Moutinho, H.R., and Al-Jassim, M.M. (2005) Electrical modification in Cu(In,Ga)Se$_2$ thin films by chemical bath deposition process of CdS films. *J. Appl. Phys.*, **97**, 053701.

67 Hanna, G., Glatzel, T., Sadewasser, S., Ott, N., Strunk, H.P., Rau, U., and Werner, J.H. (2006) Texture and electronic activity of grain boundaries in Cu(In,Ga)Se$_2$ thin films. *Appl. Phys. A*, **82**, 1.

68 Fuertes Marrón, D., Meeder, A., Sadewasser, S., Würz, R., Kaufmann, C.A., Glatzel, T., Schedel-Niedrig, T., and Lux-Steiner, M.C. (2005) Lift-off process and rear-side characterization of CuGaSe$_2$ chalcopyrite thin films and solar cells. *J. Appl. Phys.*, **97**, 094915.

69 Leendertz, C., Streicher, F., Lux-Steiner, M.C., and Sadewasser, S. (2006) Evaluation of Kelvin probe force microscopy for imaging grain boundaries in chalcopyrite thin films. *Appl. Phys. Lett.*, **89**, 113120.

70 Siebentritt, S., Sadewasser, S., Wimmer, M., Leendertz, C., Eisenbarth, T., and Lux-Steiner, M.C. (2006) Evidence for a neutral grain-boundary barrier in chalcopyrites. *Phys. Rev. Lett.*, **97**, 146601.

71 Siebentritt, S., Eisenbarth, T., Wimmer, M., Leendertz, C., Streicher, F., Sadewasser, S., and Lux-Steiner, M.C. (2007) A 3 grain boundary in an epitaxial chalcopyrite film. *Thin Solid Films*, **515**, 6168.

72 Persson, C. and Zunger, A. (2005) Compositionally induced valence-band offset at the grain boundary of polycrystalline chalcopyrites creates a hole barrier. *Appl. Phys. Lett.*, **87**, 211904.

73 Thorpe, T.P. Jr., Fahrenbruch, A.L., and Bube, R.H. (1986) Polycrystalline cadmium telluride. *J. Appl. Phys.*, **60**, 3622.

74 Durose, K., Boyle, D., Abken, A., Ottley, C.J., Nollet, P., Degrave, S., Burgelman, M., Wendt, R., Beier, J., and Bonnet, D. (2002) Key aspects of CdTe/CdS solar cells. *phys. status solidi B*, **229**, 1055.

75 Vigil-Galán, O., Valliant, L., Mendoza-Perez, R., Contreras-Puente, G., and Vidal-Larramendi, J. (2001) Influence of the growth conditions and postdeposition treatments upon the grain boundary barrier height of CdTe thin films deposited by close spaced vapor transport. *J. Appl. Phys.*, **90**, 3427.

76 Woods, L.M., Levi, D.H., Kaydanov, V., Robinson, G.Y., and Ahrenkiel, R.K. (1998) Electrical characterization of CdTe grain-boundary properties from as processed CdTe/CdS solar cells, in *2nd World Conference on Photovoltaic Solar energy Conversion* (eds J. Schmid et al.), European Commission, Vienna, Austria, p. 1043.

77 Visoly-Fisher, I., Cohen, S.R., Gartsman, K., Ruzin, A., and Cahen, D. (2006) Understanding the beneficial role of grain boundaries in polycrystalline solar cells from single-grain-boundary scanning probe microscopy. *Adv. Funct. Mater.*, **16**, 649.

78 Visoly-Fisher, I., Cohen, S.R., Ruzin, A., and Cahen, D. (2004) How polycrystalline devices can outperform single-crystal ones: Thin film CdTe/CdS solar cells. *Adv. Mater.*, **16**, 879.

79 Fuertes Marrón, D., Glatzel, T., Meeder, A., Schedel-Niedrig, T., Sadewasser, S., and Lux-Steiner, M.C. (2004) Electronic structure of secondary phases in Cu-rich CuGaSe$_2$ solar cell devices. *Appl. Phys. Lett.*, **85**, 3755.

80 Glatzel, T., Steigert, H., Sadewasser, S., Klenk, R., and Lux-Steiner, M.C. (2005) Potential distribution of Cu(In,Ga)(S,Se)$_2$-solar cell cross-sections measured by Kelvin probe force microscopy. *Thin Solid Films*, **480–481**, 177.

81 Mainz, R., Streicher, F., Abou-Ras, D., Sadewasser, S., Klenk, R., and Lux-Steiner, M.C. (2009) Combined analysis of spatially resolved electronic structure and composition on a cross-section of a thin film Cu(In$_{1-x}$Ga$_x$)S$_2$ solar cell. *phys. status solidi (a)*, **206**, 1017.

82 Bonnet, D. (1970) Preparation and properties of polycrystalline CdS$_x$Te$_{1-x}$ films. *phys. status solidi (a)*, **3**, 913.

83 Durose, K., Cousins, M.A., Boyle, D.S., Beier, J., and Bonnet, D. (2002) Grain boundaries and impurities in CdTe/CdS solar cells. *Thin Solid Films*, **403–404**, 396.

84 McCandless, B.E. (2001) Thermochemical and kinetic aspects of cadmium telluride solar cell processing, in *II–VI Compound Semiconductor Photovoltaic Materials Symposium at the MRS Spring Meeting* (eds R. Birkmire *et al.*), MRS, Warrendale, PA, p. H1.6.1.

12
Electron Microscopy on Thin Films for Solar Cells
Daniel Abou-Ras, Melanie Nichterwitz, Manuel J. Romero, and Sebastian S. Schmidt

12.1
Introduction

Electron microscopy and it related techniques, be it on bulk samples in scanning electron microscopy (SEM) or on thin specimens in transmission electron microscopy (TEM), provide the possibilities not only to image positions of interest down to the angstroms range but also to analyze microstructures, compositions, as well as electrical and optoelectronic properties of individual layers and their interfaces in thin-film solar cells. It is the motivation of the present chapter to give an overview of these various techniques applied in SEM and TEM, highlighting their possibilities and also limitations. Also, an introduction into sample preparation for electron microscopy is given, which is often underestimated but eventually decides on the quality of the image data acquired and analysis results obtained.

Figure 12.1 represents an overview of the various electron and photon emissions occurring upon interaction of the impinging electron beam with a semiconducting thin film. The analysis techniques for these emission signals will be introduced in the following subsections.

12.2
Scanning Electron Microscopy

SEM is the technique applied most frequently for imaging of thin-film solar cells. Corresponding modern microscopes provide insight into layer thicknesses, surface topographies, and various other features in solar-cell thin-film stacks with resolutions down to below 1 nm. Also very frequently, scanning electron microscopes are equipped with energy-dispersive X-ray (EDX) detectors, which are used for analyzing local elemental compositions in thin films. However, the possibilities of analysis in SEM go far beyond imaging and compositional analysis.

This chapter gives an overview of the various imaging and analysis techniques applied on a scanning electron microscope. It will be shown that imaging is divided

Advanced Characterization Techniques for Thin Film Solar Cells,
Edited by Daniel Abou-Ras, Thomas Kirchartz and Uwe Rau.
© 2011 Wiley-VCH Verlag GmbH & Co. KGaA. Published 2011 by Wiley-VCH Verlag GmbH & Co. KGaA.

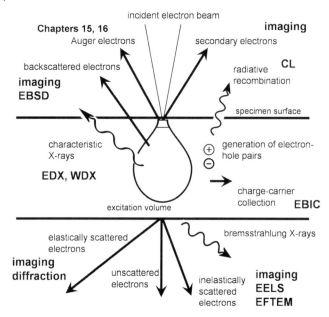

Figure 12.1 Various electron and photon emissions upon irradiation of a semiconducting specimen by an electron beam. Indicated are also the techniques making use of these emissions, that is, EBSD, EDX and WDX spectrometry, CL, EBIC measurements, EELS, and EFTEM).

into that making use of secondary electrons (SE) and of backscattered electrons (BSEs), resulting in different contrasts in the images and thus providing information on compositions, microstructures, and surface potentials. Also, it will be demonstrated how important it is to combine various techniques on identical sample positions, in order to enhance the interpretation of the results obtained from applying individual SEM techniques.

Of particular importance for all SEM techniques are the energy of the incident electron beam E_b (adjusted via the acceleration voltage) and the electron-beam current I_B, which is adjusted by the current through the filaments of thermionic electron guns such as W or LaB_6 cathodes, or by apertures for microscopes with field-emission guns. The electron-beam energy E_b defines the penetration depth R into the specimen material, which may be approximated via the empirical expression [1] $R = 4.28 \times 10^{-2} \, \mu m [\rho/(g/cm^3)]^{-1} (E_b/keV)^{1.75}$, where ρ is the average density of the specimen material. On the other hand, the electron-beam current I_B affects the number of electrons per second in the electron beam and thus the electron-injection rate into the sample.

Electrons are focused by a series of electromagnetic lenses, resulting in electron-probe diameters on the specimen surface of down to below 1 nm. Scanning coils move the focused beam across the sample surface. Further extensive introductions into SEM and its related techniques can be found in various text books (e.g., Ref. [2]).

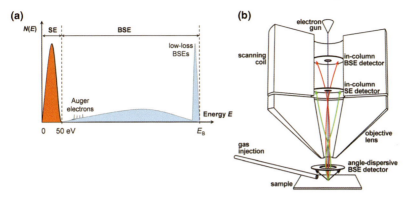

Figure 12.2 (a) Energy distribution of scattered SEs and BSEs. The spectrum features also peaks related to Auger electrons. (b) The electron trajectories in a typical scanning electron microscope, equipped with SE and BSE detectors as well as with a gas-injection system.

12.2.1
Imaging Techniques

Electrons used for imaging in a scanning electron microscope are either SEs or BSEs. Both types of electrons are emitted from the specimen upon electron irradiation (Figure 12.2). The emission spectrum also contains contributions from Auger electrons (used for surface analysis of samples, see Chapter 15 and Section 16.3), which are emitted after ionization of an inner-core shell, as alternative to characteristic X-rays (see Section 12.2.3). By convention, SEs exhibit energies between 0 and 50 eV, whereas BSEs cover the energy range between 50 eV and the energy of the primary beam E_b.

SEs are mostly collected outside of the microscope column by a positively biased collector grid and then accelerated on a scintillator layer in front of a photomultiplier (note that also low-energy BSEs are thus detected). Due to their very small exit depths of few nanometers, and also due to the fact that the SE yield depends on the tilt of a given surface element, SEs are used to image the surface topography.

SEs are retarded by a positive bias and repelled by a negative one at the sample surface. Also, these electrons are affected by the electrical field present owing to differently biased regions. Thus, negatively charged regions appear bright and positively charged areas dark. Therefore, it is possible to obtain a voltage contrast and also information on the local doping [3].

SEs may also be detected inside the microscope column, by use of annular detectors. As much as the electron beam is focused by the series of electrostatic and magnetic lenses when running from top to bottom, it is spread energetically when emitted from the specimen and traveling up the microscope column. SEs and BSEs exhibit substantially different trajectories, and therefore, two corresponding annular detectors at different vertical positions in the column may be used (Figure 12.2). Images recorded by the in-column SE detector rather contain information from the surface.

The in-column BSE detector collects mainly Rutherford-type BSEs, that is, multiple elastically scattered BSEs (by nucleus–electron interaction), emitted into rather small angles with respect to the impinging electron beam, and the energy of these BSEs depends on the atomic number Z. This is why the technique is often referred to as Z contrast imaging, that is, it allows for detecting phases-exhibiting differences in Z. Additional energy filtering of these BSEs may reveal even small variations in compositions [4].

Apart from Rutherford-type BSEs, also Mott-type (single elastically scattered) BSEs are emitted, however, at much larger angles (with respect to the impinging beam). The exit volume is therefore also smaller, that is, the BSEs are emitted close to the surface, and in consequence, the corresponding detector is positioned just beneath the end pole-piece of the microscope. Modern microscopes provide lens systems which separate the Mott BSEs, emitted at very large angles ($>60°$), from the Rutherford-type BSEs scattered into smaller angles [5]. Therefore, the contribution from Z contrast in the Mott BSE image is reduced substantially. Rather, the images exhibit a crystallographic contrast, also termed channeling contrast, resulting from the fact that the yield in BSEs is high when atomic columns in a grain are oriented (nearly) parallel to the trajectories of the BSEs, that is, the crystal forms channels for the BSEs. Then, the corresponding grain appears bright in the BSE image (else, lower intensities are obtained in the images). Sometimes, it is possible to obtain a channeling contrast also by use of a SE detector, which forms since channeled BSEs may lead to the release of SEs, carrying on the crystallographic contrast.

The use of gases injected on the specimen during imaging is an important issue, since, for example, in case of cross-section thin-film solar-cell specimens, the SEM imaging may be complicated by use of insulating materials such as glass or plastic foil substrates. The idea is to introduce a gas, for example, N_2, which is then cracked by the impinging electron beam, reducing charges present on the specimen surface. That is, the impinging electron beam is not substantially deflected, which may have considerable consequences on imaging and analysis.

12.2.2
Electron Backscatter Diffraction

BSEs may not only be used for imaging of thin films but also give information on crystal symmetry and orientation when diffracted at corresponding sets of atomic planes. This is the principle of electron backscatter diffraction [6] (EBSD), which is depicted in the schematics in Figure 12.3. When an electron beam impinges into a crystal residing within a polycrystalline thin film, BSEs are emitted in all directions, that is, some of these electrons meet the Bragg condition for a specific set of atomic planes in the crystal. Electron beams diffracted at a specific set of atomic planes form, when emerging from the specimen, a band on a planar detector, that is, a charge-coupled device (CCD) camera with a thin scintillator layer. The bands of the various sets of atomic planes in a crystal result in a diffraction pattern, which is referred to as Kikuchi or EBSD pattern.

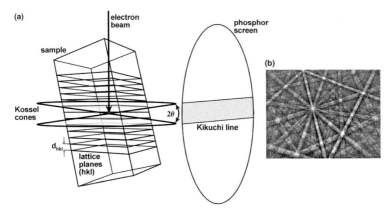

Figure 12.3 Schematics of the EBSD measurement (a) and a EBSD pattern from a Cu(In,Ga)Se$_2$ crystal (b).

It is apparent from Figure 12.3 that the sample surface is tilted substantially off the position perpendicular to the electron beam, which is done in order to reduce the absorption of BSEs by the surrounding material. The tilt angle usually applied of about 70° is a trade-off between reducing absorption of BSEs and still obtaining good resolution of the EBSD patterns. In a fully automated EBSD system, the positions (corresponding to the interplanar angles) and widths (related to the Bragg angles) of few bands are identified, and these bands are then indexed by comparison with simulated patterns from specific crystals. In order to increase the speed of evaluation, the diffraction bands in a given EBSD pattern are transformed to peaks in a radius-angle coordinate system, the so-called Hough space [6]. In present EBSD systems, EBSD patterns can be acquired and indexed at various points of a mesh, when scanning across a region of interest on the specimen, at velocities of up to above 600 patterns per second, resulting in an EBSD map.

It is important to point out that EBSD is a very surface-sensitive technique. In spite of the fact that at, for example, 20 kV, the penetration depth of the electron beam may be few micrometers for the typical absorber materials in inorganic thin-film solar cells, the exit depths for the BSEs contributing to the EBSD patterns are only few tens of nanometers. Therefore, the quality of the EBSD patterns and hence that of the EBSD maps depends mainly on the quality of the surface preparation of the specimen (see Section 12.4 for details). A very thin graphite layer (few nanometers) on top of a polished cross-section specimen has shown to aid in preserving the surface for several weeks and also in reducing the drift during the EBSD measurement.

Since the local orientation of a crystal (with respect to a reference coordinate system) at a specific measuring point in the map can be identified, the EBSD patterns can be evaluated such that the EBSD map represents the local-orientation distribution, given by colors (see Figure 12.4a and legend therein). When acquiring EBSD patterns on a large area of about of 1 mm^2 or larger, integral film textures can be extracted which compare to those from texture measurements by means of X-ray

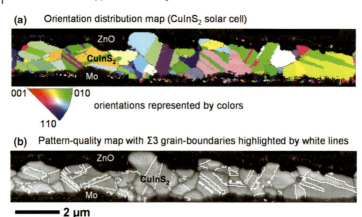

(a) Orientation distribution map (CuInS$_2$ solar cell)

orientations represented by colors

(b) Pattern-quality map with Σ3 grain-boundaries highlighted by white lines

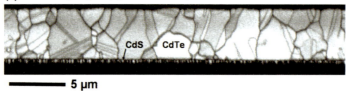

2 μm

(c) Pattern-quality map (CdTe solar cell)

5 μm

(d) Pattern-quality map (Si thin film on ZnO/glass)

5 μm

Figure 12.4 (a) Local-orientation distribution and (b) pattern-quality maps, extracted from the identical set of EBSD data, acquired on the cross-section specimen of a CuInS$_2$ thin-film solar cell (provided by J. Klaer, HZB, Berlin, Germany). (c) and (d): EBSD pattern-quality maps from a CdTe solar cell and a Si/ZnO/glass stack (also cross-section specimens; samples provided by A.N. Tiwari, EMPA, Dübendorf, Switzerland and T. Sontheimer, HZB).

diffraction (however, showing information depths of only few tens of nanometers, not several micrometers). Another way of representing EBSD data is to measure the sharpness of some diffraction bands (or peak intensities in the Hough space), which is typically expressed as gray values and termed pattern quality. The grain boundaries are visible in pattern-quality maps (see, e.g., Figure 12.4b) as dark lines since at the positions of grain boundaries, the EBSD patterns of two neighboring grains superimpose, resulting in zero solutions and therefore very low gray values.

Since the local orientations of neighboring grains are identified, also their relative orientations, that is, the so-called misorientations, can be quantified from the EBSD data. The misorientations can be parametrized by, for example, corresponding angle–axis pairs, where the point lattice of the first grain has to be rotated about the axis through the angle in order to result in the point lattice of the second grain.

There are, depending on the crystal symmetry, various symmetrically equivalent angle-axis pairs, of which the one with the lowest angle is their representative, the so-called disorientation (where the disorientation angle can be used to classify grain boundaries in an EBSD map and to extract corresponding distributions). The superposition of the point lattices of two neighboring grains results in the so-called coincidence-site lattice (CSL). The ratio formed by the volume of the CSL unit cell and that of the point lattices is termed the Σ value, which corresponds to a specific disorientation (angle–axis pair) and therefore also can be used to classify grain boundaries. Note that grain boundaries always feature five macroscopic degrees of freedom, of which three parameters are represented by the disorientation (which can be measured by means of EBSD) and two parameters are related to the normal of the grain-boundary plane (which can generally not be identified by EBSD).

A maximum misorientation angle in case of a contiguous grain is defined for neighboring pixels in an EBSD map. Thus, grain-size distributions of a polycrystalline thin film may be extracted at high accuracy. However, when acquiring an EBSD map on a region of interest, the microstructural information obtained is reduced to a two-dimensional section through the polycrystalline thin film, owing to the low information depth for EBSD of only a few tens of nanometers.

In order to overcome this obstacle, the EBSD technique may be combined with a scanning electron microscope equipped with a focused ion beam (FIB, see Section 12.4 for details). The specimen is alternately sectioned by the FIB and analyzed by EBSD. A three-dimensional (3D) data cube is then reconstructed from the resulting stack of EBSD maps. From this data cube, 3D grain-size distributions can be extracted, although a large volume has to be analyzed in order to obtain good statistics of the distributions, which is generally difficult for thin-film specimens. However, applying 3D EBSD analysis allows for imaging the 3D microstructure of a thin film at a specific region of interest, which is especially useful when studying grain boundaries by means of electron-beam-induced current (EBIC) or cathodoluminescence (CL), as will be further lined out in Section 12.2.7.

12.2.3
Energy-Dispersive and Wavelength-Dispersive X-Ray Spectrometry

When an electron impinges on a specimen, it may scatter with an inner-shell electron of a specimen atom, transferring energy sufficient for the inner-shell electron to emerge from the specimen. The empty state is then reoccupied by an electron from an elevated state, and the difference in potential energy between the two states is either transferred to a bound electron, which then can leave the atomic bond as Auger electron, or is emitted as X-ray quantum. The difference in potential energy between two electronic states in an atom and therefore the energy of the X-ray quantum is characteristic for a specific element. Thus, by recording all X-rays emitted from a specimen upon electron-beam irradiation, the chemical composition can be analyzed.

There are two ways of X-ray detection. One is based on X-ray quanta from the specimen generating charge pulses in a field-effect transistor, which then are

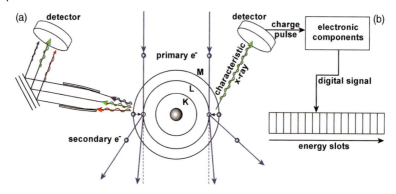

Figure 12.5 Schematics of principles of WDX (a) and EDX (b) measurements in a scanning electron microscope. While the characteristic X-rays are wavelength-dispersed by Bragg diffraction at crystals in the case of WDX, for EDX, the characteristic X-rays generate electron–hole pairs and thus current pulses in the detector, which are assigned to the corresponding energy slots in the spectrum.

processed and assigned to slots according to their energies (Figure 12.5b). This is the principle of the EDX spectrometry. The EDX spectrum consists of X-ray lines at energy positions according to the characteristic energy differences between the states of inner-shell electrons. At high beam currents of several nanoamperes, fast acquisitions of EDX line-scans or elemental-distribution maps can be performed when scanning on a line or across a specific region of interest. The reader is referred to Section 16.5 for further details.

For the other way of X-ray detection, the X-rays emitted from the specimen are Bragg diffracted at single crystals, and by using crystals with various interplanar spacings, a large wavelength range in the spectrum is covered (Figure 12.5a). This is the principle of the wavelength-dispersive X-ray spectrometry (WDX). The X-ray line resolution of WDX and also its detection limit is by an order of magnitude lower than that of EDX, and therefore, WDX is predestined for analyses of compounds where X-ray lines are very close to or even superimpose each other, and also for the analysis of rather light elements in compounds and those present at small concentrations (down to about 0.01 at.%). It should be noted that by use of present EDX evaluation software, the deconvolution procedures of two superimposing X-ray lines have demonstrated quite effective. Also, when analyzing compounds with several elements by means of WDX, often different single crystals have to be used for each element, which is time-consuming. Since WDX analyzes X-rays at rather flat angles with respect to the sample surface, owing to the geometries of analysis crystal and detector, rather high beam currents are needed in order not to suffer from long acquisition durations, for decent statistics of the measurement.

Speaking of the spatial resolution, for example, in EDX or WDX elemental-distribution maps, it is determined mainly by the excitation volume of the impinging electron beam (it is important to emphasize that in this sense, there is no difference between EDX and WDX). That is, high spatial resolution of down to about 100 nm can be achieved by reducing the acceleration voltage of the scanning electron microscope

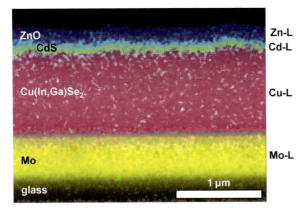

Figure 12.6 SEM cross-section image from a ZnO/CdS/Cu(In,Ga)Se$_2$/Mo/glass stack, superimposed by elemental distribution maps using Zn-L (blue), Cd-L (green), Cu-L (pink), and Mo-L (yellow) signals.

(e.g., to 5–7 keV). The spatial resolution also depends on the mean free path of the X-ray quantum of interest. For example, when regarding Cu-L X-ray quanta with energies of about 1 keV (EDX), the spatial resolutions extracted from corresponding elemental distribution maps may reach values of down to 100 nm, whereas they are substantially higher for maps using Cu-K X-rays (with energies of about 8 keV), which exhibit a much larger mean free path. In Figure 12.6, an SEM cross-section image is shown, which is superimposed by elemental distribution maps using Zn-L, Cd-L, Cu-L, and Mo-L signals. Although the CdS layer thickness is only about 50 nm, the corresponding Cd-L signals are clearly visible in between the Zn-L and Cu-L distribution maps. Also, although the energies for the Zn-L and Cu-L lines differ only by about 80 eV, the evaluation software was able to deconvolute the corresponding EDX peaks successfully.

A different scenario for EDX measurements is given when analyzing specimens with thicknesses of only about 10–50 nm in a transmission electron microscope. Then, the excitation volume may be very small, when working with a highly focused electron probe even lower than 1 nm in extension, in spite of the high acceleration voltages of 100–300 keV. The resulting spatial resolutions may be below 1 nm [7]. Whenever working with such thin specimens, however, it is important to realize the substantial reduction in count rates and therefore the deterioration in statistics.

12.2.4
Electron-Beam-Induced Current Measurements

EBIC in a scanning electron microscope is a widespread method to characterize electronic properties of thin-film solar cells. Its principle is to measure the current of the solar cell or any other charge carrier separating and collecting structure during electron beam irradiation.

Instead of light as in standard working conditions, the electron beam is used to generate free charge carriers/electron–hole pairs in the semiconducting absorber material in a defined region around the position of irradiation. These charge carriers either recombine or they are collected and in this way measurable as an external current. The probability of collection depends on the position of generation and other parameters characteristic of the solar cell as for example diffusion lengths of charge carriers in the materials involved, charge distributions, and potential drops. By means of EBIC, these properties can be studied with a high spatial resolution. A good overview of the technique is given in a review by Leamy [8].

12.2.4.1 Electron-Beam Generation

Figure 12.7 shows schematically the setup of an EBIC experiment. The solar-cell sample is irradiated by a focused electron beam and charge carriers are generated in a defined region around the position of irradiation.

The number of generated electron–hole pairs is assumed to be proportional to the band-gap energy E_g of the irradiated material [10] and the energy necessary to generate one electron–hole pair E_{eh} can be expressed as $E_{eh} = 2.1\ E_g + 1.3\ eV$. In EBIC experiments, typical electron-beam energies E_b are between 3 and 40 keV. Varying E_b changes not only the number of charge carriers generated but also the

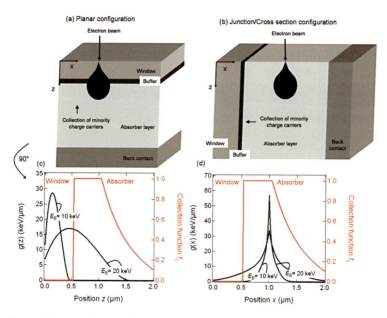

Figure 12.7 (a) Planar configuration of an EBIC experiment. (b) EBIC cross section or junction configuration. (c) Depth-dependent generation function in Cu(In,Ga)Se$_2$ according to the equations in Ref. [1] and depth-dependent collection function f_c. (d) Lateral generation function [9] and collection function perpendicular to pn-junction. f_c is the same in (c) and (d). For calculation of f_c in the quasineutral region, Eq. (12.2) was used assuming $L = 500$ nm, $S_{BC} = 1 \times 10^5$ cm/s, $D = 1$ cm^2/s.

extension of the generation region. The spatial distribution is proportional to the energy deposited by a primary electron along its path through the sample. For the description of this energy loss versus distance, different approaches have been chosen. Equations for mostly one-dimensional generation profiles in different materials have been derived experimentally by EBIC and luminescence measurements [1, 11–13], via Monte Carlo simulations [11, 14] and analytically [15]. The lateral and depth-dependent generation functions shown in Figure 12.7 are derived from equations given in Refs. [1] and [9]. In Ref. [11], a comparison of various profiles gained from different measurements and simulations is shown, which illustrates that the results vary quite significantly.

Another problem occurs when analyzing a multilayer stack such as a solar cell. The problem of different densities of the irradiated materials can be solved by normalizing the spatial coordinate to the density of the corresponding material as shown in Ref. [9], but this procedure still neglects backscattering at the interfaces between the layers. Profiles for multilayer structures including this effect were obtained via Monte Carlo simulations or by a simple cut and paste method [14].

12.2.4.2 Charge-Carrier Collection in a Solar Cell

For charge-carrier collection within the solar cell, that is, for obtaining an external voltage or current, there need to be "selective" contacts for each type of charge carrier. Upon irradiation, a gradient of the Fermi levels of electrons and holes develops, which causes an electron current to one contact and a hole current to the second one. In a "regular" pn-junction based solar cell, this is provided by the potential drop of the space-charge region of the pn-junction and ohmic (or low barrier) contacts to the external circuit.

The relevant equations describing charge-carrier transport are the continuity equations for electrons and holes in the absorber material. When assuming low-injection conditions (i.e., density of generated minority charge carriers much lower than density of majority charge carriers), the following equation for the collection function $f_c(\vec{x})$ can be derived via a reciprocity theorem [16, 17]:

$$D\Delta f_c(\vec{x}) + \mu \vec{E} \vec{\nabla} f_c(\vec{x}) - \frac{f_c(\vec{x})}{\tau} = 0 \quad (12.1)$$

D is the diffusion constant of the minority charge carriers of the corresponding material, μ their mobility, τ their lifetime, and $\vec{E}$ a possible electric field. The collection function $f_c(\vec{x})$ stands for the probability of a charge carrier generated at position $\vec{x}$ to be collected. If translation invariance in the directions parallel to the pn-junction (y and z) is assumed, a one-dimensional equation can be used. Assuming as a boundary condition that the collection probability at the edge of the space-charge region is one ($f_c(x_{SCR}) = 1$) and in case of an infinite semiconductor layer where $f_c(x) \rightarrow 0$ for $x \rightarrow \infty$, the solution for a field free absorber layer ($E=0$) is a simple exponential function $f_c(x) = \exp(-x/L)$, where L is the diffusion length of the minority charge carriers. Assuming a finite semiconductor limited by a back contact at position x_{BC}, the second boundary condition changes to $f'_c(x_{BC}) = S_{BC}/D f_c(x_{BC})$, where S_{BC} is the surface recombination velocity of the minority charge carriers at the back contact. A solution for $f_c(x)$ is

$$f_c(x) = \frac{^1/_L \cosh\left(\frac{x-x_{BC}}{L}\right) - ^{S_{BC}}/_D \sinh\left(\frac{x-x_{BC}}{L}\right)}{^{S_{BC}}/_D \sinh\left(\frac{x_{BC}-x_{SCR}}{L}\right) + ^1/_L \cosh\left(\frac{x_{BC}-x_{SCR}}{L}\right)} \quad (12.2)$$

In Figure 12.7 (red curve), an exemplary collection function of a Cu(In,Ga)Se$_2$ thin-film solar cell is shown. $f_c(x)$ is assumed to be zero in the window and buffer layers, one in the space-charge region and it is $L = 500$ nm and $S_{BC} = 1 \times 10^5$ cm/s.

12.2.4.3 Experimental Setups

In this section, an overview of the most commonly used experimental setups for EBIC measurements on thin-film solar cells is given: the junction, plan view, and edge scan configurations. In a lot of cases, it is possible and reasonable to perform EBIC measurements not only on a completed pn-junction solar cell, but on Schottky contacts produced by depositing a metal on the absorber layer. In this way, additional information about collection properties of the absorber layer is provided, especially when comparing EBIC results of Schottky contacts and pn-junctions. Since a sufficiently thin metal layer is partially transparent for electrons, it is possible to use a Schottky contact structure for all configurations described here.

Cross-Section or Junction EBIC Junction EBIC can be used to extract quantitative information about the solar cell like the width of its space-charge region and the diffusion length of the minority charge carriers in the absorber layer. In junction EBIC, the cross section of the solar cell is irradiated by an electron beam and the current signal is measured as a function of the position of irradiation a. An EBIC profile across the cross section of the solar cell perpendicular to the pn-junction can be described by integrating the collection function and the lateral generation profile over the coordinate x (see Figure 12.7):

$$I(a) = \int f_c(x) g(x-a) dx \quad (12.3)$$

It has to be taken into account that in this setup, an additional free surface is present, where enhanced recombination due to surface defects might take place. Fitting of the measured profiles to theoretical ones obtained from the equations above allows for the extraction of values for the width of the space-charge region, the back contact recombination velocity S_{BC}, and a value not for the bulk diffusion length but for an effective diffusion length L_{eff}. L_{eff} is influenced by surface recombination and, therefore, depends on the depth of generation, that is, the electron beam energy [18, 19]. Using the equations given in Refs. [18] and [19], describing the dependence of the effective diffusion length on the electron beam energy E_b, it is possible to extract values for the bulk diffusion length and the surface recombination velocity [20]. An important assumption, which has to be fulfilled for this evaluation, is that the collection function is independent of the generation function (not always the case, see Section 12.2.4.4). In this way, the junction EBIC method can be used to extract quantitative information about important solar-cell properties with a high spatial resolution compared with other techniques.

Additionally, junction EBIC can be used to determine the position of the actual p-n-junction [21, 22], to learn more about interface properties and charge and defect distributions [20], as well as to investigate the behavior of grain boundaries [22–26] and possible inhomogeneities in collection properties.

Planar EBIC from Front and Back Side In such setups, the solar cell is either irradiated from the front side like it is under standard sunlight irradiation or from the backside. The EBIC can then be expressed as (Figure 12.7) $I = \int f_c(z)g(z)dz$, where $g(z)$ is the depth-generation profile of the electron beam. Information about collection properties like the minority charge carrier diffusion length and width of the space-charge region can be obtained by measuring the collection efficiency (ratio of measured current to incident beam current) for different electron beam energies E_b and consequently changed generation profiles $g(z)$. The evaluation of this data can be performed in different ways [27–29]. Compared with junction EBIC, planar EBIC from the front side provides the advantage that there is no artificial surface, which could influence charge-carrier collection properties. The method can also be used to investigate the behavior of grain boundaries concerning charge-carrier collection [30, 31]. One difficulty is the topography of the surface irradiated, which has a huge effect on absorption of incident electrons. This has to be taken into account especially when investigating grain-boundary effects.

Edge Scan Configuration In this configuration, a charge-carrier collecting junction is deposited only on part of the surface of an absorber layer. The EBIC signal is measured in dependence of the distance of the position of irradiation to the junction [31, 32].

12.2.4.4 Critical Issues

When using EBIC in order to investigate charge-carrier collection properties of solar cells, it has to be taken into account that the generation of charge carriers in a semiconductor using an electron beam might be significantly different compared to the generation by sunlight. In the following, different scenarios are discussed.

Injection Conditions The different spatial distribution of charge carrier generation as compared to sunlight generation might influence the charge distribution in the solar cell and therefore collection properties as, for example, shown in Ref. [33].

As mentioned above, the injection level, that is, the density of excess charge carriers compared to the equilibrium doping level of the semiconductor irradiated, plays an important role. Equation (12.2) describes the collection function of the quasineutral (i.e., field free) region of an absorber layer only if low-injection conditions are present. When the density of excess charge carriers is close to or higher than the doping density, the charge distribution within the device is influenced by these excess charge carriers, which also changes collection properties [34–36]. This effect can be used to estimate the doping density with a good spatial resolution, which can for example be used to exhibit differences in the doping level of grain boundaries [26].

Temperature When using high electron beam intensities, it is possible that the sample heats up during the measurements, which may result in a modified charge-carrier collection behavior of the device. An estimate of the maximum temperature change due to electron beam irradiation is given by $\Delta T = E_b I_b r_b / K$ [37], where K is the thermal conductivity of the material irradiated, r_b is the beam radius (in our case in the order of the extension of the generation volume), E_b the electron beam energy, and I_b the electron beam current. A more precise equation for a thin film substrate structure is also given in Ref. [37].

Interactions of Sample and Electron Beam When investigating charge-carrier collection properties by means of EBIC, one has to be aware of the fact that the electrons penetrating into the sample have much higher energies compared to light of the solar spectrum. There might also be reactions taking place which are not possible under standard conditions but might influence charge-carrier collection properties. As an example, it was found that the electron beam works as a reducing agent and breaks oxygen bondings passivating compensating donors in a $Cu(In,Ga)Se_2$ layer, thereby changing its doping density [21].

Additional Interfaces/Surfaces The experimental setup might lead to additional surfaces or interfaces present in the charge carrier collecting device, which are not present under standard conditions using sunlight as excitation source. At these interfaces, there might be enhanced recombination due to an accumulation of defect states as mentioned above. A second effect might also be a different charge equilibrium due to structural reconstruction at the interface (dangling bonds, etc.). This results in band bending at the surface and, therefore, modified collection properties. As an example, it was shown that on InP, an oxide causes band bending, which assists charge-carrier collection [38]. The EBIC signal results to be nearly constant with increasing distance to the collecting junction.

In the present section, EBIC was introduced as a powerful experimental technique to gain information about electrical properties of thin-film solar cells. There are different configurations in use complementing each other and allowing a detailed insight into charge-carrier collection properties. As the experimental setup and generation conditions differ from the ones under standard conditions using sunlight for generation, interpretation of EBIC data is not always straightforward and care has to be taken. In solar-cell research, EBIC is not only used to derive values for relevant quantities like diffusion lengths or the width of the space-charge region, but also to learn about junction and interface properties as well as charge and potential distributions. In this way, it helps to develop a profound understanding of device properties and to improve solar-cell performance.

12.2.5
Cathodoluminescence

In semiconductors, the energy of impinging photons or the electron beam can promote electrons from the valence band to the conduction band, with the generation of electron–hole (e–h) pairs. The radiative recombination of these electrons and holes results in *photoluminescence* (PL, see Chapter 7) upon photon excitation and CL as a result of electron irradiation. Therefore, the radiative processes in semiconductors are fundamentally similar in both PL and CL, only differing in the excitation source. CL thus benefits from the widespread use of photoluminescence for the interpretation of the emission spectra obtained.

Luminescence spectroscopies are very useful to determine composition in semiconductor compounds, assess crystal quality, detect electronic levels associated to dopants (down to densities in the 10^{15} cm^{-3} range), and electronically active defects, but does not provide high spatial resolution. Here is where the CL mode of operation of the electron microscope in both, SEM and TEM, finds its application.

A summary of the radiative transitions that can be found in the emission spectrum of semiconductors is given in Figure 12.8. Process 1 describes the electron thermalization with intraband transition, a process that may lead to phonon-assisted luminescence or just phonon excitation. Process 2 is an interband transition involving the recombination of an electron in the conduction band and a hole in the valence band with the emission of a photon of energy $h\nu \approx E_g$. Processes 3-5 correspond to the exciton recombination, which is observable at cryogenic temperatures and, in some cases, in quantum structures at higher temperature. For the excitonic recombination, we can distinguish between free excitons (commonly denoted by *FX*) and excitons bound to an impurity ($D°X$ for a neutral donor, $A°X$ for a neutral acceptor; for the corresponding ionized impurities are D^-X and A^-X). Processes 1, 2, and 3 are *intrinsic* luminescence because they are observed in undoped semiconductors. Processes 4, 5, and 6 arise from transitions involving energy levels associated to donors and/or acceptors and they are collectively known as *extrinsic* luminescence. Process 4 represents the transition between the energy level associated to a donor and a free hole ($D°h$). Process 5 represents the transition between the free electron and the energy level associated to an acceptor ($eA°$). Donor-to-acceptor (DAP) recombination is obtained if an electron bound to the donor state recombines with a hole bound to the acceptor state, a DAP; process 6. Finally, it is worth mentioning that all these radiative recombination processes will compete with each other, with other nonradiative recombination mechanisms (process 7, representing

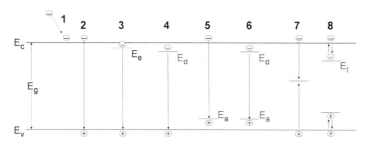

Figure 12.8 Schematic diagram of radiative transitions in semiconductors between the conduction band (E_c) and valence band (E_v) and transitions involving exciton (E_e), donor (E_d), and acceptor (E_a) levels. Nonradiative transitions via mid-gap states and traps are also shown.

recombination through a midgap state without photon emission), and with trapping levels for both electrons and holes, process 8. The contribution of each process to the overall recombination will be reflected in the local emission spectrum and is ultimately the source of contrast in the photon intensity maps acquired by CL. The possibility to control the temperature and the excitation level during the experiment makes CL ideal for investigating in detail the recombination processes in semiconductors. CL has enabled imaging of the electronic and optical properties of semiconductor structures with an ultimate resolution of about 20 nm (although 100–500 nm is a more typical value) and can provide depth-resolved information by just varying the electron beam energy (or acceleration voltage).

In general, the CL modes of the electron microscope can be divided into *spectroscopy* and *imaging*. In the *spectroscopy* mode, a spectrum is obtained over a selected area under observation in the SEM or TEM (a *point analysis* in the terminology of *X-ray microanalysis*, with the electron beam fixed over one location). In the *imaging* mode, an *image* of the photon intensity (when using a monochromator at the wavelength range of interest, a *monochromatic* image; when bypassing the monochromator, not resolved in energy, a *panchromatic* image) is acquired instead. Because *spectroscopy* and *imaging* cannot be operated simultaneously, information is inevitably lost.

Both modes can be combined in one single mode: *spectrum imaging*. The objective of the section of this chapter is introducing *spectrum imaging* as the most advanced instrumentation developed to date for CL measurements and illustrate how *spectrum imaging* can be applied to thin-film photovoltaics. Figure 12.9 shows the schematics of the instrumentation needed to setup the *spectrum imaging* mode in the SEM. The essential requirements for *spectrum imaging* are superior efficiency in the collection, transmission, and detection of the luminescence. High collection efficiency is achieved by a parabolic mirror attached to the end of a *retractable* optical guide (the collection efficiency is estimated to be about 80% when positioning the specimen at the focal point of the mirror). A hole (about 500 μm in diameter) drilled through the parabolic mirror and aligned vertically with the focal point allows the electron beam through. The light transmitted by the optical guide is focused at the entrance slit of a spectrograph by a collimating lens. For the detection, multichannel photodetectors

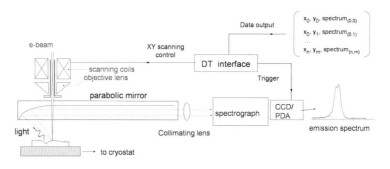

Figure 12.9 Schematics of the CL spectrum imaging setup in the SEM.

are needed. Both charged-coupled device (CCD) or photodetector array (PDA) architectures are used for multichannel spectrum acquisition, where the choice of the photodetector material (Si, GaAs, InGaAs, ...) depends on the wavelength of interest. Only at very low operating temperatures (using cryogenic liquids) is when CCDs and PDAs achieve the high sensitivity and superior performance required for this application (millisecond readout times, triggering, extremely low dark currents).

Up to this point, the described system is not yet able to perform *spectrum imaging*. The principal addition for implementing *spectrum imaging* is the digital electronics (DT interface in Figure 12.4) which (i) controls the X–Y scanning of the electron beam, (ii) sends the *triggers* to the CCD/PDA electronics for the acquisition, and (iii) process the *spectrum series* by associating each x_n, y_m pixel of the scanning with the corresponding spectrum. Thus, *spectrum imaging* combines *spectroscopy* and *imaging* in one single measurement by acquiring the emission spectrum at high speed (typically 10–100 ms) in synchronization with the scanning of the electron beam. With acquisition times of 10–20 ms by pixel on a 125×125 pixel scan, the time to acquire the entire *spectrum series* – consisting of 15,625 spectra, equivalent to more than ten million data inputs – is about 5 min. This high-speed mode is routinely used when measuring thin films at cryogenic temperatures. When a very low excitation is needed to improve the resolution, or when the emission is very low, we can increase the acquisition time per pixel (up to 500 ms to 2 s) at the cost of a much prolonged time for the measurement (hours instead of minutes). After the acquisition is complete, the *spectrum series* can be processed to

- reconstruct maps of the photon intensity, photon energy, or full-width-half maximum (FWHM) at the wavelength range of interest selected over the spectrum;
- extract the spectrum from a selected area;
- output an ASCII file with any of the calculated parameters;
- perform quantitative measurements (relative contributions of different transitions, recombination rate at extended defects, ...);
- pixel-to-pixel correlation between images is inherent to *spectrum imaging*;
- display spectrum linescans;
- run spectrum fitting routines;

and much more. Once the spectrum series is saved, it can be reexamined in the future. Even to answer one question that might be asked years after the measurements were completed.

Applications of *spectrum imaging* to the investigation of electronic properties in CdTe are presented in the remaining of the subsection. This is obviously a minor, although representative, demonstration of the results published to date and the reader is welcome to explore the available literature on this subject.

12.2.5.1 Example: Spectrum Imaging of CdTe Thin Films

Record efficiencies in CdTe solar cells have been produced using *close space sublimation* (CSS) which, simply by the thermodynamics of the process, results in a high concentration of native cadmium vacancies (V_{Cd}). The need to provide a cost-effective solution by reducing capital and operating costs results in a CdTe thin film with a very high density of defects: (i) point defects (*intrinsic*, vacancies and interstitials associated to Cd and Te; *extrinsic*, introduced by incorporation of impurities to the CdTe), and (ii) extended defects (dislocations and grain boundaries). Because of this, we can anticipate that the emission spectrum will show all the radiative (and nonradiative) transitions described above (represented schematically in Figure 12.8). CdTe thin films are then a perfect example of what to be expected in most polycrystalline semiconductors used in solar cells, in which the crystal quality is good enough to see the *intrinsic luminescence* associated to related single crystals of high purity but is otherwise strongly influenced by the presence of defects.

A *luminescence spectrum* representative of CdTe thin films is shown in Figure 12.10 ($T = 15.7$ K, $E_b = 10$ keV, $I_b = 250$ pA). At cryo temperatures, the spectrum consists of excitonic transitions (free excitons *FX* and bound excitons *BX*) and a very broad emission at lower energy, which is associated with DAP recombination. In CSS CdTe, the acceptors are complexes with participation of V_{Cd}, such as the center A or chlorine center A (after chlorine treatment) [39, 40]. The density of these complexes is so high that they interact with each other forming a band (of acceptor character) located within the band gap. The fine structure of discrete transitions associated to the DAP recombination is lost, substituted for a very broad emission. The considerable broadening of the transitions in the emission spectrum is yet another manifestation of the high density of defects in polycrystalline semiconductors. *Spectroscopy* measurements from single crystals are often required for the interpretation of the spectrum in thin films.

Spectrum imaging is very useful for the *microcharacterization* of the transitions identified in the emission spectrum and, more specifically, in determining the electronic properties of extended defects such as dislocations and grain boundaries. Results of *spectrum imaging* measurements completed on a CdTe thin film are summarized in Figure 12.11. The microstructure of the CdTe film is readily accessible to the electron microscope, as seen on Figure 12.11a, which reveals grains

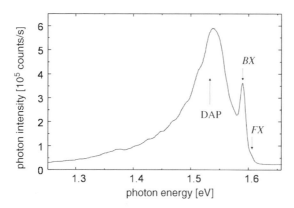

Figure 12.10 CL spectrum representative of CdTe thin films ($T = 15.7$ K, $E_b = 10$ keV, $I_b = 250$ pA). Excitonic transitions involving free excitons *FX* and bound excitons *BX*, and a very broad emission associated with DAP recombination can be resolved.

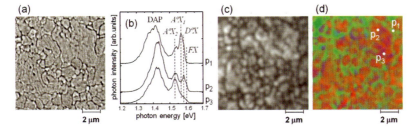

Figure 12.11 Spectrum imaging of CdTe thin films ($T = 15.7$ K, $E_b = 10$ keV, $I_b = 250$ pA). (a) Microstructure of the CdTe thin film, as seen by the SE mode of the SEM. (b) A sequence of spectra selected over different locations of Figure 12.11d, p_1, p_2, and p_3, which reveals variations in the local density of donors and acceptors. (c) Photon intensity map corresponding to the free exciton *FX*. (d) Photon energy map RGB (red: 1.530 eV, green: 1.570 eV, and blue: 1.600 eV) of the exciton recombination. The transition A^0X_1 at 1.570 eV is characteristic of grain boundaries; this is also shown in the spectrum at location p_1.

with 1–5 μm in diameter. When first examining the *spectrum series*, it becomes clear that there are large variations in the emission spectrum from location to location within the micrometer scale; an indication that the distribution of all the different electronic states present in the CdTe thin film is highly nonuniform. This is illustrated by the sequence of spectra in Figure 12.11b, selected over different locations shown in Figure 12.11a and extracted from the *spectrum imaging* file. These local variations certainly contribute to the broadening seen in the emission spectrum acquired over large areas of the film. This is worth noting, but the best capability of *spectrum imaging* is the possibility of correlating directly electronic properties at the microscale with the microstructure of the film in a completely different dimension. This is especially interesting to investigate grain boundaries. Figure 12.11c shows that from the photon intensity image for the free exciton (*FX*) it may be concluded that recombination of free excitons is preferential at grain boundaries, because there, the photon intensity associated with the FX transition is largely reduced. When examining all the excitonic transitions and then mapping the photon energy of the dominant excitonic recombination (Figure 12.11d), it becomes evident that there is a characteristic excitonic transition associated to grain boundaries. To intuitively visualize this, the photon energy is displayed in a red-green-blue (RGB) scale covering the selected spectral interval (in this case, red, 1.530 eV; green, 1.570 eV; and blue, 1.600 eV) – grain boundaries exhibit a green color in the map of Figure 12.11d. With this analysis, a *redshift* or *blueshift* is very intuitive and easy to recognize. In addition to this, and for a more quantitative analysis, an ASCII file can be produced with the energy of the transition(s) for each pixel, which can be further examined. This feature is unique to *spectrum imaging*. The distinct excitonic transition at grain boundaries (A^0X_1 at 1.570 eV) reveals that an acceptor with activation energy $\varepsilon_v + 30$ meV is preferentially located at grain boundaries, whereas an acceptor with activation energy level $\varepsilon_v + 70$ meV (A^0X_2 at 1.530 eV) is otherwise preferentially located in grain interiors. Shifting the energy level of the acceptor toward the valence band seems to be beneficial to the solar cell performance.

12.2.6
Scanning Probe and Scanning-Probe Microscopy Integrated Platform

Scanning probe microscopy (SPM) is becoming the instrument of choice in nanoscience characterization. In addition to providing excellent resolution and sensitivity, the tip (the central component of the microscope) represents de facto a *nanoprobe* that can be adapted to measure multiple properties in different modes of operation [41]. Among them, scanning tunneling microscopy (STM, see Chapter 11), atomic force microscopy (AFM, see Chapter 11), and near-field scanning optical microscopy (NSOM, see Chapter 5) are routinely used in the laboratory. On the other hand, much progress remains to be made with novel modes of operation and applications. With this motivation, we have integrated *SPM* (STM/AFM/NSOM) with SEM in one integrated platform. There are obvious benefits from this approach: (i) the tip can be manipulated with high precision under constant observation in the SEM; (ii) the tip and the electron beam are *two independent probes* that can be controlled simultaneously: one acting as *excitation probe*, the other acting as *sensing probe* – depending on the experimental configuration; (iii) SEM- and SPM-based measurements can be performed at the very same location during the same experiment. The final subsection is dedicated to introducing the modes of operation that are available to this integrated platform as well as describing its application to *luminescence spectroscopy* and *electrical measurements* in thin-film solar cells.

Figure 12.12 shows the schematics of the STM integrated inside a scanning electron microscope equipped with *CL optics (which is described in the previous section dedicated to CL spectrum imaging)*. The STM is based on a three-axis nanopositioning system (basically a digitally controlled *piezoscanner*, commercially available from various vendors) that can be mounted on top of the *mechanical stage* of the SEM. Inertial motion is used for approaching the STM tip, which is mounted on the side of

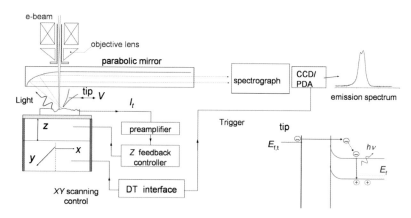

Figure 12.12 Schematics of the STM integrated inside the SEM. The STM is primarily designed to perform *tunneling luminescence* measurements. The inset shows the physics of the tunneling luminescence at low applied bias, in which minority carriers are injected by the tip in the semiconductor, recombining with available majority carriers.

the *piezoscanner*. Specialized cantilevered tips are needed to accommodate the tip between the parabolic mirror and the specimen, which is limited to 2–3 mm when the end of the tip is positioned at the focal point of the mirror. Another requirement to the geometry of the cantilevered tip is that the end of the tip must be observable in the *electron microscope*; otherwise both the electron beam (as the *excitation probe*) and/or the potential *luminescence* excited at the tip – depending on the experimental configuration – will be blocked by the tip itself. One of the advantages of this approach, though, is that the STM tip can be accurately positioned into the focal point of the parabolic mirror, improving dramatically the collection efficiency.

This setup is primarily intended to perform *tunneling luminescence microscopy*, in which tunneling electrons are responsible for the excitation of the luminescence. In semiconductors, the high localization of the tunneling electrons at the STM tip enables the controlled excitation of luminescence within a nanoscale volume underneath the tip. Therefore, *tunneling luminescence* exceeds the best spatial resolution of CL. On the other hand, the *quantum efficiency* of the tunneling luminescence is only about 10^{-4} photons/electron and, therefore, a highly efficient collection optics is absolutely critical. With the optics of the CL setup, an *external quantum efficiency* as high as 10^6 photons/nAs can be obtained, which is close to collecting every single photon. With this, the possibility of running *spectrum imaging* (*spectrum-per-pixel* measurements) of the *tunneling luminescence* is within reach. Now the *digital interface* (see Figure 12.12) controls the scanning of the STM tip and sends the *triggers* to synchronize the spectrum acquisition with the STM scanning. Because of the integration of both SEM and STM, CL and *tunneling luminescence spectroscopies* can be performed over the same location. This is very interesting for the thin films used in solar cells, in which the *surface electronics* can play a critical role in the solar cell response. Because of the small volume of excitation, the STM-based luminescence can be used to measure the electronic states present near the surface. A comparison of the tunneling luminescence and CL spectra over the same location can be useful to compare the electronic properties of the surface versus the bulk of the film.

For illustration of the capabilities of this STM–SEM platform, we present results for Cu(In,Ga)Se$_2$ thin films, in which the surface electronics is critically involved in the formation of the junction [42]. Figure 12.13a and b shows the STM image and the corresponding photon energy map for the tunneling luminescence (color-coded so that red and blue shifts in the emission spectrum are intuitive). Figure 12.13c and d shows the SEM image (over the same area) and the corresponding photon energy map for the CL excitation. This film is deposited in the regime of selenium deficiency and shows larger variations in energy from grain to grain when compared to the standard Cu(In,Ga)Se$_2$ deposited under Se overpressure, as clearly evidenced by the photon energy map of Figure 12.13d.

The most interesting result when comparing the photon energy maps (same RGB-energy scale used in both) is that variations in the luminescence are largely mitigated at the Cu(In,Ga)Se$_2$ surface (Figure 12.13b). The naturally occurring depletion of Cu makes the surface more uniform than the underlying Cu(In,Ga)Se$_2$.

Cu(In,Ga)Se$_2$ is very accessible to observation in the STM, but for other systems, such as CdTe, organic semiconductors, or silicon thin films, the intrinsic low

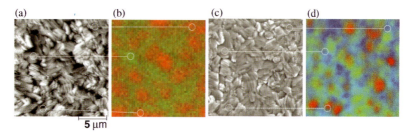

Figure 12.13 (a) STM image of a Cu(In,Ga)Se$_2$ thin film and (b) corresponding photon energy map of the tunneling luminescence ($V_t = -5$ V, $I_t = 50$ nA). (c) SEM image matching the previous STM and (d) corresponding photon energy map of the CL ($E_b = 5$ keV, $I_b = 750$ pA). The RGB scale in (b) and (d) corresponds to R: 1.17 eV, G: 1.20 eV, B: 1.22 eV.

conductivity will cause unstable operation of the STM under constant current conditions (current feedback), with the crashing of the STM tip as the most likely outcome. This can be solved by using the force feedback of the AFM instead. Figure 12.14 shows the schematics of the AFM implemented inside the SEM. The STM tip is now substituted by a force sensor consisting of an ultrasharp metallic tip attached to a self-sensing and -actuating piezo tuning fork (TF). This is the most compact force sensor that can be built and inserted in the confined space between the parabolic mirror and the specimen. The TF is driven by an oscillating voltage source near its resonance frequency $V_{tf}(\omega_r)$. The interaction between the tip and the specimen causes a reduction in the oscillation amplitude (monitored by the current $I_{tf}(\omega_r)$) and a shift in the resonance frequency and phase of the oscillation. One maintains the distance Z constant by controlling the amplitude of the TF oscillation during the scanning of the tip while applying a voltage to the tip to create pulses of tunneling current in synchronization with the TF oscillation. Tunneling current pulses as high as about 100 nA have been observed without significant degradation of

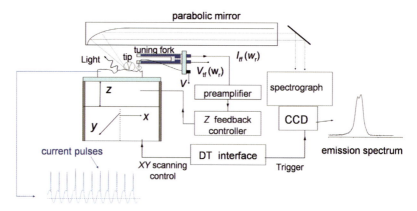

Figure 12.14 Schematics of the AFM integrated inside the SEM.

the AFM operation. In consequence, we can run tunneling luminescence measurements in both the STM or AFM configuration.

There are other benefits from using the AFM setup. First, the oscillation amplitude of the TF and the setpoint can be selected to drive the tip into the intermittent-contact or tapping mode. In this mode, the tip can establish a contact at the bottom of each oscillation. If, for example, a *p–n*-junction is present in the specimen, the tip can locally forward bias the cell and excite electroluminescence (EL). A map of the AFM-based EL will be specific to the depletion region of the solar cell and not so much to the surface electronic states, such as in tunneling luminescence measurements, which are inherently noncontact measurements.

The second, and probably best, advantage of using the AFM is decoupling the *z*-feedback from the electrical measurements that can be performed with the conductive tip (in the STM, the current is fed to the *z*-feedback electronics to follow the topography of the film). This opens up the possibility of performing conductive AFM measurements (topography vs. conductivity) and measuring local *I-V* characteristics. This is available to any commercial AFM, however.

What is unique to the combination of AFM and SEM in the characterization of thin films is the possibility to measure the lateral electron transport across grain boundaries and estimate the diffusion length locally at individual grains. In this configuration, the lateral electron transport across a single-grain boundary can be measured by maintaining the tip over one grain and measuring the difference in the current/voltage sensed by the tip when the electron–hole pairs are excited (by the e-beam) within the grain interior and the adjacent grain across the grain boundary. This is basically an EBIC measurement in which one of the terminal contacts is substituted by the AFM tip. Using this method, we have found evidence for a significant barrier for electron transport across grain boundaries in $CuGaSe_2$, which is not present in $CuInSe_2$ or $Cu(In,Ga)Se_2$. On the other hand, the exponential decay of the current (voltage) sensed by the tip when the electron beam moves away from the tip (similar to the method used by EBIC) can be applied to estimate the diffusion length locally at specific locations of interest in the film, under observation in the SEM.

Figure 12.15 illustrates the application of the AFM-based measurements of the electron transport to polycrystalline silicon films (of *n*-type polarity). Figures 12.15a and b are SEM and AFM images revealing the microstructure of the film and Figure 12.15c corresponds to an image of the current sensed by the tip at one specific location. There are two sources of contrast in the current images. First, the tip can establish a local Schottky diode and the holes excited by the electron beam (minorities in *n*-type) can be collected from regions close to the tip. This collected current, of positive polarity in our setup, extends from the tip the local (hole) diffusion length (see Figure 12.15c). Second, in this contact scheme with *remote* ground, the silicon film acts as a *current divider*. Thus, a fraction of the primary electrons from the electron beam (the specimen current) flows to the tip and the rest to the ground, depending on the relative resistance along these two different paths – this is referred to as REBIC for remote EBIC [43]. This current is of negative polarity in this EBIC image. In the case of a film of uniform resistivity, the current detected by the

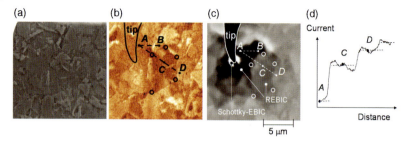

Figure 12.15 AFM-based EBIC measurements performed on polycrystalline silicon thin films. (a and b) The SEM and AFM images acquired over the same area of the film. (c) An image of the current sensed by the tip (acquired during the scanning of the electron beam) with the tip making contact over the location at the end tip. The current linescan along A–C–D on (c) is shown in (d). Electron-beam energy $E_b = 15$ keV and current $I_b = 500$ pA. In order to enhance a correlation between these images, the contours of one grain are outlined for (b) and (c) (hollow circles).

picoamplifier is a straight ohmic line but if there are grain boundaries impeding the electron transport, the ohmic baseline becomes stepped (meaning that the signal drops sharply at the boundary). This is because the slope of the detected current is proportional to the local value of resistivity. Figure 12.15c shows this sharp drop in the detected current when the electron beam moves across the grain boundaries, which is due to their high resistance to the electron flow. Although electrons flow with ease across certain grain boundaries (see, for example, A–B in Figure 12.15b and c), most of the boundaries present a very high energy barrier to the electron transport (see A–C–D in the same figure) and the current toward the tip decreases rapidly when increasing the number of boundaries to be crossed by the electrons in order to reach the tip. This illustrates how the combination of SEM and AFM can be very useful to determine the electrical behavior of individual grain boundaries in microsized grain semiconductors.

12.2.7
Combination of Various Scanning Electron Microscopy Techniques

Apart from combining SEM and AFM imaging, it is also possible to unite various other SEM techniques in order to gain information on microstructure, composition, charge-carrier collection, and radiative recombination, with high spatial resolutions of down to 20–100 nm. These are especially useful when acquired on the identical region of interest. Examples for an SE image, EBSD pattern quality and orientation distribution maps, monochromatic CL images, an EBIC image, and EDX elemental distribution maps, all acquired on the identical position on a $ZnO/CdS/CuInS_2/Mo/$ glass cross-section specimen, are given in Figure 12.16. By combination of, for example, the EBSD maps and the CL images, it is possible to relate the local orientations to the local recombination behavior. The EDX maps suggest that the strongly varying local CL intensities are not due to a change in composition, and it is apparent that the EBIC intensities are high where the CL signals are rather low.

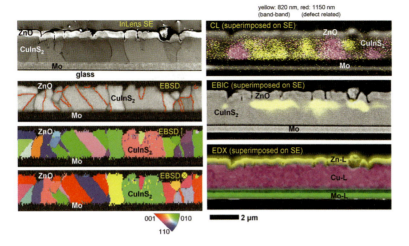

Figure 12.16 SE image (at 3 keV), EBSD pattern-quality (with Σ3 grain boundaries highlighted by red lines) and orientation distribution maps (at 20 keV), monochromatic CL images at 10 keV, 8 K, and at 820 (yellow, band–band transitions) and 1150 nm (pink, defect-related transitions), superimposed on an SE image, an EBIC image (8 keV) and EDX elemental distribution maps (7 keV) from Zn-L (yellow), Cu-L (pink), and Mo-L (green) signals. All images and maps were acquired on the identical position of a ZnO/CdS/CuInS$_2$/Mo/ glass stack (provided by courtesy of Dr. B. Marsen, HZB, Berlin, Germany; CL images in collaboration with Dr. U. Jahn, Paul-Drude Institute Berlin, Germany).

12.3
Transmission Electron Microscopy

Whenever aiming for imaging and analyses at scales of down to the angstroms range, TEM and its related techniques are appropriate tools. In many cases, also SEM techniques provide the access to various material properties of the individual layers, not requiring specimen preparation as time-consuming as TEM techniques (details on the preparation can be found in Section 12.4). Still, TEM exhibits unique possibilities, especially when studying properties of interfaces between individual layers, of grain boundaries, or of any other (sub)nanometer-scale features.

Transmission electron microscopes consist of an electron gun, emitting an electron beam, which is accelerated to energies typically ranging between 80 and 1200 keV. The way these electrons impinge on the sample differs for the conventional (CTEM) and the scanning TEM (STEM) modes (Figure 12.17). In CTEM mode, the specimen is irradiated by a (nearly) parallel electron beam, and imaging as well as electron diffraction is performed on a specific region of interest. The beam may also be focused on a spot and scanned across this region of interest, for modern microscopes with electron probe sizes of down to below 0.1 nm. This allows for imaging, electron diffraction, and compositional analysis of high spatial resolutions.

The present section will give an overview of the various techniques in CTEM and STEM modes, that is, imaging, electron diffraction, electron energy-loss

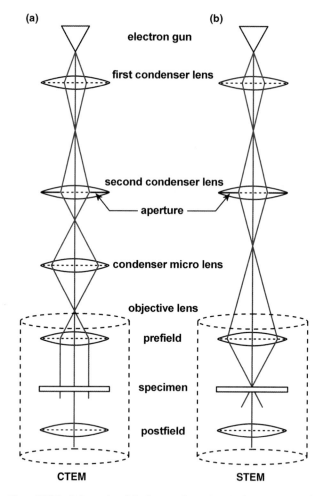

Figure 12.17 Schematics of the lens configurations in the upper part of TEM and their influences on the electron beam trajectories for CTEM (a) and for STEM (b). Note the results are a parallel beam for CTEM and a focused electron probe for STEM. Reproduced from Ref. [44].

spectrometry (EELS), and electron holography. Details on TEM-EDX can be found in Section 12.2.3. Furthermore, detailed introductions into TEM can be found, for example, in Refs. [44–46].

12.3.1
Imaging Techniques

12.3.1.1 Bright-Field and Dark-Field Imaging in the Conventional Mode

When a parallel electron beam is scattered at a crystalline specimen, a part of the electrons is diffracted at atomic planes. All electrons transmitted by the specimen are

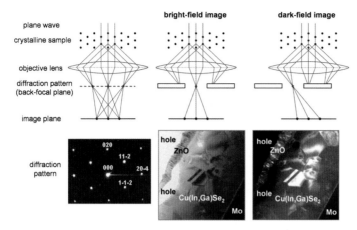

Figure 12.18 Schematics of BF and DF imaging in the CTEM mode of the microscope. The diffraction pattern is formed in the back-focal plane of the objective lens. Exemplary BF and DF images, acquired at the identical position on a cross-section specimen of a Cu(In, Ga)Se$_2$ thin-film solar cell, show the changes in contrast when selecting either direct or diffracted beams by means of the objective aperture.

imaged by the objective lens into the image plane. The electrons diffracted into same angles go through the identical reflection in the diffraction pattern, which is formed in the back-focal plane of the microscope (Figure 12.18). The center reflection of the diffraction pattern (000) is generated by the direct (undiffracted) beam, whereas the other reflections are indexed according to the *hkl* values of the set of atomic planes at which the electrons are diffracted. By use of projection lenses, either the image of the sample or the diffraction pattern can be projected on the view screen or the imaging device.

In order to enhance the contrast in a TEM image, an (objective) aperture may be introduced into the back-focal plane of the objective lens, in order to select either direct or diffracted electron beams for imaging. When the objective aperture is positioned on the central (000) reflection in the diffraction pattern, mainly the direct electron beam contributes to the image, whereas diffracted beams are (at least in part) screened. The result is that the hole in the image (Figure 12.18) appears bright, and therefore the type of imaging is termed bright-field (BF). When diffracted beams are selected by use of the objective aperture, the hole becomes dark, which is why the method is called dark-field (DF) imaging. Grains with sets of atomic planes at which the impinging electron beam is diffracted into wide angles appear dark in a BF image and bright in the DF image (where the reflection is selected by the objective aperture corresponding to the set of atomic planes). It may be concluded that BF and DF imaging makes use of diffraction or amplitude contrasts.

12.3.1.2 High-Resolution Imaging in the Conventional Mode

In the present section, the term "high-resolution" is used for imaging of the atomic lattice of polycrystalline thin films. In contrast to BF and DF imaging, where an

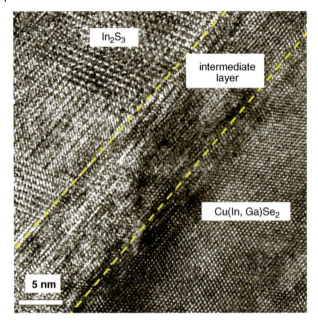

Figure 12.19 HR-TEM image from the interface between the p-type Cu(In,Ga)Se$_2$ absorber and the n-type In$_2$S$_3$ buffer. An intermediate layer of about 10 nm thickness formed between the absorber and the buffer (see Ref. [51] for further details).

objective aperture with rather small opening is applied in order to select either the direct (including also forward scattered) or diffracted beams, for high-resolution (HR) TEM, the objective aperture should be sufficiently large in order to include the direct and at least one diffracted beam [45]. HR-TEM images (as the one in Figure 12.19) must not be considered as direct images of the atomic lattice but they are interference patterns formed by phase relationships of the direct and the diffracted beams. Therefore, HR-TEM imaging is based on phase contrast. Note that in order to image an atomic lattice, corresponding atomic columns have to be oriented parallel to the impinging electron beam, which is realized by tilting the specimen in the microscope. In addition, it is important to correct the aberrations of the objective lens as well as possible.

Then, spherical aberration and defocus are used to optimize phase-information transfer, resulting in an atomic lattice image since the interference fringes in HR-TEM images are related to the atomic positions. The appearance of these fringes is affected considerably by the focus and the specimen thickness (only for uncorrected transmission electron microscopes). By varying these parameters, contrasts may be inverted, that is, white spheres on black backgrounds (as in the HR-TEM image in Figure 12.19) may be changed into black spheres on white backgrounds. Therefore, the unambiguous interpretation of HR-TEM images requires in general the acquisition of a series of TEM images from the identical region of interest at varying focus, and also the comparison of these experimental images with simulated ones [47].

12.3 Transmission Electron Microscopy

Currently, high point resolutions of down to 0.05 nm [48] have been demonstrated by use of a series of lenses correcting for the spherical [49] and chromatic [50] aberrations of the objective lens (a similar corrector for spherical aberrations may be applied for the condenser system, in order to enhance the point resolution in the STEM mode).

12.3.1.3 Imaging in the Scanning Mode Using an Annular Dark-Field Detector

Electrons transmitted through the sample in the STEM mode can be detected by use of either a BF detector, residing on the optical axis of the microscope, or a DF detector, which has the form of a ring, and is therefore termed annular DF (ADF). By varying the camera length of the ADF detector, it is possible to image using mostly Bragg diffracted electrons (large camera lengths), obtaining a crystallographic contrast, or to perform Z-contrast imaging at high scattering angles (short camera lengths, see Figure 12.20). Note that the aim of Z contrast is to achieve images dominated by single-atom scattering, which is incoherent. The contrast in these images is then formed by differences in the average atomic number Z of the position probed since the intensity in the image, I, is proportional to about Z^2, according to the Rutherford scattering cross section. Modern microscopes providing an STEM mode are equipped with high-angle ADF (HAADF) detectors which exhibit a very large central aperture.

12.3.2
Electron Diffraction

12.3.2.1 Selected-Area Electron Diffraction in the Conventional Mode

The basic setup for electron diffraction in CTEM mode with (nearly) parallel illumination of the specimen has already been given in Figure 12.18. It was shown that the electron diffraction pattern forms in the back-focal plane of the objective lens. In order to confine the diffraction pattern to a specific region of interest, an aperture

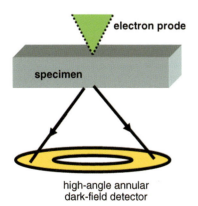

Figure 12.20 Schematics of HAADF imaging in the STEM mode of the microscope.

may be positioned on the corresponding position in the image plane. (Note that additional projection lenses – not shown in Figure 12.18 – project either the image of the sample in the image plane or the diffraction pattern in the back-focal plane on the view screen or the imaging device.) This is the concept of selected-area electron diffraction (SAED). The basic concepts of neutron and X-ray diffraction at materials, introduced in Chapter 13, apply also to electron diffraction. Thus, they shall not be addressed further at this point.

Aperture sizes of down to about 50 nm in diameter (on the specimen) may be applied. Similar as for HR-TEM, grains of interest are oriented such that atomic columns are parallel to the incident electron beam. An example of an SAED pattern acquired at the interface between a $Cu(In,Ga)Se_2$ absorber and an In_2S_3 buffer layer is given in Figure 12.21. Note that in this diffraction pattern, it is possible to identify an orientation relationship between the $Cu(In,Ga)Se_2$ and the In_2S_3 layers since the corresponding $1\bar{1}2$ and $20\bar{6}$ as well as the 004 and $2\bar{2}0$ reflections superimpose each other. (Remark: the orientations of individual crystals with respect to the incident electron beam, also termed zone axes, have the notation [uvw] (u, v, and w being the coefficients of a vector in real space), whereas atomic planes (or rather the whole set of these) is given by hkl (h, k, and l being the coefficients of a vector in reciprocal space)).

12.3.2.2 Convergent-Beam Electron Diffraction in the Scanning Mode

Although SAED is very useful for giving information on the crystal structures of individual thin films and also on the orientation relationships of two neighboring layers, the region of interest probed is in general much larger than many

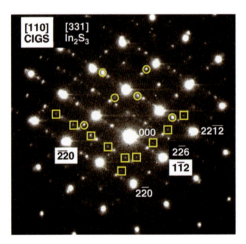

Figure 12.21 SAED pattern acquired at the interface between $Cu(In,Ga)Se_2$ absorber and In_2S_3 buffer. In addition to reflections attributed to the perfect crystals, also reflections from twin boundaries are visible, which are present in the $Cu(In,Ga)Se_2$ (circles) and In_2S_3 layers (squares).

crystalline features present in individual grains. Also, the crystallographic information obtained is in many cases not precise due to a relaxation of the Bragg conditions for a thin specimen and small grains present in the specimen. A technique overcoming these limitations is convergent-beam electron diffraction (CBED) in the STEM mode of the microscope. Using this technique, the illumination of the specimen is not parallel but convergent, with probe sizes in the range of few nanometers to few tens of nanometers, that is, sufficiently small in order to probe very small specimen volumes. The result is that instead of an array of sharp maxima as in SAED, the diffraction pattern in the back-focal plane consists of a pattern of disks of intensity. The size of these disks depends on the beam convergence angle, which is controlled by the size of the condenser aperture. The larger this angle, the larger the size of the CBED disks.

The camera length (magnification of the diffraction pattern) is also an important parameter since it defines the angular range of the pattern. At large camera lengths, mainly the central 000 disk is visible, whereas a wide area of the reciprocal space comes into sight for small camera-length values, containing also electrons scattered into high angles. By acquiring CBED patterns at varying convergence angles and camera lengths, not only small deviations in lattice parameters within individual grains may be detected, but also specimen thicknesses and crystal symmetries of unknown materials may be determined.

12.3.3
Electron Energy-Loss Spectrometry and Energy-Filtered Transmission Electron Microscopy

Two related analysis techniques in a TEM are electron energy-loss spectroscopy (EELS) and energy-filtered TEM (EFTEM, sometimes also referred to as energy-selected imaging, ESI). Both techniques are based on the loss of kinetic energy of the beam electrons due to the interaction with the specimen. The double-differential cross section of the interaction $d^2\sigma/(dEd\Omega)$ describes the angular distribution of the beam electrons after the scattering as a function of their energy loss. This distribution depends strongly on the elements and the bond types within the specimen and is a fingerprint of the material or compound under investigation. In EELS and EFTEM, the energy distribution of beam electrons after the interaction with a specimen is exploited to analyze the specimen.

12.3.3.1 Scattering Theory
A good knowledge of electron-scattering theory [52] is required, in order to interpret the acquired energy-loss spectra and energy-filtered images. In electron microscopy, beam electrons with kinetic energies E_B interact with the constituents of a specimen material via electromagnetic interactions. Two types of interactions are usually distinguished: elastic and inelastic interactions.

In elastic or quasielastic scattering, no or little energy is transferred between the beam electrons and the specimen. This is the interaction of the beam electrons with atomic nuclei in the specimen, which have a much larger rest mass than the beam electrons, and phonons. Although the energy transfer is negligible, momenta may be transferred between the beam electrons and single nuclei or phonons. This transfer can change the momentum directions of the primary beam electrons but affects their absolute value only to a negligible extent.

In inelastic scattering, a significant amount of energy and momentum is transferred between the beam electron and the specimen. This is the interaction of the beam electron with electrons within the specimen. Due to the energy transfer, these specimen electrons can be excited from a lower into a higher unoccupied electronic state. This includes single-electron excitations as well as collective excitations such as plasma resonances (plasmons). The energy distribution of the beam electrons having been scattered at the specimen is therefore strongly coupled with the electronic structure of the specimen.

12.3.3.2 Experiment and Setup

In general, both types of scattering, elastic and inelastic, occur and together determine the double-differential cross section of the interaction $d^2\sigma/(dEd\Omega)$, which is unique for each specimen. It is not (yet) convenient in TEM to record the total scattering intensity as a function of scattering angle and energy loss. This is due to technical reasons and storing capacities. For EELS and EFTEM, only electrons within a specific angular range (i.e., within a cone) are utilized (see Figures 12.17 and 12.22a). The aperture angle α_c (i.e., the collection angle of the spectrometer, see Figure 12.22a) of this cone is either limited by the lens system or by an aperture in the optical path. In order to gain access to the energy distribution of the primary beam electrons scattered at the specimen, these electrons are focused on monochromatic spots, which are spatially separated along one axis according to their kinetic energies, as shown in Figure 12.22a. Ideally, the position of these spots in the resulting spectrum would be independent of the initial momentum of the electrons and the position of the interaction within the specimen. In practice, this is realized by means of imperfect electromagnetic prism devices (i.e., spectrometers and energy filters), within or below the TEM column. An example of a post-column filter installed below the TEM column is shown in Figure 12.22b. If properly aligned in a TEM, such spectrometers exhibit an energy-dispersive plane, containing the spectrum, and an image plane where the image or diffraction pattern of the sample is formed. For a review of EELS and EFTEM instrumentation and spectrometer devices, see Refs. [52, 53].

The measured quantity in EELS and EFTEM is the electron intensity as a function of sample position and energy loss $I(x,y,\Delta E)$ (sometimes also of scattering angles and energy loss $I(\varphi,\theta,\Delta E)$). These intensity distributions $I(x,y,\Delta E)$ are obtained by applying one of the following two approaches:

- **EELS.** Energy-loss spectra (diffraction spectra) $I(\Delta E)$ are acquired on positions (x,y) within a region of interest, see Figure 12.23a. For this approach, the projective lens system projects the spectrum (energy-dispersive) plane of the spectrometer on the detecting device (e.g., a CCD). The procedure described is performed in STEM mode.

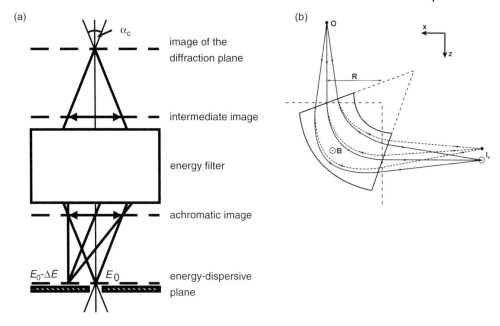

Figure 12.22 (a) The principle of an EELS. Beam electrons with energy loss ΔE are focused on spots different from those for beam electrons with the initial energy E_0. α_c denotes the collection angle of the spectrometer. (b) Principle of a post-column energy filter. The magnetic field B deflects electrons emanating from the object point O on monochromatic spots along the z-axis, where the intensity distribution I_z is formed. The lines with arrows indicate the path of electrons without energy loss, the dashed line the path of electrons that lost some energy.

- **EFTEM.** Images or diffraction patterns with intensity distributions $I(x,y)$ are acquired by using only electrons within selected energy intervals (from ΔE to $\Delta E + dE$), see Figure 12.23a. For this approach, an energy-selecting slit (with width dE) is placed into the spectrum plane of the spectrometer at position ΔE, and the projective lens system projects the image plane of the spectrometer on the detecting device (e.g., a CCD camera).

The energy resolution of the intensity distributions $I(x,y,\Delta E)$ is not only dependent on the quality of the spectrometers but also on the initial momentum and energy distribution of the beam electrons prior to the interaction with the specimen. The resolution can be improved substantially by applying monochromators to minimize the FWHM of the initial energy distribution.

12.3.3.3 The Energy-Loss Spectrum

Figure 12.23b shows an example of an EELS of a Si sample. In general, such a spectrum can be divided into several sections (corresponding to different causes for energy loss).

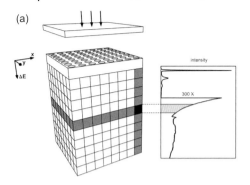
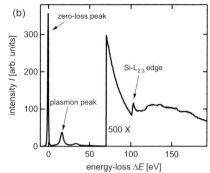

Figure 12.23 (a) Data cube of the intensity distribution $I(x,y,\Delta E)$. With EFTEM, the intensity distribution $I(x,y)$ for certain energy losses ΔE is acquired. EELS is applied to acquire the intensity distribution $I(\Delta E)$ for a sample position (x,y). (b) EEL spectrum of a Si sample. The zero-loss peak, a plasmon peak, and the Si-$L_{2,3}$ edge are indicated.

- **Zero loss (0 eV).** The peak at zero energy loss is called "zero loss" or "elastic" peak. Its intensity represents all electrons which traveled through the specimen without measurable energy loss within the resolution limit of the spectrometer, such as Bragg-diffracted electrons.
- **Low-loss region (0–50 eV).** The first few hundred millielectron volts energy loss of the spectrum contains intensity fluctuations due to phonon scattering and intraband excitation. The subsequent region up to an energy loss of 50 eV contains beam electrons that have excited plasmons, Cerenkov radiation, and interband transitions. It is, therefore, possible to extract local band-gap energies [54] by evaluating the low-loss signals.

While penetrating the specimen, beam electrons polarize the dielectric medium. In turn, the medium exerts a force on the beam electrons which results in energy loss. Hence, the energy-loss distribution is related to the polarizability of the specimen [52] and the low-loss region can be analyzed in order to determine the local complex dielectric function $\varepsilon(\Delta E)$ of the specimen [55].

- **Ionization edges (50 eV to few keV).** These edges or peaks (see Figure 12.23b) appear at energy-loss values of approximately the binding energy of electrons in core shells. Interactions with the beam electrons excite these inner-shell electrons into unoccupied electronic states or into the continuum. The heights of and the area under the edges are very sensitive to the local composition. Up to 50 eV beyond an ionization edge, the so-called energy-loss near-edge structure (ELNES) can be studied. This structure contains information on the binding state and the coordination of the involved atoms and depends on the joint density of states, which includes the density of ground and excited states. The two energy-loss spectra in Fig. 12.24 show how the shape and the onset of the Si-$L_{2,3}$ edge changes from Si to SiO_2 (Fig. 12.24). In state-of-the-art microscopes, EELS and ELNES

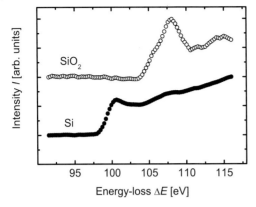

Figure 12.24 EELS of the Si-$L_{2,3}$ edge acquired for an Si and SiO_2 sample [57]. The onset and the shape of the Si-$L_{2,3}$ edge change for different adjacent atoms.

analyses can be performed down to the atomic scale [56], using the STEM mode. The extended energy-loss fine-structure (EXELFS) covers a few hundred eV beyond the ionization edge and can be analyzed to obtain information about the distance of neighboring atoms. It arises through interference between the electron wave backscattered from neighboring atoms and the outgoing wave of the excited electrons.

- **Background**. Background signals, generated by various electron energy-loss processes, additional to the ones mentioned above, are superimposed on the characteristic features. The background includes beam electrons which excited quasi-free electrons, and electrons that lost energy owing to bremsstrahlung. The local decrease of the background intensity at higher energy loss can be described well by a power law of the energy loss ΔE^{-n}. However, the exponent n may change with varying energy loss ΔE.
- **Thickness effect**. EELS is very sensitive to the collection angle (Figure 12.22a) and the thickness of the specimen. The thicker the sample, the higher the probability for multiple scattering of beam electrons in the specimen. The resulting energy loss of an electron is the sum of the energy losses of the individual scattering events.

For the evaluation of an EEL spectrum, a single-scattering distribution is favorable, where each electron has been scattered only once or not at all. It is therefore beneficial to work with specimen thicknesses smaller than the inelastic mean free-path λ of electrons in the specimen material (which is typically 50–100 nm and can be estimated by use of an algorithm given in Ref. [52]). The effect of multiple scattering on a spectrum can be compensated in part by applying signal-processing methods, for example, Fourier-log or Fourier-ratio deconvolution [52].

Although sometimes obstructive, the thickness dependence is the reason why the local thickness of specimens can be measured with EELS and EFTEM.

Assuming that the number of scattering events per beam electron follows a Poisson distribution, the specimen thickness t can be calculated from the ratio of the total beam electron intensity I_t and the intensity of the unscattered or elastically scattered electrons I_0 according to: $t/\lambda = \ln(I_t/I_0)$. A thickness-distribution map from a region of interest can be computed from corresponding unfiltered and zero-loss-filtered images.

Limitations. The energy and spatial resolutions of EELS and EFTEM are limited because of several effects. First of all, imperfect instrumentation affects these resolutions, which may be in part compensated by the use of monochromators, aberration-corrected lenses, and spectrometers. However, not only the instrumentation has an effect but also the material and thickness of the specimen. Delayed or deformed ionization edges or edges at high energy losses are difficult to be evaluated. In addition to beam damage, contamination of the specimen surface with C or Si from the residual gas in the column can change the energy-loss spectrum substantially.

12.3.3.4 Applications and Comparison with EDX Spectroscopy

With the information of the energy-loss distribution, a variety of specimen properties is accessible. However, the successful application of EELS and EFTEM requires good knowledge about the microscope and the scattering theory and may involve also modeling and simulation [58].

One main application is the analysis of the local chemical composition of elements in the specimen, for which both techniques compete with EDX spectroscopy in TEM. Sigle [59] give an overview of state-of-the-art applications of EELS and EFTEM and a detailed comparison with EDX. A summary of this comparison is given below.

- **Resolution.** EELS in the STEM mode offers a slightly better spatial resolution than EDX (sub Å), because the sample volume which provides the information can be defined by apertures confining the collection angle of the spectrometer [52]. This may compensate for beam broadening in the sample. The energy resolution, especially when using a monochromator, is much better in an EELS (<1 eV) compared with an EDX spectrum (down to few tens of electron volt, depending on the element). Therefore, EELS provides access to further specimen properties apart from composition (see above).
- **Count rate and background.** In comparison with EDX, EELS profits from a better count rate but suffers from a higher background. This is partly because the X-ray fluorescence yield is very low, especially for light elements. In addition, the scattered electrons have an angular distribution peaked in forward direction, while characteristic X-rays are emitted isotropically. In the EDX signal, the background arises mainly from bremsstrahlung X-rays, whereas in the EELS signal, there are several contributions (see paragraph on "Background" above).
- **Quantification and detection limit.** Although EELS has advantages in case of the detection of light elements, the detection limit strongly depends on the specimen material. In some cases, even single atoms can be detected [60]. The detection

limit for EDX is approximately 0.1–1 at.%. Compared with EDX, EELS allows for a standardless quantification (without reference specimen) of the local composition. However, very thin specimens are required, and modeling is needed for a quantitative specimen analysis. While EDX in a TEM is a well-established acquisition technique, EELS is more demanding for the operator during acquisition and quantification.
- EELS is usually better for the detection and quantification of **light elements**, and EDX can be advantageous for the detection and quantification of **heavy elements**.
- **Information content**. In general, the information content is higher in EELS, since the spectra contain features like ELNES, EXELFS, information about the polarizability of the specimen, and provide information about the specimen thickness.

12.3.4
Off-Axis and In-Line Electron Holography

In CTEM, the exit-plane electron wavefunction (i.e., of electrons emerging from the specimen at the bottom surface, having interacted or not) may be written in the form $a(x,y)\exp[i\varphi(x,y)]$, where a and φ are the amplitude and the phase of this wavefunction, whereas x and y are spatial coordinates. The objective lens of the microscope images this electron wave on the image plane of the objective lens. This image of the original wavefunction shall have the form $A(x,y)\exp[i\Phi(x,y)]$, where A and Φ are the corresponding amplitude and phase. The intensity distribution of a TEM image, $I(x,y)$, does not contain information on the phase of the electron wave since it can be considered the square of the corresponding amplitude distribution: $I(x,y) = A^2(x,y)$. Thus, in CTEM, the information on the spatial phase distribution $\Phi(x,y)$, and thus also that on the exit-plane wavefunction, $\Phi(x,y)$, is lost.

A way to overcome this obstacle is the application of off-axis electron holography [61]. Figure 12.25a shows a schematic representation of the experimental setup of this technique. The key principle is to make use of an unscattered reference electron wave (since the wave goes through a hole in the specimen) and the electron wavefunction of the scattered electrons (at the specimen) interfering with one another. This process results in a hologram, which contains the information on both, the amplitude and the phase distribution $A(x,y)$ and $\Phi(x,y)$. The interference of the direct (unscattered) and the scattered electron wave is realized by a biprism, which is a positively charged wire. In a second step, the amplitude and phase distribution of the exit-plane wavefunction (i.e., $a(x,y)$ and $\varphi(x,y)$) are reconstructed from the hologram. Figure 12.25b and c shows the reconstructed amplitude and phase images from a ZnO/CdS/Cu(In,Ga)Se$_2$ layer stack. The amplitude image is governed mainly by the contrasts due to Bragg diffraction (mainly visible in the ZnO and Cu(In,Ga)Se$_2$ layers) and variations in thickness.

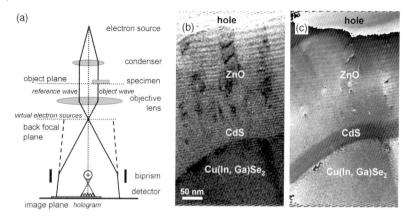

Figure 12.25 (a) Schematics of the experimental setup of off-axis electron holography (reproduced from Ref. [63]). Reconstructed amplitude (b) and phase (c) images from a ZnO/CdS/Cu(In,Ga)Se$_2$ layer stack (data by courtesy of M. Lehmann, TU Berlin/TU Dresden).

If the primary electron beam is not dynamically scattered (e.g., Bragg diffracted) in the specimen (for applying medium-resolution electron holography), the spatial phase distribution $\varphi(x,y)$ of the exit-plane electron wave is proportional to the electrostatic potential $V(x,y)$ averaged along the path through the specimen via

$$\varphi(x,y) = \sigma\, t(x,y)\, V(x,y) \tag{12.4}$$

where σ is an electron interaction constant dependent on the acceleration voltage, and t is the effective specimen thickness, taking also into account possible contamination layers on the top and bottom surfaces of the specimen. That is, the contrasts in the phase image can be attributed to a spatially varying, electrostatic Coulomb potential in the specimen which the impinging electrons experience when they travel through it. In a crystalline sample, the electrostatic potential $V(x,y)$ contains contributions from the ionic lattice of the crystal and also the space charges (via the Poisson equation). Twitchett et al. [62] demonstrated the application of phase imaging by means of off-axis electron holography at a bulk-Si p–n-junction. Since the contribution from the ionic crystal lattice to $V(x,y)$ is the same for p- and n-type Si, the contrast in the phase image is due to the distribution of space charges at the junction. However, this approach is complicated substantially when applied to a p–n-heterojunction as the one shown in Figure 12.25. It should also be noted that in general, off-axis holography (at medium resolution) is performed on specimens with thicknesses of few hundreds of nanometers, in order to provide sufficiently high signal-to-noise ratios.

Even when working at a microscope without biprism, phase-contrast images may still be obtained by the acquisition of a series of TEM images at varying focus, where the exit-plane electron wave interferes with itself. This technique (for medium resolutions) is termed Fresnel contrast analysis or in-line electron holography [64] and particularly useful at phase or grain boundaries, that is, where the phase is likely to change. The corresponding TEM images feature Fresnel fringes (Figure 12.26) at these boundaries, and the contrasts and spacings of these fringes vary with varying defocus value Δf. By use of a reconstruction algorithm [65], the spatial phase distributions $\varphi(x,y)$ of the exit-plane electron waves are calculated. From these distributions, corresponding electron-static potential images can be obtained via Eq. (12.4). The electrostatic potential is also termed mean inner potential (MIP) since the images are projections of the three-dimensional potential. Note that in-line electron holography only gives information about the changes in the phase (and would not give you absolute values, as does off-axis electron holography), and that it is based on a phase-object approximation. In this approach, the spatial amplitude distribution $a(x,y)$ is considered homogeneous around the phase or grain boundary. This situation is achieved by working on areas with constant thicknesses and by avoiding dynamical scattering conditions, that is, by screening diffracted beams, using the smallest objective aperture available (Section 12.3.1.1).

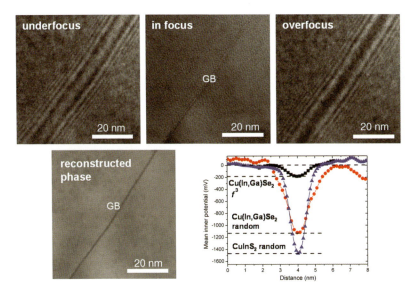

Figure 12.26 TEM images from a region around a grain boundary in a CuInS$_2$ thin film (top), acquired at varying focus value. In the images recorded at underfocus and overfocus conditions, Fresnel fringes around the grain boundary are visible. From this series of images, a phase distribution image was reconstructed (bottom). In addition, profiles across various grain boundaries in Cu(In,Ga)Se$_2$ and CuInS$_2$ absorber layers were extracted from MIP distribution images, calculated from corresponding phase distribution images by use of Eq. (12.4).

When extracting profiles from MIP spatial distributions obtained on Cu(In,Ga)(S, Se)$_2$ grain boundaries (Figure 12.26 shows an example from a CuInS$_2$ layer), the general result is potential wells with FWHM of about 1–2 nm and depths of 1–2 V. The width and depth of a potential well may be interpreted in terms of whether varying space-charge densities or changes in the ionic lattice of the crystal are responsible for the changes in phase detected. The spatial resolution of the MIP profiles exhibits values of about 0.4–0.6 nm and is limited by the size of the objective aperture.

A decent comparison of off-axis and in-line holography and also a further introduction into the techniques can be found in Ref. [66].

12.4
Sample Preparation Techniques

The last section of this chapter is dedicated to specimen preparation for electron microscopy. The position in the chapter suggests a minor importance; however, the preparation decides indeed on the quality of imaging and analyses. It is specifically the specimen surface to which one should always pay the greatest attention. All contamination, scratches, and further disturbances complicate the interpretation of images and the evaluation of the analysis data. The focus of the present section is put on the preparation of cross-sectional specimens. A more general introduction may be found in Ref. [67].

12.4.1
Preparation for Scanning Electron Microscopy

A very simple cross-section preparation for thin-film solar cells on glass substrates, but also for those on foils, is to fracture or cut the cell, which normally takes only seconds. However, since these solar cells in general consist of rather brittle materials, the result may be disappointing, yet justified when aiming for a quick glance on the sample or when surface roughness on the cross section is not a severe issue. Further more elaborated (and also more time-consuming) methods are described in the following.

- **Combination of mechanical and ion polishing.** Stripes of the solar cell are cut, which then are glued face-to-face together, best by use of epoxy glue which is suitable for high vacuum. From the resulting stack, slices are cut, of which the cross sections are mechanically polished. Care is advised not to apply too much pressure during the polishing, and that the scratches remaining on the cross-section surface are in the range of 100 nm or smaller in depth. Then, this cross-section sample is introduced in an ion-polishing machine, equipped with, for example, Ar-ion beams. In order to obtain rather flat cross-section surfaces, small incident angles of about 4° should be chosen. A thin layer of graphite (of about 5 nm in thickness) reduces drift during the measurement and also preserves the cross-section surface, often for several weeks. The graphite seems also to enhance the signal intensities for EBIC (Section 12.2.4) and CL (Section 12.2.5) measurements, probably since it reduces the surface recombination substantially.

- **FIB in a scanning electron microscope**. A focused Ga-ion beam can be used to sputter trenches into specimens, exposing cross sections at positions of interest. In a scanning electron microscope, the progress of the sputtering can be monitored by corresponding electron detectors. Also, by alternately FIB slicing and SEM imaging, a 3D image of the specimen may be reconstructed. Similarly, also EBSD and EDX/WDX may be combined with FIB slicing. It is of advantage to be able to prepare cross-section specimens and perform imaging and analyses on these specimens without breaking the vacuum, that is, reducing the effects of surface contaminations substantially. It should be pointed out that by means of FIB, cross-section specimens from thin-film solar cells may be prepared independently of the substrate, which should be particularly helpful when working with, for example, sensitive foils as substrates.
- **Plasma etching in a glow-discharge apparatus**. Apart from polishing or slicing cross sections by means of ion beams, also plasmas may be used for this purpose [68]. Particularly, instruments used for glow-discharge optical emission or mass spectrometry (see Chapter 16.1 for details) may be employed. The specimen functions as cathode in the glow-discharge setup, and the Ar plasma ignited sputters the specimen atoms. Few seconds sputtering is sufficient for reducing surface roughnesses substantially.

12.4.2
Preparation for Transmission Electron Microscopy

The ultimate goal is always to obtain specimens which are transparent for electrons, that is, which exhibit thicknesses ranging from few to several hundreds of nanometers (you would like to work, for example, with rather thin specimens for HR-TEM, and with relatively thick ones for off-axis electron holography). When aiming for very thin regions in the specimens, this is generally achieved by forming a hole, preferably at positions within the layer stacks of the thin-film solar cells. At the fringe of this hole, the layer thicknesses are rather small, which increase moving away from the hole according to the angle of the wedge formed (see Refs. [69–72] for overviews and more details). Another very helpful preparation method is FIB, providing a tool for selecting precisely the position from where a lamella is to be extracted.

- **Combining mechanical and ion polishing**. An approach applied frequently for preparing cross-section specimens from thin-film stacks is the combination of mechanical and ion polishing. There are various ways to the final TEM specimen:
  –Stripes of the solar cell are cut, which then are glued face-to-face together by use of epoxy glue (similar as for the preparation of SEM specimens). One may also glue a piece of Si single crystal to the sample stripe. The resulting stack is

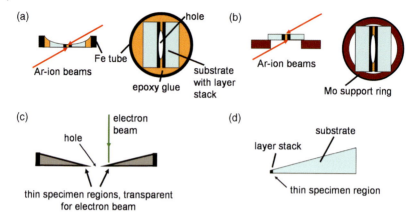

Figure 12.27 Schematics for conventional TEM preparation methods for cross-section specimens. (a) Forming a stack by gluing to two stripes of solar cells face-to-face together, and embedding the stack in a Fe tube (cross section and plan views). (b) Polishing the stack and gluing it to a Mo support ring (cross section and plan views). (c) Cross-section view of the geometry of the final TEM specimen with respect to the incident electron beam. (d) Preparation of a wedge-shaped specimen by means of a tripod. At the tip of the wedge, the specimen is transparent for the electron beam.

introduced into a small tube of about 3 mm in diameter, filled with epoxy glue. From the tube, disks are cut, which are polished on both sides down to thicknesses of about 100 μm. A dimple-grinder introduces a circular deepening in the sample so that the resulting minimum specimen thickness is about 20–30 μm. Finally, Ar-ion milling at rather small angles (3.5–6°) is performed until a hole forms, preferably at the interface between the two stripes of the solar cell (i.e., the fringe of the hole intersects the thin-film stacks, see Figure 12.27a). –From the stack formed by gluing to stripes of solar cells together face-to-face, also directly disks may be cut. These disks are polished on both sides until they exhibit thicknesses of about 10 μm. Rings made of Mo or Ni are glued on these very thin cross-section specimens, which will then support them. The Ar-ion milling process is somewhat shorter than in the procedure described above, due to the reduced thickness of the specimen (Figure 12.27b). Also for this approach, the result is a specimen with a hole at the interface between the solar-cell stripes. That is, in side view, the specimen has the shape of a wedge, with decreasing thickness toward the hole (Figure 12.27c).

- **Tripod polishing**. The stack form by gluing two stripes of solar cells face-to-face together may also be polished by use of a so-called Tripod [73], where the polishing plane is defined by three points. Two points are given by Teflon legs, which are kept coplanar, while on the third point at a corresponding leg, the sample is mounted. The pod heights are adjustable by micrometer screws. By use of a Tripod, either plan-parallel specimens may be polished, which are then glued to supporting rings (as described above), or wedge-shaped specimens may be produced (see Figure 12.27d). Often, specimens prepared by use of a Tripod are

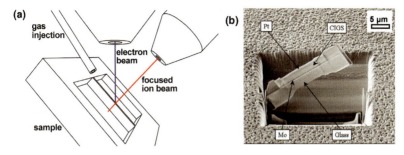

Figure 12.28 (a) Schematics of the configuration of the electron and Ga ion sources as well as of the gas injection in a focused-ion beam system within a SEM. The preparation of a TEM lamella is indicated. (b) SEM image of a TEM lamella prepared by FIB from a CdS/Cu(In,Ga)Se$_2$/Mo/glass stack.

post-treated in an Ar-ion polishing machine, in order to further reduce the thicknesses and to optimize the surface qualities.

- **Using a FIB in a scanning electron microscope**. Similarly as for SEM specimen preparation, trenches may be formed on two sides of a specific region of interest (Figure 12.28). The residual specimen lamella between the two trenches can be thinned further, until it is transparent for the electron beam. This lamella is then extracted by use of a micromanipulator needle, and usually put on a TEM grid or welded (e.g., by use of Pt gas) to an appropriate TEM holder. Surface contamination layers formed as consequences of the ion-beam bombardment can be reduced substantially by decreasing the voltage and current of the impinging ion beam. FIB preparation may be also combined with polishing of the specimen in order to reduce the FIB milling duration. Further information of specimen preparation for TEM by use of a FIB can be found in, for example, Refs. [74, 75].

References

1 Everhart, T.E. and Hoff, P.H. (1971) Determination of kilovolt electron energy dissipation vs penetration distance in solid materials. *J. Appl. Phys.*, **42** (13), 5837–5846.

2 Reimer, L. (1985) *Scanning Electron Microscopy, Physics of Image Formation and Microanalysis, Springer Series in Optical Sciences*, (ed. P.W. Hawkes) vol. 45, Springer, Berlin.

3 Perovic, D.D., Castell, M.R., Howie, A., Lavoie, C., Tiedje, T., and Cole, J.S.W. (1995) Field-emission SEM imaging of compositional and doping layer semiconductor superlattices. *Ultramicroscopy*, **58**, 104–113.

4 Jaksch, H. (2008) Low Loss BSE imaging with the EsB Detection system on the Gemini Ultra FE-SEM, in *EMC 2008, Vol 1: Instrumentation and Methods, Proceedings of the 14th European Microscopy Congress 2008, Aachen, Germany, September 1–5, 2008* (eds M. Luysberg, K. Tillmann, and T. Weirich), Springer, Berlin, pp. 555–556.

5 Jaksch, H. (2008) Strain related contrast mechanisms in crystalline materials imaged with AsB detection, in *EMC 2008, Vol 1: Instrumentation and Methods, Proceedings of the 14th European Microscopy Congress 2008, Aachen, Germany, September 1–5, 2008* (eds M. Luysberg, K. Tillmann, and T. Weirich), Springer, Berlin, pp. 553–554.

6 Schwartz, A.J., Kumar, M., Adams, B.L., and Field, D.P. (eds) (2009) *Electron Backscatter Diffraction in Materials Science*, Springer, New York.

7 Watanabe, M., Ackland, D.W., Burrows, A., Kiely, C.J., Williams, D.B., Krivanek, O.L., Dellby, N., Murfitt, M.F., and Szilagyi, Z. (2008) Improvements in the X-ray analytical capabilities of a scanning transmission electron microscope by spherical-aberration correction. *Microsc. Microanal.*, **12**, 515–526.

8 Leamy, H.J. (1982) Charge collection scanning electron microscopy. *J. Appl. Phys.*, **53** (4), R53–R80.

9 Rechid, J., Kampmann, A., and Reinecke-Koch, R. (2000) Characterising superstrate CIS solar cells with electron beam induced current. *Thin Solid Films*, **361–362**, 198–202.

10 Holt, B. (1989) *SEM Microcharacterization of Semiconductors*, Academic Press, New York.

11 Bonard, J.M., Ganiere, J.D., Akamatsu, B., Araujo, D., and Reinhart, F.K. (1996) Cathodoluminescence study of the spatial distribution of electron–hole pairs generated by an electron beam in $Al_{0.4}Ga_{0.6}As$. *J. Appl. Phys.*, **79**, 8693–8703.

12 Werner, U., Koch, F., and Oelgart, G. (1988) Kilovolt electron energy loss distribution in Si. *J. Phys. D: Appl. Phys.*, **21**, 116–124.

13 Oelgart, G. and Werner, U. (1984) Kilovolt electron energy loss distribution in GaAsP. *phys. status solidi (a)*, **85**, 205–213.

14 Mohr, H. and Dunstan, D.J. (1997) Electron-beam-generated carrier distributions in semiconductor multilayer structures. *J. Microsc.*, **187**, 119–124.

15 Kanaya, K. and Okayama, S. (1972) Penetration and energy-loss theory of electrons in solid targets. *J. Phys D: Appl. Phys.*, **5**, 43–58.

16 Donolato, C. (1989) An alternative proof of the generalized reciprocità theorem for charge collection. *J. Appl. Phys.*, **66** (9), 4524–4525.

17 Donolato, C. (1985) A reciprocity theorem for charge collection. *Appl. Phys. Lett.*, **46** (3), 270–272.

18 Donolato, C. (1983) Evaluation of diffusion lengths and surface recombination velocities from electron beam induced current scans. *Appl. Phys. Lett.*, **43** (1), 120–122.

19 Jastrzebski, L., Lagowski, J., and Gatos, H.C. (1975) Application of scanning electron microscopy to determination of surface recombination velocity: GaAs. *Appl. Phys. Lett.*, **27** (10), 537–539.

20 Kniese, R., Powalla, M., and Rau, U. (2009) Evaluation of electron beam induced current profies of $Cu(In,Ga)Se_2$ solar cells with different Ga-contents. *Thin Solid Films*, **517**, 2357–2359.

21 Matson, R.J., Noufi, R., Ahrenkiel, R.K., and Powell, R.C., (1986) EBIC investigations of junction activity and the role of oxygen in $CdS/CuInSe_2$ devices. *Solar Cells*, **16**, 495–519.

22 Romero, M.J., Al-Jassim, M.M., Dhere, R.G., Hasoon, F.S., Contreras, M.A., Gessert, T.A., and Moutinho, H.R. (2002) Beam injection methods for characterizing thin-film solar cells. *Prog. Photovoltaics: Res. Appl.*, **10**, 445–455.

23 Liang, Z.C., Shen, H., Xu, N.S., and Reber, S. (2003) Characterization of direct epitaxial silicon thin film solar cells on a low-cost substrate. *Sol. Energy Mater. Sol. Cells*, **80**, 181–193.

24 Nichterwitz, M., Abou-Ras, D., Sakurai, K., Bundesmann, J., Unold, T., Scheer, R., and Schock, H.W. (2009) Influence of grain boundaries on current collection in $Cu(In,Ga)Se_2$ thin-film solar cells. *Thin Solid Films*, **517**, 2554–2557.

25 Edwards, P.R., Galloway, S.A., and Durose, K. (2000) EBIC and luminescence mapping of CdTe/CdS solar cells. *Thin Solid Films*, **361–362**, 364–370.

26 Galloway, S.A., Edwards, P.R., and Durose, K. (1999) Characterization of thin film CdS/CdTe solar cellsusing electron

and optical beam induced current. *Solar Energy Mater. Sol. Cells*, **57**, 61–74.

27. Scheer, R. (1999) Qualitative and quantitative analysis of thin film heterostructures by electron beam induced current. *Solid State Phenom.*, **67–68**, 57–68.

28. Wu, C.J. and Wittry, D.B. (1978) Investigation of minority-carrier diffusion lengths by electron beam induced bombardment of Schottky barriers. *J. Appl. Phys.*, **49** (5), 2827–2836.

29. Sieber, B., Ruiz, C.M., and Bermudez, V. (2009) Evaluation of diffusion-recombination parameters in electrodeposited CuIn(S,Se)$_2$ solar cells by means of electron beam induced currents and modelling. *Superlattice Microstruct.*, **45**, 161–167.

30. Liang, Z.C., Shen, H., Xu, N.S., and Reber, S. (2003) Characterization of direct epitaxial silicon thin film solar cells on a low-cost substrate. *Sol. Energy Mater. Sol. Cells*, **80**, 181–193.

31. Scheer, R. and Lewerenz, H.W. (1995) Diffusion length measurements on n-CuInS2 crystals by evaluation of electron-beam induced current profiles in edge-scan and planar configurations. *J. Appl. Phys.*, **77** (5), 2006–2009.

32. Kuiken, H.K. and van Opdorp, C. (1985) Evaluation of diffusion length and surface recombination velocity from a planar collector geometry electron-beam-induced current scan. *J. Appl. Phys.*, **57** (6), 2077–2090.

33. Kniese, R., Powalla, M., and Rau, U. (2007) Characterization of the CdS/Cu(In,Ga)Se$_2$ interface by electron beam induced currents. *Thin Solid Films*, **515**, 6163–6167.

34. Schmid, D., Jäger-Waldau, G.J., and Schock, H.W. (1991) Diffusion length measurement and modeling of CuInSe$_2$-(Zn,Cd)S solar cells, in *Proceedings of the 10th European Photovoltaic Solar Energy Conference, Lisbon, Portugal, April 8–12, 1991* (eds A. Luque, G. Sala, W. Palz, G. dos Santos, and P. Helm), Kluwer Academic, Dordrecht, pp. 935–938.

35. Fossum, J.G. and Burgess, E.L. (1978) Silicon solar cell designs based on physical behaviour in concentrated sunlight. *Solid State Electron.*, **21**, 729–737.

36. Cavalcoli, D. and Cavallini, A. (1994) Evaluation of diffusion length at different excess carrier concentrations. *Mater. Sci. Eng.*, **B24**, 98–100.

37. Röll, K. (1980) The temperature distribution in a thin metal film exposed to an electron beam. *Appl. Surf. Sci.*, **5**, 388–397.

38. Hakimzadeh, R. and Bailey, S.G. (1993) Minority carrier diffusion length and edge surface recombination velocity in In P. *J. Appl. Phys.*, **74** (2), 1118–1123.

39. Castaldini, A., Cavallini, A., Fabroni, B., Fernandez, P., and Piqueras, J. (1996) Comparison of electrical and luminescence data from the A center in CdTe. *Appl. Phys. Lett.*, **69**, 3510–3513.

40. Hofmann, D.M., Omling, P., Grimmeiss, H.G., Meyer, B.K., Benz, K.W., and Sinerius, D. (1992) Identification of the chlorine A center in CdTe. *Phys. Rev. B*, **45**, 6247–6250.

41. Meyer, E., Jarvis, S.P., and Spencer, N.D. (2004) Scanning probe microscopy in materials science. *MRS Bulletin*, **29**, 443–448.

42. Turcu, M., Pakma, O., and Rau, U. (2002) Interdependence of absorber composition and recombination mechanism in Cu(In, Ga)(Se,S)$_2$ heterojunction solar cells. *Appl. Phys. Lett.*, **80**, 2598–2600.

43. Holt, D.B., Raza, B., and Wojcik, A. (1996) EBIC studies of grain boundaries. *Mater. Sci. Eng.*, **B42**, 14–23.

44. Williams, D.B. and Carter, C.B. (2009) *Transmission Electron Microscopy, A Textbook for Materials Science*, Springer Science + Business Media, LLC, Norwell.

45. Fultz, B. and Howe, J.M. (2008) *Transmission Electron Microscopy and Diffractometry of Materials*, Springer, Berlin.

46. Reimer, L. (1993) *Transmission Electron Microscopy, Physics of Image Formation and Microanalysis*, Springer, Berlin.

47. Thust, A., Coene, W.M.J., Op de Beeck, M., and Van Dyck, D. (1996) Focal-series reconstruction in HRTEM: Simulation studies on non-periodic objects. *Ultramicroscopy*, **64**, 211–230.

48 Kisielowski, C., Freitag, B., Bischoff, M., van Lin, H., Lazar, S., Knippels, G. *et al.* (2008) Detection of single atoms and buried defects in three dimensions by aberration-corrected electron microscope with 0.5-angstrom information limit. *Microsc. Microanal.*, **14**, 454–462.

49 Rose, H. (1994) Correction of aberrations, a promising means for improving the spatial and energy resolution of energy-filtering electron microscopes. *Ultramicroscopy*, **56**, 11–25.

50 Kabius, B., Hartel, P., Haider, M., Müller, H., Uhlemann, S., Loebau, U., Zach, J., and Rose, H. (2009) First application of C_c-corrected imaging for high-resolution and energy-filtered TEM. *J. Electron. Microsc.*, **58** (3), 147–155.

51 Abou-Ras, D., Rudmann, D., Kostorz, G., Spiering, S., Powalla, M., and Tiwari, A.N. (2005) Microstructural and chemical studies of interfaces between Cu(In,Ga)Se$_2$ and In$_2$S$_3$ layers. *J. Appl. Phys.*, **97** (12), 084908-1-8.

52 Egerton, R.F. (1996) *Electron Energy-Loss Spectroscopy in the Electron Microscope*, Plenum Press, New York.

53 Reimer, L. (1995) *Energy-Filtering Transmission Electron Microscopy*, Springer, Berlin.

54 Gu, L., Ozdol, V.B., Sigle, W., Koch, C.T., Srot, V., and van Aken, P.A. (2010) Correlating the structural, chemical, and optical properties at nanometer resolution. *J. Appl. Phys.*, **107**, 013501.

55 Ryen, L., Wang, X., Helmersson, U., and Olsson, E. (1999) Determination of the complex dielectric function of epitaxial SrTiO$_2$ films using transmission electron energy-loss spectroscopy. *J. Appl. Phys.*, **85**, 2828–2834.

56 Muller, D.A., Kourkoutis, L.F., Murfitt, M., Song, J.H., Hwang, H.Y., Silcox, J., Dellby, N., and Krivanek, O.L. (2008) Atomic-scale chemical imaging of composition and bonding by aberration-corrected microscopy. *Science*, **319**, 1073–1076.

57 Ahn, C.C. and Krivanek, O.L. (1983) *EELS Atlas*, Gatan, Pleasanton.

58 Verbeeck, J. and Van Aert, S. (2004) Model based quantification of EELS spectra. *Ultramicroscopy*, **101**, 207–224.

59 Sigle, W. (2005) Analytical transmission electron microscopy. *Ann. Rev. Mater. Res.*, **35**, 239–314.

60 Krivanek, O.L., Mory, C., Tence, M., and Colliex, C. (1991) EELS quantification near the single-atom detection level. *Microsc. Microanal. Microstruct.*, **2**, 257–267.

61 Lichte, H. (1986) Electron holography approaching atomic resolution. *Ultramicroscopy*, **20**, 293–304.

62 McCartney, M.R., Smith, D.J., Hull, R., Bean, J.C., Voelkl, E., and Frost, B. (1995) Direct observation of potential distribution across Si/Si p-n junctions using off-axis electron holography. *Appl. Phys. Lett.*, **65** (20), 2603–2605.

63 Lehmann, M. and Lichte, H. (2002) Tutorial on off-axis electron holography. *Microsc. Microanal.*, **8**, 447–466.

64 Dunin-Burkowski, R.E. (2000) The development of Fresnel contrast analysis, and the interpretation of mean inner potential profiles at interfaces. *Ultramicroscopy*, **83**, 193–216.

65 Bhattacharyya, S., Koch, C.T., and Rühle, M. (2006) Projected potential profiles across interfaces obtained by reconstructing the exit face wave function from through focal series. *Ultramicroscopy*, **106**, 525–538.

66 Latychevskaia, T., Formanek, P., Koch, C.T., and Lubk, A. (2010) Off-axis and inline electron holography: Experimental comparison. *Ultramicroscopy*, **110** (5), 472–482.

67 Ayache, J., Beaunier, L., Boumendil, J., Ehret, G., and Laub, D. (2010) *Sample Preparation Handbook for Transmission Electron Microscopy, volume 1: Methodology, volume 2: Techniques,* Springer, Berlin.

68 Shimizu, K. and Mitani, T. (2010) *New Horizons of Applied Scanning Electron Microscopy, Springer Series in Surface Science,* (eds G. Ertl, H. Lüth, and D.L. Mills) vol. 45, Springer, Berlin.

69 Bravman, J.C., Anderson, R.M., and McDonald, M.L. (eds) (1987) Specimen preparation for transmission electron microscopy of materials, in *Proceedings of the MRS 1987 Fall Meeting, Boston, Massachusetts, U.S.A., December 3, 1987, Materials Research Society Symposium*

70 Anderson, R. (1990) Specimen preparation for transmission electron microscopy of materials II, in *Proceedings of the MRS 1990 Spring Meeting, San Francisco, U.S.A., April 19–20, 1990, Materials Research Society Symposium Proceedings*, **199**, Materials Research Society, Pittsburgh, PA.

71 Anderson, R., Tracy, B., and Bravman, John (1991) Specimen preparation for transmission electron microscopy of materials III, in *Proceedings of the MRS 1991 Fall Meeting, Boston, Massachusetts, U.S.A., December 5–6, 1991, Materials Research Society Symposium Proceedings*, **254**, Materials Research Society, Pittsburgh, PA.

72 Anderson, R.M. and Walck, S.D. (1997) Specimen preparation for transmission electron microscopy of materials IV, in *Proceedings*, **115**, Materials Research Society, Pittsburgh, PA.

Proceedings of the MRS 1997 Spring Meeting, San Francisco, U.S.A., April 2, 1997, Materials Research Society Symposium Proceedings, **480**, Materials Research Society, Pittsburgh, PA.

73 Ayache, J. and Albarède, P.H. (1995) Application of the ion-less tripod polisher to the preparation of YBCO superconducting multilayer and bulk ceramics thin films. *Ultramicroscopy*, **60**, 195–206.

74 Giannuzzi, L.A.C and Stevie, F.A. (eds) (2005) *Introduction to Focused Ion Beams: Instrumentation, Theory, Techniques and Practice*, Springer, Berlin.

75 Mayer, J., Giannuzzi, L.A., Kamino, T., and Michael, J. (2007) TEM sample preparation and FIB-induced damage. *MRS Bull.*, **32**, 400–407.

13
X-Ray and Neutron Diffraction on Materials for Thin-Film Solar Cells

Susan Schorr, Christiane Stephan, Tobias Törndahl, and Roland Mainz

13.1
Introduction

In order to understand natural and artificially produced materials, a detailed understanding of their crystal structures is required. This information is a basis for research in physics, chemistry, biology, and materials science. Among the various experimental methods, neutron and X-ray (photon) scattering have become key techniques of choice. Both techniques are complementary. In X-ray scattering, it is almost exclusively the electrons in atoms which contribute to the scattering, whereas neutrons interact with the atomic nuclei. This has an important consequence: the response of neutrons from light atoms (such as hydrogen or oxygen) is much higher than for X-rays, and neutrons easily distinguish atoms of comparable (or even equal) atomic number (see Figure 13.1). Due to the fact that neutrons interact with atoms via nuclear rather than electrical forces and nuclear forces are very short-range (of the order of a few fermis, i.e., 10^{-15} m), the cross section for such an interaction is very small. The size of a scattering center (nucleus) is typically 10^5 times smaller than the distance between such centers. As a consequence, neutrons can travel large distances through most materials without being scattered or absorbed. Thus, neutrons penetrate matter much more deeply than X-rays.

While neutron scattering provides insights into the crystal structure with high resolution, X-ray scattering has the advantage that (due to a larger scattering cross section) measurement durations are usually much shorter, compared with neutron scattering. Additionally, lab-scale X-ray sources are broadly available.

13.2
Diffraction of X-Rays and Neutron by Matter

Most of all inorganic, solid materials can be described as crystalline. When X-rays or neutrons interact with a crystalline substance, coherent elastic scattering may occur which is also termed diffraction.

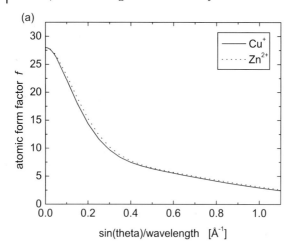

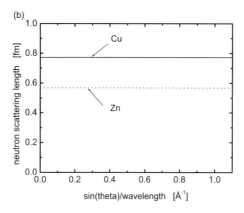

Figure 13.1 Example of scattering amplitudes of two isoelectronic cations. (a) The atomic form factor f of Cu^+ and Zn^{2+}. For better visibility, the atomic form factor of Ga^{3+} was not shown. (b) The coherent neutron scattering length b for Cu and Zn.

Elastic X-ray scattering can be described accurately in terms of classical electromagnetic theory. An electron in an alternating electromagnetic field will oscillate with the same frequency as the field. When an X-ray beam hits an atom, the electrons around the atom start to oscillate with the same frequency as the incoming beam. According to classical electromagnetic theory, an accelerated charge – here the oscillating electron – emits electromagnetic radiation. The sum of the contributions of these radiations to the scattered amplitude of all the electrons of an atomic species in the crystal is expressed by the atomic scattering factor f. At zero angle, all the scattered waves are in phase, and the scattered amplitude is the simple sum of the contributions from all Z (where Z is the atomic number of the atom) electrons, that is, $Z = f$. As the scattering angle increases, f becomes smaller than Z because of the increasing destructive interference effects between the Z-scattered waves (Figure 13.1a and b).

The scattering of neutrons by nuclei is a quantum-mechanical process. Formally, the process has to be described in terms of the wavefunctions of the neutron and the electrostatic potential caused by the nucleus. The scattering of a neutron by a single nucleus can be described using the cross section σ, measured in barns (1 barn = 10^{-28} m^2), which is equivalent to the effective area presented by the nucleus to the passing neutron. If the neutron "hits" this area, it is scattered isotropically. This is due to the fact that the range of the nuclear potential is very small compared with the wavelength of the neutron; thus, the nucleus is effectively a point scatterer. X-rays, on the other hand, are not scattered isotropically because the electron clouds around the atom are comparable in size to the wavelength of the X-rays. The amplitude of the neutron wave scattered by the nucleus depends on the strength of the interaction between the neutron and the nucleus. Because the scattered neutron wave is isotropic, its wavefunction can be written as $(-b/r)e^{ikr}$ if the scattering nucleus is in the origin of the coordinate system ($\mathbf{k}$ is the wave vector of the neutron with $k = 2\pi/\lambda$, and $\mathbf{r}$ is the position). The scattering amplitude, b, is referred to as the scattering length of the nucleus, which expresses the strength of the interaction between the neutron and the scattering nucleus. It varies in an irregular way with the atomic number (Figure 13.2). The relatively large scattering amplitudes of, for example, hydrogen and oxygen atoms in comparison with heavy metal atoms allow these atoms to be located within the unit cell. The scattering cross section σ is related to b by the simple relation $\sigma = 4\pi b^2$ [1].

The atoms in a crystal are arranged in a regular pattern. In almost all directions into which waves are scattered, destructive interference occurs. That is, the interfering scattered waves are out of phase, and not any residual energy leaves the solid sample. However, for a small fraction of these directions, the coincidence of the scattered waves results in constructive interference. The waves are in phase, and scattered X-rays or neutrons leaving the sample into various directions. The scattering amplitude of a unit cell is determined by summing the scattering amplitudes f or

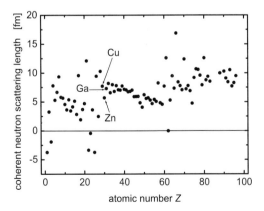

Figure 13.2 Coherent neutron scattering length in dependence on atomic number Z. Copper, zinc and gallium are marked.

b, from all atoms in the unit cell, respectively. The summation must take into account the path or phase differences between all the scattered waves and is expressed by the dimensionless number F_{hkl}, the structure factor. F_{hkl} must not only express the amplitude of scattering from a lattice plane with the Miller indices hkl but also the phase angle of the scattered wave. Therefore, F_{hkl} is represented mathematically as a complex number, that is,

$$F_{hkl} = \sum_{n=0}^{N} f_n \exp\{2\pi i(hx_n + ky_n + lz_n)\} \quad \text{structure factor for X-rays}$$

$$F_{hkl} = \sum_{n=0}^{N} b_n \exp\{2\pi i(hx_n + ky_n + lz_n)\} \quad \text{structure factor for neutrons}$$

(13.1)

In Eq. (13.1), N is the number of symmetrically nonequivalent atomic positions in the unit cell and x_n, y_n, z_n are the atomic coordinates.

The intensities I_{hkl} of scattered X-ray or neutron waves are proportional to the squares of their amplitudes, or F_{hkl} multiplied by its complex conjugate F_{hkl}^*, hence $I_{hkl} \propto |F_{hkl}|^2$

The wavelengths λ of neutrons are related to their velocity v through de Broglie's equation $\lambda = h/mv$. Neutrons emerge from a nuclear reactor with a range of velocities and hence wavelengths of which the maxima are typically in the range 1–2 Å, that is, close to X-ray wavelengths. Similarly as for X-rays, useful single wavelength beams are achieved by the use of crystal monochromators. Therefore, neutron diffraction is geometrically similar to X-ray diffraction (XRD).

For simplicity, the scattering of X-rays and neutrons from a family of lattice planes hkl is often considered as reflection from a series of parallel planes inside the crystal. The two parallel incident rays 1 and 2 make an angle θ with these planes (Figure 13.3), which have a lattice plane distance d_{hkl}. A reflected beam of maximum intensity (Bragg peak) will result if the waves represented by 1' and 2' are in phase. The difference in path length between 1 and 1' and 2 and 2' must then be an integer multiple of the wavelength λ. This relationship is expressed mathematically in

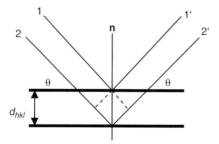

Figure 13.3 Schematic representation of the reflection of waves from lattice planes. The vector **n** is the normal of the plane and θ represents the diffraction or Bragg angle.

Bragg's law (Eq. (13.2)). The angle at which a reflected beam of maximum intensity occurs is termed Bragg angle θ

$$n\lambda = 2d_{hkl} \sin\theta \tag{13.2}$$

Polycrystalline diffraction methods may be classified as "fixed λ, varying θ'" techniques (EDXRD discussed in Section 13.5 is a notable exception). For these techniques, a sufficiently large number of more or less randomly oriented crystallites are present in the specimen such that hkl planes in some of the crystallites are oriented, by chance, at the appropriate Bragg angles for reflection. The family of planes of a given d_{hkl} interplanar spacing reflects at the same 2θ angle with respect to the direct beam. In situations where the crystallites are randomly oriented, the diffracted intensities are uniform. Else, the analyzed ensemble of crystallites exhibits a texture or preferred orientation. The analysis of preferred orientations in thin films is important since it almost invariably arises as a consequence of the processes of crystallizations and recrystallizations or sintering during the growth processes of the thin films.

13.3
Neutron Powder Diffraction of Absorber Materials for Thin-Film Solar Cells

Ternary and quaternary compound semiconductors used as absorber materials in thin-film solar cells often contain electronically similar elements, for example, Cu and Ga (chalcopyrites $CuGaS_2$, $CuGaSe_2$, $Cu(In,Ga)S_2$ or $Cu(In,Ga)Se_2$) or Cu and Zn (kesterites Cu_2ZnSnS_4 and $Cu_2ZnSnSe_4$). The cations Cu^+, Ga^{3+}, and Zn^{2+} have the identical number of electrons (28). Since atomic scattering form factors f are proportional to the atomic number Z, the positions of the unit cell atoms of similar atomic number are not easy to be determined. Hence these cations named above are not distinguishable in the atomic structure by conventional XRD. The problem can be solved using neutron diffraction, because of the different neutron scattering lengths of copper, gallium, and zinc ($b_{Cu} = 7.718(4)$ fm, $b_{Ga} = 7.288(2)$ fm, $b_{Zn} = 5.680(5)$ fm [2]).

Moreover, since atomic nuclei cross sections are very small, the destructive interference effects, which in the case of X-rays lead to a decrease in the scattering amplitude with angle, are also small and neutron-scattering amplitudes do not decrease rapidly with angle. This gives the advantage that in a neutron diffraction experiment Bragg peaks with valuable intensity can be observed also in the high-Q region. In Figure 13.4a and b, the X-ray and neutron diffraction pattern of a kesterite sample are shown for comparison.

13.3.1
Example: Investigation of Intrinsic Point Defects in Nonstoichiometric CuInSe₂ by Neutron Diffraction

The concentration of intrinsic point defects in $Cu(In,Ga)Se_2$ can be determined by Rietveld analysis of X-ray and neutron powder diffraction data using the refined cation-site occupancy values [4]. This compound crystallizes in the chalcopyrite-type

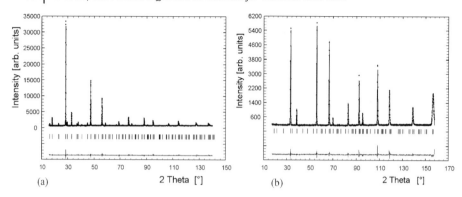

Figure 13.4 Rietveld analysis of a diffraction pattern of a kesterite sample with the chemical composition $Cu_{1.99}Zn_{0.49}Fe_{0.51}Sn_{1.01}S_4$ using X-rays (a) and neutrons (b) [3].

crystal structure with space group $I\bar{4}2d$. In general, Cu(In,Ga)Se$_2$ exhibits a non-stoichiometric composition which can be expressed with the overall formula $Cu_y(In,Ga)_{1-y}(Se,S)_{1/2+y}$ (y ranging from 0 to 1). Within the chalcopyrite-type structure, monovalent Cu is situated on the positions 4a (0 0 0) and trivalent In or Ga on the positions 4b (0 0 $^1/_2$) in the unit cell. The cations are tetrahedrally coordinated by the anions and vice versa.

In principle, 12 different types of intrinsic point defects may exist in this crystal structure, which are vacancies, anti-site, and interstitial defects. Introducing non-stoichiometry into the structure can cause various kinds of these defects. Copper deficiency in Cu(In,Ga)Se$_2$ leads to the presence of Cu-vacancies (V_{Cu}). Moreover, unoccupied 4a sites in the unit cell may be occupied by In or Ga, where an In$_{Cu}$ or Ga$_{Cu}$ antisite defect is formed. For the determination of the fraction of the amount of In, Ga, Cu, and vacancies occupying the two possible cation positions 4a and 4b, neutron diffraction is applied. The distribution of the cations within the crystal structure can be revealed by the method of average neutron-scattering length [5] on the basis of the refined site-occupancy values. The average neutron-scattering length $\bar{b}_j$ (j is either 4a or 4b) is given by

$$\begin{aligned}\bar{b}_j &= Cu_j \cdot b_{Cu} + In_j \cdot b_{In} + Ga_j \cdot b_{Ga} + V_j \\ \bar{b}_j &= Cu_j \cdot b_{Cu} + In_j \cdot b_{In} + Ga_j \cdot b_{Ga} + V_j\end{aligned} \quad (13.3)$$

where $b_{Cu} = 7.718\,(4)$ fm, $b_{In} = 4.065(2)$, and $b_{Ga} = 7.288\,(2)$ fm are the neutron-scattering lengths of Cu, In, and Ga [2], respectively, and V_j is the vacancy fraction on the corresponding position. The requirement for the calculation of defect by this method is given by

$$Cu_j + In_j + V_j = 1 \quad (13.4)$$

The procedure for calculation of defect concentrations after neutron diffraction experiment is explained using the example of nonstoichiometric $Cu_yIn_{1-y}Se_{1/2+y}$ ($y \neq 0.5$). First, the experimental average neutron scattering length $\bar{b}_j^{exp}$ for both cation positions has to be calculated using the site-occupancy factors resulting from the

Rietveld analysis (occ_{4a} and occ_{4b}). The Rietveld method is a least square fit based method for structural refinement of an experimentally determined diffraction pattern to a structure model and described in detail by Young [6]. Assuming in the Rietveld calculations that Cu is situated only on 4a positions and In only on the 4b positions, the experimental average neutron-scattering lengths $\bar{b}_{4a}^{exp}$ and $\bar{b}_{4b}^{exp}$ are defined as follows:

$$\bar{b}_{4a}^{exp} = occ_{4a} \cdot b_{Cu} \qquad \bar{b}_{4b}^{exp} = occ_{4b} \cdot b_{In} \qquad (13.5)$$

The experimental average neutron-scattering length has to be compared with a theoretical neutron scattering length ($\bar{b}^{calc}$) calculated on the basis of a cation distribution model, defining the theoretical occupancies of Cu and In on the 4a and the 4b position:

$$\bar{b}_{4a}^{calc} = occ(Cu)_{4a}^{theor} \cdot b_{Cu} + occ(In)_{4a}^{theor} \cdot b_{In} + occ(V)_{Cu}$$
$$\bar{b}_{4a}^{calc} = occ(Cu)_{4b}^{theor} \cdot b_{Cu} + occ(In)_{4b}^{theor} \cdot b_{In} + occ(V)_{In} \qquad (13.6)$$

The sum of the theoretical occupancies of Cu and In on the 4a and 4b positions ($occ(Cu)_{4a}^{theor}$ and $occ(In)_{4b}^{theor}$) should not exceed the concentrations of Cu and In within the sample, which has to be determined by quantitative, compositional analysis, as, for example, wavelength-dispersive X-ray spectrometry (WDX, see Chapter 12).

In order to determine the distribution of the cations Cu^+ and In^{3+} on the two possible positions (4a and 4b), the next step is to minimize the difference between $\bar{b}_j^{exp}$ and $\bar{b}_j^{calc}$ by building up a reasonable cation distribution model. Since $b_{Cu} > b_{In}$, a difference between experimentally determined and calculated average neutron-scattering length of the 4a site like $\bar{b}_{4a}^{exp} < \bar{b}_{4a}^{calc}$ is caused by indium occupying this site (In_{4a}) or copper vacancies (V_{Cu}) are present on this position. Thus, a dropping of $\bar{b}_{4a}^{exp}$ can be caused by copper vacancies and/or indium on the 4a position. Moreover, it has to be taken into account that in dependence of the Cu/In ratio (copper-rich or copper-poor $Cu_yIn_{1-y}Se_{1/2+y}$) both cations can also occupy interstitial positions. On the other hand, a Cu_{In} defect would elevate the value of the experimental average neutron-scattering length $\bar{b}_{4b}^{exp}$ due to $b_{Cu} > b_{In}$.

In Figure 13.5, an example for the increase and decrease in the experimental average neutron-scattering length of the two possible cation sites in the chalcopyrite-type structure ($\bar{b}_{4a}^{exp}, \bar{b}_{4b}^{exp}$) in dependence of stoichiometry in $Cu_yIn_{1-y}Se_{1/2+y}$ is shown. To proof the assumed cation distribution model, the experimental and theoretical average neutron-scattering lengths of both cation sites have to be compared. First the amount of copper necessary to lift up $\bar{b}_{4b}^{exp}$ is calculated and subtracted from the total amount of copper in the sample. The amount of copper left can be distributed on the 4a site (Cu_{4a}) and on interstitial positions (Cu_i). In the case of a copper-poor composition, the amount of copper left is required completely to achieve $\bar{b}_{4a}^{exp} = \bar{b}_{4a}^{calc}$; thus the possibility of a Cu_i defect can be excluded. This conclusion also corresponds with the high formation energy needed to form such a defect in copper-poor $CuInSe_2$ calculated by Zhang et al. [7]. The knowledge of the

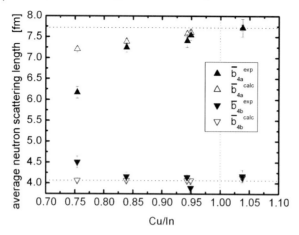

Figure 13.5 Experimentally determined average neutron scattering length of the 4a and 4b position in the chalcopyrite type structure in comparison to the neutron scattering length of copper (b_{Cu}) and indium (b_{In}).

total amount of Cu_{In}, In_{Cu}, Cu_{Cu}, and In_{In} leads to an easy identification also of the fractions of V_{Cu} and V_{In}.

13.4
Grazing Incidence X-Ray Diffraction (GIXRD)

XRD from randomly oriented polycrystalline thin films often suffer from low peak intensity and poor peak to background ratio for symmetrical XRD measurements such as θ–2θ powder diffraction techniques. The low diffraction intensity from a thin top layer in a θ–2θ scan is mainly related to the fact that the path length of the X-rays in the thin film is short. Therefore, most of the radiation instead interacts within the underlying substrate. In addition to lower peak intensity in θ–2θ scans, the thin-film reflections may also be superimposed and difficult to distinguish from substrate reflections, which complicate the evaluation of the XRD data. To be able to improve the situation for weakly diffracting thin films, low-angle XRD techniques such as grazing incidence X-ray diffraction (GIXRD) have been developed. In a symmetric θ–2θ measurement, the scattering vector is perpendicular to the sample surface and only lattice planes in parallel to the substrate surface contribute to the diffractograms. Furthermore, the angle of the incoming X-rays to the sample surface is changed during the measurement to always be equal to θ, that is, half of the scattering angle 2θ. GIXRD is an asymmetric XRD scan where the path length of the X-rays in a thin film is increased by using a fixed angle of incidence, α, for the incoming X-rays. Since the GIXRD measurement is performed at a constant angle of incidence where only the detector is moved over a 2θ range of interest, it is also called a detector scan. In contrast to a θ–2θ scan, the direction of the scattering vector changes during a grazing incidence measurement and is no longer perpendicular to the sample surface. This implies that the angle between the diffracting lattice planes and the sample surface changes during the course of a GIXRD measurement. A GIXRD measurement is best performed on a randomly oriented thin film in order to fulfill the Bragg condition for diffraction for every chosen angle of incidence, α.

A diffractogram measured by grazing incidence may in fact not show any reflections at all if the analyzed thin film is highly textured or epitaxial, because such thin layers all have their grains aligned to the underlying substrate in specific way. However, a good way for thin film analysis is to perform a symmetric θ–2θ scan for information about the film texture in combination with an asymmetric GIXRD scan for information regarding the random orientation contribution from the same film.

The attenuation of the intensity of an incoming X-ray beam in matter is defined by the Lambert–Beer law $I = I_0 e^{-\mu l}$. Here μ is the linear attenuation coefficient, l the traveled path length of the beam in the medium, and I_0 the intensity prior to entrance into the material. The linear attenuation coefficient is wavelength dependent and defined by $\mu = \mu_m \rho$, where μ_m is the mass absorption coefficient and ρ the density of the material. The mass absorption coefficient value, with the unit of $m^2 \, kg^{-1}$, is usually tabulated instead of μ. For Cu K$_\alpha$ radiation, μ is normally found in the region from 10^4–$10^6 \, m^{-1}$, which corresponds to penetration depths of 1–100 μm. The traveled path length l of the X-rays in a sample can be described by $l = 2d/\sin\theta$ for a θ–2θ scan and by $l = d(1/\sin\alpha + 1/\sin(2\theta - \alpha))$ for GI scan, respectively. The path length l is the sum of the distance traveled by the X-rays in the material prior to scattering takes place and the exit length after scattering occurs, and d is the penetration depth perpendicular to the sample surface. The problem for thin films is that the given penetration depth, d, for the X-rays is commonly much larger than the actual film thickness of the sample. Thus, the most obvious gain by using the grazing incidence technique is that the total diffraction volume is increased by increasing the path length of the X-rays within the thin film itself, which leads to higher peak intensities and overall improved statistics. For example, if a parallel beam setup with a beam width of 1 mm is used, which is typical for an X-ray mirror, the sample projection length is $(\sin\alpha)^{-1}$ mm. Thus, an α-value of 1° results in a projection length of 57 mm, which largely increases the total diffraction volume. At such an angle of incidence, less of the X-rays penetrate down into the substrate and the presence of the substrate peaks in the diffractogram is reduced. However, caution is adviced when performing small angle of incidence measurements because the projection length may extend the size of the analyzed sample. This may lead to unknown reflections in the diffractogram that originate from the sample holder.

The surface sensitivity of GIXRD may be increased further by using α-values of only a few tenths of a degree. For flat film surfaces, GIXRD measurements using small α-values cause the X-rays to be refracted at the air/sample interface. The reason for this refraction of the X-rays is that the refractive index, n, at X-ray energies is slightly smaller than 1, indicating that the X-rays traverse the air/sample interface into a less dense medium. n is approximated by $n = 1 - \delta - i\beta$, where both δ and β are in the 10^{-6} range for X-rays. One effect of refraction is that the X-rays are totally reflected if α is smaller than the critical angle for total external reflection, $α_c$. $α_c$ is roughly dependent on the density of the thin film and is usually in the region of 0.1–0.5° for Cu K$_α$ radiation. For GIXRD measurements on flat samples, where $α < α_c$, the penetration depth of the X-rays is usually below 100 Å, which makes the measurement very surface sensitive. At $α > α_c$, the penetration depth increases rapidly as it is influenced more strongly by the linear attenuation coefficient. One important effect for GIXRD measurements with α-values around the critical angle is that the measured scattering angle 2θ is different from the actual Bragg angle $2θ_B$ due to the refraction of the X-rays at the sample surface. The corresponding peak shift Δ2θ is always positive and defined by $Δ2θ = 2θ − 2θ_B$. The actual expression of Δ2θ has been derived by Toney and Brennan [8] and is expressed as

$$\Delta 2\theta \approx \alpha - 1/(2)^{1/2} \left\{ \left[(\alpha^2 - \alpha_c^2)^2 + 4\beta^2 \right]^{1/2} - \alpha_c^2 + \alpha^2 \right\}^{1/2} \quad (13.7)$$

From this expression, it follows that the maximum value of Δ2θ is obtained close to the critical angle and is $α_c − β^{1/2}$, which is $α_c$ for $β = 0$. Furthermore, the peak shift is zero in GIXRD at zero and large angles of incidence, whereas Δ2θ varies linearly with α for $α < α_c$, and by using a small angle approximation $α_c^2/2α$ for $α \gg α_c$. Due to the peak shift that occurs in GIXRD close to the critical angle, it is important to make corrections to the measured data prior to analysis. This is especially true if properties that can depend on the actual 2θ value, such as strain and film composition, are to be evaluated. If the critical angle for a material is known, the measured data can be corrected if the diffractometer is carefully aligned. However, it may be a good idea to measure the position of the film reflections at α values of a few degrees or even by a θ–2θ scan to estimate the peak shift, Δ2θ. It is clear that the average information depth of a GIXRD measurement is dependent on the angle of incidence, where changing α and performing the same detector scan for several α results in depth resolved structural information of the analyzed sample. Thus, if a multilayer sample is analyzed by GIXRD, the intensity from peaks originating from the topmost layer increases with α for $α < α_c$, toward an intensity maximum reached for $α = α_c$. Above $α > α_c$, the intensity of the reflections from the top film decreases and the intensity of the peaks from the underlying layers increases as the X-rays reach further down into the sample. An example of depth profiling by GIXRD is demonstrated by Toney et al. for iron oxide thin films [9]. The total diffraction volume at a certain angle of incidence is also dependent on the substrate surface roughness. For thin films on rough substrates, a maximum scattering intensity can sometimes be found at α values higher than that of $α_c$.

For grazing incidence measurements, a nonfocusing diffractometer geometry is used since a divergent beam, which is commonly used in θ–2θ focusing geometry setups, will not end up on the focusing circle of the diffractometer, which leads to large focusing errors (Figure 13.6). One example of a nonfocusing geometry, commonly used for GIXRD, is a parallel beam setup with a parabolic multilayer X-ray mirror on the primary side coupled with a parallel-plate collimator on the detector side. The advantage of the X-ray mirror is that it yields high intensities due to

13.5 Energy Dispersive X-Ray Diffraction (EDXRD)

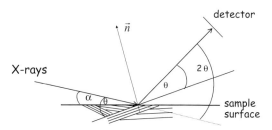

Figure 13.6 Principle geometry of a GIXRD experiment: α – incident angle, θ – Bragg angle, n – normal of the lattice plane which fulfills the Bragg equation.

reflecting a large portion of the divergent X-rays from the X-ray tube in a parallel beam, and that displacement errors or sample roughness do not lead to peak shifts in 2θ. Another option for GIXRD is to use a slit-based system on the primary side together with a parallel plate collimator and a monochromator on the detector side. A comparison of net peak intensities between a θ–2θ and two different grazing incidence setups is shown in Table 13.1, and the difference between a GI- and a θ–2θ scan for a ZnO/CuIn$_{0.5}$Ga$_{0.5}$Se$_2$/Mo/glass structure using a parallel-beam setup is displayed in Figure 13.7. In Figure 13.7, it can be seen that the intensity of the ZnO peaks is greatly enhanced in the GIXRD measurement as compared to the θ–2θ scan. Note that the intensity of all reflections from the underlying structure does not reduce to zero intensity, which may be related to the large surface roughness of the CuIn$_{0.5}$Ga$_{0.5}$Se$_2$ layer.

13.5
Energy Dispersive X-Ray Diffraction (EDXRD)

According to the Bragg equation (Eq. (13.2)), the lattice plane distance d_{hkl} can be written as a function of diffraction angle θ and wavelength λ. Therefore, to obtain a diffractogram – that is, the diffraction intensity as a function of d_{hkl} – it is sufficient to vary either the diffraction angle θ or the wavelength λ. While in angle-dispersive X-ray diffraction (ADXRD), the wavelength of the radiation is kept constant and the angle

Table 13.1 Comparison between three different diffractometer setups, A–C, for a 270-nm thick Cu$_2$O film on SiO$_2$. A: θ – 2θ, Bragg–Brentano, B: GI, $\alpha = 0.3°$, slit + 0.40° parallel-plate collimator and LiF(200) monochromator, and C: GI, $\alpha = 0.3°$, X-ray mirror + 0.15° parallel-plate collimator.

Peak	A (cps)	B (cps)	C (cps)
(110)	12	6	420
(111)	85	80	5300
(200)	9	28	2400
(211)	—	—	90
(220)	16	15	1200

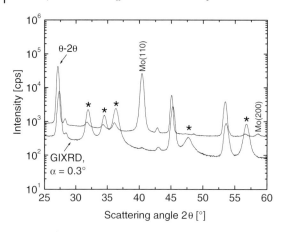

Figure 13.7 Comparison of θ–2θ and GIXRD (α = 0.3°) diffractograms for a ZnO/CuIn$_{0.5}$Ga$_{0.5}$Se$_2$/Mo structure on a glass substrate, where the individual layers are 0.05, 2, and 0.4 μm thick, respectively. Peaks denoted by (*) correspond to the wurtzite structure ZnO top layer and all unmarked reflections originate from the CuIn$_{0.5}$Ga$_{0.5}$Se$_2$. The sample is measured with an X-ray mirror and 0.27° parallel-plate collimator setup.

is varied, in energy-dispersive X-ray diffraction (EDXRD) the angle is kept constant and the wavelength of the radiation is varied. This variation of wavelength is realized by the use of polychromatic radiation (typically synchrotron radiation) in combination with an energy-dispersive detector and a multichannel analyzer. Note that for electromagnetic radiation, energy E and wavelength λ are related by $E = hc/\lambda$ (where h is Planck's constant and c is the speed of light). The Bragg equation (Eq. (13.2)) can therefore be written as

$$d_{hkl}(E) = hc/(2E\sin(\theta)) \tag{13.8}$$

Polychromatic synchrotron radiation consists of a high flux of photons with a wide range of energies [10]. Only those photons which fulfill Eq. (13.8) for a given interplanar distance d_{hkl} are diffracted in the direction of the detector at an angle of 2θ (additional to incoherent scattering). In principle, the photons of different energies are detected one after another. However, owing to the high intensity of synchrotron radiation and the high detection speed of modern detectors, it is possible to obtain a complete diffractogram within seconds with an energy range of typically 6–100 keV [11].

Since in EDXRD the angle stays constant during the measurements, no moving parts are needed. Therefore the acquisition time for a complete diffractogram is only determined by the intensity of the used radiation and the speed of the detector. This predestines EDXRD for time-resolved *in-situ* measurements. Information such as phase formation, composition gradients of alloys (e.g., in Cu(In,Ga)Se$_2$), and grain growth can be obtained as a function of time, temperature, pressure, and other parameters [12–14]. Additionally, X-ray fluorescence (XRF) signals – even of heavy

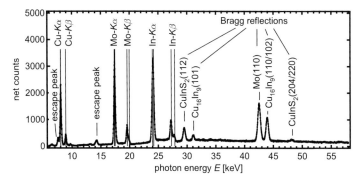

Figure 13.8 Accumulated EDXRD spectrum acquired during a growth process of a CuInS$_2$ thin film.

elements – are measured simultaneously to the Bragg reflexes. These XRF signals provide information about material adsorption and desorption as well as elemental distributions [15, 16].

Disadvantages of EDXRD compared with ADXRD are the necessity of a highly brilliant, polychromatic X-ray source (which is only available at synchrotrons) and a lower peak resolution stemming from the limited energy resolution of energy-dispersive detectors [10]. Figure 13.8 shows a typical EDXRD spectrum measured during the growth of a CuInS$_2$ film by sulfurizing a Cu–In precursor layer on Mo-coated glass. Fluorescence signals of the elements Cu, Mo, and In as well as Bragg reflexes of CuInS$_2$, the metallic Cu$_{16}$In$_9$ phase, and Mo are visible within a single spectrum.

The main purpose of *in-situ* measurements is the time-resolved observation of sample properties such as crystal structure, elemental composition, and morphology. Usually, a change of sample properties is brought about by changing the external parameters such as temperature, atmosphere, and pressure. The general principle of an experimental setup for *in-situ* EDXRD is outlined in Figure 13.9. In order to ensure well-defined reaction conditions, a chamber which isolates the sample environment from the laboratory environment is necessary. Such a chamber needs X-ray windows for the incoming and the diffracted radiation. The choice of window material

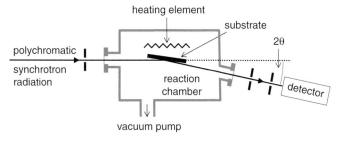

Figure 13.9 Schematic drawing of a general setup for *in-situ* EDXRD experiments.

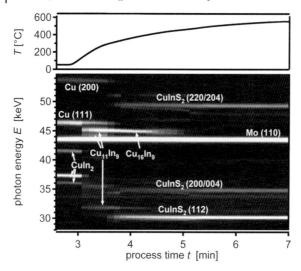

Figure 13.10 Time-resolved EDXRD spectra during RTP synthesis of CuInS$_2$. The complete reaction of the metallic precursor with sulfur takes place within a few minutes [17].

depends on the energy range of interest and the process conditions. Al and graphite can cope with high temperatures and have sufficient transparencies for energies above 15 keV. If detection of lower energies is desired (e.g., in order to detect fluorescence signals of Cu or lighter elements), beryllium, polyimide, or diamond windows are more appropriate. (However, beryllium is connected with potential safety risks and polyimide windows exhibit low temperature stabilities.)

In the following, the potential of *in-situ* EDXRD will be demonstrated by some application examples, all connected to the study of reactions playing a role during the growth of CuInS$_2$ thin films. Figure 13.10 presents time-dependent EDXRD spectra during the synthesis of CuInS$_2$ by rapid thermal processing (RTP) [17]. Here, the signal intensity of the spectra is gray-scale coded, where black represents low intensities and white represents high intensities. A complete spectrum was measured every 7 s, providing a reasonable time resolution for the phase formation during the process, which in total took only a few minutes.

Despite the mentioned low peak resolution, *in-situ* EDXRD has successfully been used to monitor grain sizes in polycrystalline materials as a function of time and temperature by carefully analyzing the profile of the Bragg reflexes [14]. To be precise, it is only the domain size that can be seen. Domain refers to a region with undisturbed crystal lattice. Therefore the domain size is a lower limit for the grain size since at least at grain boundaries the crystal lattice is interrupted. The profile of a Bragg reflex of a homogeneous crystalline material measured by EDXRD is a convolution of a Gaussian and a Cauchy profile. The breadth β^C of the Cauchy contribution of a reflex is according to the Scherrer equation proportional to the inverse of the domain size D [18]

$$\beta^C_{hkl} = (k^{\text{Scherrer}} hc)/(D \sin\theta) \qquad (13.9)$$

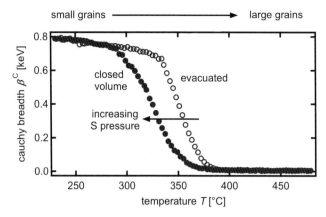

Figure 13.11 Temperature dependence of the evolution of the Cauchy breadth of the 112-reflex of a CuInS$_2$ layer measured by EDXRD during annealing in different sulfur pressure conditions. The decrease in the Cauchy breadth correlates with an increase in grain sizes within the CuInS$_2$ layer from some 10 nm to 1 μm. The temperature range at which the decrease in Cauchy breadth takes place shifts to lower temperatures when the sulfur pressure is increased [14].

where $k^{Scherrer}$ is the Scherrer constant, h the Planck's constant, c the speed of light in vacuum, and 2θ the diffraction angle. As a consequence, the evolution of the Cauchy breadth is a measure for the evolution of domain size or the lower limit of grain size. Figure 13.11 shows the evolution of Cauchy breadth with time and temperature during a recrystallization process in a CuInS$_2$ thin film under two different sulfur pressure conditions.

A major advantage of the EDXRD method is that it provides fluorescence signals additionally to Bragg reflexes which are contained within a single spectrum (as seen in Figure 13.8). The intensity of a fluorescence signal for a specific element is in principle determined by the intensity of the incoming synchrotron radiation, the attenuation of the incoming radiation, the density of the elements in the sample, the probability for the transition from an incidence photon to a fluorescence photon of that element, and the attenuation of the emitted fluorescence radiation. The attenuation of incidence and fluorescence radiation in the sample itself depends on the depth distribution of the elements. Therefore, if all other parameters are known, the fluorescence signals provide information about the evolution of depth distributions of elements within the sample as a function of time. Since the number of fluorescence signals that are measured under one or more fixed angles is limited, the depth distributions have to be parameterized such that the number of parameters does not exceed the number of measured fluorescence signals. It can be demonstrated that a simple parameterization (where the knowledge of phase formation gained from the simultaneously measured Bragg reflexes are taken as additional constrains) can be used to obtain *in-situ* depth distributions by a quantitative analysis of the fluorescence signals [16]. Figure 13.12 presents a series of depth distributions during a CuInS$_2$ synthesis process. The parameterized depth distributions have been

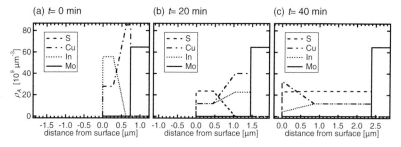

Figure 13.12 Series of modeled depth distributions during $CuInS_2$ synthesis by sulfurization of a Cu–In precursor (on a Mo-coated substrate). The figures represent the atomic density ρ_A of the elements in the layer as a function of the distance from the layer surface. (a) At the beginning of the process ($t=0$), the precursor layer consists of pure Cu and $CuIn_2$. (b) Eventually, the sulfurization starts and $CuInS_2$ forms on top of the metallic layer consisting of $Cu_{16}In_9$. (c) Finally, the sulfurization is complete. Excess Cu forms $Cu_{2-x}S$ on top of the $CuInS_2$ layer [16].

modeled by comparing calculated to measured fluorescence intensities. The results are in good agreement with *ex-situ* analyzed samples from break-off experiments [19]. The resolution of the method can be increased by using more than one detector to simultaneously measure fluorescence signals under different exit angles.

References

1 Lovesey, S.W. (1984) *Theory of Neutron Scattering from Condesed Matter*, vol. 2, Claredon Press, Oxford.
2 Sears, V.F. (1992) Neutron scattering lengths and cross sections. *Neutron News*, 3 (3), 26–37.
3 Schorr, S., Höbler, H.-J., and Tovar, M. (2007) A neutron diffraction study of the stannite–kesterite solid solution series. *Eur. J. Mineral.*, 19 (1), 65–73.
4 Furrer, A., Mesot, J., and Straessle, T. (2009) *Defects in Solids in Neutron Scattering in Condensed Matter Physics*, 4th edn (eds J.L. Finney and D.L. Worcester), World Scientific, Singapore, pp. 226–229.
5 Schorr, S., Tovar, M., Stuesser, N., Sheptiakov, D., and Geandier, G. (2006) Where the atoms are: Cation disorder and anion displacement in (DX^{VI})-$(A^{I}B^{III}X^{VI}_2)$ semiconductors. *Physica B – Condens. Matter*, 385, 571–573.
6 Young, R.A. (2000) *The Rietveld Method*, Oxford Science Publications London, UK.
7 Zhang, S.B., Wei, S.H., Zunger, A., and Katayama-Yoshida, H. (1998) Defect physics of the $CuInSe_2$ chalcopyrite semiconductor. *Phys. Rev. B*, 57 (16), 9642–9656.
8 Toney, M.F. and Brennan, S. (1989) Observation of the effect of refraction on X-rays diffracted in a grazing-incidence asymmetric Bragg geometry. *Phys. Rev. B*, 39 (11), 7963–7966.
9 Toney, M.F., Huang, T.C., Brennan, S., and Rek, Z. (1988) X-ray depth profiling of iron oxide thin films. *J. Mater. Res.*, 3 (2), 351–356.
10 Buras, B. and Gerward, L. (1989) Application of X-ray energy-dispersive diffraction for characterization of materials under high pressure. *Prog. Cryst. Growth Charact.*, 18, 93–138.
11 Genzel, C., Denks, I.A., Gibmeier, J., Klaus, M., and Wagener, G. (2007) The materials science synchrotron beamline EDDI for energy-dispersive diffraction

analysis. *Nucl. Instrum. Methods Phys. Res., A*, **578**, 23–33.

12 Rissom, T., Mainz, R., Kaufmann, C., Caballero, R., Efimova, V., Hoffmann, V., and Schock, H.W. (2010) Examination of growth kinetics of copper rich Cu(In,Ga)Se$_2$-films using synchrotron energy dispersive X-ray diffractometry. *Sol. Energy Mater. Sol. Cells*. doi: 10.1016/j.solmat.2010.05.007

13 Mainz, R., Klenk, R., and Lux-Steiner, M. (2007) Sulphurisation of gallium-containing thin-film precursors analysed *in-situ*. *Thin Solid Films*, **515** (15), 5934–5937.

14 Rodriguez-Alvarez, H., Mainz, R., Marsen, B., Abou-Ras, D., and Schock, H.W. (2010) Recrystallization of Cu–In–S thin films studied *in situ* by energy-dispersive X-ray diffraction. *J. Appl. Crystallogr.*, **43** (5), 1053.

15 Weber, A., Mainz, R., and Schock, H.W. (2010) On the Sn loss from thin films of the material system Cu–Zn–Sn–S in high vacuum. *J. Appl. Phys.*, **107** (1), 013516.

16 Mainz, R. and Klenk, R. In-*situ* analysis of elemental depth distributions in thin films by combined evaluation of synchrotron X-ray fluorescence and diffraction. To be published.

17 Rodriguez-Alvarez, H., Koetschau, I.M., Genzel, C., and Schock, H.W. (2009) Growth paths for the sulfurization of Cu-rich Cu/In thin films. *Thin Solid Films*, **517**, 2140–2144.

18 Otto, J.W. (1997) *J. Appl. Crystallogr.*, **30**, 1008–1015.

19 Calvo-Barrio, L., Perez-Rodriguez, A., Alvarez-Garcia, J., Romano-Rodriguez, A., Barcones, B., Morante, J., Siemer, K., Luck, I., Klenk, R., and Scheer, R. (2001) Combined in-depth scanning auger microscopy and Raman scattering characterization of CuInS$_2$ polycrystalline films. *Vacuum*, **63**, 315–321.

14
Raman Spectroscopy on Thin Films for Solar Cells

Jacobo Álvarez-García, Víctor Izquierdo-Roca, and Alejandro Pérez-Rodríguez

14.1
Introduction

In this chapter, the capabilities of Raman spectroscopy for the advanced characterization of thin films for solar cells are reviewed. Raman spectroscopy is an optical nondestructive technique based on the inelastic scattering of photons with elemental vibrational excitations in the material. The line shape and position of the Raman bands are determined by the crystalline structure and chemical composition of the measured samples, being sensitive to the presence of crystalline defects, impurities, and strain. Presence of peaks characteristic of different phases also allows for the identification of secondary phases that are strongly related to the growth and process conditions of the films. All this gives a strong interest to the analysis of the Raman spectra, providing a powerful nondestructive analytical tool for the crystalline and chemical assessment of the films. In addition, the combination of a Raman spectrometer with an optical microscope also allows for achieving a high spatial resolution (below 1 μm), thereby enabling mapping and surface and in-depth resolved homogeneity analysis.

The chapter is structured in four main sections: the two first ones are devoted to a revision of the fundamentals of Raman spectroscopy (Section 14.2) and vibrational modes in crystalline materials (Section 14.3). Section 14.4 deals with the main experimental considerations involved in the design and implementation of a Raman scattering setup. This is followed by a detailed description of the application of Raman scattering for the structural and chemicophysical analysis of thin-film photovoltaic materials (Section 14.5), with the identification of crystalline structure and secondary phases, evaluation of film crystallinity, analysis of chemical composition of semiconductor alloys, characterization of nanocrystalline and amorphous layers, and stress effects. This includes the description of representative state of the art and recent case examples that illustrate the capabilities of the technique for the advanced characterization of layers and process monitoring in thin-film photovoltaic technologies.

Advanced Characterization Techniques for Thin Film Solar Cells,
Edited by Daniel Abou-Ras, Thomas Kirchartz and Uwe Rau.
© 2011 Wiley-VCH Verlag GmbH & Co. KGaA. Published 2011 by Wiley-VCH Verlag GmbH & Co. KGaA

14.2
Fundamentals of Raman Spectroscopy

Raman spectroscopy relies on the analysis of the electromagnetic radiation inelastically scattered by a material. The macroscopic theory, introduced in this section, provides a simple and consistent description of the underlying physical processes. Nevertheless, in order to fully describe the light–matter interaction process, it is necessary to take into account quantum mechanical considerations. A detailed description of light-scattering theory can be found in Ref. [1]. Further detailed information on the fundamentals of Raman scattering and its application for the analysis of semiconductor materials is given in Refs. [2–4].

Transmission, reflection, refraction, and *absorption* constitute the fundamental processes that describe most of the photon–matter interactions. However, a small fraction of light undergoes different interaction mechanisms resulting in the radiation of electromagnetic energy. These processes are known as light scattering and typically account for less than 1% of the photon–matter interactions. When a photon is scattered by a medium, it may either retain its energy (*elastic scattering*) or loose part of it (*inelastic scattering*).

In the case of the interactions between light and semiconducting materials, light scattering is a straightforward consequence of the dielectric properties of the medium. In the presence of an externally applied electric field, atoms within a dielectric material redistribute their electrical charge, creating an internal electric field (polarization), which tends to compensate the external field. In linear and isotropic materials, the polarization response $\vec{P}$ is linear with the electric field of the incoming radiation, according to

$$\vec{P} = \chi \cdot \vec{E}_I = \chi \cdot \vec{E}_I^0 \cdot \cos(\vec{k}_I \cdot \vec{r} - \omega_I \cdot t) \tag{14.1}$$

In this equation, $\vec{E}_I^0$ is the amplitude of the electric field from an electromagnetic incident wave with wavevector $\vec{k}_I$ and frequency ω_I. The factor χ is the electrical susceptibility of the medium, and may be interpreted as the density of electric dipoles per unit volume, which is a characteristic of the material. When the applied field is oscillating, such as the one associated with the electromagnetic radiation, the induced polarization field is also time dependent, and therefore, according to the basic electrodynamic theory, it radiates energy. Consequently, light scattering may be interpreted as a result of the radiative processes associated with the existence of electric dipoles at the atomic scale, distributed throughout the material.

While the previous description provides a simple picture of elastic light scattering, Raman scattering involving inelastic interactions requires a more accurate analysis of the polarization mechanisms. Polarization is caused by the redistribution of charges within the atoms forming a solid. However, in a semiconducting crystal at temperatures above 0 K, atoms do not occupy static positions, but they oscillate around their equilibrium positions with a characteristic frequency Ω. Such periodic oscillations modulate the dielectric response of the medium, leading to the occurrence of radiative polarization fields oscillating at a frequency different to the frequency of the external electric field.

To illustrate this argument mathematically, one may consider a perfect crystal in which all the atoms are oscillating at a frequency Ω. Notice that for the sake of simplicity, we will restrict this discussion to the analysis of a single collective oscillation mode in a crystal (phonon), characterized by a frequency Ω and wavevector $\vec{q}$ – in a more general approach, an expansion in terms of Fourier components may be considered. Under these assumptions, the amplitude of the oscillations of an atom in the crystal may be expressed as

$$\vec{X}(\vec{r},t) = \vec{X}^0(q,t) \cdot \cos(\vec{q} \cdot \vec{r} - \Omega \cdot t) \tag{14.2}$$

In terms of this frequency-dependent amplitude, the susceptibility of the crystal can be expressed as a first-order Taylor expansion

$$\chi(\vec{q},\Omega) = \chi_0 + \left(\frac{\partial \chi}{\partial X^j}\right) \cdot X^j(\vec{q},\Omega) \tag{14.3}$$

In the previous expression, the index j runs over the three crystallographic directions. Using Eqs. (14.1) and (14.3), and basic trigonometric relations, the polarization vector may be expressed as

$$\vec{P} = \chi \cdot \vec{E}_I^0 \cos(\vec{k}_I \cdot \vec{r} - \omega_I \cdot t) + \left(\frac{\partial \chi}{\partial X^j}\right) \cdot X^j(\vec{q},\Omega) \cdot \vec{E}_I^0 \cdot \cos(\vec{k}_{AS} \cdot \vec{r} - \omega_{AS} \cdot t)$$

$$+ \left(\frac{\partial \chi}{\partial X^j}\right) \cdot X^j(\vec{q},\Omega) \cdot \vec{E}_I^0 \cdot \cos(\vec{k}_S \cdot \vec{r} - \omega_S \cdot t)$$

where

$$\begin{aligned}\vec{k}_S &= \vec{k}_I - \vec{q} & \omega_S &= \omega_I - \Omega \\ \vec{k}_{AS} &= \vec{k}_I + \vec{q} & \omega_{AS} &= \omega_I - \Omega\end{aligned} \tag{14.5}$$

The first term of Eq. (14.4) corresponds to a wave equation with the same frequency and wavevector as the incoming field (zero-order term), and is responsible for the elastic contribution to the scattering radiation, sometimes called *Rayleigh scattering*. The second and third terms in Eq. (14.4) contain a factor proportional to the first derivative of the susceptibility, and account for the two terms responsible for the Raman scattering processes. This factor makes the inelastic contribution much less intense than the elastic term. Typically, the Raman scattering intensity is 10^6–10^9 times less intense than the Rayleigh signal.

On the other hand, it is worth to notice that the Raman scattering signal consists of two contributions, symmetrically shifted in frequency with respect to the frequency of the excitation light ω_I. The magnitude of this shift (ω), known as "*Raman shift*," is directly equal to the vibration frequency of the crystal (Ω). The Raman shift is usually

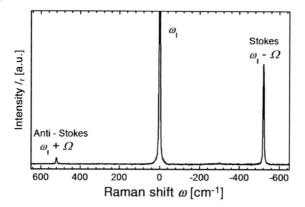

Figure 14.1 Scattering spectrum of crystalline Si showing two inelastic contributions shifted to higher (anti-Stokes) and lower (Stokes) energy. The energy shift is the energy of the phonon involved in the photon–phonon interaction. The observed peak corresponds to a triple degenerated phonon band at 520 cm^{-1}, as described in Section 14.3.

expressed as a wave number (inverse wavelength), and it is related to the excitation (λ_I) and scattered (λ_{scatt}) wavelengths by

$$\omega\,(\mathrm{cm}^{-1}) = \frac{1}{\lambda_{scatt}(\mathrm{cm})} - \frac{1}{\lambda_I(\mathrm{cm})} \tag{14.6}$$

For practical analytical purposes, the information provided by the two inelastic contributions is redundant, and usually only the low-frequency signal (Stokes component) is analyzed, since its intensity is higher than the high-frequency component (anti-Stokes band), as shown in Figure 14.1. Normally, Raman spectra are plotted as a function of the absolute value of the Raman shift.

The frequency, intensity, and band shape of the Raman bands provide useful information in relation to the way in which atoms vibrate in a crystalline lattice. These parameters, in turn, indirectly depend on a number of physical and chemical factors, including the chemical composition and crystallographic structure of the material, the presence of impurities, stress, or crystalline defects. Such interdependence between the Raman spectra and different physical properties allows inferring basic materials properties based on the analysis of the spectra, and constitute the fundamentals of the technique.

14.3
Vibrational Modes in Crystalline Materials

In the previous section, we had assumed a simplified case in which all the atoms in a crystal oscillate at a unique frequency. However, in real crystals, the lattice dynamics is characterized by the existence of multiple collective oscillation modes (phonons), which can be described in terms of their frequency (Ω) and wavevector ($\vec{q}$) [5]. As a result, the band structure of Raman spectra of real materials is usually complex, particularly in those cases in which the crystal symmetry is low. Nevertheless, there exists an important restriction that must be taken into account, which significantly reduces the complexity of the Raman spectra. In a perfect crystalline material, any inelastic photon–phonon interaction must fulfill two fundamental relations, given by Eq. (14.5). The first relation stands for the energy conservation law, which requires that the energy balance in the scattering process must be properly preserved. Likewise, the conservation of the crystalline quasimomentum requires also that the difference between the wavevectors of the scattered and incoming photon must be equal to $\pm\vec{q}$ (wavevector of the phonon involved in the scattering process). Furthermore, since the energy of visible photons is about three orders of magnitude greater than the typical energy of lattice oscillations, inelastic scattering at practical experimental conditions occurs without a significant change in the photon energy, and accordingly, without any substantial change in its wavevector module. Consequently, the conservation of the crystalline quasimomentum requires that phonons involved in inelastic scattering processes must have a wavevector module close to zero: $|\vec{q}| = 0$. According to these conservation laws, the Stokes process can be interpreted as the process of creation of a phonon with frequency (Ω) and wavevector $(\vec{q})$ by the incident photon, and this results in a scattered photon with lower energy, while the anti-Stokes process would correspond to the annihilation of a phonon with frequency (Ω) and wavevector $(\vec{q})$ by the incident photon, resulting in a scattered photon with higher energy.

The previous argument related to the energy and quasimomentum conservation laws allows predicting the number of optically active bands for any given crystalline structure. Raman active phonons correspond necessarily to long-wave oscillation modes (or zone-center phonons, in relation to its position in the Brillouin zone), which involves the in-phase displacement of all the equivalent atomic positions in the crystal. Therefore, the number of optical vibrational modes in three-dimensional crystalline structures is always $3(N - 3)$, where N is the number of atoms in the crystal base (notice that this formula takes into account the fact that three normal modes necessarily correspond to pure translations in the crystal).

At this point, it is worth to mention that the symmetry characteristics of a particular crystalline structure may lead to additional restrictions in the characteristics of the Raman spectrum. A first restriction deals with the fact that some of the optically active modes may be symmetrically equivalent, thus leading to Raman modes with the same frequency. When this happens, the corresponding Raman mode is said to be "degenerated." On the other hand, a second noticeable particularity of some lattice vibrations is its lack of Raman activity. Non Raman-active modes necessarily involve the displacement of atoms in the lattice in such a way that the terms in the derivative of the electrical susceptibility are null, or they cancel each other. From Eq. (14.4), it is clear that under this condition, the Raman intensity is zero.

In order to illustrate the concepts described in the previous paragraphs, we shall consider now the crystalline structure of Si as a practical example. Si crystallizes in a diamond-like face-centered cubic structure, in which two Si atoms occupy the nodes of the crystal unit cell, at positions given by $\vec{u}_1 = (0, 0, 0)$ and $\vec{u}_2 = (1/4, 1/4, 1/4)$

(space group $Fd\bar{3}m$). Therefore, one may expect three Raman-active phonons for this structure, which in principle may occur at different frequencies. In this case, due to the high-symmetry characteristics of the Si structure, the frequency of these three active modes is the same (the mode is said to be triple-degenerated), and therefore, one single band is observed in the (Stokes) Raman spectrum of crystalline Si, as shown in Figure 14.1.

The previous simple example already reveals that a complete comprehension of the Raman spectrum of a material necessarily requires a detailed analysis of the symmetry properties of the crystal structure. Nevertheless, in addition to the previous considerations, it must be taken into account also the fact that in polar oscillation modes (those involving an asymmetric displacement of ions in the lattice, which result in a net electric field over the crystal), the photon/phonon coupling results in a splitting of the phonons with propagation vector parallel (longitudinal) and perpendicular (transverse) to the electric field. Such splitting is known as "*LO–TO splitting*," and from a practical point of view, it results in the occurrence of two bands in the spectrum for each polar mode, occurring at different frequencies. As a representative example of LO–TO splitting, we may consider the case of the GaAs crystal. The crystalline structure of GaAs is essentially the same as the one of Si, but replacing each of the two atoms in the atomic base by one atom of Ga and As. Although this modification results in a different symmetry space group ($F\bar{4}3m$), it is possible to show that the three optical modes in this crystalline structure are also degenerated. However, unlike in the case of the Si crystalline structure, these modes involve the displacement of ions with different charges, which result in a net electric field (polar modes) that is responsible for the LO–TO splitting of the mode. The experimentally measured frequencies for these two bands are 268 and 292 cm^{-1}, corresponding to the TO and LO modes, respectively.

14.4
Experimental Considerations

From the experimental point of view, Raman spectroscopic measurements require a relatively simple setup, consisting of a laser source, focusing and collection optics, and a sensitive spectrophotometer. Figure 14.2 shows a schematic diagram of the experimental setup required for these measurements. From the different elements constituting the system, the spectrometer unit is the most critical one.

The following subsections intend to introduce the most important considerations in the selection of an adequate Raman setup.

14.4.1
Laser Source

Continuous gas lasers with powers ranging from some tens of milliwatts to a few watts have been traditionally used as light sources in Raman spectroscopy. Recently, highly

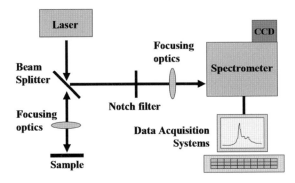

Figure 14.2 Schematic diagram of Raman scattering experimental setup.

temperature stabilized solid-state lasers have become also a good alternative to conventional gas lasers.

Since the intensity of the Raman signal linearly increases with laser power, laser power should be kept as high as possible. Nevertheless, over a certain power threshold, thermal effects become observable in the spectra. Spectral changes may be evident in the case of thermally unstable films, or materials sensitive to oxidation. However, even at moderate powers thermal effects may cause subtle changes in the spectra, which may lead to wrong interpretations of the spectra. Asymmetric band broadening and red shift of the Raman bands are usually the first observable thermal effects. These effects can be especially important when working with a Micro-Raman configuration, because in this case the incident excitation power is concentrated in a very small (micrometric) volume, which can lead to very high power densities.

Inelastic Raman scattering occurs regardless of the excitation wavelength. In this sense, laser wavelength is not a critical parameter in many applications. Nevertheless, some considerations should be kept in mind when selecting a proper wavelength. First, the Raman cross section increases with a ω_i^4 factor. Thus, shorter wavelengths are normally expected to yield better Raman signals. However, this is not always the case due to the fact that the Raman scattering intensity is also proportional to the scattering volume, which normally decreases exponentially with light frequency due to light absorption. Therefore, light penetration is a relevant factor, which has not only an impact on the Raman intensity, but also on the volume probed by the measurement. This is interesting in the analysis of semiconductors with high optical absorption in the visible region, as those involved in thin-film solar cells: in this case the direct comparison of spectra measured with different excitation wavelengths (and, hence, with different penetration depths of the scattered volume) provides a simple procedure to analyze the in-depth homogeneity of the layers, even if this is normally achieved at an estimative low-depth resolution level.

A second consideration in relation to the selection of the laser wavelength deals with the possible existence of *resonant Raman* effects [6]. Resonance occurs when the excitation wavelength matches the energy gap of the material. Under these conditions, a strong enhancement of the Raman signal occurs, normally accompanied by other spectral effects. Due to the complexity of the resonant spectra, these are not necessarily the optimum conditions for thin film characterization, and the adequacy of operating under resonance should be evaluated in each case. Finally, it is worth to mention that near-infrared (NIR) or ultraviolet (UV) excitation wavelengths may be a particularly adequate choice in cases in which under visible excitation light, the Raman signal cannot be resolved due to the existence of a strong background fluorescence signal. In those cases, the use of excitation wavelengths far away from the electronic absorption bands of the material (as those at NIR or UV regions) constitutes an effective means to reduce background fluorescence.

14.4.2
Light Collection and Focusing Optics

In many cases, Raman spectra are acquired within the so-called *backscattering* configuration, in which the light scattered at 180° with respect to the incident laser beam is analyzed. Optical systems based either on conventional optical discrete elements or fiber optics are used for light focusing and collection purposes, commonly in conjunction with beam-splitting optics. A particular useful arrangement is based on the use of a metallographic microscope, which allows attaining diffraction-limited spatial resolution (micro-Raman spectroscopy, also known as Raman microprobe configuration) [4]. This has very interesting applications for the development of mapping and uniformity analysis measurements down to the submicrometre range.

14.4.3
Spectroscopic Module

Due to the narrow bandwidth of the Raman bands, particularly in the case of semiconducting materials, high spectral resolution (wavelength resolution $\Delta\lambda < 0.1$ nm) is desirable. Besides, the low intensity of the Raman signal requires the use of sensitive CCD detectors, which are usually actively cooled to minimize dark current noise. The excellent performance of currently available multichannel CCD detectors makes them the preferred choice for most applications in front of photomultiplier-based systems, which require using scanning monochromators.

Special attention should be paid to the performance of the selected optical system in terms of stray light rejection. Since the Raman bands appear at spectral positions relatively close to the much more intense laser excitation light, stray light must be necessarily considered. For this reason, double or triple grating spectrometers are often the best choice for laboratory systems. In the case of single grating spectrometers, it is necessary to use a notch filter in order to minimize the intensity of the laser light directly going into the spectrometer.

14.5
Characterization of Thin-Film Photovoltaic Materials

14.5.1
Identification of Crystalline Structures

As earlier discussed, bands in the Raman spectra of thin-film crystalline materials essentially provide the phonon energy of the zone-center phonons in the corresponding crystalline structure. The analysis of the frequency and number of bands in the spectra allow determining the crystalline structures present in the film. However, unlike with other techniques such as X-ray diffraction, simulation techniques for predicting the Raman spectrum of a given material are significantly complex, and more often, crystalline structure and phase identification is accomplished by comparing the spectra with data from reference materials. In any case, in order to avoid wrong interpretations of the spectra, it is always desirable to analyze cautiously the reference spectrum of the material under investigation. Once this is done, Raman spectroscopy can be effectively used to investigate the crystallographic properties of thin-film materials, providing useful information on the existence of secondary phases with a submicron spatial resolution. Even though the required integration time depends on the experimental setup, the film characteristics, and the type of analysis to be performed, in general it is possible to perform a measurement within a few seconds to a few minutes. Moreover, when combined with a motorized stage, the technique can be used in order to make a chemical mapping of the surface of the film ("Raman mapping").

We shall now introduce the potential of Raman spectroscopy as an analytical tool by considering a particular example in some detail. We will discuss the case of the chalcopyrite-type structure, which is characteristic of many thin-film photovoltaic materials, including $CuInSe_2$, $CuInS_2$, $CuGaSe_2$, and related alloys [7, 8]. From the crystallographic point of view, the chalcopyrite-type structure is significantly more complex than IV-type or III–V materials, and correspondingly, the Raman spectrum reflects the existence of more modes in which atoms in the lattice can oscillate. The unit cell in the chalcopyrite-type structure is occupied by two formula units, leading to a total of 21 vibrational modes. These modes are classified according to the symmetry of the atomic displacements, and expressed in group theory notation as

$$\Gamma_{opt} = A_1 \oplus 2A_2 \oplus 3B_1 \oplus 3B_2 \oplus 6E \tag{14.7}$$

Equation (14.7) is interpreted in the following way. The zone-center phonon spectrum in the chalcopyrite-type structure contains one A_1 symmetry mode, two A_2 modes, and so on. By convention, double-degenerate modes are designed by the letter E, while letters A and B are reserved to totally symmetric and antisymmetric nondegenerate vibrational modes, respectively. Therefore, in this case, the 21 vibrational modes are grouped into nine nondegenerate modes and six double-degenerate modes. By analyzing in more detail the symmetry of each vibration, it can

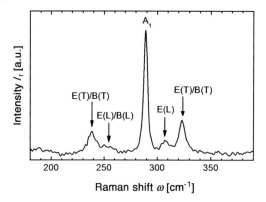

Figure 14.3 Raman spectrum of a CuInS$_2$ film, showing characteristic chalcopyrite-type Raman bands.

be shown that the two A_2 modes are not active by Raman, while B_2- and E-type modes involve polar displacements, resulting in LO–TO splitting of these modes. The experimental spectrum of a representative chalcopyrite compound, CuInS$_2$, is shown in Figure 14.3.

The spectrum of CuInS$_2$ is characterized by a dominant band at 290 cm^{-1}, corresponding to the A_1 symmetry phonon mode. This is in fact a common characteristic to all chalcopyrite-type materials, for which the A_1 mode involving only the displacement of the lighter anions while the heavier cations remain at rest yields higher scattering intensity. On the other hand, due to the relatively poor intensity of the other bands and the fact that they overlap, their identification necessarily requires a more accurate spectral analysis based on the use of polarized light and special light excitation/collection geometries. In this way, it is possible to figure out particular experimental configurations in which Raman scattering is only allowed for selected symmetry phonons. The dependence of the scattering intensity with the polarization and the scattering geometry is a consequence of the symmetry characteristics of the derivative susceptibility tensor, and gives rise to configurations in which certain symmetry-type phonons are forbidden ("selection rules"). Even though these methodologies are extremely useful for fundamental studies, they have limited applicability in the case of thin-film polycrystalline materials, in which multiple randomly oriented crystals are excited simultaneously.

Once the basic spectral properties of chalcopyrite-type materials have been introduced, we will consider now the case of more complex structures in which multiple crystalline phases may occur. Figure 14.4 shows the Raman spectrum or a CuInSe$_2$ precursor grown by electrodeposition. Single-step electrodeposition of CuInSe$_2$ followed by recrystallization under sulfurizing conditions allows the fabrication of high crystalline quality CuIn(S,Se)$_2$ absorbers [9]. Electrodeposition-based processes have interest because of their higher potential to achieve a significant reduction in fabrication costs in relation to standard physical vapor deposition (PVD) processes.

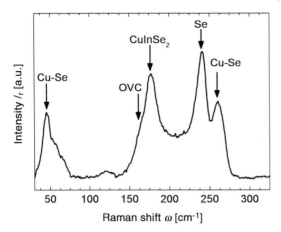

Figure 14.4 Raman spectrum of a CuInSe$_2$ electrodeposited precursor, in which different secondary phases are identified.

This spectrum shows the presence of modes characteristic of both Cu-rich and Cu-poor secondary phases, which coexist with the desired chalcopyrite-type phase [10]. The existence of chalcopyrite-type CuInSe$_2$ can be inferred from the observation of a characteristic A_1 band at 173 cm^{-1}. Nevertheless, additional bands in the spectra appear, in relation to the precipitation of elementary Se and binary Cu$_x$Se phases, as indicated in Figure 14.4. Presence of these phases is related to the used electrodeposition conditions that lead to precursor layers with overall Se-rich and Cu-rich composition, and they play a significant role in the recrystallization of the layers during the sulfurizing step. Moreover, special attention should be paid to the low-frequency shoulder observable in the region of 153 cm^{-1}. This shoulder is associated with the presence of an ordered vacancy compound phase (OVC), with chemical composition CuIn$_3$Se$_5$ [11]. A more detailed analysis of the crystallographic structure of this OVC phase reveals that it is closely related to the chalcopyrite-type structure. Essentially, the OVC structure can be derived by randomly introducing in the chalcopyrite-type lattice complex defects in the form of (In$_{Cu}$) antisites and Se vacancies (V_{Se}^{2-}), and imposing the preservation of the charge neutrality in the lattice. For example, the CuIn$_3$Se$_5$ OVC structure may be readily obtained by introducing in a defect-free chalcopyrite lattice a combination of two In$_{Cu}$ and one V_{Se}, for every three molecular formula units (CuInSe$_2$). Likewise, other combinations of defects in the primary chalcopyrite structure lead to a variety of OVC structures characterized by different stoichiometries. By using group theory methods combined with simple lattice dynamic models, it is possible to show that the spectrum of the OVC structure should be also characterized by a totally symmetric A_1 band, in which the displacement of the anions is equivalent to those involved in the A_1 mode of the chalcopyrite-type structure. Nevertheless, in good agreement with the experimental findings, the expected frequency of the A_1 band for the OVC phase is lower, due to the fact that the introduced (V_{Se}^{2-}) vacancies lead to an effective weakening of the average anion–cation bonding constant.

Another example of identification of secondary phases can be found in the CdTe system. In this case, Raman scattering is very sensitive to the presence of Te aggregates that can be present in CdTe thin films depending on their growth and processing conditions [12]. Tellurium has characteristic vibrational modes at 123 (E) and 141 cm^{-1} (A_1). The frequency of the A_1 mode coincides with that of the TO mode from CdTe, that has also an LO mode at 167 cm^{-1}. Observation of the modes related to Te aggregates in the film is favored by the fact that Te has at least two orders of magnitude larger scattering cross section compared to that of CdTe.

In the previous examples, Raman spectroscopy provided particularly useful information in relation to the identification of phases present in the film. In particular, detection of OVC phases in chalcopyrites constitutes a good alternative to X-Ray diffraction analysis, due to the strong overlapping between the different X-Ray diffraction peaks from the different crystalline structures. Furthermore, in many occasions the formation of secondary phases occurs primarily at the interface regions, which makes its detection by means of other techniques not necessarily straightforward. In relation to this aspect, we will consider now a second example in which the high spatial resolution provided by micro-Raman spectroscopy allowed investigating the chemical characteristics of the interface of a photovoltaic absorber. Figure 14.5 shows a spectrum measured with a Raman microprobe with the laser spot directly focused on the cross section of a CuInS$_2$ film grown onto Mo-coated glass. The spectrum measured with the laser probe located on the region close to the CuInS$_2$/Mo interface shows the presence of two bands at about 380 and 410 cm^{-1}, which correspond to the main vibrational modes characteristic of the hexagonal structure of MoS$_2$ [13].

MoS$_2$ and MoSe$_2$ interfacial phases have been observed at the interface region between the Mo back contact and the absorber layer in both CuInS$_2$- and CuInSe$_2$-based solar cells. Presence of such phases has a relevant effect on the characteristic of the devices, as they prevent from the formation of a Schottky barrier at the back region

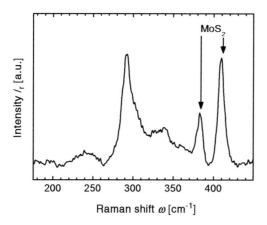

Figure 14.5 Cross-section Raman spectrum of a CuInS$_2$ film grown onto Mo-coated glass, revealing the formation of a MoS$_2$ phase at the CuInS$_2$/Mo interface.

of the cell. On the other hand, they might be also responsible in some cases for film adhesion failure [13–15].

14.5.2
Evaluation of Film Crystallinity

The crystalline quality of a photovoltaic absorber is important in order to ensure its adequate optoelectronic properties. The presence of crystalline defects may result in changes in the doping levels, density of states (DOS), and electronic transport properties. Crystalline defects embrace several modifications of the crystalline lattice of the material, which result in the breakage of the symmetry of the crystal. Point defects, such as vacancies, interstitials, or antisites, are zero-dimensional defects affecting only ions in the lattice. Other crystalline defects, such as dislocations or grain interfaces, are physically extended over one- or two-dimensions in the space.

Some crystalline defects result in characteristic changes in the Raman spectrum. In particular, the introduction of impurities in a crystal often results in characteristic localized vibrational modes (LVMs) that may be active by Raman. This is the case, for example, for N atomic impurities in GaAs lattice [16], which give rise to an LVM at about 475 cm^{-1}. Furthermore, the relative intensity of the LVM can be used to determine the impurity concentration in the host lattice. Usually, impurity concentrations above 1% are required in order to be able to resolve LVM.

Raman spectroscopy is also extremely sensitive to lattice disorder. For example, variations in the cation arrangement in ternary tetrahedral compounds often result in additional Raman bands, which are often described as "disorder activated modes." In some cases, these bands arise from noncenter phonons, which become active by the relaxation of the quasicrystalline momentum conservation law, caused by the breakage of the crystalline symmetry. In general, either point or extended defects can be responsible for such activation of noncenter phonons because of the breakage of the momentum conservation law. These effects determine a characteristic change in the shape of the main peaks in the spectra, with the appearance of a shoulder at the low- or high-frequency side that is due to the contribution of the modes at the vicinity of the center of the Brillouin zone [17–20]. The location of this contribution in relation to the frequency of the Raman peak is determined by the dispersion curves of the phonons involved in the process in the vicinity of the $q=0$ point: in Si, this contribution appears at the low-frequency side of the Raman peak [19], while semiconductors as CuInS$_2$ and CuInSe$_2$ show this disorder-induced contribution at the high-frequency side of the main Raman mode [20]. The resulting shape of the band can be modeled assuming a correlation length model that is described in Section 14.5.4. Estimation of the correlation length allows obtaining a quantitative estimation of the degree of disorder in the crystals related to the presence of a high density of defects.

Unfortunately, in many cases Raman spectroscopy cannot be used to quantify or even identify the presence of specific defects. Nevertheless, the analysis of the Raman spectra of a polycrystalline material always provides useful information in relation to the film crystallinity. When either point or extended defects are introduced into the crystal lattice, the density of phonon states is also modified with respect to the perfect crystal lattice, resulting in a reduction in the intensity of the Raman bands. However, one may keep in mind that even though the reduction in the intensity of the Raman bands in the spectrum of a thin film may be indicative of a higher density of defects, the band intensity also strongly depends on the experimental conditions. For this reason, the absolute intensity of the Raman bands is rarely used as a crystalline quality indicator. Alternatively, the analysis of the Raman bandwidth provides a much more reliable method to evaluate the defect density in the material. The bandwidth of a Raman band is inversely proportional to the phonon lifetime, which obviously depends on the density of defects that promote phonon scattering. Therefore, a thin-film material with a high density of defects is characterized by the increased width of their bands, with respect to the natural bandwidth of the defect-free crystal. In the case of chalcopyrite-type $CuInS_2$ and $CuInSe_2$ photovoltaic absorbers, the increased bandwidth of the A_1 Raman band has been confirmed to be correlated with a degradation of the film crystallinity, as corroborated by a comparative study using X-ray diffraction and electron microscopy [13, 21]. Furthermore, a correlation between the full width at half maximum of the A_1 band and photovoltaic parameters, such as the open-circuit voltage and the fill factor [22, 23], was found.

14.5.3
Chemical Analysis of Semiconducting Alloys

Alloying obviously leads to changes in the lattice dynamics of the crystal, which manifest in the Raman spectra. In the simplest case, alloying results in a gradual change in the force constants and effective mass that causes a Raman band to shift its frequency. Although it is not always the case, the shift of the Raman band is often linear with the degree of alloying. Even if it is not linear, the frequency of the Raman band can still be used to unambiguously determine the composition of the alloy. $Cu(In,Ga)Se_2$ and $Cu(In,Ga)S_2$ are representative examples of this mode behavior [24–27]. In these cases, the A_1-dominant band in the spectrum is observed to shift linearly from the reference position of the $CuInSe_2(S_2)$ compound ($CuInSe_2$, 173 cm^{-1}; $CuInS_2$, 290 cm^{-1}) to the frequency of the band at the $CuGaSe_2(S_2)$ crystal ($CuGaSe_2$, 183 cm^{-1}; $CuGaS_2$, 310 cm^{-1}). Nevertheless, in some cases, two bands instead of one single band are observed for a given vibrational mode. In such a case, the alloy is said to present a "bimodal behavior" – in opposite to the previous "one-mode behavior." A representative photovoltaic material presenting this behavior is $CuIn(S,Se)_2$. In this case, the A_1 band splits into two modes, which are referred to as "Se-like" and "S-like" bands, since they appear closer to the respective positions of the reference ternary compounds at 173 and 290 cm^{-1}, respectively [28]. Moreover, the dependence of the frequency of both modes with the degree of alloying is significantly different. While the "Se-like" mode is clearly affected by the variation of the Se/S ratio in the quaternary alloy, the "S-like" band is nearly unaffected by the change in the film stoichiometry. This can be seen in Figure 14.6, which corresponds to the plot of Raman spectra from S-rich alloys with different composition.

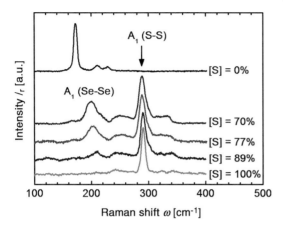

Figure 14.6 Raman spectrum from CuIn(S,Se)$_2$ alloys with different composition. For comparison, spectra from CuInSe$_2$ ([S] = 0%) and CuInS$_2$ ([S] = 100%) layers are also shown. Notice that the Se-like mode significantly shifts with the alloy composition, while the frequency of the S-like band remains nearly unaffected.

The dependence of these modes on the degree of alloying can be used for the quantitative determination of the chemical composition of thin-film quaternary photovoltaic absorbers. In addition, it provides an effective means of establishing a compositional mapping of the film, and allows also investigating the existence of in-depth resolved composition inhomogeneities. This can be done by Raman microprobe measurements performed with the laser probe focused at different positions from the cross section of the layers [29–31]. Figure 14.7 shows an example of such measurements performed on a CuIn(S,Se)$_2$ layer synthesized with a gradual change in the S to Se content ratio [31]. In Figure 14.7, the spectra from the surface region are characterized by a dominant S-like mode, which points out the existence of an S-rich surface region. Moving the laser spot toward the back region leads to a decrease in this peak and a corresponding increase in the Se-like mode, which correlates with the existence of a Se-rich alloy at the back region of the layer. In addition, the spectra measured close to the interface with the back Mo contact show the peaks characteristic of the interfacial MoSe$_2$ phase.

This kind of in-depth resolved measurements is of interest because in this case there is no need for a special preparation of the specimens before their observation. However, their in-depth resolution is limited by the relatively large size of the laser spot (of the order of 0.7–1 μm, depending on the objective numerical aperture and the excitation wavelength). For a compositionally inhomogeneous layer, this implies that adjacent regions of different composition in the layer might contribute to the spectrum measured with the laser spot at a given position. Higher in-depth resolution (about 400 nm) has been achieved by the combined AFM/Raman microprobe mapping of cross sections of samples prepared in the form of standard TEM specimens [32]. Another option is the use of an Ar$^+$ ion beam for the selective sputtering of the layer to different depths [33]: this allows us to obtain a depth

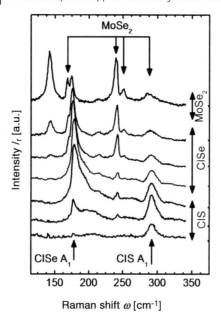

Figure 14.7 Micro-Raman spectra measured with the laser spot at different positions through the cross section of a CuIn(S,Se)$_2$ layer with graded composition on an Mo-coated glass substrate.

resolution of the order of 100 nm because of the high light absorption of these semiconductors. However, in this case, the choice of the sputter conditions requires special care in order to minimize the presence of damage-induced effects in the spectra.

14.5.4
Nanocrystalline and Amorphous Materials

In the previous sections, we have restricted our discussion to the analysis of crystalline and polycrystalline materials. For practical purposes, a defect-free nanocrystal with diameters above 100–200 nm may be regarded as an infinite crystal. This is due to the fact that the phonon frequency and Raman scattering cross section are mostly dependent on the short-range ordering of the crystal lattice, up to a few atomic neighbors. In this sense, it is worth to highlight that Raman spectroscopy is much more sensitive to the local environment of the atoms in the crystal than X-ray diffraction techniques. As discussed in Section 14.5.2, a remarkable consequence of this characteristic is the activation of LVM caused by the introduction of impurities in the lattice.

When the effective correlation length over which the material can be considered crystalline is further reduced over this limit (100–200 nm), spectral effects become measurable. In the case of nanocrystalline materials, the phonon correlation length may be physically limited by the size of the crystal itself, but more generally, it may be restricted by the presence of crystalline defects resulting in a higher phonon scattering probability. When the phonon correlation length can no longer be considered infinite, the breakage of the crystalline symmetry leads to a relaxation of the quasi-momentum conservation law, and the activation of noncenter phonon modes with $\vec{q} \neq 0$. As a result, the band shape of the Raman bands is modified, resulting in an effective asymmetric broadening and shift with respect to the crystalline band. In this case, the modeling of the spectrum allows estimating the effective correlation band of the material, which in the case of nanostructured films, can be associated with the grain size.

Phonon confinement effects in nanocrystalline Si have been extensively studied [17, 19]. Moreover, the Raman spectrum of nanocrystalline Si can be accurately modeled based on the simplifying assumption that the phonon dispersion curve can be described by a parabola. Then, the Raman-band shape of the Si band may be expressed according to the equation

$$I(\omega) \propto \int_0^{\frac{2\pi}{a_0}} \frac{e^{-\frac{q^2 L^2}{8}} 4\pi q^2 dq}{[\omega - \omega(q)]^2 + [\Gamma_0/2]^2} \tag{14.8}$$

which for any particular case allows extracting the correlation length (L) by fitting the experimental spectrum.

The validity of the correlation length model is conditioned by the assumption that the material has some degree of crystalline ordering. When the phonon correlation length is further reduced, additional considerations must be taken into account. In particular, when a solid material progressively loses its crystalline order, any phonon can contribute to inelastic light scattering. Therefore, the resulting spectrum is conditioned by the phonon DOS, which determines which interactions are more probable. In general, the spectra of amorphous materials are characterized by the presence of broad bands resembling the phonon DOS.

An important practical application of Raman spectroscopy in the field of photovoltaic materials deals with the determination of the amorphous fraction in Si thin films. This parameter has a significant importance from the technological point of view, since it affects the open circuit voltage of the cell, and at the same time, is strongly dependent on the film-deposition conditions. Figure 14.8 shows characteristic Raman spectra from layers exhibiting a variable content of amorphous (a-Si) and microcrystalline (μc-Si) phases [19].

As may be appreciated in Figure 14.8, the presence of an a-Si phase in the films leads to a broad band in the region of 480 cm^{-1}, for which the Si phonon DOS has a maximum. The integral intensity of this band in relation to the intensity of the crystalline peak at 520 cm^{-1} provides a direct indication of the amorphous/microcrystalline ratio. Moreover, the following model may be used for quantification purposes:

$$\phi_c = \frac{I_c}{I_c + \gamma I_a} \tag{14.9}$$

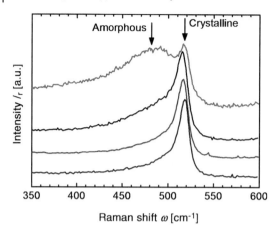

Figure 14.8 Raman spectra of Si films containing a mixture of amorphous and microcrystalline phases.

where the parameter γ is an empirical constant, ϕ_c is the crystalline/amorphous fraction, and I_c and I_a are the intensity of the crystalline and amorphous bands, respectively.

14.5.5
Evaluation of Stress

In a crystalline material, the frequency of the Raman bands is a characteristic inherent to the crystalline structure, and the chemical composition of the material. However, the phonon frequency may be modified by means of introducing an external force able to modify the internal energy of the crystal. When a mechanical stress is applied to a crystal, it results in microscopic atomic displacements, resulting in a net strain. Such strain modifies the interatomic distance, and consequently, the ionic and covalent bonding forces between neighbor atoms. This mechanism is responsible for the variation of the phonon frequency. In principle, even though one should take into account the tensor nature of stress and strain magnitudes, a compressive stress results in a reduction in the interatomic distance, increasing the interatomic forces, which leads to a positive ("blue") shift of the phonon frequency. When a tensile stress is applied, the phonon frequency is decreased (or "red shifted").

The existence of stress in thin-film materials can be assessed by Raman spectroscopy by means of accurately determining the frequency of the Raman modes. In the case of crystalline Si, and for a uniaxial or biaxial stress, the frequency shift can be related to the magnitude of the stress by the following expression [34]:

$$\sigma\,(\text{MPa}) = -434\,\Delta\omega\,(\text{cm}^{-1}) \tag{14.10}$$

Notice that the magnitude of the Raman shift associated with typical stress values is in the range of 1 cm^{-1}. Therefore, the performance of these measurements requires a

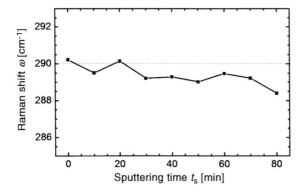

Figure 14.9 Frequency of the A_1 chalcopyrite-type characteristic mode in a $CuInS_2$ film after different sputtering times.

high-resolution spectrometer. Furthermore, special attention should be paid to the stabilization of the sample temperature, and the avoidance of laser-induced local heating. In the case of Si, the magnitude of the thermally induced Raman shift of the band is in the range of $0.024\,cm^{-1}/K$, which leads to the fact that a local temperature increase of just $20\,°C$ may lead to an error in the determination of the stress magnitude of about 210 MPa.

We shall consider an example of the application of this method to the evaluation of stress in a thin photovoltaic material. Figure 14.9 shows the variation of the phonon frequency of the A_1 chalcopyrite-type band in a $CuInS_2$ photovoltaic absorber, obtained by sequential Raman measurements performed after sputtering to different depths with an Ar^+ ion gun.

In Figure 14.9, a progressive red shift of the band is observed toward the back interface. This shift was related to the existence of residual stress in the layer, as confirmed by X-ray diffraction analyses, possibly caused by the particular growth conditions and the difference between the thermal expansion coefficients of the absorber and the metallic back contact. The maximum band was estimated from the Raman shift to correspond to a tensile stress in the range of 360 MPa [26].

14.6 Conclusions

This chapter gives an overview of the capabilities of Raman spectroscopy for the advanced characterization of thin-film photovoltaic materials. The applicability of the technique for the microstructural and chemicophysical analysis of the films is illustrated by the description of state-of-the-art and recent examples that corroborate the potential of Raman scattering for these applications. This includes features related to the identification of crystalline structures and secondary phases, the evaluation of the film crystallinity and presence of defects, the measurement of the composition of semiconductor alloys, the quantification of the amorphous to microcrystalline fraction in partially amorphous films, and the analysis of stress. Recent developments of Raman microprobe based techniques for the advanced in-depth resolved characterization of the films with high-depth resolution are also described. The relevant impact of all these features on the characteristics of the solar cells gives a strong interest to Raman spectroscopy not only for the fundamental characterization of the films, but also as a nondestructive process monitoring tool applicable in a production line of thin-film cells and modules.

References

1 Cardona, M. (1975) *Light Scattering in Solids*, vol. I, Springer, Heidelberg, New York.

2 Ferraro, J.R., Nakamoto, K., and Brown, Ch.W. (2003) *Introductory Raman Spectroscopy*, 2nd edn, Academic Press, New York.

3 Smith, E. and Dent, G. (2005) *Modern Raman Spectroscopy: A Practical Approach*, John Wiley & Sons, Inc., New York.

4 Jawhari, T. (2000) Micro-Raman spectroscopy of the solid state: applications to semiconductors and thin films. *Analusis*, **28**, 15–21.

5 Yu, P. and Cardona, M. (1996) *Fundamentals on Semiconductors*, Springer, Berlin.

6 Paillard, V., Puech, P., Laguna, M.A., Temple-Boyer, P., Caussat, B., Couderc, J.P., and de Mauduit, B. (1998) Resonant Raman scattering in polycrystalline silicon thin films. *Appl. Phys. Lett.*, **73**, 1718–1720.

7 Álvarez-García, J., Barcones, B., Pérez-Rodríguez, A., Romano-Rodríguez, A., Morante, J.R., Janotti, A., Wei, S.-H., and Scheer, R. (2005) Vibrational and crystalline properties of polymorphic $CuInC_2$ (C=Se, S) chalcogenides. *Phys. Rev. B*, **71**, 054303–054312.

8 Rincón, C. and Ramírez, F.J. (1992) Lattice vibrations of $CuInSe_2$ and $CuGaSe_2$ by Raman microspectrometry. *J. Appl. Phys.*, **72**, 4321–4324.

9 Lincot, D., Guillemoles, J.F., Taunier, S., Guimard, D., Sicx-Kurdi, J., Chaumont, A., Roussel, O., Ramdani, O., Hubert, C., Fauvarque, J.P., Bodereau, N., Parissi, L., Panheleux, P., Fanouillere, P., Naghavi, N., Grand, P.P., Benfarah, M., Mogensen, P., and Kerrec, O. (2004) Chalcopyrite thin film solar cells by electrodeposition. *Sol. Energy*, **77**, 725–737.

10 Izquierdo-Roca, V., Saucedo, E., Ruiz, C.M., Fontané, X., Calvo-Barrio, L., Álvarez-García, J., Grand, P.-P., Jaime-Ferrer, J.S., Pérez-Rodríguez, A., Morante, J.R., and Bermúdez, V. (2009) Raman scattering and structural analysis of electrodeposited $CuInSe_2$ and S-rich quaternary $CuIn(S, Se)_2$ semiconductors for solar cells. *phys. status solidi (a)*, **206**, 1001–1004.

11 Xu, Ch.-M., Xu, X.-L., Xu, J., Yang, X.-J., Zuo, J., Kong, N., Huang, W.-H., and Liu, H.-T. (2004) Composition dependence of the Raman A_1 mode and additional mode in tetragonal Cu–In–Se thin films. *Semicond. Sci. Technol.*, **19**, 1201–1206.

12 Rai, B.K., Bist, H.D., Katiyar, R.S., Chen, K.-T., and Burger, A. (1996) Controlled micro oxidation of CdTe surface by laser irradiation: A micro-spectroscopy study. *J. Appl. Phys.*, **80**, 477–481.

13 Álvarez-García, J., Pérez-Rodríguez, A., Romano-Rodríguez, A., Morante, J.R., Calvo-Barrio, L., Scheer, R., and Klenk, R. (2001) Microstructure and secondary phases in coevaporated $CuInS_2$ films: Dependence on growth temperature and chemical composition. *J. Vac. Sci. Technol. A*, **19**, 232–239.

14 Kohara, N., Nishiwaki, S., Hashimoto, Y., Negami, T., and Wada, T. (2001) Electrical properties of the $Cu(In,Ga)Se_2/MoSe_2/$Mo structure. *Sol. Energy Mater. Sol. C.*, **67**, 209–215.

15 Abou-Ras, D., Kostorz, G., Bremaud, D., Kälin, M., Kurdesau, F.V., Tiwari, A.N., and Döbeli, M. (2005) Formation and

characterisation of MoSe$_2$ for Cu(In,Ga)Se$_2$ based solar cells. *Thin Solid Films*, **480–481**, 433–438.

16 Panpech, P., Vijarnwannaluk, S., Sanorpim, S., Ono, W., Nakajima, F., Katayama, R., and Onabe, K. (2007) Correlation between Raman intensity of the N-related local vibrational mode and N content in GaAsN strained layers grown by MOVPE. *J. Cryst. Growth*, **298**, 107–110.

17 Fauchet, P.M. and Campbell, I.H. (1988) Raman spectroscopy of low-dimensional semiconductors. *Crit. Rev. Solid State*, **14**, S79–S101.

18 Gouadec, G. and Colomban, P. (2007) Raman spectroscopy of nanomaterials: How spectra relate to disorder, particle size and mechanical properties. *Prog. Cryst. Growth Charact.*, **53**, 1–56.

19 Garrido, B., Pérez-Rodríguez, A., Morante, J.R., Achiq, A., Gourbilleau, F., Madelon, R., and Rizk, R. (1998) Structural, optical and electrical properties of nanocrystalline silicon films deposited by hydrogen plasma sputtering. *J. Vac. Sci. Technol. B*, **16**, 1851–1859.

20 Camus, C., Rudigier, E., Abou-Ras, D., Allsop, N.A., Unold, T., Tomm, Y., Schorr, S., Gledhill, S.E., Köhler, T., Klaer, J., Lux-Steiner, M.C., and Fischer, Ch.-H. (2008) Phonon confinement and strain in CuInS$_2$. *Appl. Phys. Lett.*, **92**, 101922–101925.

21 Izquierdo-Roca, V., Pérez-Rodríguez, A., Morante, J.R., Álvarez-García, J., Calvo-Barrio, L., Bermudez, V., Grand, P.P., Parissi, L., Broussillon, C., and Kerrec, O. (2008) Analysis of S-rich CuIn(S,Se)$_2$ layers for photovoltaic applications: Influence of the sulfurisation temperature on the crystalline properties of electrodeposited and sulfurised CuInSe$_2$ precursors. *J. Appl. Phys.*, **103**, 123109–123116.

22 Rudigier, E., Enzenhofer, T., and Scheer, R. (2005) Determination of the quality of CuInS$_2$-based solar cells combining Raman and photoluminescence spectroscopy. *Thin Solid Films*, **480–481**, 327–331.

23 Izquierdo, V., Pérez-Rodríguez, A., Calvo-Barrio, L., Álvarez-García, J., Morante, J.R., Bermudez, V., Ramdani, O., Kurdi, J., Grand, P.P., Parissi, L., and Kerrec, O. (2008) Raman scattering microcrystalline assessment and device quality control of electrodeposited CuIn(S,Se)$_2$ based solar cells. *Thin Solid Films*, **516**, 7021–7025.

24 Tanino, H., Deai, H., and Nakanishi, H. (1993) Raman spectra of CuGa$_x$In$_{1-x}$Se$_2$. *Jpn. J. Appl. Phys.*, **32**, 32–33, 436–438.

25 Papadimitriou, D., Esser, N., and Xue, C. (2005) Structural properties of chalcopyrite thin films studied by Raman spectroscopy. *phys. status solidi (b)*, **242**, 2633–2643.

26 Álvarez-García, J. (2002) Characterisation of CuInS$_2$ Films for Solar Cell Applications by Raman spectroscopy, http://www.tdx.cesca.es/TDX-0122103-094011/index.html, University of Barcelona.

27 Caballero, R., Izquierdo-Roca, V., Fontané, X., Kaufmann, C.A., Álvarez-García, J., Eicke, A., Calvo-Barrio, L., Pérez-Rodríguez, A., Schock, H.W., and Morante, J.R. (2010) Cu deficiency in multi-stage co-evaporated Cu(In,Ga)Se$_2$ for solar cells applications: Microstructure and Ga in-depth alloying. *Acta Mater.*, **58**, 3468–3476.

28 Bacewicz, R., Gebicki, W., and Filipowicz, J. (1994) Raman scattering in CuInS$_{2x}$Se$_{2(1-x)}$ mixed crystals. *J. Phys.: Condens. Matter*, **6**, L777–L780.

29 Takei, R., Tanino, H., Chichibu, S., and Nakanishi, H. (1996) Depth profiles of spatially-resolved Raman spectra of a CuInSe-based thin-film solar cell. *J. Appl. Phys.*, **79**, 2793–2795.

30 Witte, W., Kniese, R., Eicke, A., and Powalla, M. (2006) Influence of the Ga content on the Mo/Cu(In,Ga)Se$_2$ interface formation. Proceedings of the Fourth IEEE World Conference on Photovoltaic Energy Conversion, IEEE, Piscataway, NJ, pp. 553–556.

31 Izquierdo-Roca, V., Fontané, X., Álvarez-García, J., Calvo-Barrio, L., Pérez-Rodríguez, A., Morante, J.R., Ruiz, C.M., Saucedo, E., and Bermúdez, V. (2009)

Electrochemical synthesis of $CuIn(S,Se)_2$ alloys with graded composition for high efficiency solar cells. *Appl. Phys. Lett.*, **94**, 061915–061918.

32 Schmid, T., Camus, C., Lehmann, S., Abou-Ras, D., Fischer, Ch.-H., Lux-Steiner, M.C., and Zenobi, R. (2009) Spatially resolved characterization of chemical species and crystal structures in $CuInS_2$ and $CuGa_xSe_y$ thin films using Raman microscopy. *phys. status solidi (a)*, **206**, 1013–1016.

33 Fontané, X., Izquierdo-Roca, V., Calvo-Barrio, L., Pérez-Rodríguez, A., Morante, J.R., Guettler, D., Eicke, A., and Tiwari, A.N. (2009) Investigation of compositional inhomogeneities in complex polycrystalline $Cu(In,Ga)Se_2$ layers for solar cells. *Appl. Phys. Lett.*, **95**, 261912–261915.

34 Wu, X., Yu, J., Ren, T., and Liu, L. (2007) Micro-Raman spectroscopy measurement of stress in silicon. *Microelectron. J.*, **38**, 87–90.

15
Soft X-Ray and Electron Spectroscopy: A Unique "Tool Chest" to Characterize the Chemical and Electronic Properties of Surfaces and Interfaces

Marcus Bär, Lothar Weinhardt, and Clemens Heske

15.1
Introduction

In view of the complexity of thin-film solar cells, which comprise a multitude of layers, interfaces, surfaces, elements, impurities, and so on, it is critically important to characterize and understand the chemical and electronic properties of these components. As a first step, systems (that have generally been *empirically* optimized) need to be characterized, and a detailed understanding of the inner workings and limitations (optimization barriers) needs to be achieved. In a second step, concepts and process modifications to overcome the barriers need to be developed and implemented, and the results obtained in the first step form the experimental basis to help in such "brainstorming" processes. In step three, finally, the characterization approaches will need to ascertain whether the implemented process modifications were successful and whether, from a chemical and electronic perspective, the modifications indeed modified the targeted aspects of the chemical and electronic structure derived in the first step.

To fulfill such demands, a combination of various techniques, illuminating different aspects of the electronic and chemical structure, needs to be employed. It is the purpose of this chapter to discuss and demonstrate the power of a "tool chest" comprising soft X-ray and electron spectroscopies to illuminate the electronic and chemical properties from a variety of different perspectives. These techniques are in part very old (in fact, almost as old as the initial discovery of X-rays by W.C. Röntgen in 1895) and in part brand new, with the to-date most comprehensive approach to study the electronic bulk structure of a material (the so-called resonant inelastic (soft) X-ray scattering (RIXS) map approach [1–5]) being developed in the late 2000s.

The techniques in the tool chest require unique experimental environments. Some of them can be performed in ultra-high vacuum surface-science systems in a single-investigator laboratory; others require high-brightness X-ray radiation from a third-generation synchrotron radiation source. These techniques need to be coupled with optimally selected sample series: Either custom-designed to address a specific question, from a world-record (or lab-record) batch to establish a benchmark picture,

Advanced Characterization Techniques for Thin Film Solar Cells,
Edited by Daniel Abou-Ras, Thomas Kirchartz and Uwe Rau.
© 2011 Wiley-VCH Verlag GmbH & Co. KGaA. Published 2011 by Wiley-VCH Verlag GmbH & Co. KGaA

taken directly from industrial preparation processes to shed light on the peculiarities of the "real world" systems, or, simply, from failed cells to investigate failure mechanisms (e.g., after prolonged stress testing). The answers that can be derived from the spectroscopic results will be as good as the questions that are being asked, and thus experiment planning is a further central component to a successful approach to study, understand, and optimize the chemical and electronic structure of materials, surfaces, and interfaces for thin-film solar cells.

The chapter is arranged as follows: Section 15.2 will describe the here-demonstrated characterization techniques of the "soft X-ray and electron spectroscopy tool chest," together with two short examples on band-gap determination and the RIXS map approach. In Section 15.3, a detailed example will be given that demonstrates the unique capabilities of the tool chest for characterizing the *chemical* properties of surfaces in thin-film solar cell devices. Section 15.4 describes examples of *electronic* interface structure investigations, and Section 15.5 summarizes the main messages of this chapter.

15.2
Characterization Techniques

When a soft (ca. 50–1500 eV) X-ray photon impinges on a material surface, it interacts with the sample by exciting an electron from an occupied to an unoccupied electronic state. Numerous secondary electronic processes can follow, including the excitation of secondary electrons by inelastic scattering of the initially excited electron and the relaxation of electrons with lower binding energy into the hole created by the initial excitation. The energy gained in the latter process can be used to emit a secondary photon or electron, which, in turn, can undergo inelastic scattering to produce further secondary electrons and/or photons. Additional processes occur when the sample is exposed to electrons (as opposed to photons), giving rise to a similar (but different) cascade of secondary processes. The spectroscopic techniques discussed in this chapter make use of these cascades of events in various different ways.

Figure 15.1 presents an overview of the various spectroscopic techniques that probe specific aspects of the interaction cascade. Table 15.1 lists the techniques together with the corresponding depth sensitivity and the probed electronic states. In photoelectron spectroscopy (PES [6]), the kinetic energy of the originally excited electron is probed. A typical PES spectrum thus records the number (intensity) of emitted electrons as a function of kinetic energy E_{kin}, which is then conveniently converted into a "binding energy" E_B scale by using Einstein's well-known equation (all energies relative to the Fermi energy)

$$E_{kin} = h\nu - E_B \tag{15.1}$$

When excited with soft X-rays from a lab source, PES is generally denoted as "XPS," while UV excitation gives spectra denoted as "UPS" spectra. Due to the short inelastic mean free path (IMFP) of electrons [7, 8] at kinetic energies attainable with soft X-ray excitation, PES spectra give a surface-sensitive view of the occupied core and valence

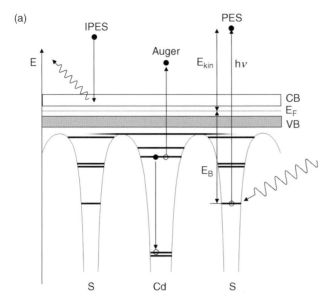

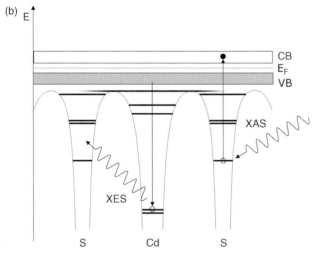

Figure 15.1 Schematic energy diagram of CdS, illustrating (a) the various electron ("PES," "IPES," and "Auger,") and (b) X-ray ("XES" and "XAS,") spectroscopies discussed in this chapter. For definition of the abbreviations, see text.

level states present at the surface of the sample. This, in turn, gives detailed insights into the electronic structure (e.g., the band structure of the valence band) and the chemical properties (e.g., the local chemical environment of a specific atomic species, giving rise to a distinct binding energy).

As PES, X-ray emission spectroscopy (XES) probes the *occupied* states and hence can be used to investigate the chemical and electronic structure of samples.

Table 15.1 Overview of the spectroscopic methods in the "soft X-ray and electron spectroscopy tool chest".

Measurement technique		Excited by	Detection of	Depth sensitivity	Probed electronic states
UPS	UV photoelectron spectroscopy	UV photons	Electrons	Surface	Occupied valence levels
XPS	X-ray photoelectron spectroscopy	X-ray photons	Electrons	Surface	Core levels
IPES	Inverse photoelectron spectroscopy	Electrons	UV photons	Surface	Unoccupied valence levels
XAES	X-ray excited Auger electron spectroscopy	X-ray photons	Electrons	Surface	Occupied valence levels and core levels
XES	X-ray emission spectroscopy	X-ray photons	X-ray photons	Near-surface bulk	Occupied valence levels and core levels
XAS	X-ray absorption spectroscopy	X-ray photons	X-ray photons	Near-surface bulk	Unoccupied valence levels and core levels
			Sample current	Surface	
			Auger electrons	Surface	
RIXS	Resonant inelastic X-ray scattering	X-ray photons	X-ray photons	Near-surface bulk	Occupied and unoccupied valence levels and core levels
UV-Vis		UV-visible photons	UV-visible photons	Bulk	Occupied and unoccupied valence levels

For XES, a core-level electron is excited by (soft) X-ray photons. The created core hole is filled by a lower binding-energy (most often valence) electron, causing the emission of a photon, which is detected by a wavelength-dispersive X-ray spectrometer. The probability for this process (compared with that of, e.g., direct PES) is very small, demanding the experiments to be performed at high-brightness synchrotron radiation sources. Because the radiative relaxation process involves a localized core hole, XES is an excellent tool to probe the *local* structure near select atomic species. In contrast to PES, XES is a photon-in–photon-out (PIPO) technique, and thus the information depth is determined by the $1/e$-attenuation length of soft X-rays in matter [9], which is material-dependent and roughly two orders of magnitude larger than the IMFP in PES. Consequently, XES probes the near-surface *bulk* of the sample with typical attenuation lengths of 20–200 nm.

The overwhelming majority of the core holes created by the initial excitation process (giving rise to PES) are filled by Auger transitions, as sketched in the center of the scheme in Figure 15.1a. In the Auger process (here denoted XAES for "X-ray-excited Auger electron spectroscopy"), a characteristic electron is emitted that reveals atom-specific information about the chemical state of the emitting atom, as in the case of PES in a surface-sensitive fashion. Due to different initial and final states involved in the Auger process compared with the PES process, the chemical sensitivity is complementary to that of the PES process. Auger transitions are generally named according to the principal quantum numbers of the three involved electronic levels. For example, sodium Auger transitions that involve a 1s core hole and 2p electrons for the relaxing and emitting electrons, respectively, are denoted as "Na KLL." For a complete description of the nomenclature see Ref. [10].

While PES, XES, and XAES probe occupied electronic states, X-ray absorption spectroscopy (XAS [11]) and inverse photoelectron spectroscopy [12] (IPES) investigate *unoccupied* electronic states. XAS is an element-specific technique, sensitive to the bonding environment and geometry. It involves the excitation of a core electron into an unoccupied state of the conduction band. For this, the excitation energy is scanned across the absorption edge of the investigated core level (CL), and secondary processes such as, for example, soft X-ray fluorescence due to radiative decay (fluorescence yield, FY), the current drawn by the sample to maintain charge neutrality (total electron yield, TEY), the number of electrons emitted above a certain threshold (partial electron yield, PEY), or the number of Auger electrons emitted in the corresponding Auger emission process (Auger electron yield, AEY) are used as detection channels. The information depth of XAS is determined by the detection mode (yield) and varies from bulk sensitivity to surface sensitivity in the order given above. If an entire XES spectrum is recorded for each (or a select number of) point(s) in an XAS spectrum near the absorption edge, the combination of XAS and XES is referred to as RIXS. A RIXS map, as shown in Figure 15.2 for CdS, thus gives the most complete spectroscopic information about both, the occupied and unoccupied electronic states. RIXS can be used to "test" calculated band structures [13–15].

In IPES, a slow electron is directed at the surface, and a UV photon that stems from the electronic relaxation into a lower unoccupied electronic state is detected. By scanning the initial electron energy and detecting UV photons of fixed photon energy, information on the conduction band (including, in particular, the position of the conduction band minimum (CBM) with respect to the Fermi energy) can be obtained. Alternatively, the electron energy can be kept constant, and the detected photon energy is varied using a grating spectrometer.

In combining PES and IPES, or XES and XAS, the band gap of the surface (PES/IPES) or near-surface bulk (XES/XAS) of the sample can be determined. Such data can be complemented by optical (UV-Vis) absorption, which gives insight into the bulk band-gap energy of the material. Apart from the differences in depth sensitivity, this band-gap information may also include additional aspects. For example, the PES/IPES band-edge positions can be given with respect to the Fermi energy, while this is not the case for XES/XAS and UV-Vis (because these approaches measure the energetic *difference* between two electronic states without relation to the Fermi energy). In XES/XAS, in contrast, the band gap is derived near the atoms of a specific species (due to the involvement of a CL), while PES/IPES and UV-Vis are

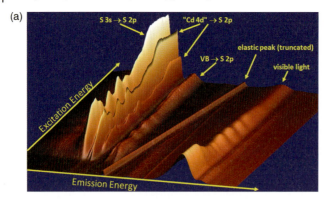

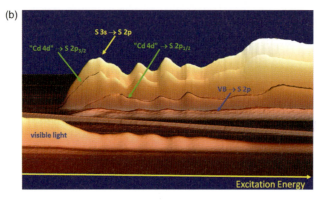

Figure 15.2 RIXS map of CdS. XES spectra are represented along the "emission energy" (detection energy) axis (a), while XAS spectra are represented along the "excitation energy" (beamline energy) axis (b). The peak heights correspond to photon intensity, as recorded by a soft X-ray spectrometer. All labels pertain to the relaxation transitions in CdS, giving rise to photon emission (the "visible light" line is due to visible fluorescence from the sample).

not element-specific. An example for a band-gap determination with these techniques is given in Figure 15.3, which illustrates the band-gap widening of chalcopyrite $Cu(In,Ga)(S,Se)_2$ thin-film solar cell absorbers toward the surface [16, 17], as seen in the band-gap energy progression as the experimental techniques become more surface-sensitive.

It should be noted that none of these techniques (or, for that matter, *any* experimental technique) exclusively probe(s) the ground state of the system. Rather, any experimental probe will influence the system, in our case by measuring the transition probability between an initial state and a final state under the influence of a quantum-mechanical operator (Fermi's golden rule) that describes the photon field (for PES, XES, XAS, UV-Vis, and IPES), the Coulomb interaction (for Auger transitions), or two-photon processes (described by the Kramers–Heisenberg formalism, for RIXS). While the ground state might be involved as the initial state

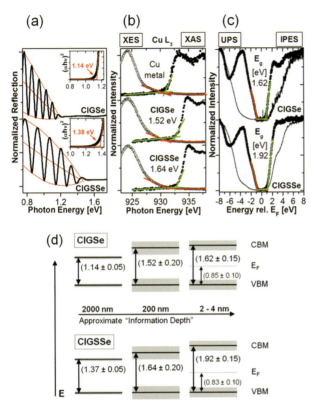

Figure 15.3 (a) Optical (UV-Vis) reflection spectra of Cu(In,Ga)Se$_2$ (CIGSe) and Cu(In,Ga)(S,Se)$_2$ (CIGSSe) thin films. Insets: E_g determination (± 0.05 eV) using an approximated absorption coefficient. (b) Cu L$_3$ XES and XAS spectra of CIGSe, CIGSSe samples, and metallic Cu. The lower bound for the ground-state band-gap values are determined by linear extrapolation of the leading edges (± 0.20 eV). (c) UPS and IPES for the as-introduced (thin solid lines) and cleaned samples (dots). Band-gap error bars are ± 0.15 eV. (d) Schematic representation of the derived band gaps as a function of the approximate experimental information depth (keeping in mind the exponentially decaying information profile of the various experimental techniques) – modified from Ref. [18].

(only for PES and RIXS), this is not the case in general. Even more importantly, it is not directly evident how the best approximation of the initial state can be obtained, especially in the presence of core (XES, XAS, RIXS) or valence (UV-Vis) excitons, but a variety of "best-practice" approaches has been established. These include the use of extrapolation of spectral leading-edges [19–21] and direct comparison with theoretically derived spectra [14, 22, 23] (i.e., using Fermi's golden rule or the Kramers–Heisenberg-formalism).

In the following two sections, the here-described techniques will be put to work. In Section 15.3, we will use the example of wet-chemical treatments of chalcopyrite thin-film solar cell absorbers to demonstrate the chemical sensitivity of the characterization techniques. In Section 15.4, we will discuss determinations of the electronic structure at thin-film solar cell interfaces.

15.3
Probing the Chemical Surface Structure: Impact of Wet Chemical Treatments on Thin-Film Solar Cell Absorbers

In this section, the prominent example of wet-chemical treatments of chalcopyrite thin-film solar cell absorbers will be used to demonstrate the capabilities of combining PES (here, XPS), XAES, and XES to reveal treatment-induced modifications of the absorber surface.

Today, the standard ZnO/buffer/Cu($In_{1-X}Ga_X$)(S_YSe_{1-Y})$_2$/Mo/glass structure of chalcopyrite thin-film solar cells contains a thin CdS buffer layer (~50 nm), usually prepared by chemical bath deposition (CBD). Record efficiencies both on a laboratory scale (>20.0% [24, 25]) and for large-area commercial modules (13.5% for 3459 cm^2 [26]) have thus been achieved. However, there is a strong impetus to replace CdS by other compounds, for ecological as well as economical reasons. The simple omission of the buffer layer, however, does not result in high-efficiency solar cells [27]. This situation can be improved by a (pre)treatment of the Cu($In_{1-X}Ga_X$)(S_YSe_{1-Y})$_2$ (CIGSSe) absorber in an aqueous solution of ammonium hydroxide and Cd^{2+}-ions (i.e., indentical to the CBD solution, but without the S source) prior to ZnO deposition. This leads to an increase in short circuit current density, open circuit voltage, and fill factor [28]. Such a Cd^{2+}/NH_3-treatment was first proposed for $CuInSe_2$ thin-film solar cell absorbers by Ramanathan et al. [27] and later successfully employed also by several other groups [27, 29, 30]. Even though Cd is still involved, this process is advantageous, since the amount of Cd-containing waste is largely reduced because, in contrast to the batch-type CBD, the solution can be reused repeatedly. In the following it will be shown how a combination of XPS, XAES, and XES is capable of giving a detailed insight into the modifications of the chemical surface structure induced by the Cd^{2+}/NH_3-treatment.

XPS survey spectra for the untreated CuIn(S,Se)$_2$ absorber and after Cd^{2+}-treatment at two concentrations (3 and 4.5 mM) are exemplarily shown in Figure 15.4. A wealth of information can be derived from such spectra. For example, the strong decrease of the intensity of the Na 1s line at ~1070 eV shows that surface Na (stemming from the glass substrate and diffusing through the CuIn(S,Se)$_2$/Mo during the absorber preparation at about 550 °C) is removed by the Cd^{2+}/NH_3-treatments. Furthermore, a slight reduction of the "native" O content at the surface of the (air-exposed) CuIn(S,Se)$_2$ absorbers is found when using Cd^{2+}-concentrations of up to 3 mM, while concentrations of 4.5 mM and above (not shown) lead to a pronounced *enhancement* of the O signal. After all Cd^{2+}/NH_3-treatments, Cd is detected on the CuIn(S,Se)$_2$ surface (see Ref. [31] for a more detailed discussion of peak intensities).

Before demonstrating how the soft X-ray and electron spectroscopies can help to identify the Cd species, first the surface composition of the untreated CuIn(S,Se)$_2$

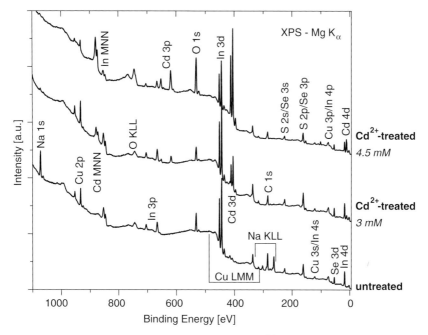

Figure 15.4 XPS survey spectra of an untreated and two Cd^{2+}/NH_3-treated $CuIn(S,Se)_2$ thin-film solar cell absorbers (3 and 4.5 mM Cd^{2+}-concentration). The different photoemission and Auger lines are identified as indicated.

absorber, especially the $S/(S + Se) = Y$ ratio, is determined. In Figure 15.5, the corresponding detail XPS spectrum of the S 2p and Se 3p region is shown. To compute the $[S]/([S] + [Se])$ ratio, the peak areas were determined by fitting them with Voigt line shapes (which is a convolution of Gauss- and Lorentz-shaped contributions and represents the best description of CL peaks for materials with a significant band gap) and a linear background, coupling the line widths of the two (spin-orbit split) S 2p and Se 3p lines, respectively. Furthermore, the spin-orbit splitting of S 2p (1.2 eV [32, 33]) and Se 3p (5.7 eV [32, 33]) and the 2:1 peak ratio (which corresponds to the multiplicity $2j + 1$) of the $j = 3/2$ and $j = 1/2$ components were kept constant. No peak area correction for the energy-dependence of the analyzer transmission and the IMFP of the electrons is necessary, since the S 2p and Se 3p electrons have very similar kinetic energies. The surface $[S]/([S] + [Se])$ ratio can then simply be derived from the fit-derived areas, corrected by photoionization cross sections from [34]. Here it is found to be 0.75 (i.e., significantly S-rich).

The Cd $M_{45}N_{45}N_{45}$ Auger emission line is shown in Figure 15.6. As indicated, the Cd "deposition" takes place in two different concentration regimes, up to 3 mM Cd^{2+}, and 4.5 mM Cd^{2+} and above. For the higher concentration regime, the Cd Auger line is shifted to lower kinetic energies, in parallel with a strong increase in peak intensity. The dashed and dotted lines represent a spectral decomposition of the 4.5 mM spectrum, revealing two different Cd species in the high-concentration

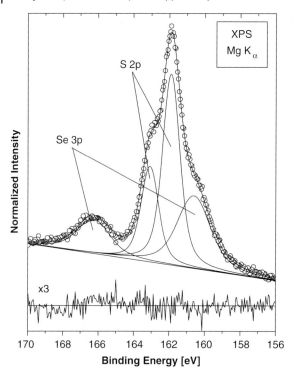

Figure 15.5 S 2p and Se 3p XPS detail spectrum of the investigated CuIn(S,Se)$_2$ thin-film solar cell absorber. The continuous lines show the corresponding fits, together with their (threefold magnified) residuum.

regime. In parallel, an enhancement in the O 1s and Cd 3d XPS intensity is found. Further information can be obtained by computing the "modified Auger parameter" [α^*(Cd) = binding energy (Cd 3d$_{5/2}$) + kinetic energy (Cd M$_4$N$_{45}$N$_{45}$)]. The value of 785.3 eV in the high-concentration regime suggests the formation of Cd(OH)$_2$ [32, 35]. As mentioned, a second Cd species in addition to the now-identified Cd(OH)$_2$ is present, as found for lower Cd^{2+} concentrations. While XPS and XAES can clearly distinguish the two different species and even give strong evidence for a Cd(OH)$_2$ deposition in the high-concentration regime, the identification of the second species is more difficult. XPS and XAES line positions and Auger parameters point toward CdS and/or CdSe, which are not easily distinguishable [32, 35]. To resolve this problem, an additional spectroscopic viewpoint is needed, and thus XES experiments for an untreated and a Cd^{2+}/NH$_3$-treated (1.5 mM) CuIn(S,Se)$_2$ absorber were performed.

The S L$_{2,3}$ XES spectra shown in Figure 15.7 reveal the changes in local chemical bonding undergone by the S atoms upon Cd^{2+}/NH$_3$-treatment (1.5 mM Cd^{2+} concentration). A detailed discussion of the various features included in this spectrum can be found in Ref. [36]. As already discussed in Section 15.2, XES, as a "PIPO" technique, is more bulk-sensitive than XPS and XAES. Here, the majority

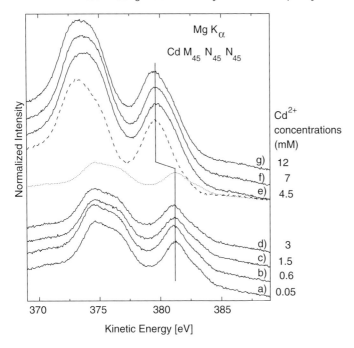

Figure 15.6 Cd $M_{45}N_{45}N_{45}$ XAES spectra of CuIn(S,Se)$_2$ absorbers after Cd^{2+}/NH$_3$-treatment with different Cd^{2+}-concentrations (0.05–12 mM, spectra a–g), indicating two different Cd species in the different concentration regimes. The dashed and dotted spectra below (e) are discussed in the text [31].

of the spectrum is associated with approximately the upper 100 nm of the CuIn(S,Se)$_2$ film. To get information about the changes induced by the Cd^{2+}/NH$_3$-treatment at the absorber *surface*, the spectrum of the untreated absorber (a) was subtracted from that of the Cd^{2+}/NH$_3$-treated film (b) in Figure 15.7, yielding the (magnified) difference spectrum (c). Comparison with a CdS reference (e) suggests the formation of CdS at the absorber surface, particularly by noting the two peaks at 150.4 and 151.5 eV (marked by the vertical lines (2)). These peaks correspond to valence electrons with strong Cd 4d character decaying into the S 2p$_{1/2}$ and 2p$_{3/2}$ core holes, respectively, and thus indicate S–Cd bonds. Additionally, the upper valence band of CdS can be observed at about 156 eV. Consequently, for spectrum (d) in Figure 15.7, a suitable amount (96%) of spectrum (a) was subtracted from (b). This subtraction assumes that 96% of the signal is due to S atoms in the CuIn(S,Se)$_2$-environment, and 4% to S atoms in a different environment. The derived difference spectrum (d) in Figure 15.7 now very closely resembles the CdS reference spectrum with the exception of a "dip" at 159 eV, which is associated with Cu 3d-derived states and hence gives a direct probe of S–Cu bonds [36]. In the present case, the Cd^{2+}/NH$_3$-treatment apparently induces the breaking of some of these bonds in favor of the formation of S–Cd bonds. This is in agreement with the experimental finding of Cu in used Cd^{2+}/NH$_3$-treatment solutions after extended absorber

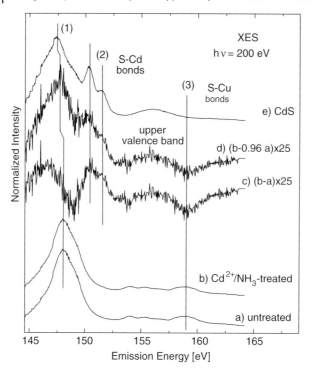

Figure 15.7 S $L_{2,3}$ X-ray emission spectra of (a) an untreated and (b) a Cd^{2+}-treated CuIn(S,Se)$_2$ absorber (1.5 mM Cd^{2+} concentration). Spectrum (c) shows the enlarged (×25) difference of (a) and (b). The spectral features at lines (2) indicate the presence of S–Cd bonds. Spectrum (d) was obtained by subtracting 96% of spectrum (a) from spectrum (b). For comparison, spectrum (e) shows a CdS-reference film [31].

soaking [27] and can be explained by an ion exchange via the corresponding Cd- and Cu-ammine complexes in the Cd^{2+}/NH$_3$-treatment solution [37].

The assumed 4% of CdS signal intensity correspond to approximately one monolayer of CdS. Using the IMFP of the electrons (λ) calculated by the QUASES-IMFP-TPP2M code [38], this is in agreement with the respective layer thickness determined from the attenuation of XPS and XAES substrate signals [31]. Hence, the dominant Cd species formed in the low-concentration regime is CdS, and a significant amount of Cd in a Se-environment can be excluded (in agreement with the S-*rich* nature of the investigated CuIn(S,Se)$_2$ surface). XES measurements of Cd^{2+}/NH$_3$-treated CuIn(S,Se)$_2$ samples treated in high-concentration solutions also indicate the formation of S–Cd bonds [39], which together with the XAES finding of the formation of two Cd species (see Figure 15.6) for these Cd^{2+}-concentrations ($\geq$4.5 mM) is indicative for the formation of a Cd(OH)$_2$/CdS bilayer in that concentration regime.

In summary, soft X-ray and electron spectroscopies allow us to derive the following picture of the Cd^{2+}/NH$_3$-treatment process. In the low-concentration regime, that is, for Cd^{2+}-ion concentrations up to 3 mM, the formation of S–Cd bonds can be found on the S-rich chalcopyrite absorber surface. The deposited amount of "CdS" is nearly independent of the Cd^{2+} concentration. The S atoms forming the S–Cd bonds stem from the absorber surface, as evidenced by the breaking of S–Cu bonds. The latter can be explained by an ion exchange via the corresponding Cd- and Cu-ammine complexes in the Cd^{2+}/NH_3-treatment solution [37]. Without additional diffusion processes, the "CdS" amount is limited to the equivalent of one monolayer, that is, to a state in which all S surface atoms are bound to Cd atoms, as corroborated by the intensity behavior of XPS and XAES signals and by the spectral deconvolution of XES spectra. For concentrations of 4.5 mM and above, a "CdS"/$Cd(OH)_2$ bilayer is formed.

It is the combination of the various spectroscopic techniques, using different aspects of chemical sensitivity and different information depths, together with a suitably designed sample series, and a well-posed initial question, that leads to deep insights into the chemical properties of thin-film solar cell absorber surfaces and the chemical modifications induced by a "real-world" chemical treatment. Such insights are extremely valuable for a better understanding of empirically optimized processes and can lead the way to performance enhancements in real-world solar cell systems.

15.4
Probing the Electronic Surface and Interface Structure: Band Alignment in Thin-Film Solar Cells

The electronic structure, in particular the alignment of the transport levels (i.e., the CBM and the valence band maximum (VBM)), plays a crucial role for the charge transport in every electronic device and thus also for solar cells. In principle, there are three general ways how the conduction bands may align at an interface, as depicted in Figure 15.8. The situation shown in Figure 15.8a, that is, a downward step in the conduction band and hence a negative conduction band offset (CBO, also

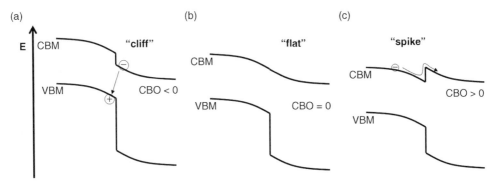

Figure 15.8 Possible (conduction) band alignments at a semiconductor heterointerface: (a) downward step ("cliff"), (b) vanishing band offset ("flat"), and (c) upward step ("spike").

called "cliff") can lead to a reduced recombination barrier (as indicated by the arrow in Figure 15.8), with potential consequences for the performance of the associated cell [40, 41]. The alignment with an upward step in the conduction band and hence a positive CBO (also called "spike"), which is shown in Figure 15.8c, might also be undesirable since the electron transport across the interface may be impeded [42]. Finally, the transition point between a cliff and a spike, that is, a flat conduction band alignment, is shown in Figure 15.8b.

In the following, we will show how the CBO of an interface can be measured, using the example of CdS/Cu(In,Ga(S,Se)$_2$ thin-film solar cells. Particularly important for a correct determination are the complex properties of the different layers of the Cu(In,Ga(S,Se)$_2$ cell. For example, high-efficiency Cu(In,Ga(S,Se)$_2$ films generally exhibit different structural and chemical properties at the surface (or back-surface) than in the bulk of the film (e.g., the band-gap energy, as mentioned in Section 15.2). Thus, great care has to be exercised in determining the band alignments at interfaces in such complex systems, analyzing both, the conduction and valence bands separately, and taking the chemical structure of the interface into account.

As a first example, we will now discuss the interface band alignment of CdS/CuInSe$_2$ (i.e., with a Ga- and S-free absorber surface), discussed in detail in Ref. [17]. The samples were taken from the (former) Shell Solar pilot line, prepared by a rapid thermal annealing of elemental layers on Mo-coated soda-lime glass. CdS films were deposited by CBD. The CuInSe$_2$ layers also contain Ga, but it is localized near the back contact of the cell and thus only trace amounts of Ga can be found at the CuInSe$_2$ surface.

The determination of the band alignment is now done in two steps:

1) In the first step, the band-edge position of the CuInSe$_2$ surface as well as of the CdS surface on the CuInSe$_2$ absorber are determined by UPS and IPES.
2) The values derived in the first step are corrected by the changes in band bending that occur due to the formation of the interface. This can be done by using core-level lines measured by XPS, as will be discussed below.

Figure 15.9 shows the UPS and IPES spectra of the absorber surface (a through f) and of a thick CdS film. Since the samples were taken from a large-scale industrial production line, they were exposed to air for a (minimized) period of time prior to the measurements and thus exhibit surface contaminants. As visible in Figure 15.9, most prominently in the UPS spectra, these contaminants obscure the measurement with very surface-sensitive techniques like UPS or IPES. For this reason the samples were cleaned by using low-energy (500 eV) Ar$^+$ ions at a current density of 1 µA/cm^2 and low and varying angles. More recently, it was found that ion energies as low as 50 eV are very efficient at removing surface contaminants from chalcopyrite surfaces while effectively avoiding any sputter-induced damage to the surface (see also discussion below) [43].

For a variety of reasons [19–21], a linear extrapolation of the leading edge in the spectra can be used to approximate the positions of VBM and CBM in the spectra. While the as-introduced sample (Figure 15.9a) is dominated by the surface contaminations and thus results in an artificially increased "band-gap energy" of 2.7 eV,

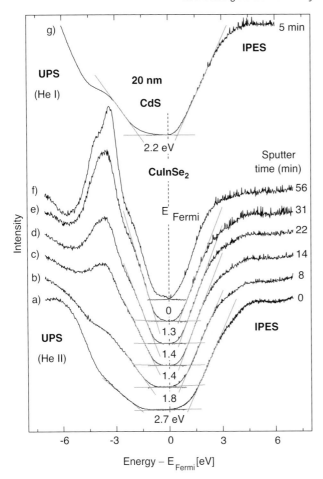

Figure 15.9 UPS and IPES spectra of a CuInSe$_2$ thin-film solar cell absorber and of a 20 nm CBD–CdS/CuInSe$_2$ interface sample. The spectra were acquired after subsequent Ar$^+$-ion sputter steps, as listed on the right-hand side. Gray lines indicate the linearly extrapolated band edges; the determined band gaps are given in the center [17].

a band-gap energy of 1.4 eV can be derived for the sample surface at intermediate stages of the cleaning process (Figure 15.9c and d). Note that it is found that the surface contaminants are also removed during the CBD process [31, 44]. Hence, the surface band-gap energy derived after the sputter treatment is expected to be a good approximation of the CuInSe$_2$ band gap directly at the CdS/CuInSe$_2$ interface. Further Ar$^+$ ion bombardment leads to a further reduction of the derived values to 1.3 eV and, finally, to the emergence of a Fermi edge after prolonged sputter treatment, indicating the formation of metallic surface species (Figure 15.9f). This is consistent with earlier investigations, in which metallic species (both Cu and In) are found after prolonged sputter-treatment of the CuInSe$_2$ surface [45–48]. The such-derived surface band-gap energy of the CuInSe$_2$ film of 1.4 (±0.15) eV is in accordance with the surface band-gap widening described in Section 15.2, presumably due to Cu depletion at the surface of the film [49–51]. The derived band-edge positions are −0.8 (±0.1) eV for the VBM and 0.6 (±0.1) eV for the CBM, respectively.

The band-edge positions and the surface band-gap energy of the CdS buffer layer were determined in a similar way (a sputter time of 5 min was sufficient to clean the surface). Values of −1.8 (±0.1) eV for the VBM and 0.4 (±0.1) eV for the CBM were derived. It is interesting that the resulting surface band gap of 2.2 (±0.15) eV is below that of a CdS bulk film (2.4 eV). This reduction of the band gap can be ascribed to a pronounced intermixing of S and Se at the interface [52, 53], which leads to the formation of a Cd(S,Se) film.

In the first (crude) approximation, the determined values indicate a CBM of the $CuInSe_2$ absorber 0.2 eV above that of the CdS buffer layer, which would indicate a small cliff. However, this value has to be corrected for changes in bend bending occurring during the interface formation, as is illustrated in Figure 15.10. The UPS and IPES measurements described above measure the energetic levels at the surface of the respective films, that is, at the $CuInSe_2$/vacuum and CdS/vacuum interface. These positions are marked with open circles in Figure 15.10.

The deposition of an overlayer on a substrate will generally induce a charge transfer both localized at the interface (i.e., the formation of chemical bonds) and a longer range charge transfer to equilibrate the electrochemical potentials. This may lead to the formation of an interface dipole (or, more precisely, a modification of the previously present surface dipole), and a change in band bending in the absorber film. This is indicated in Figure 15.10 by the difference of the dotted line (without

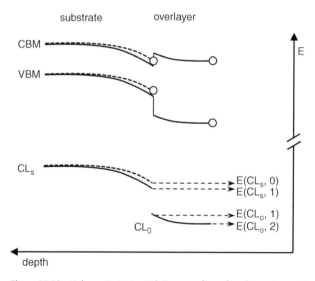

Figure 15.10 Schematic (potential) impact of interface formation on the position of the electronic levels in the substrate. The energy positions of the core levels can be used to quantify this "interface-induced band bending."

overlayer) and the continuous line (with overlayer). Further band bending may occur in the overlayer, as also shown in Figure 15.10. Both "shifts" can be determined by measuring CL energies by means of XPS and by computing the interface-induced band bending (IBB) correction, as will be described in the following. Since the described shifts are caused by surface/interface charges basically generating an electric field, all electronic levels (VBM, CBM, and all CLs) will shift by the same amount. If (at least) one sample with a sufficiently thin overlayer is available, such that both CL signals from the substrate and from the overlayer can be detected, then the actual positions of the VBMs and CBMs of substrate and overlayer at the interface and thus the band alignment can be determined. The necessary IBB correction can be computed with the positions of a substrate core level CL_S with energies $E(CL_S,0)$ without overlayer and $E(CL_S,1)$ with thin overlayer, and those of an overlayer core level CL_O with energies $E(CL_O,1)$ with thin overlayer and $E(CL_O,2)$ with thick overlayer, as follows:

$$IBB = E(CL_S, 0) - E(CL_S, 1) + E(CL_O, 1) - E(CL_O, 2) \tag{15.2}$$

Using this value and the VBM and CBM positions, we can compute the VBO and the CBO energies:

$$\begin{aligned} VBO &= E(VBM_O) - E(VBM_S) + IBB \\ CBO &= E(CBM_O) - E(CBM_S) + IBB \end{aligned} \tag{15.3}$$

where CBM_O (VBM_O) are the energies of the CBM and VBM of the overlayer and CBM_S (VBM_S) the corresponding energies of the substrate.

In the discussed case of the CdS/CuInSe$_2$ interface, IBB values using five different CLs of the substrate and three different CLs of the overlayer were computed and averaged to minimize the influence of chemical shifts occurring during the interface formation. An average value of IBB = 0.2 (±0.1) eV was found. The resulting band-alignment picture is summarized in Figure 15.11. It is characterized by a flat conduction band alignment, that is, a CBO of 0.0 (±0.2) eV, and a VBO of 0.8 (±0.2) eV. Due to the strong intermixing of S and Se at the interface mentioned above, we can assume that the offset is not abrupt but rather "smeared out" over an interfacial region, as indicated in Figure 15.11 by the dashed and curved line in the valence band region.

The result of a flat conduction band alignment is of large relevance for the understanding of Cu(In,Ga)(S,Se)$_2$-based thin-film solar cells. As discussed above, such an alignment is an intermediate case between a barrier for electron transport (a spike) and cliff-like arrangement that could lead to enhanced recombination. Experiments on interface structures involving absorbers with increased bulk band gap (e.g., a Cu(In,Ga)S$_2$ absorber) show that such cliffs indeed exist, even for optimized solar cell systems, as shown in Figure 15.12 [43]. Such findings may explain why the very high efficiencies expected for wide-gap chalcopyrite absorbers have so far been elusive. In general, the gain in open-circuit voltage of

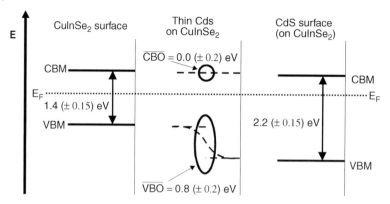

Figure 15.11 Schematic summary of the experimental band alignment results at the CdS/CuInSe$_2$ interface. The left and right parts of the figure represent the level alignment at the surfaces of the CuInSe$_2$ and CdS films, respectively (solid lines). The central part displays the level alignment directly at the interface (dashed lines), which, in addition to the energy positions far away from the interface, also takes an interface-induced band bending as well as interdiffusion into account [17].

these devices is smaller than what would be expected by the increase of the bulk band gap, which is in accordance with the finding of a sizeable (unfavorable) cliff in the conduction band alignment.

In summary, thus, the results demonstrate the intricate details of interfaces and their electronic structure in real-world thin-film solar cells, in particular in view of the band offsets between different heterojunction partners. Since such offsets are

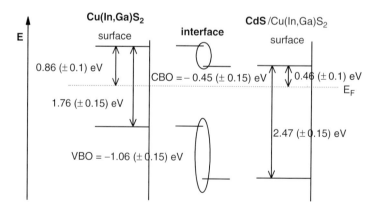

Figure 15.12 Schematic summary of the experimental band alignment results at the CdS/Cu(In,Ga)S$_2$ interface. The left and right parts of the figure represent the level alignment at the surfaces of CuInSe$_2$ and CdS films, respectively. The central part displays the level alignment directly at the interface, which, in addition to the energy positions far away from the interface, also takes an interface-induced band-bending correction into account [43].

fundamental properties influencing the charge carrier transport and thus performance of solar cells, a detailed knowledge of the electronic structure is of crucial importance for insight-based optimization approaches. The here-described combination of experimental methods, namely PES with UV and X-ray excitation, coupled with IPES, allow a detailed assessment of both the occupied and unoccupied electronic states as a function of overlayer thickness, and thus can contribute significant insights into the complicated nature of high-efficiency, real-world thin-film solar cells.

15.5
Summary

In order to aid in the development of high-efficiency thin-film solar cell devices, a unique tool chest of characterization techniques is available that uses soft X-rays and electrons to probe the electronic and chemical properties of surfaces and interfaces. By applying a variety of different viewpoints, for example, by varying the depth-sensitivity and by combining methods that probe occupied and unoccupied electronic states, a comprehensive picture of the complex processes at empirically optimized real-world semiconductor surfaces and heterojunctions can be drawn. Such insights can be used to propose novel optimization approaches and, furthermore, be employed to monitor the success of deliberately introduced modifications. By analyzing the electronic properties on the most microscopic scale, for example, using localized core holes and their wave function overlap with delocalized bands, a detailed understanding of both, long-range electronic properties and local chemical bonding, can be obtained.

As an outlook, it should be mentioned that some of the here-described methods are PIPO techniques. Due to their $1/e$-attenuation lengths in the order of a few tens to a few hundreds of nanometers, PIPO techniques have recently been successfully employed to investigate samples under non-vacuum conditions [3, 4, 54–56], and thus, in the future will lend themselves to monitor the electronic and chemical properties of thin-film solar cell interfaces *while they are being formed*. This promises new and exciting insights into the interface formation dynamics and will open up a new avenue for deliberate optimization of thin-film solar cells.

References

1 Weinhardt, L., Fuchs, O., Fleszar, A., Bär, M., Blum, M., Weigand, M., Denlinger, J., Yang, W., Hanke, W., Umbach, E., and Heske, C. (2009) Resonant inelastic soft X-ray scattering of CdS: A two-dimensional electronic structure map approach. *Phys. Rev. B*, **79**, 165305.

2 Fuchs, O., Weinhardt, L., Blum, M., Weigand, M., Umbach, E., Bär, M., Heske, C., Denlinger, J., Chuang, Y., McKinney, W., Hussain, Z., Gullikson, E., Jones, M., Batson, P., Nelles, B., and Follath, R. (2009) High-resolution, high-transmission soft X-ray spectrometer

for the study of biological samples. *Rev. Sci. Instrum.*, **80**, 63103.

3 Fuchs, O., Zharnikov, M., Weinhardt, L., Blum, M., Weigand, M., Zubavichus, Y., Bär, M., Maier, F., Denlinger, J., Heske, C., Grunze, M., and Umbach, E. (2008) Comment on "Isotope and Temperature Effects in Liquid Water Probed by X-ray Absorption and Resonant X-ray Emission Spectroscopy" – Fuchs et al. reply. *Phys. Rev. Lett.*, **100**, 249802.

4 Weinhardt, L., Fuchs, O., Blum, M., Bär, M., Weigand, M., Denlinger, J., Zubavichus, Y., Zharnikov, M., Grunze, M., Heske, C., and Umbach, E. (2010) Resonant X-ray emission spectroscopy of liquid water: Novel instrumentation, high resolution, and the "map" approach. *J. Electron Spectrosc. Relat. Phenom.*, **177**, 206.

5 Blum, M., Weinhardt, L., Fuchs, O., Bär, M., Zhang, Y., Weigand, M., Krause, S., Pookpanratana, S., Hofmann, T., Yang, W., Denlinger, J.D., Umbach, E., and Heske, C. (2009) Solid and liquid spectroscopic analysis (SALSA) – a soft X-ray spectroscopy endstation with a novel flow-through liquid cell. *Rev. Sci. Instrum.*, **80**, 123102.

6 Hüfner, S. (2003) *Photoelectron Spectroscopy*, Springer, Berlin.

7 Seah, M.P. and Dench, W.A. (1979) Quantitative electron spectroscopy of surfaces: A standard data base for electron inelastic mean free paths in solids. *Surf. Interf. Anal.*, **1**, 2.

8 Tanuma, S., Powell, C.J., and Penn, D.R. (1994) Calculations of electron inelastic mean free paths. V. Data for 14 organic compounds over the 50–2000 eV range. *Surf. Interface Anal.*, **21**, 165; QUASES-IMFP-TPP2M code for the calculation of the inelastic electron mean free path, Version2.2 http://www.quases.com/.

9 Henke, B.L., Gullikson, E.M., and Davis, J.C. (1993) *At. Data Nucl. Data Tables*, **54**, 181; http://www-cxro.lbl.gov/optical_constants/atten2.html.

10 Briggs, D. and Seah, M.P. (1990) *Practical Surface Analysis – Auger and X-ray Photoelectron Spectroscopy*, 2nd edn, Wiley Interscience, New York.

11 Stöhr, J. (1992) *NEXAFS Spectroscopy*, Springer, Berlin.

12 Smith, N.V. (1988) Inverse photoemission. *Rep. Prog. Phys.*, **51**, 1227.

13 Carlisle, J.A., Shirley, E.L., Hudson, E.A., Terminello, L.J., Callcott, T.A., Jia, J.J., Ederer, D.L., Perera, R.C.C., and Himpsel, F.J. (1995) Probing the graphite band structure with resonant soft-X-ray fluorescence. *Phys. Rev. Lett.*, **74**, 1234.

14 Eich, D., Fuchs, O., Groh, U., Weinhardt, L., Fink, R., Umbach, E., Heske, C., Fleszar, A., Hanke, W., Gross, E., Bostedt, C., Von Buuren, T., Franco, N., Terminello, L., Keim, M., Reuscher, G., Lugauer, H., and Waag, A. (2006) Resonant inelastic soft X-ray scattering of Be chalcogenides. *Phys. Rev. B*, **73**, 115212.

15 Lüning, J., Rubensson, J., Ellmers, C., Eisebitt, S., and Eberhardt, W. (1997) Site- and symmetry-projected band structure measured by resonant inelastic soft X-ray scattering. *Phys. Rev. B*, **56**, 13147.

16 Schmid, D., Ruckh, M., Grunwald, F., and Schock, H.W. (1993) Chalcopyrite/defect chalcopyrite heterojunctions on the basis of $CuInSe_2$. *J. Appl. Phys.*, **73**, 2902.

17 Morkel, M., Weinhardt, L., Lohmüller, B., Heske, C., Umbach, E., Riedl, W., Zweigart, S., and Karg, F. (2001) Flat conduction-band alignment at the $CdS/CuInSe_2$ thin-film solar-cell heterojunction. *Appl. Phys. Lett.*, **79**, 4482.

18 Bär, M., Nishiwaki, S., Weinhardt, L., Pookpanratana, S., Fuchs, O., Blum, M., Yang, W., Denlinger, J., Shafarman, W., and Heske, C. (2008) Depth-resolved band gap in $Cu(In,Ga)(S,Se)_2$ thin films. *Appl. Phys. Lett.*, **93**, 244103.

19 Eich, D., Ortner, K., Groh, U., Chen, Z.H., Becker, C.R., Landwehr, G., Fink, R., and Umbach, E. (1999) Band discontinuity and band gap of MBE grown HgTe/CdTe(001) heterointerfaces studied by k-resolved photoemission and inverse photoemission. *phys. status solidi (a)*, **173**, 261.

20 Gleim, T., Heske, C., Umbach, E., Schumacher, C., Faschinger, W., Ammon, C., Probst, M., and Steinrück, H.-P. (2001)

Reduction of the ZnSe/GaAs(100) valence band offset by a Te interlayer. *Appl. Phys. Lett.*, **78**, 1867.

21 Gleim, T., Heske, C., Umbach, E., Schumacher, C., Gundel, S., Faschinger, W., Fleszar, A., Ammon, C., Probst, M., and Steinrück, H.-P. (2003) Formation of the ZnSe/(Te/)GaAs(100) heterojunction. *Surf. Sci.*, **531**, 77.

22 Weinhardt, L., Fuchs, O., Umbach, E., Heske, C., Fleszar, A., and Hanke, W. (2007) Resonant inelastic soft X-ray scattering, X-ray absorption spectroscopy, and density functional theory calculations of the electronic bulk band structure of CdS. *Phys. Rev. B*, **75**, 165207.

23 Weinhardt, L., Fuchs, O., Fleszar, A., Bär, M., Blum, M., Weigand, M., Denlinger, J., Yang, W., Hanke, W., Umbach, E., and Heske, C. (2009) Resonant inelastic soft X-ray scattering of CdS: A two-dimensional electronic structure map approach. *Phys. Rev. B.*, **79**, 165305.

24 Repins, I., Contreras, M.A., Egaas, B., DeHart, C., Scharf, J., Perkins, C.L., To, B., and Noufi, R. (2008) Short communication: Accelerated publication 19·9%-efficient ZnO/CdS/CuInGaSe$_2$ solar cell with 81.2% fill factor. *Prog. Photovoltaics: Res. Appl.*, **16**, 235.

25 Dimmler, B. (2010) CIS: The Mainstream Technology in a Professional Energy Market, presented at the 5th World Conference on Photovoltaic Energy Conversion, Valencia, Spain, September 6–10.

26 Green, M.A., Emery, K., Hishikawa, Y., and Warta, W. (2009) Solar cell efficiency tables (Version 34). *Prog. Photovolt.: Res. Appl.*, **17**, 320.

27 Ramanathan, K., Bhattacharya, R.N., Granata, J., Webb, J., Niles, D., Contreras, M.A., Wiesner, H., Hasoon, F.S., and Noufi, R. (1997) Properties of Cd and Zn partial electrolyte treated CIGS solar cells. Proc. 26th IEEE PVSC, p. 319.

28 Bär, M., Muffler, H., Fischer, C., Zweigart, S., Karg, F., and Lux-Steiner, M.C. (2002) ILGAR-ZnO Window Extension Layer: An adequate substitution of the conventional CBD-CdS buffer in Cu(In,Ga) (S,Se)$_2$ based solar cells with superior device performance. *Prog. Photovoltaics: Res. Appl.*, **10**, 173.

29 Wada, T., Hayashi, S., Hashimoto, Y., Nishiwaki, S., Sato, T., Negami, T., and Nishitani, M. (1998) High efficiency Cu(In,Ga)Se$_2$ (CIGS) solar cells with improved CIGS surface. Proc. 2nd World Conf. of Photovolt. Energy Conv., p. 403.

30 Canava, B., Guillemoles, J.-F., Yousfi, E.-B., Cowache, P., Kerber, H., Loeffl, A., Schock, H.-J., Powalla, M., Hariskos, D., and Lincot, D. (2000) Wet treatment based interface engineering for high efficiency Cu(In,Ga)Se$_2$ solar cells. *Thin Solid Films*, **361–362**, 187.

31 Weinhardt, L., Gleim, T., Fuchs, O., Heske, C., Umbach, E., Bär, M., Muffler, H., Fischer, C., Lux-Steiner, M., Zubavichus, Y., Niesen, T., and Karg, F. (2003) CdS and Cd(OH)$_2$ formation during Cd treatments of Cu(In,Ga)(S,Se)$_2$ thin-film solar cell absorbers. *Appl. Phys. Lett.*, **82**, 571.

32 NIST X-ray Photoelectron Spectroscopy Database, NIST Standard Reference Database 20, Version 3.4, http://srdata.nist.gov/xps/.

33 Wagner, C.D., Riggs, W.M., Davis, L.E., and Moulder, J.F. (1979) *Handbook of X-Ray Photoelectron Spectroscopy* (ed. G.E. Muilenberg), Perkin-Elmer Corporation, Eden Prairie, MN.

34 Scofield, J.H. (1976) Hartree-Slater subshell photoionization cross-sections at 1254 and 1487 eV. *J. Electron. Spectrosc.*, **8**, 129.

35 Wagner, C.D. (1990) *Practical Surface Analysis*, vol. **1** (eds D. Briggs and M.P. Seah), John Wiley & Sons Inc., New York, p. 595.

36 Heske, C., Groh, U., Fuchs, O., Umbach, E., Franco, N., Bostedt, C., Terminello, L., Perera, R., Hallmeier, K., Preobrajenski, A., Szargan, R., Zweigart, S., Riedl, W., and Karg, F. (2001) X-ray emission spectroscopy of Cu(In,Ga)(S,Se)$_2$-based thin film solar cells: Electronic structure, surface oxidation, and buried interfaces. *phys. status solidi (a)*, **187**, 13.

37 Bär, M., Weinhardt, L., Heske, C., Muffler, H., Lux-Steiner, M., Umbach, E., and Fischer, C. (2005) Cd/NH$_3$ treatment

of Cu(In,Ga)(S,Se)$_2$ thin-film solar cell absorbers: A model for the performance-enhancing processes in the partial electrolyte. *Prog. Photovolt.*, **13**, 571.

38 Tougaard, S., QUASES-IMFP-TPP2M code for the calculation of the inelastic electron mean free path, Version 2.2 http://www.quases.com/.

39 Bär, M. (2004) Novel Cd-free window structure for chalcopyrite-based thin film solar cells Ph.D. thesis, TU Berlin http://opus.kobv.de/tuberlin/volltexte/2004/687/.

40 Hengel, I., Neisser, A., Klenk, R., and Lux-Steiner, M.C. (2000) Current transport in CuInS$_2$:Ga/CdS/ZnO-solar cells. *Thin Solid Films*, **361–362**, 458.

41 Klenk, R. (2001) Characterisation and modelling of chalcopyrite solar cells. *Thin Solid Films*, **387**, 135.

42 Liu, X. and Sites, J.R. (1996) Calculated effect of conduction-band offset on CuInSe$_2$ solar-cell performance. Proceedings of the AIP Conference, Lakewood, p. 444.

43 Weinhardt, L., Fuchs, O., Gross, D., Storch, G., Umbach, E., Dhere, N., Kadam, A., Kulkarni, S., and Heske, C. (2005) Band alignment at the CdS/Cu(In,Ga)S$_2$ interface in thin-film solar cells. *Appl. Phys. Lett.*, **86**, 062109.

44 Kylner, A. (1999) The chemical bath deposited CdS/Cu(In,Ga)Se$_2$ interface as revealed by X-ray photoelectron spectroscopy. *J. Electrochem. Soc.*, **146**, 1816.

45 Heske, C., Fink, R., Umbach, E., Riedl, W., and Karg, F. (1996) Surface Preparation Effects of polycrystalline Cu(In,Ga)Se$_2$ thin films studied by XPS and UPS. *Cryst. Res. Technol.*, **31**, 919.

46 Niles, D.W., Ramanathan, K., Hasoon, F., Noufi, R., Tielsch, B.J., and Fulghum, J.E. (1997) Na impurity chemistry in photovoltaic CIGS thin films: Investigation with X-ray photoelectron spectroscopy. *J. Vac. Sci. Technol. A*, **15**, 3044.

47 Otte, K., Lippold, G., Frost, F., Schindler, A., Bigl, F., Yakushev, M.V., and Tomlinson, R.D. (1999) Low energy ion beam etching of CuInSe$_2$ surfaces. *J. Vac. Sci. Technol. A*, **17**, 19.

48 Weinhardt, L., Morkel, M., Gleim, Th., Zweigart, S., Niesen, T.P., Karg, F., Heske, C., and Umbach, E. (2001) Band alignment at the CdS/CuIn(S,Se)$_2$ heterojunction in thin film solar cells. Proc. 17 European Photovolt. Solar Energy Conf., Munich, p. 1261.

49 Schmid, D., Ruckh, M., Grunwald, F., and Schock, H.W. (1993) Chalcopyrite/defect chalcopyrite heterojunctions on the basis of CuInSe$_2$. *J. Appl. Phys.*, **73**, 2902.

50 Dullweber, T., anna, G.H., Rau, U., and Schock, H.W. (2001) A new approach to high-efficiency solar cells by band gap grading in Cu(In,Ga)Se$_2$ chalcopyrite semiconductors. *Sol. Energy Mater. Sol. Cells*, **67**, 145.

51 Klein, A. and Jaegermann, W. (1999) Fermi-level-dependent defect formation in Cu-chalcopyrite semiconductors. *Appl. Phys. Lett.*, **74**, 2283.

52 Heske, C., Eich, D., Fink, R., Umbach, E., van Buuren, T., Bostedt, C., Terminello, L.J., Kakar, S., Grush, M.M., Callcott, T.A., Himpsel, F.J., Ederer, D.L., Perera, R.C.C., Riedl, W., and Karg, F. (1999) Observation of intermixing at the buried CdS/Cu(In,Ga)Se$_2$ thin film solar cell heterojunction. *Appl. Phys. Lett.*, **74**, 1451.

53 Weinhardt, L., Pookpanratana, S., Morkel, M., Niesen, T., Karg, F., Ramanathan, K., Contreras, M., Noufi, R., Umbach, E., and Heske, C. (2010) Sulfur gradient-driven Se diffusion at the CdS/CuIn(S,Se)$_2$ solar cell interface. *Appl. Phys. Lett.*, **96**, 182102.

54 Heske, C., Groh, U., Fuchs, O., Weinhardt, L., Umbach, E., Schedel-Niedrig, T., Fischer, C., Lux-Steiner, M., Zweigart, S., Niesen, T., and Karg, F. (2003) Monitoring chemical reactions at a liquid–solid interface: Water on CuIn(S,Se)$_2$ thin film solar cell absorbers. *J. Chem. Phys.*, **119**, 10467.

55 Heske, C. (2004) Spectroscopic investigation of buried interfaces and liquids with soft X-rays. *Appl. Phys. A*, **78**, 829.

56 Fuchs, O., Zharnikov, M., Weinhardt, L., Blum, M., Weigand, M., Zubavichus, Y., Bär, M., Maier, F., Denlinger, J., Heske, C., Grunze, M., and Umbach, E. (2008) Isotope and temperature effects in liquid water probed by X-ray absorption and resonant X-ray emission spectroscopy. *Phys. Rev. Lett.*, **100**, 027801.

16
Elemental Distribution Profiling of Thin Films for Solar Cells

Volker Hoffmann, Denis Klemm, Varvara Efimova, Cornel Venzago, Angus A. Rockett, Thomas Wirth, Tim Nunney, Christian A. Kaufmann, and Raquel Caballero

16.1
Introduction

The present chapter will give an overview about the techniques glow discharge-optical emission (GD-OES) and glow discharge-mass spectroscopy (GD-MS), secondary ion mass spectroscopy (SIMS) and sputtered neutral mass spectroscopy (SNMS), Auger electron spectroscopy (AES), and X-ray photoelectron spectroscopy (XPS), which are methods broadly available for depth-resolved information about the matrix and trace element concentrations in multilayer stacks as those used for thin-film solar cells. In addition, also energy-dispersive X-ray spectrometry (EDX) in a scanning electron microscope may be applied on fractured cross-section specimens of these multilayer stacks, providing a quick way of analyzing the spatial distribution of the matrix elements, which is why this technique is also included in this chapter.

AES, XPS, SIMS, and SNMS have a good lateral resolution, and depth-resolved information may be obtained by the investigation either of cross-section specimens or (most simply) of a fractured surface (see Section 16.6.3.1). All of these methods are in general performed in combination with a sputtering process when aiming for depth-resolved elemental distribution analysis. Thereby, ions are accelerated toward the surface of the sample and collide with the atoms in the region near the surface. The resulting collision cascade leads to the emission of sample material from the surface and forms a more or less flat crater. While AES and XPS only use this process to expose the region of interest, SIMS and SNMS analyze the sputtered material. Because of practical reasons, different sputtering rates are employed at times, that is, fast sputtering for removing sample material and slow sputtering during the analysis.

XPS, AES, and SIMS work under ultrahigh vacuum conditions and must be performed at high-acceleration voltages in order to obtain high sputtering yields. These techniques make use of a scanning ion beam for sputtering, which allows the discrimination of crater edge effects. For the SIMS analysis, only an electronically gated area is used for the analysis. Similarly, in AES and XPS measurements, only

---

Advanced Characterization Techniques for Thin Film Solar Cells,
Edited by Daniel Abou-Ras, Thomas Kirchartz and Uwe Rau.
© 2011 Wiley-VCH Verlag GmbH & Co. KGaA. Published 2011 by Wiley-VCH Verlag GmbH & Co. KGaA.

secondary electrons from the inner, flat part of the crater contribute to the analysis. In order to reduce the sputtering-enhanced roughness, sample rotation may be applied, where ultimate depth resolution is needed [1].

The sputtering process may change the properties of the material owing to ion damage to the sample surface. At acceleration voltages of several tens of kV, significant mixing of the material occurs, which deteriorates the depth resolution.

This effect is less pronounced for techniques, which use lower voltages, such as GD-OES and GD-MS. After several collisions, the energy of the sputtering particles is reduced to about 50 eV, and only a small number of atom layers participate at the collision process. These collisions also cause a nearly uniform angle distribution of the sputtering particles, which reduces the sputtering-enhanced roughness. The higher current density in the high-pressure (several hPa) sputtering instruments (GD-OES and GD-MS) results also in a higher thermal load on the sample. Therefore, pulsed discharges are preferred in the case of heat-sensitive samples. Even when applying pulsed discharges, the erosion rate is much higher for GD-OES and GD-MS, than for XPS, AES, and SIMS.

Last but not least, it must be mentioned that all sputtering techniques suffer from preferential sputtering, which means that sputtering of light elements is preferred in comparison with heavier elements. For techniques which analyze the sputtered material itself, such as SIMS or SNMS, after a short time equilibrium is achieved. XPS and AES, however, analyze the modified surface. Therefore, if the sample consists of elements with very different sputtering yields, preferential sputtering must be taken into account during quantification. Sputtering yields vary with varying surface-binding potential, which is often different between different phases or crystal orientations at the surface, which may lead to preferential sputtering and roughening of the sample surface.

Most of the examples given in the present chapter are analytical results from $Cu(In,Ga)(S,Se)_2$ thin films applied as solar absorber materials. The In and Ga distributions across these layers are in general not homogeneous and depend strongly on the method used for the film deposition, the process parameters involved and aging/storage conditions [2, 3].

Although in an ideal single-junction solar-cell device, the band-gap energy of the absorber is considered constant across the absorber layer, a designed band-gap profile may yield a considerable efficiency enhancement in a device affected by recombination loss at the various internal interfaces. For $Cu(In,Ga)Se_2$ thin-film solar cell devices, the in-depth Ga distributions in $Cu(In,Ga)Se_2$ thin films are proportional to those of the corresponding band-gap energies [2, 4]. The measurement of the Ga and In distributions is, hence, a key issue to be considered for device optimization.

Impurities, such as Na or Fe, in chalcopyrite thin films, are of importance for the quality of the final device [5, 6]. Their influence on the growth process and the electronic properties of the semiconductor material, and also their final distribution within the thin film may have a serious impact on the electronic quality of the material grown and hence also on the solar-conversion efficiency of the final device. Appropriate analysis tools for the concentrations and distributions of these trace elements are therefore needed, as those described in the present chapter.

16.2
Glow Discharge-Optical Emission (GD-OES) and Glow Discharge-Mass Spectroscopy (GD-MS)

GD-OES and GD-MS are among the most important techniques for the direct analysis of solids. The short time for the analysis (typically 5 min (OES) – 50 min (MS) for a thin film of about 5 μm thickness) is caused by the high erosion rate – for OES typically 1–6 μm/min, and for MS 0.1–1 μm/min. Also, the capability to analyze many elements, including the light elements, simultaneously makes GD-OES and GD-MS powerful techniques. The introduction of the radio frequency (rf) mode in OES expanded the application to the analysis of nonconducting materials [7, 8].

When the sputtering conditions are optimized, depth profiling is possible with a depth resolution of about 5–10% of the achieved depth. Since the development of fast and cheap personal computers for data acquisition, nearly all GD-OES instruments are suitable and used both, for bulk analysis and for depth profiling. Also modern GD-MS instruments are prepared for depth profiling, but still only in direct current (dc) mode for conducting samples. Currently, rf-GD-MS instruments only exist as prototypes in the research environment.

16.2.1
Principles

Glow discharges work in a noble-gas (generally Ar) atmosphere at pressures ranging from about 100 to 1000 Pa. All GD instruments work under obstructed conditions (when no positive column is build), provided by a short distance between the cathode and the anode with the analytical sample of interest serving as the cathode. During the discharge, the cathode is bombarded by positive noble gas ions and atoms. Material is sputtered from the sample, which is then partly ionized and excited to emit photons corresponding to the characteristic spectral lines. The glow discharge gives information on the elemental composition of the analytical material removed layer by layer from the cathode. It can be estimated that the number of Ar atoms in the discharge exceeds the number of sputtered sample atoms approximately by a factor of 1000. This explains why GD calibration curves are usually linear over a wide range of elemental concentration and matrix effects are less important. The separation of sputtering and excitation in space and time is also essential for the low matrix effects. The atoms and ions from the sample are highly dissolved in the argon gas and have no more information about former neighbors in the solid sample.

16.2.2
Instrumentation

16.2.2.1 Plasma Sources
Most GD instruments use the Grimm-type design (see Figures 16.1 and 16.2), where the anode has a ring shape, whose annular front is spaced by only 0.1–0.2 mm from the surface of the sample [9]. Thus, the discharge is restricted to a sample area which

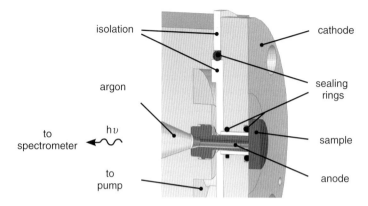

Figure 16.1 Diagram of a typical glow discharge source used for dc GD-OES analysis.

is equal to the inner aperture of the anode tube (typically 2–10 mm). This source configuration is not only suitable for flat and vacuum-tight samples, but also special sample holders exist for cylindrical, spherical, or pin-type samples.

In GD-MS, many instruments are of type VG9000, which exhibit a source design for low power and low gas-flow conditions and with possible cryogenic cooling of sample and source. Most of these instruments are only equipped with pin-geometry source. Later, also flat-type sources were introduced, therefore, GD-MS became available to thin-layer analyses. The currently only commercial GD-MS instrument is using a Grimm-type source with a fast flow concept. Additionally, developments on the ElementGD introduced μs-pulsed discharge by a pulsed voltage generator [10]. This approach allows much tighter control on sputter rates and therefore also GD-MS analyses in the low μm and nm range.

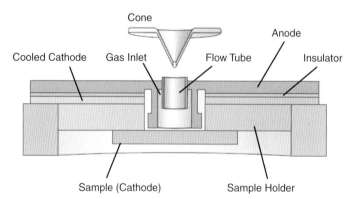

Figure 16.2 Diagram of the glow discharge source of ElementGD (reprinted with permission from Thermo Fisher Scientific (Bremen) GmbH).

16.2.2.2 Plasma Conditions

With constant discharge conditions, the spectral line intensities and ion currents are proportional to the number densities of sputtered atoms in the plasma and hence, the element concentration in the sample used. In order to maintain reproducible excitation and ionization conditions in the glow discharge source, the discharge conditions (i.e., the argon pressure, voltage, and current) are controlled carefully. The current–voltage characteristics, which describe the discharge, depend on pressure and matrix composition of the thin film analyzed. Thus, for different matrix compositions, it is only possible to keep two of the three discharge parameters constant. It turns out that the pressure has the lowest influence on the light emission [11]. Therefore, voltage and current are kept constant in dc mode. Standard conditions for a 4 mm source in GD-OES are 700 V and 20 mA, and a VG9000-type instrument with a 10 mm source uses 1000 V and 3–10 mA in most cases. The ElementGD is mostly used in the range from 30 to 60 mA at 500–800 V.

Conductive material can be analyzed with constant potential across cathode and anode (dc mode). If the potential varies at rf discharges, nonconductive materials such as thin-film solar cells on glass substrates may be also analyzed. Both, dc and rf excitation, can be applied continuously or in a pulsed mode. However, only pulsed rf generators are offered commercially with repetition rates of up to 10 kHz. In rf mode, the measurement of the plasma current is difficult and therefore, other parameters, such as constant power and constant voltage, are preferred.

16.2.2.3 Detection of Optical Emission

The optical radiation of the excited sample atoms is detected end-on through a window with an appropriate transmittance. MgF_2 is the most commonly used as window material.

Apart from the glow discharge source with its associated gas and power supplies, vacuum pumps, and controls, the optical system is the most important part of a GD instrument. In order to make use of the analytical capability of the discharge source, the spectrometer must exhibit sufficient resolving power and an adequate spectral range. Commercial optical GD spectrometers can cover the wavelength range from 110 to 800 nm, which contains the most sensitive lines of all elements. The high-spectral resolution is required in order to avoid spectral interference with lines of other elements and molecules, in particular of the discharge gas Ar. A monochromator can be used for sequential analysis or as additional wavelength channel.

In most common polychromator systems for GD-OES, each detection channel is equipped with a photomultiplier tube (PMT) and a corresponding high-voltage supply. The rapid development in the performance of charge-coupled devices (CCDs) and related detection techniques makes CCD detection systems very attractive. These systems are available for bulk and depth-profile analysis and allow flexible selection of nearly all lines and simultaneous measurement of signal and background. However, they still suffer from lower sensitivity and speed of data acquisition in comparison with PMT systems.

16.2.2.4 Mass Spectroscopy

Most installed GD-MS instruments are equipped with a double-focusing, reversed Nier–Johnson geometry mass spectrometer [12]. This geometry provides high mass resolution measurement (with $m/\Delta m$ up to 10 000), which is crucial for the determination of concentrations of trace elements in complex matrices. The ions are extracted through an exit orifice by the pressure difference into the mass spectrometer. Signal detection in GD-MS is performed by means of ion-counting detectors (secondary electron multipliers or Daly detectors) and analog ion current detectors (secondary electron multipliers or Faraday cups). In combination, both detection principles (analog and counting) are able to measure the ion current with 12 orders of linear dynamic range. The disadvantage of these mass spectrometric systems (spectrometer with detector) for thin-layer analyses is that they record the mass spectrum sequentially, that is, they are considered as slow in comparison with simultaneously recording instruments.

Time-of-flight (TOF) mass spectrometers [13] have the advantage of very fast simultaneous detections of full-range mass spectra, and therefore deliver the capability to analyze even thinner layers than by use of double-focusing MS. On the other hand, they suffer from lower dynamic range than double-focusing instruments.

16.2.3
Quantification

16.2.3.1 Glow Discharge-Optical Emission Spectroscopy

Because of the complex nature of the discharge conditions, GD-OES is a comparative analytical method. Therefore, standard reference materials must be used in order to establish a relationship between the measured line intensities and the elemental concentration. In quantitative bulk analysis, which has been developed to very high standards, the calibration is performed with a set of calibration samples of compositions similar to those of the unknown samples. Normally, an element present at high concentration in the sample is used as reference, and the internal standard method is applied [14]. This method is not generally applicable in depth-profile analysis of a multilayer stack, because the various layers on which a depth profile was acquired often comprise largely different types of materials, that is, a common reference element is not available [15].

The quantification algorithm, most commonly used in dc GD-OES depth profiling, is based on the concept of constant emission yield R_{ik} [11]. It was found that the number of photons with wavelength k, emitted per atom or ion of element i in a plasma (i.e., emission yield R_{ik}) is almost matrix- and pressure-independent. The statement is only valid if the source is operated under constant excitation conditions, that is, mainly at constant voltage and current. The given line intensity I_{ik} depends on the concentration c_i of element i in the sample j and on the sputtering rate q_j, which is the total sputtered mass of sample j per time:

$$I_{ik} = c_i q_j R_{ik} \tag{16.1}$$

16.2 Glow Discharge-Optical Emission (GD-OES) and Glow Discharge-Mass Spectroscopy (GD-MS)

From this equation, the emission yields of elements can be determined by the analysis of standard samples. For a standard sample, the concentrations of all elements and the sputtering rate must be known. When line intensities of all elements present in the sample were measured and the corresponding emission yields are known, the element concentrations of unknown samples and the sputtering rate can be determined using Eq. (16.1). These concentrations can be converted into the depth of the crater, if we assume that the density ϱ of the sample is an average of the density of the pure element densities ϱ_i

$$\varrho = \sum (c_i \varrho_i) \tag{16.2}$$

$$z = \sum \left(\frac{q}{\varrho} A \Delta t \right) \tag{16.3}$$

where q is the sputtering rate of the sample, Δt is sputtering duration, and A is the sputtering area.

16.2.3.2 Glow Discharge-Mass Spectroscopy

GD-MS is a unique quantitative technique, exhibiting low matrix dependence and very high linear range compared with other solid-state analytical techniques. In addition, the relative sensitivity factors (RSFs) are matrix independent over a wide range of concentrations, which is an advantage, especially, when determining very low concentrations. The RSF – a term used by VG9000 users and software – simply means the reciprocal slope of a virtual calibration curve (relative sensitivity coefficient, RSC, is therefore a more correct term), based on the theory that the calibration curve is linear and does not have an intercept [16]. All of these facts allow for performing GD-MS analyses routinely by using typical RSFs determined on various matrices. Such an RSF data set is called the standard RSF. RSF data in this set are normalized to Fe, that is, the RSF for Fe equals 1. Because the RSFs for various elements vary only within about one order of magnitude, semiquantitative analysis becomes possible even without any standard and calibration.

$$\text{RSF}_{x/y} = \frac{I_y}{I_x} \cdot c_{x/y} \tag{16.4}$$

where I_x is the ion intensity (corrected for ion abundance) of element x, and $c_{x/y}$ is the certified concentration of element x in matrix y of the calibration sample. The normalization of the analyte- and matrix-specific RSF is performed in the following way:

$$\text{StdRSF}_x = \frac{\text{RSF}_{x/y}}{\text{RSF}_{\text{Fe}/y}} \tag{16.5}$$

where StdRSF_x is the standard RSF of element x normalized to Fe, that is, $\text{RSF}_{\text{Fe}} = 1$. The standard RSF set contains a large list of standard RSFs for many elements from experimental results. Finally, for the quantification of an analytical result, the standard RSFs are applied as follows (for the example of determining the Na concentration in a Si matrix):

$$c_{Na/Si} = \frac{I_{Na}}{I_{Si}} \times \frac{StdRSF_{Na}}{StdRSF_{Si}} \quad (16.6)$$

16.2.4
Applications

16.2.4.1 Glow Discharge-Optical Emission Spectroscopy

Figure 16.3a shows GD-OES in-depth elemental distribution profiles from a Cu(In, Ga)Se$_2$ thin film on Mo-coated glass substrates acquired by means of pulsed-rf glow discharge. In Figure 16.3b, these profiles are given quantified according to the procedure described in Section 16.2.3.1. Layers of CuInSe$_2$ and CuGaSe$_2$ on Mo-coated glass substrates were used as standards. The results of the quantification were confirmed by means of XRF results (measuring the integral compositions of the Cu (In,Ga)Se$_2$ layer). The gradients of Ga and In in the layer show a good agreement with energy-dispersive X-ray diffraction [17] (see Section 13.5 for an introduction) and AES measurements (see Section 16.4.7).

16.2.4.2 Glow Discharge-Mass Spectroscopy

Figure 16.4 shows a system of Si layers on and SiC substrate acquired with continuous dc at 800 V and 6 mA on a VG9000 instrument with concentration distributions from part per billion range for impurities and doping elements together with matrix concentrations of Si and C.

Figure 16.5 shows an example of the capabilities of a pulsed-dc GD-MS source on an ElementGD instrument (900 V, 1.0 kHz, 50 µs pulse width, 295 ml/min Ar) for matrix (a) and trace element distributions (b) in two 2.5 µm thick layers on Mo-coated

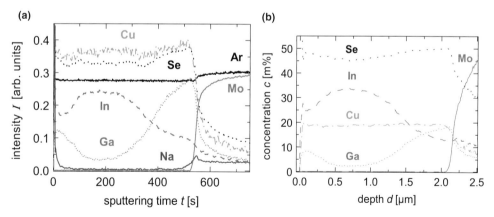

Figure 16.3 Measured (a) and quantified (b) in-depth elemental distribution profiles from a Cu(In,Ga)Se$_2$/Mo/glass stack, using U$_{peak-peak}$ 1000 V, 3.5 hPa, 350 Hz, 7% duty cycle.

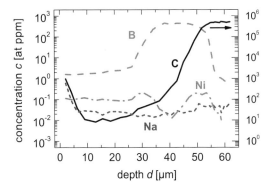

Figure 16.4 Trace and matrix concentrations of a Si-layer system on a SiC-substrate (C concentration scaled by the right axis).

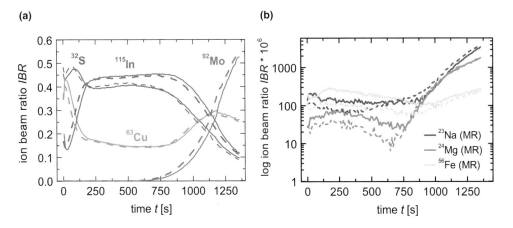

Figure 16.5 (a) Matrix element distributions, given in terms of ion-beam ratio concentrations, from two CuInS$_2$/Mo/glass stacks (solid and dotted lines) analyzed by pulsed dc-GD-MS and acquired at mass resolution $m/\Delta m = 4000$). (b) Trace element distributions of the identical multilayer stacks. Note that the time t is related to the depth of the sputtering crater.

glass substrates from different CuInS$_2$ deposition runs. These two CuInS$_2$ layers exhibit not a large difference in matrix-element but large differences in trace element distributions. Since only dc-GD-MS was used, analyses of the Mo layer and of the glass substrate were not possible.

Figure 16.6 shows the analysis results obtained on a Si thin layer (100 nm) on a glass substrate, measured by a prototype rf-GD-TOF system [18], exhibiting also signals from the interface between Si and glass, as well as from the glass substrate. Increased intensities of Zn and Na signals are found in the interface region between Si and glass. The analysis of this transition into the nonconductive substrate glass and of the glass itself is only possible since rf-GD could be used.

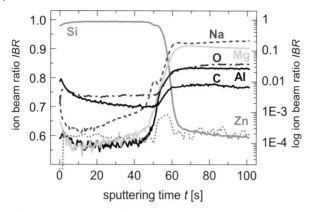

Figure 16.6 Rf-GD-TOF data from a 100 nm Si layer on glass substrate, ion beam ratio of trace elements to the sum of all measured elements (matrix element Si: linear scale, all other elements: log scale), 50 W, 1.8 ms pulse length, 45% duty cycle.

16.3
Secondary Ion Mass Spectrometry (SIMS)

SIMS [19, 20] is one of the most sensitive composition depth-profiling techniques available. In many cases, ion count rates may exceed 10^9 counts/s for pure elements with background signals of less than 1 counts/s. Thus, it is possible under these analysis conditions to detect part per trillion level impurities in some cases. Sensitivities in the part per million range are routine for almost all elements. Even difficult species such as N can be detected in the form of molecular ions, and inert gases such as Ar can also be measured. Furthermore, SIMS can provide quantitative depth scales with good accuracy based on sputtered crater depth measurements. Because virtually every ion emitted from the surface can be detected, very slow sputtering rates can still permit useful analyses. The result is that SIMS can also be operated in a "static" mode in which there is effectively no sputtering of the surface (only a fraction of a monolayer is removed). This allows near surface analysis without alteration of the composition. Many higher end instruments also provide the ability to map the location of individual atoms on the probed surface with micron-scale (and sometimes submicron) resolution. SIMS is best for analyzing trace impurity depth profiles in known matrices or measuring relative changes in a matrix constituent in the absence of major changes in the chemistry of the matrix.

16.3.1
Principle of the Method

SIMS is a method in which ionized species are sputtered from a sample surface into vacuum and accelerated by an electric field. Energy and mass of these species are then analyzed and detected with high efficiency. Given the high efficiency of the ion detector and the probability of an atom leaving the surface as an ion, it is possible to

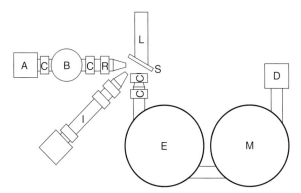

Figure 16.7 A schematic illustration of a typical SIMS instrument. The various parts shown are as follows. (A) gas source, (B) optional mass analyzer for the primary-ion beam that sputters the sample, (C) magnetic or electrostatic lenses to focus the primary beam, (D) ion-detection electronics (usually a multichannel plate, electron multiplier, or Faraday cup), (E) energy analyzer [usually electrostatic], (I) optional depth-profiling sputter gun (for time-of-flight instruments), (L) sample introduction load-lock, (M) mass analyzer [time-of-flight, magnetic sector, or quadrupole mass spectrometer], (R) raster plates to deflect the primary ion beam across the sample, and (S) sample.

detect very small impurity concentrations in a solid. The technique can often provide an image of the surface from the detected ions, and can measure molecular species, in some instruments to very high masses. A typical SIMS instrument is shown schematically in Figure 16.7.

The method for removing species from the sample surface is sputtering, generally by use of an ion beam rastered across the surface. This beam can be composed of electropositive (e.g., Cs^+), electronegative (e.g., O_2^+) or inert gas (e.g., Ar^+) ions. The electronegativity of the sputtering ion increases the yield of ions of the opposite electronegativity. An exemplary SIMS depth profile from a $Cu(In,Ga)Se_2$ layer on GaAs single-crystal substrate is shown in Figure 16.8.

The primary methods for measuring the sputtered species mass include TOF, magnetic sector, or quadrupole mass spectrometers. In all cases kinetic energy is measured first, for example by deflection by an electric field through an aperture. Once the kinetic energy of the ion is known, the charge-to-mass ratio can be determined. In the TOF method, the mass is found based on the ion velocity determined by the kinetic energy, E, as $E = 1/2\, mv^2$, where m is the ion mass and v is the velocity of the ion. The energy is determined by the secondary ion accelerating field. From the time required for the ion to emerge from the sample to the detector, its mass can thus be calculated. For this method, it is necessary that the analysis ion beam is pulsed very rapidly such that the time at which an ion is emitted from the sample can be determined very precisely. This requires a very high-performance ion gun operating on a very small duty cycle. Therefore, a second ion gun is applied for the sputtering, removing sample ions down to a fixed depth (forming a crater), and no analysis takes place during that sputtering duration. Note that TOF SIMS operates essentially as static SIMS between sputtering steps.

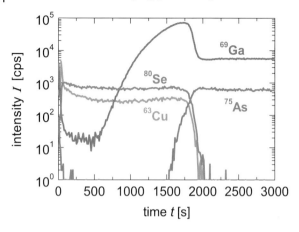

Figure 16.8 A typical SIMS profile for a single-crystal epitaxial layer of Cu(In,Ga)Se$_2$ on GaAs. Note the change in signal intensity for Ga between the Cu(In,Ga)Se$_2$, which contains only about 5 at.% Ga, and the GaAs which is 50 at.% Ga. The change is because the ion yield for Ga from GaAs is much lower than for Ga in Cu(In,Ga)Se$_2$.

The magnetic-sector mass spectrometer uses the cyclotron radius due to the Lorentz force of the ion moving at a known velocity in a known magnetic field. Thus, by detecting ions at a given angle, the mass-to-charge ratio is determined. In such an instrument, the magnetic field is typically swept to select a mass while keeping the ion energy constant. Finally, a quadrupole mass spectrometer includes an electromagnetic field that permits ions of a given mass and energy to pass through an rf electromagnetic field while rejecting other ions. The TOF instrument has a high-mass resolution (sufficient to distinguish between ^{30}Si^{1}H at a mass of 30.9816 atomic mass units (AMUs) and ^{31}P at 30.97376 AMUs) and the advantage of recording the entire mass spectrum for every analysis pulse. For the magnetic sector approach, high resolutions may be achieved, similar as for the TOF instruments with in average very high signal levels and consequently excellent detection limits. The lowest mass resolution and the highest detection limit exhibit the quadrupole mass spectrometer.

A variant technique is secondary neutral mass spectrometry (SNMS) [21–23]. In this method, a highly ionized gas plasma is generated near to the sample surface, for the purpose of ionizing neutral species sputtered from the sample. In case of SIMS, only a small fraction (a few percent at most) of the sputtered species leaves the surface as ions. Also, ion yield fluctuates substantially in a broad range from 10^{-5} to few percent. By ionizing sputtered neutral species, the fraction of species ionized is increased to nearly 100% and thus, the sensitivity of the technique is enhanced correspondingly. Furthermore, if the ionization probability were independent of the composition of the sample surface, many of the artifacts of conventional SIMS associated with ion yield fluctuations may be avoided. While SNMS is an attractive approach, it is important to realize that the ionization probability in a plasma changes considerably when metal ions of various species are introduced into the gas. Therefore, while many ions are produced, the ion yield has strong matrix effects, and the yield changes based on what was recently sputtered from the sample surface. At the same time, the transmission of ions through the gas phase and into the detector is not as high as for SIMS, that is, the sensitivity of SNMS is not necessarily improved.

16.3.2
Data Analysis

Analysis of SIMS data begins with the instrument setup, in order to eliminate instrumental artifacts. Some of these artifacts include changes in secondary ion yield, interferences, and degradation of depth resolution by displacement of atoms during sputtering. A brief summary of some of these issues is presented in the following.

Two typical mass spectra obtained by SIMS from a Cu(In,Ga)Se$_2$ thin film are shown in Figure 16.9. Comparison of the two mass spectra in Figure 16.9 reveals that there are many signals that result from molecular interferences. For example, the large number of peaks beyond In at masses 113 and 115 AMUs are due to molecular species. One may take advantage of the fact that an atom sputtered from the sample surface can have up to a few tens of eV of kinetic energy in addition to the energy provided by the extraction lens, which accelerates the ions away from the sample surface. Molecular species exhibit much smaller kinetic energies because attempting to sputter a molecule with a high kinetic energy generally leads to its fragmentation. Therefore, by closing the spectrometer apertures (field and contrast diaphragms and energy slit) the detection of atomic species can be favored over molecular ones species. However, this also reduces sensitivity. Examples of energy spectra for an atomic and a molecular species are shown in Figure 16.10. Adjusting the energy

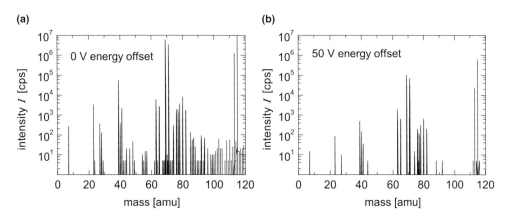

Figure 16.9 Two mass spectra from Cu(In,Ga)Se$_2$ thin films. The mass spectrum (a) was acquired under conditions maximizing sensitivity, while the acquisition of spectrum (b) was optimized for eliminating molecular interferences. The additional peaks in figure (a) are mostly compounds of elements in spectrum (b) with Na and O.

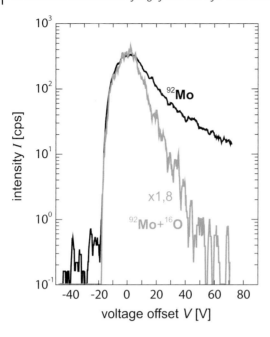

Figure 16.10 Energy spectra for the 92 AMU atomic species ^{92}Mo (gray) and the 108 AMU molecular species ^{92}Mo + ^{16}O (scaled such that the maximum signals are roughly equal). Note that at about 60 V offset, the signal intensity from atomic species is increased by a factor of 100 relative to that of the molecular species.

analyzer to the energy level greater than what the extraction lens provides is selective for atomic species. This method is not necessary with a TOF instrument because it has such a high-mass resolution that molecular species can be easily separated from atomic species.

The output of a SIMS analysis is quite simple, plotting number of counts of ions per second as the sample is sputtered. However, the secondary ion yield depends significantly on the chemistry of the surface from which the ion is emitted, the sputtering rate depends on the matrix being analyzed, and complex changes can take place that can locally alter the sputtering yield at an interface. In the exemplary depth profile shown in Figure 16.8, the profiles from the volume of the Cu(In,Ga)Se$_2$ thin film are relatively smooth for each matrix element. However, close to the interface between and the GaAs substrate, one can see the Ga signal at the back of the Cu(In,Ga)Se$_2$ film, well above the steady-state level in the GaAs. This is the case in spite of the fact that the Cu(In,Ga)Se$_2$ film contains at most 10 at.% Ga, while the substrate is 50 at.% Ga. The reason for the change is the ion yield, which is much higher for Ga in Cu(In,Ga)Se$_2$ than for Ga in GaAs. There is no way to eliminate such ion yield changes. One may correct for such a change if the ion yield is determined separately in each matrix. By calculating the fraction of the two matrices present, one can adjust the count rate in order to reduce the effect.

Because the ion yield depends upon the number of electronegative and electropositive species, sputtering with one species or the other changes the yield [24, 25]. A clean surface that has not been analyzed contains few of the sputtering beam species. As sputtering begins and large concentrations of sputtering atoms are implanted into the surface the ion yields will change [26]. The onset of sputtering also typically amorphizes the sample, which alters ion yields. Likewise, there are strong ion-yield effects typically found in the presence of variable concentrations electronegative or electropositive species in the sample. For example, O in samples profiled using a Cs^+ ion beam affects ion yields, typically increasing the yield of electropositive species and decreasing the yield of electronegative species. The details of how to detect and avoid such interferences are too lengthy to describe here and in any case are an art rather than a science. However, altering the depth-profile conditions, for example, by changing the primary ion species and the polarity of the detected species, and by looking for common elements that cause yield changes such as O, are a big help. In order to really quantify the data, one requires a relatively constant composition area in a sample and a known standard profiled immediately before or after the sample in question for comparison.

16.3.3
Quantification

Quantitative analysis of SIMS data can, in theory, be conducted based on tabulated ion yields. However, this is generally unreliable. As described above, the ion yield varies considerably from one matrix to another, with the presence of impurities (especially electropositive or electronegative elements), and with the details of the analysis condition. This may even extend to the location within a sample holder in which a sample is mounted (because of fluctuations in the secondary-ion accelerating field). Therefore, it is necessary for quantitative analysis to perform all measurements under as nearly identical conditions as possible. This often means measuring a standard sample immediately before or after measuring a test specimen. Even so, most SIMS operators will be lucky to obtain quantitative compositions to within a factor of two. The ideal way of quantifying low-concentration impurity elements is to work with a standard sample into which a known concentration of this impurity element has been implanted. This approach provides quite a reliable ion yield standard for a given analysis. An example of an implant profile from a $Cu(In,Ga)Se_2$ thin-film sample is shown in Figure 16.11. Both, atomic concentrations and depth, were quantified.

From the dose of the implant the height of the Gaussian can be calculated in atoms per cm^3 and the data thus quantified. The depth scale can also be determined from the peak of the implant, as done in the example here, although it is easier to measure the crater depth with a profilometer. Note that the depth scale ends in the $Cu(In,Ga)Se_2$ thin film because the sputtering rate of Mo is different from that of $Cu(In,Ga)Se_2$ and must be calibrated separately. This is best done by measuring craters of several depths, some ending in the $Cu(In,Ga)Se_2$ thin film and some in the Mo layer.

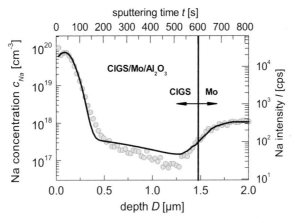

Figure 16.11 A SIMS composition depth profile (circles) of Na in a polycrystalline Cu(In,Ga)Se$_2$ thin film, and a fit to the data based on a Gaussian curve and a baseline count rate in the bulk of the sample.

16.3.4
Applications for Solar Cells

SIMS is used extensively in solar-cell manufacture in order to detect the presence of impurities, to determine the doping profiles, to analyze interdiffusion at heterojunctions, and to acquire compositional depth profiles of matrix elements. Considering the application of SIMS in some of the major solar-cell technologies, we may expect the following. In crystalline Si solar cells, the primary application is determination of the in-depth elemental distributions of dopants. During process development, impurity detection is important, but usually Si processing can (must) not introduce significant numbers of impurities. O and C are common trace elements in Si and are easily detected in SIMS in a mode detecting negative secondary ions. In amorphous Si solar cells, impurity detection and characterization of dopant distributions, studies of the movement of dopants during processing and profiling of the complete structure of multijunction structures are important. Research has shown that in extremely high-efficiency solar cells made by chemical vapor deposition, in some cases, the growth process may result in dopant diffusion and interdiffusion of layers. Proper choice of growth process may reduce these effects substantially. SIMS is an important method used to study such changes in dopant and matrix element distributions. In CdTe devices, SIMS is used for characterizing the results of processes such as CdCl$_2$ treatment. Finally, SIMS has been used regularly to study elemental distributions in Cu(In,Ga)Se$_2$ solar absorbers on Mo/glass substrates (see Figure 16.12). SIMS can also be used to resolve questions concerning, for example, diffusion of beneficial impurities such as Na [27, 28] and harmful impurities such as Fe [29] into the absorber from substrates such as soda-lime glass and stainless steel.

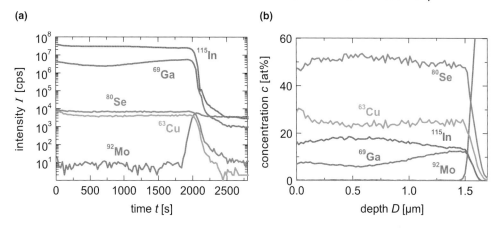

Figure 16.12 (a) Raw SIMS depth profiles from a Cu(In,Ga)Se$_2$ polycrystalline film on Mo-coated glass and (b) this SIMS data as quantified depth profiles. The ion yields for the various elements were quantified by comparison with energy-dispersive X-ray and XPS measurements, giving the integral concentrations of the matrix elements in the sample.

16.4
Auger Electron Spectroscopy (AES)

16.4.1
Introduction

AES is an analytical technique applied in order to determine the elemental composition and, in any case, to gather also information on chemical bonding from the surface region of a solid material. Compositional depth profiles can be obtained by combining AES with ion-beam sputtering. AES is broadly used for an extensive variety of materials applications, especially those requiring surface sensitivity and high spatial resolution.

16.4.2
The Auger Process

The emission of electrons as a result of the ionization process shown in Figure 16.13 were discussed in 1922 by Meitner [30] as well as were described in 1925 by Auger [31] who observed their tracks in a Wilson cloud chamber and explained correctly their origin. Figure 16.13 shows the atom immediately after K-shell ionization by an incident primary particle. The Auger process starts when an electron from an elevated energy state occupies the empty state in the K shell (as shown in Figure 16.13 for an L electron). The energy released by this transition is either emitted as a photon or given to another L electron. If this energy is sufficient, an electron can emerge from the atom. The process described above is termed a KLL Auger transition. The kinetic

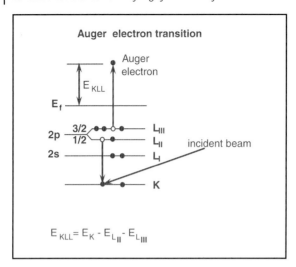

Figure 16.13 Schematics of an Auger electron emission process.

energy of an Auger electron is equal to the energy difference of the singly ionized initial state and the doubly ionized final state. For an atom of atomic number Z undergoing an Auger process involving the levels WXY, the Auger electron kinetic energy is given by the difference in the binding energies of energy levels W, X, and Y. Therefore, the resulting kinetic energy E_{KLL} is described by $E_{KLL} = E_K - E_{L_{II}} - E_{L_{III}}$ whereas all energies are referenced to the Fermi level E_f. The energies and relative intensities of WXY Auger transitions (KLL, LMM, and MNN) show a systematic behavior with increasing atomic number Z. In heavier elements, which exhibit more energy levels, more Auger transitions are possible.

Auger emission can be initiated bombarding a sample by electrons, X-rays, or ions. However, Auger instruments are typically equipped with electron guns because electron beams can be focused to very small areas.

16.4.3
Auger Electron Signals

As outlined above, Auger electrons are characterized by their kinetics energies. They can be detected by measuring the energy distribution $N(E)$ of secondary electrons, where E is the electron energy. The Auger electrons appear as prominent features together with small loss peaks on the $N(E)$ curve among the various secondary electrons ejected. Most loss peaks obtained in secondary electron spectra are either plasmon or ionization losses (plasmons are the result of collective oscillations of valence electrons against the positive atom cores).

Auger electrons may be emitted from all elements except for from H and He, and therefore, the element may be identified by its Auger spectrum, that is, by the energy positions of the peaks attributed to the transitions WXY mentioned above. The concentration of an element is related to the intensity of its peaks. Due to the dependence of the atomic energy levels, the energy loss spectrum, and the valence band structure on the local bonding, the energy positions and the shapes of Auger peaks are a result of the chemical environment of the atom. These features in the Auger electron spectrum may be interpreted accordingly.

16.4.4
Instrumentation

AES is performed in an ultrahigh vacuum system in order to minimize surface contamination during analysis. Gas molecules colliding with the surface would deposit roughly one atomic monolayer per second at 1.3×10^{-4} Pa, assuming all the molecules stick to the surface. However, sticking probabilities are usually much lesser than 1, especially after adsorption of the first monolayer. Therefore, Auger analysis should be performed at pressures $p < 1 \times 10^{-6}$ Pa, in order to inhibit reactions with adsorbates, which may alter the surface composition.

In 1967, Palmberg [32] and, separately, Tharp and Scheibner [33] reported on the measurement of Auger electrons by using a low-energy electron diffraction system. Harris [34, 35] showed the advantage of using the first derivative of the direct spectrum in order to identify Auger peaks on the large secondary electron background. In 1969, Palmberg et al. [36] introduced the cylindrical mirror analyzer (CMA) and obtained Auger spectra with improved sensitivity and with subsecond data acquisition duration.

Nowadays, an Auger system is equipped with an electron gun emitting the primary electron beam for Auger electron excitation, an electron energy analyzer and detector measuring the emitted electrons from the specimen, a secondary electron detector acquiring secondary electron images, an ion gun for surface cleaning and for performing depth-profile analysis, and a sample manipulator to locate the area of interest at the analyzer focal point.

The electrons are emitted either from a LaB_6 cathode (older systems) or from a Schottky field emitter. Schottky field emission sources may provide 1 nA of current into probe diameters of about 10 nm (electron beam energies $E_P = 10$ keV) which is sufficient for Auger analysis if channel plates for signal detection are used.

The most commonly used energy analyzers are the CMA shown in Figure 16.14 or the concentric hemispherical analyzer (CHA), which is well described in Section 16.5.2.

The basic construction of the CMA consists of a pair of coaxial cylinders with entrance and exit slits and a detector. An electric field between the two cylinders deflects electrons to the detector to an extent dependent on their kinetic energy. Scanning the electric field sweeps electrons of different energies across the detector, thus generating the spectrum. The electron gun is located coaxially within the CMA. The advantage of this arrangement is a minimization of shadowing in case of rough surfaces. The CMA is a suitable tool investigating samples with rough surfaces. The intensity response of a CMA is relatively constant as a function of sample angle over a wide range of angles.

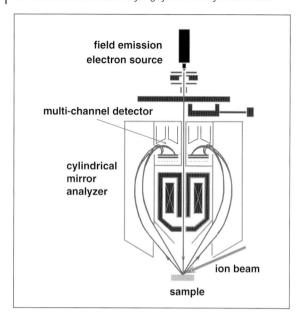

Figure 16.14 Scheme of the CMA (reprinted with permission of Physical Electronics GmbH).

An important parameter of all energy analyzers is the resolution ΔE, which is generally determined by the full width at half maximum (FWHM). ΔE varies proportionally with the energy E and for a given analyzer geometry, the relative resolution, $\Delta E/E$, is constant. The relative energy resolution of CMAs is between 0.3 and 1.2%. As mentioned above, the FWHM of peaks measured by the CMA increase with E, however, the angular acceptance of electrons is defined by the input slit geometry and remains constant with energy. Therefore, the measured spectrum cannot be expressed by $N(E)$ but is described by the product $E\ N(E)$.

The detector is usually a single-channeltron electron multiplier or a microchannel-plate electron multiplier. More detailed additional information about the CMA as well as other energy analyzers is available in the literature [37].

In Figure 16.15a the $E\ N(E)$ surface spectrum of the active layer of a solar cell is presented. As mentioned above (Section 16.4.3), the various Auger peaks appear on a varying background. Figure 16.15b shows the same spectrum after differentiation, which may be carried out either electronically (older systems) or by applying a suitable algorithm such as the Savitzky–Golay filter [38]. The Auger peaks in the filtered spectrum (Figure 16.15b) appear as characteristic features on a low varying background. The peak intensity can be easily determined from the distance between peak maximum and peak minimum. Identification of Auger peaks will be commonly carried out by comparing peak energies and peak shapes with the standard spectra collected in Ref. [39].

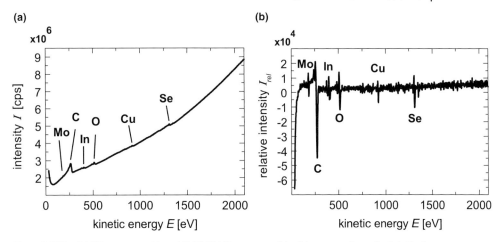

Figure 16.15 $E N(E)$ spectrum (a) and $E dN(E)/dE$ spectrum (b) of the top surface of a CuInSe$_2$/Mo/glass stack (exciting primary-electron energy was 5 keV at 20 nA primary electron beam current).

16.4.5
Auger Electron Signal Intensities and Quantification

Figure 16.16 illustrates the effects within the sample created by the incident electron beam. Size and shape of the interaction volume depend on the energy of exciting primary electrons and the atomic number of the sample material. Auger electrons are created both in the near surface region and within greater depths of the analyzed sample. However, only those electrons which are excited within the near surface region can escape the sample. Electrons that are scattered by interactions with the electrons of the sample atoms incurring energy losses form loss structures on the low energy side of the Auger peaks and electrons experiencing random and multiple loss processes form the continuous background. The distance traveling

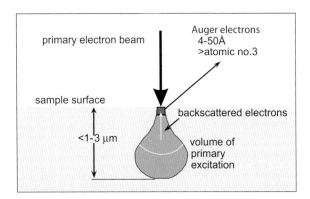

Figure 16.16 Electron beam–sample interaction (reprinted with permission of Physical Electronics GmbH).

an electron without energy loss is λ the inelastic mean free path. It specifies the detected volume from which Auger electrons can escape the sample. λ depends on the material and the energy E_{WXY} and is independent of the energy of the primary electrons. Experimentally determined values are presented in Ref. [40]. The smallest λ values are near 75 eV. Between 100 and 2000 eV, λ increases approximately as $E^{1/2}$ but rises more rapidly thereafter and below about 75 eV, λ again increases. Low-energy electrons can escape from only the first several atom layers of a surface because they have short mean free paths in solids, for example, they are strongly absorbed by even a monolayer of atoms. This property gives AES its high surface sensitivity.

Two further important factors are the ionization cross section, σ, and the backscattering factor, r. σ depends on the energy of the primary electrons E_p, on the critical energy for ionization E_W, and on the ionized species. Maximum ionization should occur near to about $E_p/E_W = 3$. However, extra ionization caused by backscattered electrons increases the ratio to about $E_p/E_W = 6$. The backscattering factor r depends on E_p and on the atomic number Z. For increased E_p and Z values, an enhanced ionization contribution is obtained due to the effect from backscattered electrons.

Surface roughness is also an important issue. The escape probability of electrons from rough surfaces is smaller than for flat ones. Electrons emerging from a rough surface may be reabsorbed by the surrounding material. Using an analyzer with a large acceptance angle such as a CMA, this roughness effect is minimized [1].

16.4.6
Quantification

Various procedures exist for quantifying Auger peak intensities. A quantification approach commonly used in the AES community introduces an atomic sensitivity factor, S_i, [39] of element i. The atomic concentration, c_A, of a given element A in a sample can be expressed as

$$c_A = (I_A/S_A)/\Sigma_i(I_i/S_i) \tag{16.7}$$

where I_A = Auger intensity of WXY transition of element A and I_i = Auger intensity of WXY transition of element i.

Equation 16.7 can be applied if only minor variations of r, λ, and the Auger transition probability, γ, of the WXY transition appear, such as for homogeneous samples. For these samples, a set of RSFs may be used, determined for each beam voltage, normalized to a reference material [39].

However, in case of alloys or compounds, sensitivity factors determined on pure elemental standard samples may lead to incorrect results. The atomic density of element i, N_i, r, and λ can sometimes vary considerably, and also changes in peak shape, induced by the changed chemical environment of the emitting atom, can appear. Either sensitivity factors determined on samples of the same composition or correction factors, as proposed in the literature [41, 42], have to be applied for quantification.

16.4.7
Application

If the electron beam is scanned across the sample surface along with a secondary electron detector, corresponding images can be acquired, which is useful when localizing the region of interest on the sample surface. Auger line scans or maps can be recorded by measuring the intensity of a certain Auger peak while scanning the electron beam across a region of interest. In Figure 16.17a–c, a secondary electron image from a cross-section sample of a glass/Mo/Cu(In,Ga)Se$_2$/CdS/ZnO stack and the corresponding linescans across this stack are presented (the distribution profiles shown in Figure 16.17c are details of the ones given in Figure 16.17b. A primary electron beam of 10 keV at 1 nA was applied.

The ZnO layer appears on the right hand side of the linescan (Figure 16.17b) followed by the Cu(In,Ga)Se$_2$ thin film, the Mo layer and the glass substrate. Since the glass substrate is an electrically insulating material, charging effects caused by the primary electron beam appear. Charging can be reduced substantially either by applying low primary beam energies and currents, by tilting the sample to a grazing-incidence angle of the primary electron beam or by neutralizing charging by low-energy ion bombardment. In the present case, the sample was tilted (60° to the surface normal). Additionally, the surface was flooded by low-energy ions (70 eV).

When applying ion-beam sputtering on the identical thin-film stack, in-depth elemental distribution profiles were obtained. Commonly, alternating AES and sputtering is performed. The resulting profile is shown in Figure 16.17d. The analysis was performed applying 5 keV at 5 µA Ar$^+$ ion sputtering as well as 5 keV primary electron beam excitation ($I = 20$ nA). Intensity quantification was not possible since no appropriate sensitivity factors for the Cu(In,Ga)Se$_2$ layer were available. The depth was calculated from the sputtering time by introducing the sputtering rate which was determined from in-depth elemental distribution profiles of a standard layer of known thickness.

The in-depth profile agrees well with the linescan acquired on the cross-section sample, shown in Figure 16.17c. While Cu and Se signal intensities are constant throughout the layer, those of Ga and In signals exhibit gradients already found by GD-OES and SIMS (see Sections 16.2.4.1 and 16.3.4). The decrease of the In, Cu, Se, and Ga signals at the surface is caused by the covering C contamination layer.

Substantial roughness may be found at the bottom of the sputter crater. The results are a decrease of the depth resolution and intensity structures as well as a broadening of signals at interfaces in the thin-film stack. Furthermore, strong roughness leads to a decrease of intensities. These effects can be clearly seen in Figure 16.17d, where the structured course of In, Cu, Se, and Ga signals is smoothed and their intensities are lower than in the linescan analysis and the signals at the Cu(In,Ga)Se$_2$/Mo as well as Mo/glass interfaces are slightly broadened. In this context, it should be noted that roughness may be minimized by rotating the sample during sputtering.

16 Elemental Distribution Profiling of Thin Films for Solar Cells

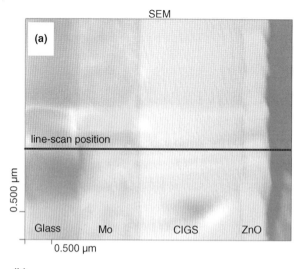

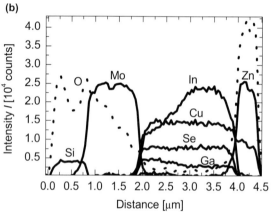

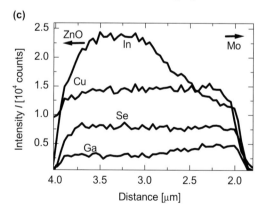

Figure 16.17 (a) Scanning electron micrograph from a cross-section of a glass/Mo/Cu(In,Ga)Se$_2$ (CIGS)/CdS/ZnO stack. (b) Linescan across this stack by means of scanning AES. (c) Detail of the linescan presented in (a). (d) Auger in-depth elemental distribution profiles of the layer of the identical stack, performed at 5 keV Ar$^+$ ion sputtering.

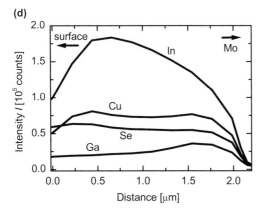

Figure 16.17 (Continued)

16.5
X-Ray Photoelectron Spectroscopy (XPS)

16.5.1
Theoretical Principles

XPS or electron spectroscopy for chemical analysis (ESCA) is probably one of the most widely used surface science techniques. It utilizes X-ray photons to ionize atoms and analyzes the kinetic energies of the ejected photoelectrons. As the core electron binding energies of the elements are distinctive, photoelectron spectroscopy can be a valuable tool in determining material composition. Furthermore, chemically inequivalent atoms of the same element are known to show measurably differing binding energies. This "chemical shift" information can be most useful in investigating surface reactions or changes in the state of the surface, as induced by, for example, oxidation.

At its heart, XPS is based on the photoelectric effect outlined by Einstein in 1905 [43] and was developed by Siegbahn et al. in the 1950s and 1960s [44]. The Einstein relationship relates the kinetic energy of the emitted photoelectron E_{KE} excited by a photon of energy $h\nu$, to the binding energy of the electrons in the atom E_{BE} referenced to the vacuum level. For photoemission from solids, the binding energy is referenced to the Fermi level of the material, resulting in Eq. (16.8):

$$E_{KE} = h\nu - E_{BE} - \phi \qquad (16.8)$$

where ϕ is the workfunction of the solid. Equation (16.8) assumes that the photoelectrons suffer no change in energy between emission from the surface and detection in the spectrometer, that is, the process is elastic. In practice, the loss of energy due to the workfunction of the spectrometer also needs to be included in Eq. (16.8). This offset is typically handled by the instrument data system internally. If the experiment is conducted by use of incident photons of much greater energy than the workfunction, the kinetic energy spectrum will correlate directly to the discrete binding energies of the electrons in the solid. However, there are a finite number of electrons which lose energy due to inelastic collisions in the solid, which results in the photoelectron peaks detected against a background increasing from high-to-low kinetic energy [45].

Once an incident photon has interacted with a solid, and a photoelectron has been ejected, the solid is left with a vacancy in a core electron level. This is generally occupied by an electron from a higher energy level, and the subsequent de-excitation of the atom may occur by either emission of an Auger electron or X-ray fluorescence. The resultant spectrum is a combination of peaks attributed to photoelectrons and Auger electrons.

Figure 16.18a shows a typical wide scan (or "survey") XPS spectrum, and Figure 16.18b a high-resolution scan of the Si2p region for a Si wafer coated with a thin oxide layer. The area under each peak in the spectrum is directly related to the amount of the substance present in the irradiated area. This allows standard-less, quantitative analysis of the surface or thin film to be performed.

XPS is a surface sensitive technique. The sensitivity toward surface atoms is due to the relatively short path that a photoelectron can travel before it is subject to an inelastic scattering process, that is, spectra are the product of atoms in the outer ten nanometers of the sample surface. For the analysis of thin films, cycles of ion bombardment followed by analysis are performed, extending the depth that can be investigated to a few microns, with accurate determination of compositional changes

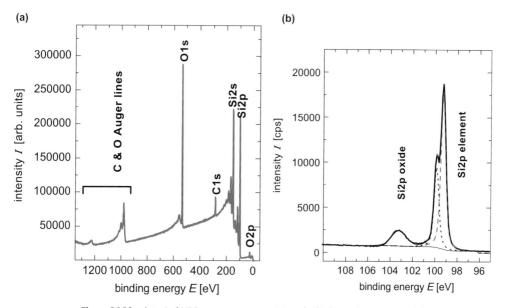

Figure 16.18 A typical XPS survey spectrum (a) and a high-resolution scan of the Si$_2$p region (b) from a Si wafer sample, showing the chemical state information available.

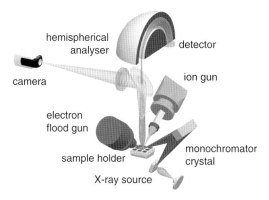

Figure 16.19 Schematic of a typical XPS instrument configuration. (*The schematic of the K-Alpha XPS instrument used by kind permission of Thermo Fisher Scientific*).

at interfaces between layers, which is possible due to the narrow electron escape depth.

16.5.2
Instrumentation

A schematic of the typical internal components of an XPS system required for thin-film solar cell analysis is shown in Figure 16.19. XPS analyses have to be performed in high or ultrahigh vacuum. Primarily this is to prevent the emitted photoelectrons colliding with gas molecules, which would significantly attenuate the signal. The vacuum also assists in increasing the time before an impurity-free surface adsorbs gas molecules from the background in the chamber sufficiently to influence the analysis.

The X-ray photon source used for the XPS experiment varies according to the desired resolution and power required, from simple, low-power X-ray "guns" to the use of synchrotron sources. Lab instruments generally have either a monochromated X-ray source, which are usually configured to expose the sample to only Al-Kα X-rays, or a nonmonochromated source. Modern instruments tend to have monochromated sources which offer better spectral resolution, good sensitivity, and the possibility of focusing the beam on a smaller area. The X-rays are monochromated by reflecting the beam from a quartz crystal, which has a suitable Bragg constant to transmit the Al-Kα X-rays only.

Most instruments include a CHA for the accurate analysis of the kinetic energy distribution of the photoelectrons. Electrons are focused on the entrance of the analyzer by a series of lenses. A potential difference, called the pass energy, is applied across the two hemispherical plates, which allows electrons within a narrow kinetic energy range to be transmitted through the analyzer. Electrons are collected at the exit by a detector, either channeltrons or a multichannel plate. XPS analyzers typically run in a constant energy mode, that is, the pass energy is fixed, and a retarding potential is applied before the analyzer slit in order to select the kinetic energy. By scanning the energy range, the spectrum is generated.

No special sample preparation is required for XPS. Both conducting and insulating samples can be analyzed. For conducting samples, the ejected photoelectrons are replaced by conduction from ground through the sample holder. Insulating samples, such as thin-film solar cells deposited on a glass substrate, require an external source of electrons to prevent an excessive build-up of charge at the surface, which distorts the appearance of the spectrum. Electrons are typically supplied from a low energy flood source, which generates a beam slightly larger than the analysis area with the aim of either neutralizing any build-up of charge, or by overcompensating gas ions to eliminate excess negative charge which may be present on the sample and deflect the flood gun beam.

The final component required is an ion source for removing surface material, forming a crater in which the chemical composition is measured. Ar^+ is the most commonly used ion, with typical acceleration energies in the range of 0.1–5 kV. A typical depth-profile experiment is carried out by alternating spectral acquisition and periods of ion sputtering. The sample is usually rotated during the etching cycle in order to minimize sputter induced roughness in the direction of the ion beam and to reduce the possibility of heavier elements migrating into lower layers, both of which can cause a significant reduction in depth resolution [1].

The main limitation of XPS is that the best possible spatial resolution using a lab instrument is around 3 μm. This is due to the twin difficulties of focusing the X-ray beam below 10 μm, and the low signal level when operating the analyzer to give the ultimate spatial resolution of 3 μm. Acquisition times can also be large, with several hours required for depth-profile experiments on large areas of interest or thick layers.

16.5.3
Application to Thin Film Solar Cells

XPS can be applied to analysis of the surface chemistry, layer compositions, and compositional gradients within thin-film solar cells. Figure 16.20 shows a depth profile through a complete device, from the ZnO front to the Mo back contact. By initially considering only the variation in each elemental signal it is possible to obtain direct quantification of the composition of each layer. The signals are quantified by measuring the area of a peak due to the emission electron from a particular orbital, and normalizing it via a RSF based on the ionization potential of that electron in that element. There are two main groups of sensitivity factors, based either on calculation (Scofield [46]) or experimental observation (Wagner [47]). The Scofield set is more widely used, as it covers the entire periodic table. This is of particular interest in the $Cu(In,Ga)Se_2$ layer, where the In and Ga elemental distributions can be measured with good depth resolution, which is of the order of a few tens of nanometers, and can be less than 3 nm (in this case the structure of each of the interfacial layers is quite rough, and so the interface resolution measured in this experiment is in the range 50–80 nm. Some variation in the ZnO stochiometry is also evident from the elemental profile. These two examples

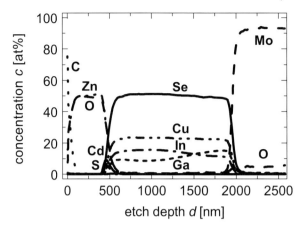

Figure 16.20 XPS depth profile of a Cu(In,Ga)Se$_2$ solar cell. The data were collected using a 150 μm X-ray spot (Al-Kα 1486.6 eV) and a 180° hemispherical analyzer fitted with a 128-channel detector. The sample was sputtered using 2 kV/2 μA Ar$^+$ ions, over a 2 × 4 mm area, and it was rotated during etching. The total acquisition time was 6 h.

demonstrate the rich information that can be obtained from a simple level of data processing.

By close examination of the spectra, it is possible to identify changes in chemical state. For example, at the interface of two layers, it is possible that there could be an interaction between components that influences the performance of a device. This interaction layer could be only a few nanometers in thickness, but would be evident in several levels of the XPS depth profile.

Figure 16.21 shows In3d spectra from an area near the CdS layer and one near the Mo back contact. The small variation in peak position indicates that the chemistry at

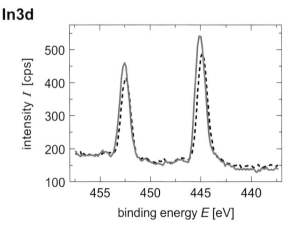

Figure 16.21 In3d spectra for regions near the CdS layer (dashed line), and near the Mo layer (solid line).

the two positions is slightly different. This is most likely to be due to the change in the [In]/[Ga] ratio between these two regions. Information such as this can indicate problems in the manufacturing process, such as unwanted oxidation (*c.f.* the presence of a low level of oxygen detected in the Mo layer in Figure 16.20).

XPS depth-profiling analysis is destructive, and accurate calibration of the etching rate through multilayered samples is challenging. Often depth scale calibration relies on another technique (e.g., an SEM cross section) or is based on the etch rate measured using a standard of known thickness, such as Ta_2O_5/Ta or SiO_2/Si. The limit on the spatial resolution means that identifying the precise location of submicron surface defects is impossible, and that rough interfaces will have an average depth resolution measured by the experiment.

Despite these limitations, XPS has proven to be an important technique for the investigation of elemental distributions in solar cell devices, failure analysis, and quality assurance applications. Advances in instrument design and software capability have enabled significant reductions in acquisition time and detection limits. The ability to analyze samples with no special preparation steps required permits both initial research developments and finished devices to be measured. In summary, the ability to directly quantify the concentrations and subtle variations in the chemistry of material present, with extreme surface selectivity, makes XPS a useful addition to the armory of the materials analyst.

16.6
Energy-Dispersive X-Ray Analysis on Fractured Cross Sections

In order to detect small traces of possibly light elements, such as, for example, Na, EDX is certainly not an appropriate technique to be used as the detection limit in particular for light elements is not sufficient. But if in need of a fast method with virtually no sample preparation involved, which can provide valuable information regarding the in-depth matrix-element distributions of the deposited thin film (e.g., the In and Ga distributions in $Cu(In,Ga)Se_2$ thin films), EDX in a scanning electron microscope may be the method of choice.

16.6.1
Basics on Energy-Dispersive X-Ray Spectrometry in a Scanning Electron Microscope

As part of the inelastic interaction of sample atoms with the electron beam in a SEM, X-rays characteristic for these atoms are emitted (Section 12.2.3). These X-rays can be used to identify constituent elements of the sample under investigation. Depending on the detector in use, both, the wavelength or the energy of the emitted X-rays, can be used for a compositional analysis [48]. Since sample drift can be an issue when recording elemental distribution profiles on a solar cell cross section, the use of high acquisition rates is often advantageous.

When investigating homogeneously composed samples, EDX is a well suited method for the quantitative determination of the sample composition [49]. It is, however, important to realize that, performing SEM–EDX on multilayer stacks in a cross-sectional configuration, the spatial extension of the interaction volume of the electron beam with the specimen always comprises inhomogeneous material distributions. This will render efforts to obtain reliable quantified information futile in the vast majority of cases. However, in general the qualitative distribution of a matrix element that is obtained by referring to the net counts may be transformed into a corresponding concentration distribution by relating the average net count of a specific element to its average concentration, which may be determined, for example, by means of X-ray fluorescence [50]. For good comparability of various measurements, it is necessary to either ensure identical measurement conditions or, alternatively, to perform normalization to an element of which ideally the concentration is assumed to be constant throughout the depth of a particular layer.

In order to ensure a favorable ratio of the characteristic X-ray line signal and the background, the kinetic energy of the exciting electrons should – as a rule of thumb – be at least twice as high as the maximum line energy used for analysis. Hence, the position of the characteristic X-ray lines for the relevant elements and their possible overlap determine the minimum acceleration voltage to be applied for the SEM–EDX analysis.

Figure 16.22 shows a net-count line scan from the fractured cross section of a ZnO/CdS/Cu(In,Ga)Se$_2$/Mo/glass stack, in which the Cu(In,Ga)Se$_2$ layer was deposited by standard multistage coevaporation [3]. The acceleration voltage was 7 kV, using a beam current of 250–300 pA. The value for the acceleration voltage was chosen according to the excitation energy for the In-L line (3.4 keV). The line scan consists of

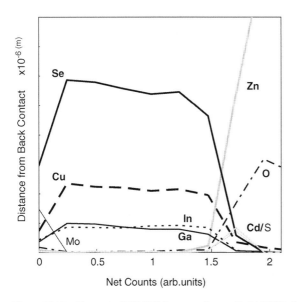

Figure 16.22 Net-count SEM-EDX line scan from a ZnO/CdS/Cu(In,Ga)Se$_2$/Mo/glass stack, recorded using acceleration voltage of 7 keV and a beam current of 200 pA.

10 measurement points between the Mo back and the ZnO front contact of the solar cell device. Although the thickness of the CdS buffer layer is only in the range of about 50 nm, it is clearly visible. Apparently, Ga and In are not homogeneously distributed. The corresponding elemental distributions exhibit in-depth gradients.

16.6.2
Spatial Resolutions

Several approaches have been published [48, 51–53], which estimate the extension R of the interaction volume of an electron beam with a compound, considerations which are relevant for all electron-beam-related techniques (see also Chapter 12). In SEM–EDX, the value R represents an intuitive and reliable limit for the spatial resolution. For $Cu(In,Ga)Se_2$ thin-film solar cell devices R varies from 130 to 270 nm, depending on the material hit by the electron beam. A definition of the spatial resolution similar to the one for analytical transmission electron microscopy [54] is difficult to obtain in the case of SEM–EDX due to the usually large specimen thickness, that is, the comparably large interaction volume of the electron beam with the sample. The signal quality and also possible superpositions of characteristic lines with those of the background have a certain impact on the minimum expansion of the patterns to be resolved. It is hence highly recommended to determine the lateral resolution experimentally for the material system and measurement conditions used. Other error sources to be considered are the measurement system error, the error introduced by sample drift during signal acquisition, the error caused by the roughness of the surfaces and interfaces of and within the sample that is analyzed, and, in connection with the sample surface roughness and sample orientation, errors introduced by reabsorption of emitted characteristic X-rays and subsequent fluorescence processes [48].

16.6.3
Applications

Modern SEM–EDX systems are often equipped with the possibility to record spectral images of an area of interest. That is, EDX spectra are recorded on a defined, spatially resolved pattern and can be analyzed after the measurement. This enables the acquisition of elemental maps and/or "wide" linescans that provide for a certain lateral averaging. Figure 16.23 shows a cross-sectional scanning electron micrograph from a $ZnO/CdS/Cu(In,Ga)Se_2/Mo/glass$ stack. Also indicated is the region of interest in which EDX data acquisition was performed using the identical measurement set-up, with 250–300 pA and with 4–5 nA electron beam current.

A net-count line scan along the arrow was extracted from the collected EDX data and is displayed in Figure 16.24a for the In-L, Ga-L and Se-L lines. As the acquisition durations were comparable in range, the number of net counts is considerably higher for the measurement at 4–5 nA. The $Cu(In,Ga)Se_2$ thin film depicted in Figure 16.23 is fabricated at a nominal substrate temperature of 330 °C. Standard

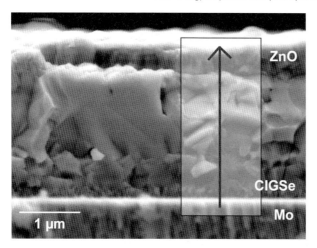

Figure 16.23 Cross-sectional SEM image from a ZnO/CdS/Cu(In,Ga)Se$_2$/Mo/glass stack, for which the Cu(In,Ga)Se$_2$ (CIGSe) absorber was deposited at the rather low nominal substrate temperature of 330 °C; the acquisition area for spectral imaging is indicated by shaded area.

processes are normally performed at temperatures above 500 °C, and a comparison with Figure 16.22 shows that at the low process temperature, a much stronger Ga gradient forms in the Cu(In,Ga)Se$_2$ layer.

It was already mentioned above that, in order to improve comparability of individual measurements, the normalization to an element with a constant concentration throughout the layer may be of use. In the case of Cu(In,Ga)Se$_2$ thin films fabricated by coevaporation, the Se concentration is assumed to comply with this requirement [50].

Figure 16.24b gives Ga and In signals normalized to the Se concentration. Due to the roughness of the investigated surface the effect of reabsorption, which is higher for lower beam energies needs to be considered. It is important to be aware of that a possible error introduced by reabsorption of Se-L X-rays is enhanced when normalization to the Se-L line is carried out for high-energy signals. On the other hand, it may also have a compensating effect for low-energy lines. After normalization, the line scans which were performed with the different beam currents show a good match. Both, Ga-L and Ga-K, signal distributions indicate features with spatial extensions as low as approximately 150 nm. These are the two distinct peaks, which designate a particularly high Ga content for distances below 1 μm away from the back contact and a peak in the Ga signal above 2 μm of distance. All of these features can be ascribed to particular stages of the Cu(In,Ga)Se$_2$ deposition process [3]. It is interesting to note the slight difference in resolution for the line scan recorded at 4–5 nA when compared with the line scan acquired at 250–300 pA. The peak width seen in the normalized spectra is slightly increased for the use of a beam current of 4–5 nA. Also, at distances larger than 2.6 μm from the back contact, the line scan at 4–5 nA exhibits an increase of the Ga signal while the line scan at 250–300 pA does not. This may be the

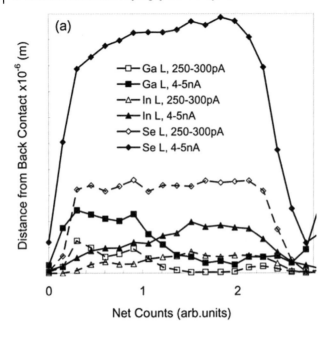
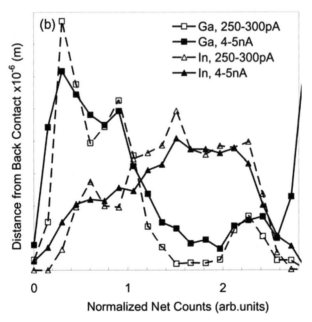

Figure 16.24 SEM-EDX line scans from the ZnO/CdS/Cu(In,Ga)Se$_2$/Mo/glass stack given in Figure 16.23: (a) net counts as recorded, (b) net counts normalized to Se-L.

consequence of the proximity of the Zn-L and Ga-L lines and an EDX signal misinterpretation due to the larger interaction volume for the higher beam current. However, the use of higher beam currents allows for a much reduced measurement duration, which reduces the risk of drift and still shows a reasonable lateral resolution.

16.6.3.1 Sample Preparation

Cross-section specimens for SEM–EDX can simply be prepared by fracturing or cutting of the sample. Care should be taken regarding contacting of the sample in order to avoid electrical charging during the measurement as this may lead to additional drift. The use of finished solar cell devices for depth profiling of the absorber layer is one way to circumvent more laborious methods of increasing the samples surface conductivity like, for example, the deposition of a very thin carbon layer. At times, it may also be advantageous to prepare polished cross-section specimens (see Chapter 12) in order to minimize errors that may be introduced by rough cross-sectional surfaces.

Acknowledgement

Volker Hoffmann, Denis Klemm, Varvara Efimova (IFW Dresden), and Cornel Venzago from AQura GmbH gratefully acknowledge the financial support from the FP6 Research Training Network GLADNET (No. MRTN-CT-2006-035459). The group from IFW Dresden thanks the Spectruma Analytik GmbH and HZB, Berlin for good collaborations. Cornel Venzago wishes to thank Agnès Tempez from Horiba Jobin Yvon for collaboration with rf-GD-TOF measurements. Christian A. Kaufmann and Raquel Caballero are grateful to Jürgen Bundesmann for technical support.

References

1 Reniers, F. (2009) In-depth analysis: Methods for depth profiling, in: *Handbook of Surface and Interface Analysis: Methods for Problem-Solving* (eds J.C. Riviere and S. Myhra) Taylor & Francis Group LLC, London, p. 273.

2 Gabor, F.A.M., Tuttle, J.R., Bode, M.H., Franz, A., Tennant, A.L., Contreras, M.A., Noufi, R., Jensen, D.G., and Hermann, A.M. (1996) Band-gap engineering in Cu(In,Ga)Se$_2$ thin films grown from (In,Ga)$_2$Se$_3$ precursors. *Sol. Energy Mater. Sol. C.*, **41–42**, 247–260.

3 Kaufmann, C.A., Caballero, R., Unold, T., Hesse, R., Klenk, R., Schorr, S., Nichterwitz, M., and Schock, H.-W. (2009) Depth profiling of Cu(In,Ga)Se$_2$ thin films grown at low temperatures. *Sol. Energy Mater. Sol. C.*, **93**, 859–863.

4 Gloeckler, M. and Sites, J.R. (2005) Band-gap grading in Cu(In,Ga)Se$_2$ solar cells. *J. Phys. Chem. Solids*, **66**, 1891–1894.

5 Sakurai, K., Shibata, H., Nakamura, S., Yonemure, M., Kuwamore, S., Kimura, Y., Ishizuka, S., Yamada, A., Matsubara, K., Nakanishi, H., and Niki, S. (2005) Properties of Cu(In,Ga)Se$_2$: Fe thin films for solar cells. *Mater. Res. Soc. Symp. Proc.*, **865**, 417–421.

6 Caballero, R., Kaufmann, C.A., Eisenbarth, R., Cancela, M., Unold, T., Eicke, A., Klenk, R., and Schock, H.W. (2009) The influence of Na on low

temperature growth of CIGS thin film solar cells on polyimide substrates. *Thin Solid Films*, **517**, 2187–2190.

7 Payling, R., Delwyn, J.G. and Bengtson, A. (eds) (1997) *Glow Discharge Optical Emission Spectrometry*, John Wiley & Sons, Inc., New York, Weinheim, Brisbane, Singapore, Toronto.

8 Marcus, K.R. (1993) *Glow Discharge Spectroscopies*, Plenum, New York.

9 Grimm, W. (1968) Eine neue Glimmentladungsquelle für die optische Emissionsspektralanalyse. *Spectrochim. Acta Part B*, **23** (7), 443–454.

10 Voronov, M., Hofmann, T., Smid, P., and Venzago, C. (2009) Microsecond pulsed glow discharge applied to a sector-field mass-spectrometer. *J. Anal. At. Spectrom.*, **24**, 676–679.

11 Bengtson, A. and Nelis, T. (2006) The concept of constant emission yield in GDOES. *Anal. Bioanal. Chem.*, **385** (3), 568–585.

12 Adams, F., Gijbels, R., and Van Grieken, R.E. (eds) (1988) *Inorganic Mass Spectrometry*, John Wiley & Sons, Inc., New York, Chichester, Brisbane, Toronto, Singapore.

13 Hohl, M., Kanzari, A., Michler, J., Nelis, T., Fuhrer, K., and Gonin, M., (2006) Pulsed r.f.-glow-discharge time-of-flight mass spectrometry for fast surface and interface analysis of conductive and non-conductive materials. *Surf. Interface Anal.*, **4**, 292–295.

14 Nelis, T. (1997) Calibration, in *Glow Discharge Optical Emission, Spectrometry* (eds R. Payling, D.G., Jones, and A. Bengtson) John Wiley & Sons, Inc., New York, Weinheim, Brisbane, Singapore, Toronto, pp. 413–417.

15 Bengtson, A. (1994) Quantitative depth profile analysis by glow discharge. *Spectrochim. Acta Part B*, **49** (4), 411–429.

16 Vieth, W. and Huneke, J.C. (1991) Relative sensitivity factors in glow discharge mass spectrometry. *Spectrochim. Acta*, **46B** (2), 137–153.

17 Rissom, T., Mainz, R., Kaufmann, C.A., Caballero, R., Efimova, V., Hoffmann, V., and Schock, H.-W. (2010) Examination of growth kinetics of copper rich Cu(In,Ga)Se$_2$-films using synchrotron energy dispersive X-Ray diffractometry. *Sol. Energy Mater. Sol. C.*, **95**, 250–253. doi: 10.1016/j.solmat.2010.05.007.

18 Lobo, L., Pisonero, J., Bordel, N., Pereiro, R., Tempez, A., Chapon, P., Michler, J., Hohl, M., and Sanz-Medel, A. (2009) A comparison of non-pulsed radiofrequency and pulsed radiofrequency glow discharge orthogonal time-of-flight mass spectrometry for analytical purposes. *J. Anal. At. Spectrom.*, **24**, 1373–1381.

19 Wilson, R.G., Stevie, F.A., and Magee, C.W. (eds) (1989) *Secondary Ion Mass Spectrometry: A Practical Handbook for Depth Profiling and Bulk Impurity Analysis*, Wiley, New York.

20 Benninghoven, A., Werner, H.W., and Rüdenauer, F.G. (eds) (1987) *Secondary Ion Mass Spectrometry: Basic Concepts, Instrumental Aspects, Applications and Trends*, Wiley, New York.

21 Szymczak, W. and Wittmaack, K. (2003) Ion-to-neutral conversion in time-of-flight secondary ion mass spectrometry. *Appl. Surf. Sci.*, **203–204**, 170–174.

22 Veryovkin, I.V., Calaway, W.F., Moore, J.F., Pellin, M.J., Lewellen, J.W., Yuelin, Li., Milton, S.V., King, B.V., and Petravic, M. (2004) A new horizon in secondary neutral mass spectrometry: Post-ionization using a VUV free electron laser. *Appl. Surf. Sci.*, **231–232**, 962–966.

23 Oechsner, H., Getto, R., and Kopnarski, M. (2009) Quantitative characterization of solid state phases by secondary neutral mass spectrometry. *J. Appl. Phys.*, **105** (6), 063523.

24 Magee, C.W. and Frost, M.R. (1995) Recent successes in the use of secondary ion mass spectrometry in microelectronics materials and processing. *Int. J. Mass Spectrom. Ion. Process*, **143**, 29–41.

25 Williams, P. and Baker, J.E. (1980) Quantitative analysis of interfacial impurities using secondary-ion mass spectrometry. *Appl. Phys. Lett.*, **36** (10), 842–845.

26 Bryan, S.R., Linton, R.W., and Griffis, D.P. (1987) Characterization and removal of ion yield transients in the near surface region of secondary ion mass spectrometry depth profiles. *J. Vac. Sci. Technol. A*, **5** (1), 9–14.

27. Rockett, A., Granath, K., Asher, S., Al Jassim, M.M., Hasoon, F., Matson, R., Basol, B., Kapur, V., Britt, J.S., Gillespie, T.S., and Marshall, C. (1999) Na incorporation in Mo and CuInSe$_2$ from production processes. *Sol. Energy Mater. Sol. Cells*, **59**, 255–264.
28. Bodegård, M., Granath, K., Stolt, L., and Rockett, A. (1999) The behaviour of Na implanted into Mo thin films during annealing. *Sol. Energy Mater. Sol. Cells*, **58**, 199–208.
29. Batchelor, W.K., Beck, M.E., Huntington, R., Repins, I.L., Rockett, A., Shafarman, W.N., Hasoon, F.S., and Britt, J.S. (2002) Substrate and back contact effects in CIGS devices on steel foil. Conference Record of the 29th IEEE Photovoltaics Specialists Conference, New Orleans, Louisiana, U.S.A., May 19–24, 2002. IEEE, Piscataway, NJ, pp. 716–719.
30. Meitner, L. (1922) Über die β-Strahl-Spektra und ihren Zusammenhang mit der γ-Strahlung. *Zeitschrift für Physik A Hadrons and Nuclei.*, **11**, 35–54.
31. Auger, P. (1925) Sur l'effet photoélectrique composé. *J. de Phys. et Le Radium*, **6**, 205–208.
32. Palmberg, P.W. (1967) Secondary emission studies on Ge and Na-covered Ge. *J. Appl. Phys.*, **38**, 2137–2147.
33. Tharp, L.N. and Scheibner, E.J. (1967) Energy spectra of inelastically scattered electrons and LEED studies of tungsten. *J. Appl. Phys.*, **38**, 3320–3330.
34. Harris, L.A. (1968) Analysis of materials by electron-excited Auger electrons. *J. Appl. Phys.*, **39** (3), 1419–1427.
35. Harris, L.A. (1968) Some observations of surface segregation by Auger electron emission. *J. Appl. Phys.*, **39** (3), 1428–1431.
36. Palmberg, P.W., Bohn, G.K., and Tracy, J.C. (1969) High sensitivity Auger electron spectrometer. *Appl. Phys. Lett.*, **15**, 254–255.
37. Young, V.Y. and Hoflund, G.B. (2009) Photoelectron Spectroscopy (XPS and UPS), Auger Electron Spectroscopy (AES), and Ion Scattering Spectroscopy (ISS), in *Handbook of Surface and Interface Analysis - Methods for Problem-Solving*, 2nd edn (eds J.C. Rivière and S. Myhra), Taylor & Francis Group, LLC, pp. 46–55.
38. Savitzky, A. and Golay, M.J.E. (1964) Smoothing and differentiation of data by simplified least squares procedures. *Anal. Chem.*, **36** (8), 1627–1639.
39. Childs, K. D., Carlson, B. A., LaVanier, L. A., Moulder, J. F., Paul, D. F., Stickle, W. F., Watson, D. G. (1995) *Handbook of Auger Electron Spectroscopy*, 3rd edn (ed. C.L. Hedberg), Physical Electronics, Minnesota.
40. Czanderna, A.W. (1975) Methods of surface analysis, in *Methods and Phenomena*, vol. **1** (eds S.P. Wolski and A.W. Czanderna) Elsevier Scientific Publishing Company, Amsterdam, Oxford, New York, p. 164.
41. Mathieu, H.J. and Landolt, D. (1981) Experimental study of matrix effects in quantitative Auger analysis of binary metal alloys. *Surf. Interface Anal.*, **3** (4), 153–156.
42. Wirth, T. (1992) Quantitative auger electron spectroscopy of silicides by extended matrix correction using $dN(E)/dE$ spectra. *Surf. Interface Anal.*, **18** (1), 3–12.
43. Einstein, A. (1905) On a heuristic viewpoint concerning the production and transformation of light. *Ann. Phys.-Berlin*, **17**, 132–148.
44. Siegbahn, K., Nording, C.N., Fahlman, A., Nordberg, R., Hamrin, K., Hedman, J., Johansson, G., Bermark, T., Karlsson, S.E., Lindgren, I., and Lindberg, B. (1967) *ESCA: Atomic, Molecular and Solid state Structure Studied by Means of Electron Spectroscopy*, Aimqvist and Wiksells, Uppsala, Sweden.
45. Watts, J.F. and Wolstenholme, J. (2003) *An Introduction to Surface Analysis by XPS and AES*, Wiley, Chichester, UK.
46. Scofield, J.H. (1976) Hartree–Slater subshell photoionization cross-sections at 1254 and 1487eV. *J. Electron Spectrosc.*, **8**, 129–137.
47. Wagner, C.D. (1981) Empirical atomic sensitivity factors for quantitative analysis by electron spectroscopy for chemical analysis. *Surf. Interface Anal.*, **3**, 211–225.

48 Brümmer, O., Heydenreich, J., Krebs, K.H., and Schneider, H.G. (eds) (1980) *Handbuch Festkörperanalyse mit Elektronen, Ionen und Röntgenstrahlen, Friedr*, Vieweg & Sohn, Braunschweig.

49 Colby, J.W. (1968) Quantitative microprobe analysis of thin insulating films. *Adv. X-Ray Anal.*, **11**, 287–305.

50 Abou-Ras, D., Kaufmann, C.A., Schöpke, A., Eicke, A., Döbeli, M., Gade, B., and Nunney, T. (2008) Elemental distribution profiles across Cu(In,Ga)Se$_2$ solar-cell absorbers acquired by various techniques. (eds M. Luysberg, K., Tillmann, and T. Weirich) Proceedings of the 14th European Microscopy Congress 2008, September 1–5, 2008, Aachen, Germany. Springer. "EMC 2008", Vol 1: Instrumentation and Methods, pp. 741–742.

51 Grün, A.E. (1957) Lumineszenz-photometrische messungen der energieabsorption im strahlungsfeld von elektronenquellen – eindimensionaler fall in luft. *Zeitschrift für Naturforschung*, **12a** (2), 89–95.

52 Rechid, J., Kampmann, A., and Reineke-Koch, R. (2000) Characterising superstrate CIS solar cells with electron beam induced current. *Thin Solid Films*, **361–362**, 198–202.

53 Mohr, H. and Dunstan, D.J. (1997) Electron-beam-generated carrier distributions in semiconductor multilayer structures. *J. Microsc.*, **187** (2), 119–124.

54 Michaels, J.R. and Williams, D.B. (1987) A consistent definition of probe size and spatial resolution in the analytical electron microscope. *J. Microsc.*, **47** (3), 289–303.

17
Hydrogen Effusion Experiments
Wolfhard Beyer and Florian Einsele

17.1
Introduction

In an effusion experiment, a sample is heated to high temperatures and released (effused) gases are detected. In particular for incorporated hydrogen in thin films, effusion measurements have been established as a fairly fast method for measurement of hydrogen concentration and the kinetics of its release, and thus on its incorporation stability [1]. Knowledge about hydrogen incorporation and stability is of particular interest for silicon-based thin-film solar cell materials like hydrogenated amorphous silicon (a-Si:H), microcrystalline silicon (μc-Si:H), related (alloy) materials, and amorphous and microcrystalline germanium (a-Ge:H and μc-Ge:H, respectively). These materials contain typically 10 at.% of hydrogen. More recently, this class of materials has gained interest also for application as undoped surface passivation layers and doped contact layers in amorphous/crystalline silicon heterojunction solar-cell technology [2]. The process of hydrogen effusion may also be of importance in the technology of crystalline silicon on glass (or on SiO_2) if plasma-grown a-Si:H films are used as a starting material [3]. Furthermore, hydrogen incorporation and its release is of interest (and has been studied) for transparent conducting oxide (TCO) materials like tin oxide (SnO_2) and zinc oxide (ZnO), which are widely applied in thin-film solar-cell technology. Although the incorporated hydrogen concentration in these TCO films is typically much smaller than in a-Si:H, hydrogen is the topic of intensive research [4] since it may act as a dopant [5]. Furthermore, during a-Si:H deposition on TCO-coated glass substrates, hydrogen may reduce the oxygen-bonded Sn or Zn atoms to the metallic phase [6].

Gas effusion measurements in general have turned out to provide a fast structural characterization of thin films, since the effusion of gas atoms and molecules from the film interior depends on the atomic density and is strongly modified by the presence of empty spaces (voids). This is particularly true for effusion of rare (inert) gas atoms like He, Ne, and Ar, which do not react with the host materials [7]. These gas atoms can be brought into the material during deposition or by ion implantation.

The aim of this chapter is to review some effusion measurement techniques, the data analysis, and selected results. Hydrogen effusion (also termed hydrogen evolution) measurements were first applied in thin-film silicon technology by Triska et al. [8] in 1975. Gas effusion measurements in general, however, go back to much earlier work, in particular in surface science where this technique has been widely used for characterizing the desorption of atomic or molecular species. Here, the technique is often termed temperature-programmed desorption (TPD) or thermal desorption spectroscopy (TDS). Basic work has been published in extensive review articles like those by Redhead [9] and Pétermann [10].

17.2
Experimental Setup

Two different experimental setups are commonly applied for hydrogen effusion measurements, the "closed" and the "open" effusion systems. In the "closed" system, a sample is heated inside a closed vacuum container and the hydrogen pressure is recorded as a function of time [11–15]. Provided that readsorption of hydrogen to the sample and adsorption/desorption of hydrogen at the walls of the container are negligible and the gas temperature remains constant, according to the gas equation

$$pV = \nu RT = NkT \qquad (17.1)$$

(with the pressure p, the gas volume V, the number of gas atoms/molecules $N = \nu N_A$, the Avogadro number N_A, the gas constant R, and the Boltzmann constant k), the time derivative of the pressure is proportional to the effusion rate dN/dt.

In an "open" system, the vacuum container where the sample is heated is constantly evacuated by a turbomolecular pump [16]. Due to the constant pumping speed of turbomolecular pumps over a wide pressure range, the hydrogen pressure at the pumping port is a measure of the effusion rate, if the adsorption/desorption at the container walls is negligible and the hydrogen flow is not inhibited by the size of the vacuum lines.

In many cases, quadrupole mass analyzers (QMA) are applied for partial pressure detection so that not only hydrogen but other desorbing gases can be measured. Such instruments typically require pressures $\leq 10^{-3}$ mbar. In particular at this point, the "open" system has some advantages to the "closed" system, as it is less sensitive to the total amount of effused gases. Both setups allow, in principle, line-of-sight mass spectrometry where effusing gas atoms or molecules can proceed without collision directly into the orifice of the quadrupole instrument. By line-of-sight mass spectrometry, desorbing gas species can clearly be identified while upon collision, different gas species may form. Since quadrupole instruments only measure ions, usually at the entrance port of the quadrupole instrument a heated filament generates free electrons, which are accelerated by an applied voltage and cause the ionization of atoms or molecules. Typically voltages up to 100 V are applied and positive ions are measured. Under these conditions, molecules are partially decomposed showing up in the mass spectra as characteristic cracking patterns, tabulated in literature [17]. For example, water (molecular weight 18) gives a cracking pattern with the masses 18, 17, 16, 2, and 1, neglecting some presence of deuterium. Thus, even for effusion of several hydrogen-bearing molecule species at the same time, the identification of the effusion of the individual molecule species may be possible. As a heating schedule, most commonly a linear increase in the sample temperature as a function of time is used, but the application of a linear variation of the reciprocal temperature has also been discussed [9].

In USA and Europe, effusion setups are commonly self-assembled and there is no standard commercial product. In Japan, there is a commercial product for basic measurements by ESCO Company (ESCO EMD-WA1000S). In the following, we describe the experimental procedures of effusion measurements for the setup realized in the Forschungszentrum Jülich. This latter apparatus has been applied successfully for thin-film characterization for almost 30 years. In Figure 17.1, the setup is shown schematically. It is an "open" system and allows absolute calibration of the effusion rate of various gases. In this effusion system, the samples are placed in a quartz tube (16 mm outer diameter), which is sealed at one side and at the other side is constantly pumped by a turbomolecular pump at a nominal pumping speed of 150 l/s. Within the quartz tube, the samples can be moved using a small thin metal boat made of nickel that can be manipulated from the outside of the quartz tube via a small magnet. For the measurement, a sample is placed in the end part of the quartz tube which is heated by an oven. The remaining samples are stored in the cold parts of the quartz tube outside the heated range. The oven is of relatively low mass in order to ensure a fast heating and cooling and is only 10 cm in length. It has a maximum power consumption of about 0.4 kW and allows heating rates >40 K/min up to a maximum temperature of 1200 °C. At this maximum oven temperature, the sample within the quartz tube attains a temperature of about 1050 °C. While a maximum temperature of about 1000 °C is sufficient for measurement of hydrogenated silicon, we note that for alloys of silicon with nitrogen or carbon higher temperatures may be

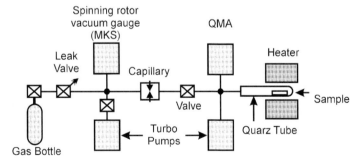

Figure 17.1 Schematic illustration of hydrogen effusion setup employed at Jülich. Samples are placed in a quartz tube which is evacuated by a turbopump. Upon heating, effusing gases are detected by a QMA. The left part of the effusion setup (separated by a valve) is used for calibration (see text).

required for all hydrogen to leave the sample. Higher temperatures (by 150–200 °C) are no problem for the quartz tube, but for the oven a special construction may be necessary. Typical (linear) heating rate is 20 K/min (applied for most effusion measurements in this article).

For the QMA, a Q-200 (with Faraday cup detector) by Leybold–Heraeus was employed in the original Jülich apparatus. This device is no longer produced but high-quality apparatus (e.g., Hiden HAL 301/3F) is on the market. For the effusion experiment, various masses (at Jülich up to about 20) are measured sequentially by the quadrupole instrument while the temperature rises. Measurements of different masses require a time delay in order to account for the RC time constant of the Faraday cup signal amplifier. The data are recorded via PC. The apparatus described so far can provide relative measurements only. In order to achieve absolute measurements, a calibration system, as also shown schematically in Figure 17.1 has been implemented.

Calibration is achieved by inserting well-known flows of calibration gases into the apparatus and by comparing the quadrupole reading of the calibration gas flow with that from the sample. In order to establish the calibration gas flow, a capillary is used with dimensions chosen such that a pressure p on the higher pressure side of the capillary of $p \approx 10^{-3}$–10^{-2} mbar results in a gas flow of about 10^{16} atoms/s through the capillary. Since such a flow causes only a weak rise of pressure (up to 10^{-5}–10^{-4} mbar) in the main turbopumped chamber, the pressure drop within the capillary exceeds two orders of magnitude. Then the gas flow through the capillary dN/dt given by the Knudsen formula for molecular flow

$$dN/dt = C_1 \, p(M)^{-1/2} \tag{17.2}$$

depends only on the gas type (molecular mass M), the capillary dimensions, and the pressure p at the higher pressure side of the capillary. The latter is measured at Jülich by a highly accurate spinning rotor vacuum gauge (MKS).

The constant C_1 can be determined by establishing a well-known flow dN/dt of a calibration gas of molecular mass M_0, for example, by using a well-known volume as a source of the calibration gas, and monitoring the pressure drop in this volume as a function of time. While this pressure drop yields with Eq. (17.1) the rate dN/dt, with the pressure p measured at the higher pressure side of the capillary and $M = M_0$, the constant C_1 in Eq. (17.2) is obtained.

Once C_1 is known, the flows dN/dt of other gases can be calculated according to Eq. (17.2), using the measured pressure p and the known molecular mass M. With this setup, only the flows of highly pure gases can be calibrated which means that highly pure calibration gases are required (like supplied, e.g., in the MiniCan system by Linde, the system must be highly leak-tight and when switching from one calibration gas to another, the calibration system needs to be pumped to low pressure (typically 5×10^{-7} mbar). At Jülich, a turbopump with a nominal pumping speed of 50 l/s is employed in the calibration system.

For an accurate measurement of a quadrupole signal for a given mass, one needs to account for two sources of signal drifts. One possibility of an error arises because the signal height of a given mass is not measured at the peak center of the quadrupole signal but rather in the flanks. This drift is usually fairly long termed and it has proven sufficient to optimize the measuring window to the peak maximum once per week. The other possible error arises because the sensitivity of a quadrupole may change (it typically decreases) as a function of time, likely due to changes in the degree of ionization of the incoming gas species. The sensitivity factors of different gases were found to vary only slightly relative to each other. To account for this effect, we determine in calibration procedures prior to and after the actual effusion experiments the sensitivity factor S_X^R of a given gas molecule X by comparing the quadrupole reading with the gas flow calculated by Eq. (17.2). In addition to the gases arising from the effusion experiment, a reference gas with the sensitivity factor S_Y^R (Y typically He, Ne, or Kr) is also calibrated. From these calibration runs, relative sensitivity factors S_X^R/S_Y^R for given molecules X are obtained. To determine the actual sensitivity S_X^A during an effusion run, the reference gas is kept flowing and the actual value of S_Y^A is monitored. The actual value of S_X^A is then determined by $S_X^A = S_Y^A (S_X^R/S_Y^R)$ [18]. By monitoring a reference gas flow during an effusion run, irregularities of the effusion system like nonconstant pumping speed or sudden changes in quadrupole sensitivity (which sometimes occur when effused gases like fluorine contaminate the QMA system) are also detected. Besides the calibration described above, effusion measurements at Jülich have often been calibrated by measuring a well-known dose of, for example, implanted H or He ions or by comparison the effusion results with impurity concentrations measured otherwise.

Important for successful H effusion measurements is also the choice of the film substrate which should affect the effusion process as little as possible. In early work, often aluminum foils or rocksalt substrates were employed which were removed by hydrochloric acid or water prior to effusion [12, 15]. Metal substrates are usually not well suited as they often contain dissolved hydrogen leading to a high hydrogen background signal. Moreover, at the interface to amorphous silicon layers, silicide formation or metal-induced crystallization may occur in certain temperature ranges leading to modifications of the effusion spectra. Glass often leads to a high-background hydrogen signal and problems may arise as it gets soft and melts at elevated temperature. We obtained best results with crystalline silicon, germanium (both only moderately doped), or sapphire platelets as substrates. In these cases, the hydrogen background signal is low, arising primarily from surface adsorbed water. However, for certain deposition conditions, the films on the latter substrates tend to form blisters and pinholes [19] typically near 300–400 °C causing modifications of the effusion spectra [16]. The reason is presumably that these latter substrates have a low solubility for molecular hydrogen so that H_2 precipitates at the film-substrate interface. The effect can be avoided by a thin coating with SiO_2, which provides a higher H_2 solubility and diffusivity. Fused quartz substrates usually do not show pinhole formation but hydrogen in-diffusion at 500–650 °C and out-diffusion above 700 °C have been reported [16].

The detection limit for effusion of molecular hydrogen (molecular mass 2) of the Jülich system is near 10^{13} molecules/s. It is caused primarily by a background mass 2 signal due to H_2O desorption from parts of the apparatus which are not heated out (like, e.g., the quartz-metal connection) when pumping down the effusion apparatus. Since the hydrogen isotope deuterium (present in natural hydrogen at about 0.015 at.%) has a much lower background and thus a lower detection limit, deuterated materials are often prepared and measured. Note that for hydrogen effusion from silicon, the results for the two different hydrogen isotopes, namely H_2 proper and D_2 (deuterium), are quite similar, that is, there is no significant isotope effect visible in hydrogen effusion [20].

Recently it was demonstrated by Kherani et al. [21] that effusion measurements of radioactive tritium (T) from a-Si:H:T films are possible without high vacuum. Here, effusing tritium (primarily HT molecules) moves in an argon atmosphere to an ionization chamber where the radioactive decay of T is monitored. The effusion spectra for tritium from amorphous silicon were not found to differ significantly from those for hydrogen or deuterium. Due to the high sensitivity of tritium decay measurements, this method can give a much lower detection limit for effusing hydrogen than obtained for the nonradioactive isotopes of hydrogen.

17.3
Data Analysis

If hydrogen effusion measurements are performed to measure the absolute hydrogen concentration, the effusion rate is integrated over the time and temperature, that is,

$$N_H = \int (dN/dt)\, dt. \qquad (17.3)$$

Additional measurement techniques like secondary ion mass spectrometry (SIMS), infrared absorption, and others may be necessary to assure that after completion of the experiment all hydrogen has left the material. As mentioned in Section 17.2, a relatively simple check of calibration is the measurement of a well-known implantation dose of a given gas species.

If the aim of the experiment is the study of hydrogen stability and processes of its release, some knowledge is required on the distribution of hydrogen within the thin film and on its bonding state, namely if hydrogen is predominantly present as a molecule or if it is bonded to atoms of the amorphous network. The hydrogen depth profile can be conveniently measured, for example, by SIMS profiling (see Section 17.2); the bonding state can be evaluated, for example, by infrared absorption measurements which give information on the concentration of hydrogen in bonds to amorphous network atoms (Si, C, etc.).

In the following, we describe the data analysis assuming a constant hydrogen concentration for the as-deposited material throughout the film depth, the bonding of hydrogen primarily to atoms of the host material (little H_2), and the measurement of hydrogen effusion in a high-vacuum system applying a constant heating rate.

Hydrogen effusion involves then surface desorption of hydrogen which can be described for a single process by the rate equation [9, 10]

$$d(N/N_0)/dt = (kT/h)(1-(N/N_0))^n \exp(-\Delta G/kT) \qquad (17.4)$$

(with N_0 and N denoting the original and effused hydrogen concentrations, k and h the Boltzmann and Planck constants, respectively, and n the order of reaction). For simplicity, the transmission coefficient [10] is set equal unity in Eq. (17.4). For a temperature T rising as $T = T_0 + \beta t$ (β the linear heating rate), Eq. (17.4) describes an effusion rate dN/dt first rising and then falling as a function of temperature and time. However, this surface desorption formula will only describe the measured effusion curves or part of them as long as surface desorption is the rate-limiting step for effusion. As we are dealing with thin films, diffusion of hydrogen as an atom or a molecule to the film surface may also be a rate-limiting step. Therefore, the analysis of the effusion data requires the identification of rate-limiting processes and the identification of the diffusing species, molecular or atomic hydrogen. Since in an effusion measurement a sample is heated to high temperatures, structural changes may occur which will also influence the effusion curves, in particular by changing the diffusion process.

17.3.1
Identification of Rate-Limiting Process

Methods for analysis of effusion data and for the identification of rate-limiting effusion processes are demonstrated in the following part of Section 17.3 for hydrogenated amorphous silicon and germanium films. Typical hydrogen (deuterium) effusion transients (also termed effusion spectra [9]) for such material are shown in Figure 17.2a and b [18, 22]. These films were deposited by plasma

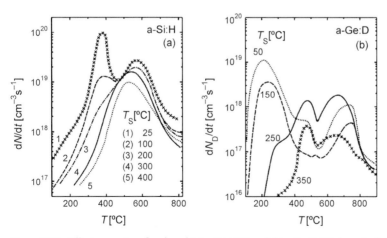

Figure 17.2 Effusion spectra of undoped (a) a-Si:H [22] and (b) a-Ge:D [18] deposited at various substrate temperatures T_S.

decomposition of SiH$_4$ and GeD$_4$ gases, respectively, at different substrate temperatures T_S. Film thickness was approximately 1 μm. For this deposition method, the hydrogen (deuterium) concentration in the films decreases significantly with increasing T_S so that films deposited near room temperature have hydrogen concentrations >20 at.%, decreasing to below 5 at.% for the highest substrate temperatures shown.

For identification of the rate-limiting process of hydrogen effusion, the effusing species must be known. For a-Si:H, line-of-sight mass spectrometry [23] demonstrates that only molecular hydrogen desorbs, that is, no atomic H and no SiH$_x$ species ($x = 1$–4) were seen. The situation for a-Ge:H is likely to be similar. This means that the primary rupture of Si–H or Ge–H bonds and the diffusion of molecular or atomic hydrogen to the film surface could be rate-limiting steps. In case of diffusion of atomic hydrogen, a surface recombination step of atomic hydrogen to form hydrogen molecules could also limit the hydrogen effusion rate. In addition, as mentioned above, changes in effusion rate with the result of effusion maxima or minima can arise if the structure of a material changes in a narrow temperature range. In particular, the H diffusion processes may be affected by structural changes. A major structural change of amorphous films occurs upon crystallization.

According to differential thermal analysis measurements performed at heating rates close to the present one, crystallization of a-Si:H is known to take place at $T = 700$–800 °C [24] and of a-Ge:H near 500 °C [25]. In Figure 17.2, crystallization is particularly visible in the effusion spectra of a-Ge:D by the effusion dip near 500 °C. In case of a-Si:H, crystallization is often barely visible since at high temperatures $T > 700$ °C, little hydrogen is commonly present.

Up to the crystallization temperature, the effusion spectra of a-Si:H and a-Ge:D (see Figure 17.2a and b) show basically one or two effusion peaks with maximum effusion rates for a-Si:H near 400 °C and 600 °C and for a-Ge:H near 200 °C and 450 °C. At high substrate temperatures ($T_S \geq 200$ °C for a-Si:H and $T_S \geq 250$ °C for a-Ge:H), only the high-temperature (HT) peaks remain.

Several experiments then allow the identification of the rate-limiting effusion process for each effusion peak. An effusion process limited by diffusion of hydrogen atoms or hydrogen molecules can be detected (i) by measuring hydrogen depth profiles by SIMS as a function of annealing or (ii) by measuring the effusion spectra of identical films of different thickness. In the first case, a decrease in hydrogen concentration toward the film surface and for many substrates also toward the film–substrate interface is expected. In the second case, the (average) diffusion length for hydrogen species required to leave the film rises with increasing film thickness. Accordingly, the effusion peak will shift to higher temperature [13, 16]. In Figure 17.3, the temperature T_M of maximum effusion rate of the two effusion peaks is plotted for a-Si:H films as a function of film thickness d [22]. It is seen that the low-temperature (LT) peak is practically independent of film thickness, while the high-temperature peaks shift to higher temperature.

Accordingly, it can be concluded that in the LT peaks in Figure 17.2a, hydrogen effusion from a-Si:H is not limited by diffusion but by the rupture of Si–H bonds. In contrast, in the HT peaks of Figure 17.2a, diffusion is the rate limiting step. For both

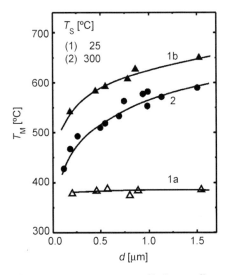

Figure 17.3 Temperature T_M of hydrogen effusion maximum versus film thickness d for plasma-grown a-Si:H [22]. Curves 1a and 1b refer to the LT and HT effusion peaks of $T_S = 25\,°C$ material, respectively, curve 2 to the effusion peak of $T_S = 300\,°C$ material.

a-Si:H and a-Ge:H, the analysis of the results shown in Figure 17.4 leads to the same conclusion for both the materials.

Here, the effusion rate of molecular H_2, HD, and D_2 is plotted as a function of temperature for sandwich structures of deuterated (or partly deuterated) material embedded into two layers of hydrogenated material. Note that in these samples, as verified by SIMS depth profiling, hydrogen effuses both at the actual surface and at

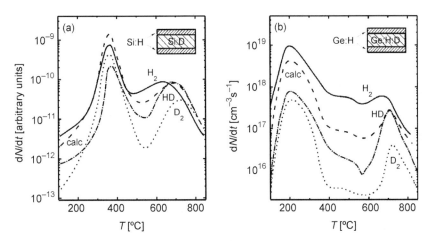

Figure 17.4 H_2, D_2, and HD effusion rates versus temperature T for sandwich samples of (a) undoped a-Si:H/a-Si:D/a-Si:H (substrate temperature $T_S = 25\,°C$) [22] and (b) a-Ge:H/a-Ge:H: D/a-Ge:H ($T_S = 50\,°C$) [18]. Dashed curves give the HD effusion rate calculated according to Eq. (17.5).

the film substrate interface. The fact that the effusion maxima of H_2, HD, and D_2 in the LT peak lie all nearly at the same temperature demonstrates that effusion is here not limited by diffusion. In the HT peaks, on the other hand, the HD and D_2 peaks show up at higher temperature than the H_2 peaks demonstrating that the effusion rate of hydrogen is limited by diffusion.

17.3.2
Analysis of Diffusing Hydrogen Species from Hydrogen Effusion Measurements

The effusion curves in Figure 17.4 can also be used for the identification of the diffusing hydrogen species. If the diffusing species is *molecular* hydrogen (deuterium), little effusion of HD is expected which may arise only from the interface areas between hydrogenated and deuterated layers and from interaction of diffusing *molecular* hydrogen (deuterium) with bound deuterium (hydrogen). In the other case of diffusion of *atomic* hydrogen (deuterium), a large effusion rate of HD is expected as HD is then formed at the film surface by a recombination process. If dN_H/dt and dN_D/dt are the effusion rates of H_2 and D_2, respectively, the effusion rate dN_{HD}/dt of HD is given by

$$dN_{HD}/dt = 2((dN_H/dt)(dN_D/dt))^{1/2} \quad (17.5)$$

This effusion rate of HD follows from the rate equation (Eq. (17.4)) for $n = 2$, yielding

$$dN_H/dt + dN_{HD}/dt + dN_D/dt = \text{constant } (N_H + N_D)^2 \exp(-\Delta G/kT) \quad (17.6)$$

The results in Figure 17.4 thus demonstrate that in the LT effusion peaks, primarily molecular hydrogen is diffusing for both a-Si:H and a-Ge:H. The fact that the effusion maxima of H_2, HD, and D_2 for this LT peak occur at the same temperature (implying that H_2 effusion is not limited by diffusion) shows that diffusion of molecular hydrogen in these materials is very fast. In the HT peaks, on the other hand, the measured and (according to Eq. (17.5)) calculated HD signals coincide, suggesting diffusion by atomic H. Since the deuterated peaks occur at higher temperature than the hydrogenated ones, effusion is apparently limited by diffusion. Thus, in this case, surface recombination of atomic hydrogen is not limiting the effusion rate. Similar results as for the HT peaks of the low substrate temperature samples of Figure 17.4a and b are obtained for H/D/H sandwich samples deposited at higher substrate temperature and showing one (HT) effusion peak only.

Thus hydrogen effusion measurements provide an important structural information, namely that for plasma grown a-Si:H and a-Ge:H films of low substrate temperature and high hydrogen content, hydrogen molecules are diffusing very fast, while for high substrate temperatures and low hydrogen content apparently a more dense material is present where molecular hydrogen (with an atomic diameter of about 2.5 Å) cannot diffuse and H diffusion takes place at much slower rate by atomic hydrogen.

17.3.3
Analysis of H$_2$ Surface Desorption

For an effusion peak limited by surface desorption, Eq. (17.4) is applicable for describing the effusion curves. The order of reaction can be determined by an analysis of the peak shape [9, 14] if there is one fixed free energy of desorption ΔG only. One expects $n=1$ if immobile neighboring hydrogen atoms form H$_2$ molecules during desorption, whereas for a surface recombination process of mobile atomic hydrogen, the order of reaction should be $n=2$. According to Eq. (17.4), a plot of log $(d(N/N_0)/dt)$ $(1-(N/N_0))^{-n}$ versus $1/T$ should yield (for a single ΔG) for $n=1$ a straight line if the order of reaction is unity. The results of Figure 17.5 [22] demonstrate for the LT effusion of a-Si:H that this is indeed the case.

Since $\Delta G = \Delta H - T\Delta S$ (ΔH the enthalpy and ΔS the entropy of the reaction), from the slope of the straight line in Figure 17.5 the enthalpy ΔH can be determined. From the intercept I of the straight line at $1/T=0$, the (experimental) entropy ΔS follows as $I=(kT/h)\exp(\Delta S/k)$ and thus ΔG ($\Delta G \approx 1.95$ eV for a-Si:H) is obtained.

An alternative way to determine ΔH, ΔS, and thus ΔG is the measurement of the effusion peak as a function of the heating rate β. According to Beyer and Wagner [26], for a linear heating rate β the formula

$$\ln(\beta/kT_M^2) = \ln[n(kT_M/h)(\exp \Delta S/k)(1-N_M/N_0)^{n-1}/\Delta H] - \Delta H/kT_M \quad (17.7)$$

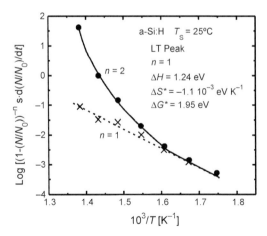

Figure 17.5 Analysis of LT hydrogen effusion peak of undoped a-Si:H ($T_S = 25\,°C$) in terms of Eq. (17.4) [22].

is valid. Here, N_M is the hydrogen concentration effused at the temperature T_M. In case of first-order reaction, this formula simplifies and leads to the approximation formula [9]

$$\Delta G/kT_M = \ln(v_1 T_M/\beta) - 3.64 \qquad (17.8)$$

with β the linear heating rate, $T = T_0 + \beta T$ and the characteristic frequency $v_1 = kT/h$ approximately equal to the phonon frequency, that is, $v_1 \approx 10^{13}\,\mathrm{s}^{-1}$.

The analysis of the effusion curves in terms of kinetic parameters becomes difficult or impossible when dealing with several overlapping processes (with different ΔG) rather than with a single process. In particular, a peak-shape analysis like the one using Eq. (17.4) will become rather meaningless. In this case, Eqs. (17.7) and (17.8) may still be applicable describing then the dominant effusion process only.

17.3.4
Analysis of Diffusion-Limited Effusion

If the effusion process is limited by diffusion, the effusion peak temperature T_M is related to the diffusion coefficient near this latter temperature. The correlation between effusion and diffusion parameters is obtained by solving the diffusion equation for out-diffusion of hydrogen from a film of thickness d at constant heating rate β. If out-diffusion toward both surface and film-substrate interface side of a film is assumed, the relation

$$\begin{aligned}\ln(D/E_D) &= \ln(d^2\beta/\pi^2 k T_M^2)\\ &= \ln(D_0/E_D) - E_D/kT_M\end{aligned} \qquad (17.9)$$

is obtained [26]. Also assumed for this relation is that the hydrogen-diffusion coefficient D can be expressed by the Arrhenius dependence

$$D = D_0 \exp(-E_D/kT) \qquad (17.10)$$

with D_0 the diffusion prefactor and E_D the diffusion energy. The diffusion parameters D_0 and E_D can be obtained according to Eq. (17.9) by plotting $\ln(d^2\beta/\pi^2 k T_M^2)$ versus $1/T_M$ either for series of identical films prepared with different thickness d or of samples measured with different heating rate β.

Two examples for application of Eq. (17.9), in order to determine the hydrogen-diffusion parameters, are shown in Figure 17.6a and b. A close agreement with results for hydrogen diffusion from SIMS hydrogen-deuterium interdiffusion measurements ($D_0 = 1.17 \times 10^{-2}\,\mathrm{cm^2/s}$, $E_D = 1.53\,\mathrm{eV}$ [27]) is observed as long as the material is not changing its structure significantly between the annealing temperature of the SIMS diffusion measurements (typically 300–450 °C) and the temperature of the evaluated H effusion peak (typically 600 °C). This condition is fulfilled for curve 2 of Figure 17.6a and for all samples in Figure 17.6b. The dashed line in Figure 17.6b shows the results of hydrogen and deuterium interdiffusion

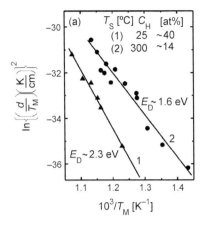
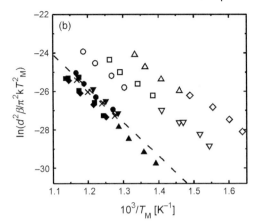

Figure 17.6 (a) Plot of $\ln(d/T_M)^2$ versus $1/T_M$ for (diffusion limited) effusion data of Figure 17.3 [22]. (b) Plot of $\ln(d^2\beta/\pi^2 kT_M^2)$ versus $1/T_M$ for samples measured at different heating rate β. Closed symbols undoped; open symbols boron-doped, x: phosphorus doped [26], dashed curve: data by Ref. [27].

measurements by Carlson and Magee [27] for undoped material demonstrating the close agreement for the H diffusion data obtained by the different methods. The results give for undoped a-Si:H deposited between 280 °C and 300 °C hydrogen-diffusion prefactors between 8×10^{-3} and 1.2×10^{-2} cm^2/s and diffusion energies E_D between 1.49 and 1.54 eV.

One may ask to what degree the experimental effusion curves are fitted by a single H-diffusion process, that is, by a single Arrhenius dependence. Using $D_0 = 1.1 \times 10^{-2}$ cm^2/s and $E_D = 1.49$ eV determined by the procedure employed in Figure 17.6b for a given sample, the corresponding effusion rate was calculated [22] and normalized to fit the experimental data in the effusion maximum. The results of experimental and calculated effusion curves are shown in Figure 17.7. A good agreement is found in a wide temperature range $T \leq T_M$. A deviation between experimental and calculated curves occurs primarily at $T > T_M$ (600 °C) suggesting that in this HT range diffusion processes with higher diffusion energy and/or material changes are involved.

A much stronger structural change than for the material of Figure 17.7 occurs, however, for the low substrate temperature a-Si:H material (curve 1 in Figure 17.6a). This material was analyzed in Sections 17.3.1 and 17.3.2 to be void rich up to annealing temperatures of 400 °C and dense for annealing temperatures of 500–600 °C. It is likely that due to material reconstruction (upon release of much hydrogen in the LT effusion peak) the HT effusion peak is here retarded, causing the changed diffusion parameters of curve 1 in Figure 17.6a, as discussed also by Kherani et al. [21].

If measurements of the maximum effusion rate T_M as a function of film thickness or heating rate are unavailable, the diffusion energy E_D may be roughly estimated

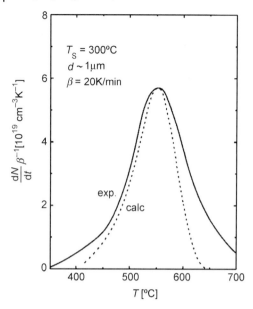

Figure 17.7 Calculated and experimental hydrogen effusion rate of plasma-grown a-Si:H (dense material) versus temperature [22].

from T_M by using Eq. (17.9) and assuming, for example, for D_0 the theoretical diffusion prefactor

$$D_0 = (1/6)v_1 a^2 \approx 10^{-3} \text{ cm}^2\text{s}^{-1} \tag{17.11}$$

for a-Si:H (a is the interatomic distance) [28]. For the data of Figure 17.7, $E_D \approx 1.33$ eV is obtained.

17.3.5
Analysis of Effusion Spectra In Terms of Hydrogen Density of States

Jackson et al. [29] proposed to analyze hydrogen-effusion measurements in terms of a hydrogen density of states, in analogy to deep-level transient spectroscopy (DLTS) measurements giving the electronic density of trap states. They relate the effusion rate dN/dt at a particular time t of the effusion experiment with the position of the hydrogen chemical potential $\mu(t)$ relative to a surface barrier energy (for hydrogen desorption) or transport energy E for hydrogen diffusion (k is the Boltzmann constant)

$$dN_H/dt = C \exp(-(E-\mu(t))/kT(t)). \tag{17.12}$$

For the hydrogen density of states $g_H(E)$, the relation

$$g_H(\mu) \approx \partial N_H/\partial \mu \tag{17.13}$$

is assumed to be valid. According to Eq. (17.12) and for a fixed prefactor C, the experimental effusion rate gives the time dependence of the chemical potential $\mu(t)$ relative to the energy E. Equation (17.13) then yields the hydrogen density of states near the H chemical potential μ. The authors assume $C = 10^{27}$ cm^{-2} s^{-1} for hydrogen surface desorption and $C \approx 10^{25}$–10^{26} cm^{-2} s^{-1} for diffusion-limited effusion for a sample of 0.5 μm film thickness. The results for sputtered deuterated a-Si:H films gave for hydrogen surface desorption (LT effusion peak) a relatively sharp H density of states distribution at an energy of about 1.85 eV. This energy corresponds to the value of $\Delta G \approx 1.9$ eV characterizing the LT hydrogen effusion peak of undoped hydrogen rich a-Si:H (see Section 17.3.3). For the diffusion-limited effusion, the authors evaluated a relatively broad hydrogen density of states extending between energies of about 1.6 eV to about 2.2 eV below the hydrogen transport path.

Drawbacks of the method are that the determined $g_H(E)$ involves any change of H effusion due to structural changes, like crystallization, densification, or formation of bubbles. Actually such structural changes are obscured, as such changes are recognized in the effusion spectra at characteristic temperatures (like the crystallization temperature) but appear in the H density of states distribution at certain energies defined by the value of C used. Furthermore, in particular at low hydrogen content, hydrogen (mass 2) signals arising from H_2O and CH_4 effusion will also contribute to the "hydrogen density of states." The analysis, furthermore, is questionable for diffusion-limited effusion as hydrogen atoms (at some depth L from the next film surface) set free to move in the transport path related to a decrease in H chemical potential will not immediately be desorbed. Due to the diffusion-limited effusion process, this hydrogen will desorb after the time $t = L^2/D(t,T)$, approximately. Here, $D(t,T)$ is the H-diffusion coefficient at the given time and temperature. Thus, it appears unlikely that this method could detect any particular H density of states feature except for the case where the samples are extremely thin.

17.3.6
Analysis of Film Microstructure by Effusion of Implanted Rare Gases

In the past few years, it was demonstrated that the effusion of rare (noble gases) like He, Ne, or Ar provide a fast characterization of thin films [7]. Since these atoms do not bind to atoms of the host material, the effusion curves are affected primarily by the material density and by the presence of various types of voids. These atoms can be brought into the material either during deposition or by ion implantation. In the latter case, the dose needs to be as small as possible to avoid ion bombardment damage, which usually densifies a given material, but may also lead to formation of bubbles (isolated voids) filled with the rare gas. In general, the implantation dose should be $\leq 10^{16}$ cm^{-2} [7, 30]. The implantation energy is typically chosen such that the maximum of the implanted atom distribution lies near half of the film thickness. Diffusion of rare gases in a solid has been described by the process of doorway diffusion, first applied to the diffusion of rare gases in glass [31]. Considering rare gas atoms sitting in cages of a host material (see Figure 17.8a), diffusion of a given atom is

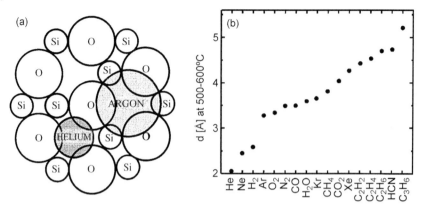

Figure 17.8 (a) Schematic illustration of helium and argon incorporation in quartz (from Ref. [31]). (b) Diameter d of atoms and molecules ordered according size from gas viscosity measurements [32].

considered to proceed if, by thermal vibrations, the doorway of a given cage opens wide enough so that the atom can move from one cage to another.

By classical modeling, Anderson and Stuart [31] derived the formula

$$E_D = C\,G\,r_D(r-r_D)^2 \qquad (17.14)$$

with a constant C, a shear modulus G, a doorway radius r_D, and a radius r of the diffusing atoms. In Figure 17.8b, approximate sizes of various atoms and molecules are given, determined by viscosity measurements [32]. Note that these measurements give temperature-dependent sizes and that the diameters shown here refer to $T \approx 500\text{--}600\,°C$. Note also that the size of Ne and H_2 is quite similar so that Ne out-diffusion can be used to predict H_2 diffusion, which is of great importance for hydrogen stability issues [28].

According to Eq. (17.14), a strong dependence of the diffusion energy E_D on the size of the diffusing atoms and molecules is expected and the corresponding effusion peaks often show such a dependence, as depicted in Figure 17.9a. Note that also the widths of the rare gas effusion peaks give structural information on the material, namely about its homogeneity, since broader effusion structures indicate the presence of more inhomogeneous materials or of material structure which is more strongly changing with rising annealing temperature. If we assume for the diffusion coefficients of the rare gases Arrhenius dependences $D = D_0 \exp(-E_D/kT)$, the diffusion energy E_D can be estimated from T_M by Eq. (17.9), using the theoretical diffusion prefactor D_0 (Eq. (17.11)). According to Eq. (17.14), a plot of the atomic radius r versus $(E_D)^{1/2}$ should give a straight line with the intercept r_D at $E_D = 0$. In Figure 17.9b, the atomic radius r is plotted versus $(E_D)^{1/2}$ for the effusion results of Figure 17.9a. A nearly straight line is obtained with an intercept $r_D \approx 0.2\,\text{Å}$. It must be noted, however, that this analysis for the present a-Si:O:H material is not expected to give accurate values for r_D. The broad hydrogen-effusion spectrum in Figure 17.9a indicates that structural changes occur with rising annealing temperature so that the

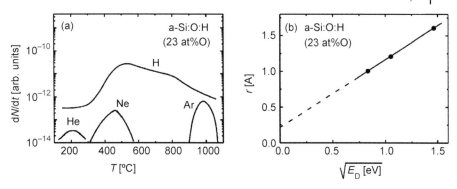

Figure 17.9 (a) Effusion spectra of hydrogen and implanted He, Ne, and Ar for plasma-deposited a-Si:O:H (oxygen concentration ≈ 23 at.%). (b) Atomic radius r of diffusing rare gas atoms in a-Si:O:H of Figure 17.9a versus square root of diffusion energy.

assumption of a constant r_D up to the annealing temperature of about 1000 °C appears quite unlikely.

For characterization of microstructure, the analysis of helium effusion peaks turned out to be highly useful [7]. This is demonstrated in Figure 17.10a and b, showing the effusion of implanted He for plasma grown a-Si:H and a-Ge:H films, deposited at different substrate temperatures T_S.

For higher substrate temperatures T_S, the results show He-effusion maxima near $T_M = 400\,°C$ for a-Si:H and near 370 °C for a-Ge:H. Similar peak effusion

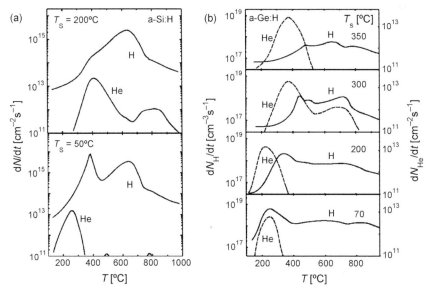

Figure 17.10 Effusion spectra of hydrogen and (implanted) helium for undoped (a) a-Si:H and (b) a-Ge:H, deposited at different substrate temperatures T_S.

temperatures are obtained when He is implanted in crystalline silicon or germanium and in fact, from literature data of the diffusion of helium in crystalline Si and Ge [33], similar effusion peak temperatures are calculated according to Eq. (17.9). In the picture of doorway diffusion, the difference between silicon and germanium may be related to the different bond length of about 2.35 Å for silicon and about 2.45 Å for Ge [34]. With decreasing substrate temperature, the He-effusion temperature decreases, indicating a more open structure (increasing doorway radius in Eq. (17.14)). Qualitatively, this structural change agrees with the results of hydrogen effusion (see Section 17.3.1). Another interesting feature in Figure 17.10a and b is the appearance of a second He-effusion peak at high temperatures ($T > 500\,°C$) for substrate temperatures near 200 °C for a-Si:H and 300 °C for a-Ge:H. Such peaks have been explained by the presence of isolated voids or bubbles [7]. If He enters such bubbles, it can leave only (except that the voids/bubbles disappear by structural changes) if according to the gas equation (Eq. (17.1)) the pressure within the bubble gets such high that He returns into the network. Thus the temperature of this HT He peak will depend primarily on the bubble size and the amount of He. Since this HT He effusion usually occurs when the material has crystallized (for a-Si:H and a-Ge:H), this void size has unlikely much relation to the original material and is not evaluated. Instead, the ratio F^{HT} of HT to total He effusion is analyzed. The concept is that F^{HT} describes the probability for diffusing He atoms to get trapped in isolated voids, that is, F^{HT} is a measure of the concentration of isolated voids, present from the deposition process. In Figure 17.11a and b, the results for T_M and F^{HT} for plasma-

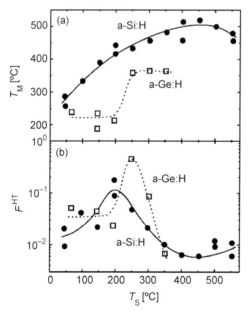

Figure 17.11 (a) Temperature T_M of (lower temperature) He effusion peak and (b) ratio F^{HT} of a-Si:H and a-Ge:H films as a function of substrate temperature T_S [30].

grown a-Si:H and a-Ge:H films are compared. It is seen that the decrease in T_M with decreasing substrate temperature T_S (and increasing hydrogen content) takes place quite gradually for a-Si:H while it occurs in a rather narrow temperature range between 200 °C and 250 °C for a-Ge:H under the applied deposition conditions. Another interesting result is the enhanced formation of isolated voids near $T_S = 200$ °C for (plasma-grown) a-Si:H and near $T_S = 250$ °C for a-Ge:H, while isolated voids show up in much smaller concentration at low substrate temperatures ($T_S < 200$ °C) and high T_S (>300 °C). This formation of isolated voids has been explained by diffusion effects of hydrogen in the subsurface region (close to the surface where the network is still fairly flexible) at the temperature of deposition, resulting in hydrogen precipitation and thus in the formation of voids [7, 30]. Thus, no isolated voids are expected when the material is permeable to diffusion of H_2 at low substrate temperatures. At high substrate temperatures, the effect presumably disappears for both materials as the concentration of incorporated hydrogen decreases.

17.4 Discussion of Selected Results

17.4.1 Amorphous Silicon and Germanium Films

17.4.1.1 Material Density versus Annealing and Hydrogen Content

In the previous section, the effusion results of amorphous silicon and germanium films have been used to illustrate the evaluation of hydrogen- and helium-effusion measurements. The general picture obtained from such measurements is that at low substrate temperatures, a hydrogen-rich material grows with interconnected voids (or decreased density) such that molecular hydrogen (H_2) can move rapidly without diffusion processes delaying H_2 release. After annealing to about 400 °C and thus after having released a high amount of hydrogen, the same film appears as a rather dense material where diffusion of hydrogen takes place by hydrogen atoms. Thus a strong structural change must have taken place during annealing near $T_A = 400$ °C. Indeed, such changes can be monitored by various methods, like measurements of the film density [35] or of changes in He effusion [7]. In Figure 17.12, some results are shown.

In Figure 17.12a, it is seen that the film thickness d of low T_S plasma-grown a-Si:H decreases (and thus the material density increases) considerably in the temperature range of LT hydrogen effusion at about 300–400 °C. In Figure 17.12b, shifts of the He-effusion peak to higher temperature show that plasma-grown a-Si:H deposited at lower substrate temperature T_S densifies much stronger than material deposited at higher T_S. Still, some densification effects are observed also for a-Si:H grown at $T_S = 300$ °C. It is also seen in Figure 17.12b that the concentration of isolated voids increases upon annealing, as the fraction F^{HT} of HT He effusion rises. This annealing effect has been attributed primarily to the change of interconnected voids to isolated ones as the material shrinks upon annealing [7].

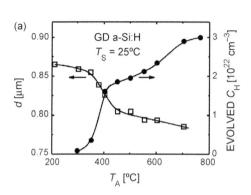
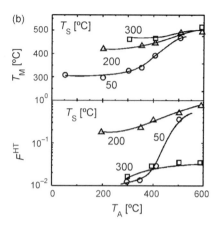

Figure 17.12 (a) Thickness d and hydrogen concentration C_H of (undoped) plasma-deposited a-Si:H of low substrate temperature and high hydrogen content as a function of annealing temperature T_A. Annealing time was 5 min [35]. (b) Temperature T_M of (LT) He effusion peak and fraction of HT (related to isolated voids) He effusion F^{HT} as a function of annealing temperature T_A (annealing time 5 min) [7].

17.4.1.2 Effect of Doping on H Effusion

Significant changes of hydrogen effusion by doping were first noted by Beyer et al. [16]. It was found that in particular by boron doping, both LT and HT hydrogen-effusion peaks of plasma-grown a-Si:H are shifted to lower temperatures. This is illustrated in Figure 17.13a for low substrate temperature material. It was demonstrated later

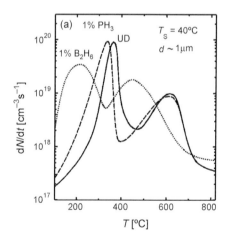
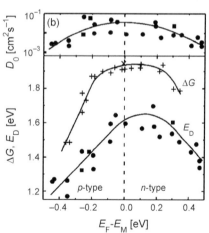

Figure 17.13 (a) Hydrogen effusion rate for a-Si:H, deposited at $T_S = 40\,°C$, undoped (UD) and doped with 1% diborane (B_2H_6) and 1% phosphine (PH_3) [1]. (b) Hydrogen free energy of desorption ΔG, hydrogen diffusion prefactor D_0, and hydrogen diffusion energy E_D (the latter two determined by HD interdiffusion experiments measured by SIMS) versus Fermi energy E_F relative to midgap E_M [1].

[1, 20] that the effect originates from a Fermi-level dependence of both the hydrogen desorption free energy ΔG and the hydrogen-diffusion energy, as shown in Figure 17.13b. Note that due to defect generation during (LT) hydrogen effusion, the Fermi level in the doped material may shift toward midgap so that the HT effusion may not always show a doping dependence.

17.4.2
Amorphous Silicon Alloys: Si-C

Alloys of amorphous silicon with carbon, nitrogen, oxygen, or germanium can be fairly easily prepared by plasma deposition by addition of, for example, methane, ammonia, carbon dioxide, or germane, respectively, to silane. Thus the bandgap can be varied from near 1 eV (a-Ge:H) to more than 2.5 eV (a-Si:C:H, a-Si:N:H, and a-Si:O:H). However, when alloy atoms are added to silicon, the tendency to grow material with an interconnected void structure is enhanced. Typical results of hydrogen effusion of a-Si:C:H films are shown in Figure 17.14a and b.

The results show that already upon addition of 15% carbon, an LT hydrogen effusion peak appears. The temperature T_M of both H effusion maxima shifts rapidly with increasing carbon content to higher temperature. These effects can be understood by an increase in (average) hydrogen-binding energy as carbon is incorporated. While the effusion spectra in Figure 17.14a look similar to those in Figure 17.2a, it must be noted that the effusion temperature in the HT peak of a-Si:C:H is often not found to be diffusion limited (on the scale of film thickness), in contrast to typical plasma-grown a-Si:H films. This may be attributed to the presence of a granular material in a-Si:C:H. For a-Si:C:H alloys with carbon concentrations exceeding about 20 at.%, a significant hydrogen effusion in the form of hydrocarbon molecules was

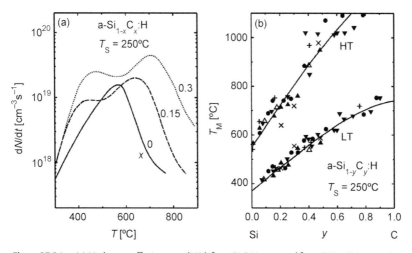

Figure 17.14 (a) Hydrogen effusion rate dN/dt for a-Si:C:H prepared from SiH_4–CH_4 gas mixtures at $T_S = 250\,°C$ [1]. (b) Temperature T_M of LT and HT effusion peaks versus film composition y of a-Si$_{1-y}$C$_y$:H films ($T_S = 250\,°C$) prepared from various gas mixtures [36].

observed, predominantly in the temperature range where LT hydrogen effusion takes place. By mass spectrometric analysis, the hydrocarbon molecules methane, ethylene, and propene were identified as the primary desorbing species [36].

17.4.3
Microcrystalline Silicon

Microcrystalline silicon films have a much stronger tendency to show microstructure effects (voids at grain boundaries) than a-Si:H films. However, deposition conditions (usually at conditions close to the amorphous-microcrystalline transition) can be found to prepare fairly dense material with diffusion-limited hydrogen effusion [37]. Figure 17.15 shows effusion results for a series of samples deposited at different substrate temperatures. Under the conditions used, the samples of $T_S = 150\,°C$ and $200\,°C$ are void-rich; those deposited at $250\,°C$ and $300\,°C$ compact with similar hydrogen-diffusion coefficients as in a-Si:H. The nature of the various LT hydrogen effusion peaks is not fully clarified.

The microstructure effects visible in hydrogen effusion are also seen by effusion measurements of implanted He and Ne. The results shown in Figure 17.16 (for μc-Si:H deposited at somewhat different deposition conditions as compared to the samples of Figure 17.15) demonstrate the growth of relatively dense μc-Si:H at higher ($T_S = 200\,°C$) substrate temperatures (He effusion near $400\,°C$ and near $900\,°C$, hydrogen effusion maximum near $600\,°C$ and neon effusion above $700\,°C$ only). In contrast, at $T_S \approx 50\,°C$ material with interconnected voids grows as indicated by a He effusion peak below $300\,°C$, a hydrogen effusion maximum near $500\,°C$ and Ne effusion starting near $200\,°C$, but extending up to $900\,°C$ and higher. Note that this latter very broad effusion process suggests the presence of a rather inhomogeneous material, as noted in Section 17.3.6.

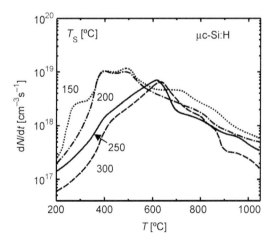

Figure 17.15 Hydrogen effusion spectra for a series of microcrystalline Si:H films deposited at different substrate temperatures T_S [37].

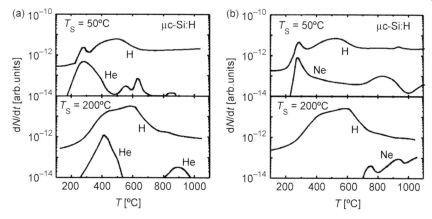

Figure 17.16 Effusion spectra of hydrogen and (a) implanted He and (b) implanted Ne for microcrystalline Si:H films deposited at $T_S = 50\,°C$ and $200\,°C$ (W. Beyer (2009) unpublished).

17.4.4
Zinc Oxide Films

Hydrogen effusion measurements of zinc oxide crystals and thin films were performed by Nickel et al. [4, 38] and, more recently, by Beyer et al. [39]. Nickel derived hydrogen-diffusion energies by effusion measurements, varying film thickness and heating rate. Hydrogen density of states measurements were presented but the drawbacks discussed in Section 17.3.5 apply.

In Figure 17.17a–c, results of effusion of implanted He and Ne are presented for LPCVD grown, sputter (SP)-grown and single crystalline ZnO, respectively. The results show rather void-rich material in the first case, more dense material in the second and highly dense material (no visible neon effusion) in the third case. The width of the He and Ne effusion peaks suggests a rather inhomogeneous material (see Section 17.3.6) in the first case and the most homogeneous material (as expected) in the last case.

17.5
Comparison with Other Experiments

As discussed in Section 17.3, the characterization of thin-film material should never rely on effusion measurements alone, since there are chances for erroneous conclusions for poorly defined material. For example, SIMS depth profiling is often indispensable for assuring that hydrogen is evenly distributed within the film. Interdiffusion experiments of hydrogen and deuterium can be of great importance for verifying the evaluation of effusion measurements in terms of the H diffusion coefficient. Infrared absorption measurements are necessary to assure the presence of, for example, silicon-bonded hydrogen in contrast to molecular hydrogen. Furthermore, a comparison of microstructure data from effusion measurements with

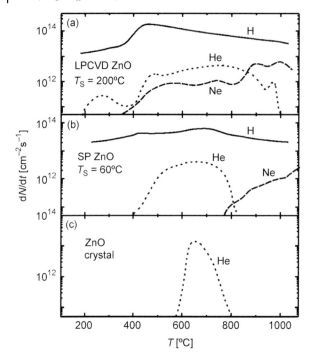

Figure 17.17 Effusion spectra of implanted He and Ne for (a) LPCVD grown, (b) sputter (SP)-grown and (c) single crystal ZnO. Also shown is the hydrogen effusion rate versus temperature of the thin-film ZnO material [39].

those from small-angle scattering is highly recommended for assuring important issues with regard to material structure.

17.6
Concluding Remarks

While the technique of hydrogen effusion was primarily developed for the quantification of incorporated hydrogen and of the thermal stability of hydrogen in amorphous Si:H films, the more recent successful structural characterization of various materials by rare gas effusion opens a wide range of further application of the method, namely for microstructure analysis of thin films in general. For amorphous silicon and related technology, effusion measurements have been applied widely and a wealth of information was obtained. However, further work like the modeling of diffusion of hydrogen atoms and molecules, rare gas atoms, and other species in thin films is highly welcome. Drawbacks of the method involve its destructive nature, the special substrates required, the costly high vacuum equipment, and the limited throughput which is a few samples per day, not counting the effort of rare gas implantations. Part of these obstacles may get overcome, in future, by, for example, confining to small spots heated by a laser, if advanced hydrogen or rare gas detection methods are developed and applied. In any case, a further successful development of thin-film silicon solar cells appears hardly thinkable without applying sophisticated microstructure analysis like the gas-effusion method.

Acknowledgments

The authors wish to thank D. Lennartz and Dr. P. Prunici for valuable technical support.

References

1. Beyer, W. (1991) Hydrogen effusion: a probe for surface desorption and diffusion. *Physica B*, **170**, 105–114.
2. Tanaka, M., Taguchi, M., Matsuyama, T., Sawada, T., Tsuda, S., Nakano, S., Hanafusa, H., and Kuwano, Y. (1992) Development of new a-Si/c-Si heterojunction solar cells: ACJ-HIT (Artificially Constructed Junction-Heterojunction with intrinsic thin-layer). *Jpn. J. Appl. Phys.*, **31**, 3518–3522.
3. Matsuyama, T., Terada, N., Baba, T., Sawada, T., Tsuge, S., Wakisaka, K., and Tsuda, S. (1996) High-quality polycrystalline silicon thin film prepared by a solid phase crystallization method. *J. Non-Cryst. Solids*, **198–200**, 940–944.
4. Nickel, N.H. (2006) Hydrogen migration in single crystal and polycrystalline zinc oxide. *Phys. Rev. B*, **73**, 1952041–1952049.
5. Van de Walle, C.G. (2000) Hydrogen as a cause of doping in zinc oxide. *Phys. Rev. Lett.*, **85**, 1012–1015.
6. Beyer, W., Hüpkes, J., and Stiebig, H. (2007) Transparent conducting oxide films for thin film silicon photovoltaics. *Thin Solid Films*, **516**, 147–154.
7. Beyer, W. (2004) Characterization of microstructure in amorphous and microcrystalline Si and related alloys by effusion of implanted helium. *Phys. Status Solidi (c)*, **1**, 1144–1153.
8. Triska, A., Dennison, D., and Fritzsche, H. (1975) Hydrogen content in amorphous Ge and Si prepared by r.f. decomposition of GeH_4 and SiH_4. *Bull. Am. Phys. Soc.*, **20**, 392.
9. Redhead, P.A. (1962) Thermal desorption of gases. *Vacuum*, **12**, 203–211.
10. Pétermann, L.A. (1972) Thermal desorption kinetics of chemisorbed gases. *Prog. Surf. Sci.*, **3**, 1–61.
11. Brodsky, M.H., Frisch, M.A., Ziegler, J.F., and Lanford, W.A. (1977) Quantitative analysis of hydrogen in glow discharge amorphous silicon. *Appl. Phys. Lett.*, **30**, 561–563.
12. Fritzsche, H., Tanielian, M., Tsai, C.C., and Gaczi, P.J. (1979) Hydrogen content and density of plasma-deposited amorphous silicon-hydrogen. *J. Appl. Phys.*, **50**, 3366–3369.
13. Biegelsen, D.K., Street, R.A., Tsai, C.C., and Knights, J.C. (1979) Hydrogen evolution and defect creation in amorphous Si:H alloys. *Phys. Rev. B*, **20**, 4839–4846.
14. McMillan, J.A. and Peterson, E.M. (1979) Kinetics of decomposition of amorphous hydrogenated silicon films. *J. Appl. Phys.*, **50**, 5238–5241.
15. Oguz, S. and Paesler, M.A. (1980) Kinetic analysis of hydrogen evolution from reactively sputtered amorphous silicon–hydrogen alloys. *Phys. Rev. B*, **22**, 6213–6221.
16. Beyer, W., Wagner, H., and Mell, H. (1981) Effect of boron-doping on the hydrogen evolution from a-Si:H films. *Solid State Commun.*, **39**, 375–379.

17 Cornu, A. and Massot, R. (1966) *Compilation of Mass Spectral Data*, Heyden and Son, London.

18 Beyer, W., Herion, J., Wagner, H., and Zastrow, U. (1991) Hydrogen stability in amorphous germanium films. *Philos. Mag. B*, **63**, 269–279.

19 Shanks, H., Fang, C.J., Ley, L., Cardona, M., Demond, F.J., and Kalbitzer, S. (1980) Infrared spectrum and structure of hydrogenated amorphous silicon. *Phys. Status Solidi (b)*, **100**, 43–56.

20 Beyer, W., Herion, J., Mell, H., and Wagner, H. (1988) Influence of boron doping on hydrogen diffusion and effusion in a-Si:H and a-Si alloys. *Mater. Res. Soc. Symp. Proc.*, **118**, 291–296.

21 Kherani, N.P., Liu, B., Virk, K., Kosteski, T., Gaspari, F., Shmayda, W.T., Zukotynski, S., and Chen, K.P. (2008) Hydrogen effusion from tritiated amorphous silicon. *J. Appl. Phys.*, **103**, 0249061–0249067.

22 Beyer, W. (1985) Hydrogen incorporation in amorphous silicon and processes of its release, in *Tetrahedrally-Bonded Amorphous Semiconductors* (eds D. Adler and H. Fritzsche) Plenum, New York, pp. 129–146.

23 Matysik, K.J., Mogab, C.J., and Bagley, B.G. (1978) Hydrogen evolution from plasma-deposited amorphous silicon films. *J. Vac. Sci. Technol.*, **15**, 302–304.

24 Tsai, C.C. and Fritzsche, H. (1979) Effect of annealing on the optical properties of plasma deposited amorphous hydrogenated silicon. *Sol. Energy Mater.*, **1**, 29–42.

25 Jones, S.J., Lee, S.M., Turner, W.A., and Paul, W. (1989) Substrate temperature dependence of the structural properties of glow discharge produced a-Ge:H. *Mater. Res. Soc. Symp. Proc.*, **149**, 45–50.

26 Beyer, W. and Wagner, H. (1982) Determination of the hydrogen diffusion coefficient in hydrogenated amorphous silicon from effusion experiments. *J. Appl. Phys.*, **53**, 8745–8750.

27 Carlson, D.E. and Magee, C.W. (1978) A SIMS analysis of deuterium diffusion in hydrogenated amorphous silicon. *Appl. Phys. Lett.*, **33**, 81–83.

28 Beyer, W. (2003) Diffusion and effusion of hydrogen in hydrogenated amorphous and microcrystalline silicon. *Sol. Energy Mater. Sol. C.*, **78**, 235–267.

29 Jackson, W.B., Franz, A.J., Jin, H.C., Abelson, J.R., and Gland, J.L. (1998) Determination of the hydrogen density of states in amorphous hydrogenated silicon. *J. Non-Cryst. Solids*, **227–230**, 143–147.

30 Beyer, W. (2004) Microstructure characterization of plasma-grown a-Si:H and related materials by effusion of implanted helium. *J. Non-Cryst. Solids*, **338–340**, 232–235.

31 Anderson, O.L. and Stuart, D.A. (1954) Calculation of activation energy of ionic conductivity in silica glasses by classical methods. *J. Am. Ceram. Soc.*, **37**, 573–580.

32 Eucken, A. (ed.) (1950) *Landolt-Börnstein, Atom- und Molekularphysik I*, Springer, Berlin, Germany, p. 325 and 369.

33 Van Wieringen, A. and Warmoltz, N. (1956) On the permeation of hydrogen and helium in single crystal silicon and germanium at elevated temperatures. *Physica*, **22**, 849–865.

34 Kennard, O. (1978) Bond lengths between carbon and other elements, in *CRC Handbook of Chemistry and Physics*, 58th edn (ed. R.C. Weast) CRC Press, West Palm Beach, FL, USA, pp. F–215.

35 Beyer, W. and Wagner, H. (1983) The role of hydrogen in a-Si:H- results of evolution and annealing studies. *J. Non-Cryst. Solids*, **59–60**, 161–168.

36 Beyer, W. and Mell, H. (1987) Composition and thermal stability of glow discharge a-Si:C:H and a-Si:N:H alloys, in *Disordered Semiconductors* (eds M.A. Kastner, G.A. Thomas, and S.R. Ovshinsky) Plenum, New York, pp. 641–658.

37 Beyer, W., Hapke, P., and Zastrow, U. (1997) Diffusion and effusion of hydrogen

in microcrystalline silicon. *Mater. Res. Soc. Symp. Proc.*, **467**, 343–348.
38 Nickel, N.H. and Brendel, K. (2003) Hydrogen density-of-states distribution in zinc oxide. *Phys. Rev. B*, **68**, 1933031–1933034.
39 Beyer, W., Breuer, U., Hamelmann, F., Hüpkes, J., Stärk, A., Stiebig, H., and Zastrow, U. (2009) Hydrogen diffusion in zinc oxide thin films. *Mater. Res. Soc. Symp. Proc.*, **1165**, M05-24, 1–6.

Part Four
Materials and Device Modeling

18
Ab-Initio Modeling of Defects in Semiconductors

Karsten Albe, Péter Ágoston, and Johan Pohl

18.1
Introduction

The characterization of semiconductors applied in photovoltaic devices by means of *ab-initio* methods has become an emerging field over recent years. This is not only because of the increasing availability of more powerful computers, but mostly due to the development of refined numerical and theoretical methods.

Explaining and predicting solid-state properties requires an understanding of the behavior of electrons in solids. First-principles or *ab-initio* calculations are those that start directly at the Hamiltonian of the Schrödinger equation containing the kinetic energy operators and the potential energy due to electrostatic interactions of electrons and nuclei. The value of *ab-initio* methods lies in the fact that material properties can be calculated from scratch without experimental input or empirical parameters. Obtaining materials data from *ab-initio* calculations may serve various purposes. Firstly, they allow for validating experimental results and therefore contribute to a better understanding of material properties. Secondly, calculated data can be used for identifying new properties or mechanisms. Thirdly, by calculating specific properties for a large database of various structures, new materials with optimized properties may be designed. In semiconductor research, *ab-initio* methods are now frequently used for calculating structural, thermomechanical and electronic properties. In practice, however, the feasibility of an *ab-initio* approach always depends on the problem at hand and may be limited for conceptual reasons. Most importantly, the computational costs which scale with the system size, that is, the number of electrons necessary to calculate a certain property of a material, restrict the applicability of *ab-initio* methods. Moreover, it is often very crucial to balance accuracy and computational cost by choosing the appropriate method for the problem at hand. The most accurate approach of treating a complex system consisting of many interacting particles, such as a semiconducting crystal, is *ab-initio* methods solving the many-body Schrödinger equation directly. This includes quantum Monte Carlo [1], configuration interaction (CI), and coupled cluster (CC) methods [2], which are computationally expensive and therefore more often applied for studying small

Advanced Characterization Techniques for Thin Film Solar Cells,
Edited by Daniel Abou-Ras, Thomas Kirchartz and Uwe Rau.
© 2011 Wiley-VCH Verlag GmbH & Co. KGaA. Published 2011 by Wiley-VCH Verlag GmbH & Co. KGaA

systems, like molecules and clusters, rather than solids. The most established *ab-initio* methods for describing electrons in solids are based on density functional theory (DFT), of which we give a brief exposition in the following. We then describe exemplarily how the defect chemistry of a semiconductor may be determined from *ab-initio* calculations and present results on the point defect chemistry of ZnO as a case study.

18.2
Density Functional Theory and Methods

In DFT, the complexity of a many-body system of N interacting electrons in the presence of ionic cores is reduced to the optimization of an electron density, which depends only on three spatial coordinates independent of the number of electrons. Hohenberg and Kohn showed in 1964 [3] that the energy E of an interacting electron gas in an external potential $V(\mathbf{r})$ due to ionic cores and external fields can be expressed by a functional $F[n(\mathbf{r})]$ of the electron density $n(\mathbf{r})$ and that the ground state energy is equivalent to $E_{GS} = \min\{\int V(\mathbf{r})n(\mathbf{r})d\mathbf{r} + F[n(\mathbf{r})]\}$. The proof itself, however, does not provide a recipe of how to construct an explicit functional $F[n(\mathbf{r})]$. Later, Kohn and Sham showed that the problem of many interacting electrons may be transformed into a problem of noninteracting particles moving in an effective potential, which may in principle include all many-body effects including exchange and correlation, and is only a functional of the electron density [4]. Within the Kohn–Sham method, the ground-state total energy E for a collection of electrons interacting with one another and with an external potential V_{ext} (including the interaction of the electrons with the nuclei) can be written as

$$E[n] = \sum_{\alpha} f_\alpha \left\langle \psi_\alpha \left| \left(\frac{1}{2}\nabla^2 + V_{\text{ext}} + \frac{1}{2}\int \frac{n(\mathbf{r'})d\mathbf{r'}^3}{|\mathbf{r}-\mathbf{r'}|} \right) \right| \psi_\alpha \right\rangle + E_{\text{xc}}[n] + E_{\text{II}}$$

where f_α is the occupation of a single-particle state ψ_α usually taken from a Fermi function, while E_{II} describes the ion–ion interaction. All many-body effects are hidden in the exchange-correlation $E_{\text{xc}}[n]$ term.

By construction, DFT is an exact theory given that the exchange-correlation functional is known. In praxis, however, the functional has to be approximated. Since the foundations of DFT have been completed with the Hohenberg–Kohn theorem and the Kohn–Sham ansatz subsequent progress in the theory has focused on efficient numerical implementations and improved exchange-correlation functionals.

18.2.1
Basis Sets

Finding a mathematically convenient basis set representing the Kohn–Sham orbitals within DFT calculations is essentially a balance of computational performance and accuracy. Any basis can be made more complete and therefore more accurate by adding more basis functions. For a plane-wave or real-space basis this can be systematically done by simply increasing the spectral range. Usually, core electrons, for which the wavefunctions exhibit strong oscillations, are excluded and replaced by so-called pseudopotentials representing the combined potential of the ionic core and core electrons. Exclusion of the core electrons results in an improved performance, especially for plane-wave basis sets, and is often a prerequisite for treating extended systems. Using a localized basis set, such as for example a Gaussian or numerical atomic orbital basis function, may have advantages concerning the scalability of the method with respect to the system size. However, in that case, more effort is needed to check the completeness of the basis set. A combination of a localized basis and a plane-wave representation is the full potential linearized augmented plane wave (FP-LAPW) method, which represents an accurate yet efficient all-electron approach [5].

18.2.2
Functionals for Exchange and Correlation

18.2.2.1 Local Approximations

In the local density approximation (LDA), the exchange-correlation energy as a functional of the density is described by a homogeneous electron gas. The correlation energy of the homogeneous electron gas can be obtained by highly accurate quantum Monte Carlo simulations [6], while the exchange energy is known analytically. The LDA is sufficiently accurate for many problems in solid-state physics, especially for describing electrons in simple metals which behave very much like free electrons. Because of its simplicity and limited computational costs, the LDA is very useful for describing large systems as well as for obtaining approximate energies and wavefunctions that can then be refined by more sophisticated levels of theory. An improvement of the LDA has been achieved by recognizing that the exchange-correlation energy can be described more accurately by including gradients of the electron density. In contrast to LDA, different variants exist of how the gradient is included. Nowadays, however, the generalized gradient approximation (GGA) by Perdew *et al.* [7] (PBE-GGA) is most widely used.

Although improving on many calculated material properties in comparison with the local approximation, the GGA still has deficiencies in various aspects. Most importantly, both the LDA and the GGA fail to reasonably describe the fundamental band gap, which is generally underestimated (sometimes by more than 50%) in both approximations. This issue is known as the band-gap problem in DFT. Approaches to improve the exchange-correlation energy by including higher derivatives of the electron density have failed, and it is now generally accepted that improvements require to properly account for nonlocality.

18.2.2.2 Functionals Beyond LDA/GGA

A direct extension of the LDA and the GGA functional is obtained by adding an on-site Coulomb repulsion term to specific orbitals representing the electrostatic repulsion between localized electrons. This approach is motivated by the lattice-based Hubbard model [8], which describes the repulsion of electrons within a narrow band, that is, electron correlation, by an empirical parameter U. Similarly, a density functional that partially corrects specific localized orbitals for the electron correlation can be designed by introducing the repulsion parameter U (LDA + U or GGA + U) [9].

The +U method has become a popular approach, which allows for a partial correction of the correlation and self-interaction error of the LDA and significantly improves the description of Mott insulators, such as transition metal oxides. Nowadays, the method is widely applied due to its relative accuracy at very reasonable computational costs. As a byproduct, the calculated band gaps of semiconductors can be improved by using the +U approach, because the increased repulsion between specific orbitals leads to an enhanced localization. However, LDA + U and GGA + U still underestimate the band gap. In addition, it is often not a priori clear how and for which orbitals U parameters should be applied. Further details on the LDA + U method can be found in a review by Anisimov [10].

The LDA and the GGA functional considerably underestimate the exchange energy, but properly account for the correlation energy. The Hartree–Fock method by definition properly accounts for the exchange interaction, but in turn cannot access the correlation energy. Therefore, one may conclude that the true answer to the problem must lie somewhere in between. This is the starting point for the development of hybrid functionals. The basic idea for hybrid functionals is thus to mix the exchange-correlation energy of the traditional LDA or GGA functionals with a fraction of exact or Hartree–Fock exchange. In recent years, it turned out that hybrid functionals are indeed able to give a much better description of the exchange and the correlation energy than traditional functionals. Also, the band gaps are significantly improved. This, however, comes at a price: hybrid functionals are generally at least two orders of magnitude more computationally expensive than their local or semilocal counterparts. Typical examples of hybrid functionals that have been engineered in such a way are B3LYP [11], PBE0 [12], and HSE06 [13, 14], just to name the most prominent ones. The hybrid functional PBE0 should clearly be distinguished from PBE, which is a semilocal GGA functional (see Section 18.2.1.1). A review on hybrid functionals applied to solids and an assessment of their accuracy in comparison to experiment has recently been given by Marsman *et al.* [15]. Another review focusing more on the general theoretical framework has been authored by Kümmel and Kronik [16], who also discuss the issue of the band-gap problem in the case of hybrid functionals.

For a more detailed general introduction into *ab-initio* total-energy calculations, we refer the reader to a review paper by Payne *et al.* [17]. A highly recommendable full treatise on electronic structure calculations including recent developments in DFT methods can be found in a recent book by Martin [18]. The different approximative functionals in DFT may be ordered into a hierarchy starting with the most basic, the local density approximation and then increasing in accuracy. Such a hierarchy is sometimes called the Jacobs ladder of DFT. A table which puts the different *ab-initio* methods and functional approximations into such a scheme is shown in Table 18.1.

Table 18.1 Overview of *ab-initio* total energy and electronic structure methods.

Method	Description	Reference
Density functional theory		
Local and semilocal functionals		
LDA	A local functional describing exchange-correlation energy that depends only on the local electron density and is obtained from the solution for the homogeneous electron gas	[6]
GGA	The exchange-correlation energy also depends on the gradient of the electron density.	[7]
Orbital-dependent and hybrid functionals		
LDA + U/GGA + U	An on-site Coulomb repulsion term, represented by the parameter U is included for specific orbitals. Description of exchange and correlation is improved if U is chosen properly.	[9]
Hybrid functionals (HSE06, PBE0, B3LYP, etc.)	Improved accuracy of the exchange-correlation energy by mixing local or semilocal functionals with a fraction of exact exchange or Hartree–Fock exchange. Provides a good alternative to local and semilocal functionals, but is computationally more expensive. In certain cases, hybrid functionals alleviate the band-gap problem.	[15, 16]
Many-body approaches		
GW	Allows for an improved description of charged excited states by treatment of the screened Coulomb interaction. It delivers a good description of band structures, but it is difficult to obtain total energies.	[19]
BSE	Explicit treatment of neutral excitations within the framework of MBPT.	[20]
Time-dependent DFT	Explicit treatment of neutral excitations within the framework of DFT.	[20]
QMC	True many-body wavefunction approach, highly accurate but also computationally expensive, currently forces are still hard to obtain	[21]

18.3
Methods Beyond DFT

More sophisticated methods that potentially alleviate the band-gap problem of DFT need to take explicitly into account the many-body interactions of excited states, in particular the Coulomb screening and the interaction of electrons and holes. The three main approaches to this problem are the GW [1]) method [19],

---

1) in GW, the G stands for Green's function and the W stands for the screened Coulomb interaction, which both enter the electron self-energy Σ.

methods based on the Bethe–Salpeter equation (BSE) [20], and time-dependent density functional theory (TD-DFT) [20]. While being very useful for the calculation of excited state properties, all of these methods suffer from their extreme computational cost, which finally limits their applicability to the study of very small systems.

GW is the method of choice when realistic electron addition and removal energies, such as measured in direct and inverse photoemission experiments, are of interest. The GW method is thus able to give much more accurate energies for charged excited states than standard DFT. GW is based on a perturbative treatment of standard DFT calculations and is therefore also referred to as a many-body perturbation theory (MBPT) approach. GW provides an approximation to the electron self-energy Σ based on a perturbative evaluation with Kohn–Sham orbitals. Pioneered already in 1965 with the work of Hedin [22] it was not until the 1980s that calculations for silicon showed that the GW method has the potential to alleviate the band-gap problem for real materials.

For an accurate treatment of neutral excitations in semiconductors such as electron–hole pairs (excitons), one has to resort to either the BSE or to TD-DFT. These methods are able to predict, for example, optical and electron energy loss spectra. As with DFT, the accuracy of TD-DFT is limited by the accuracy of the available functionals. For example, optical spectra calculated by TD-DFT within the adiabatic local density approximation (ALDA), which is the time-dependent analog of the LDA, show very good agreement with experiments for molecular materials. For solids, better agreement has been found using BSE approaches. Due to their computational complexity, however, BSE approaches are limited to relatively simple materials. The efficient description of electron–hole excited states is therefore still considered to be an unsolved problem. A review comparing GW, BSE, and TD-DFT is given in Ref. [20].

A method that does not quite fit the scheme of Table 18.1 is the Quantum Monte Carlo (QMC) method [21]. QMC is the only method presented here, which is completely independent of DFT. It is a ground-state method based solely on the many-body wavefunction as the basic object. Therefore, it takes a fundamentally different approach to the problem of electron correlation compared to all the methods described above, which are all in some way based on the electronic density. The basic idea is to sample the many-body wavefunction using random numbers. Quantum Monte Carlo is in principle able to give the correct total energy and therefore an exact treatment of exchange and correlation is possible. However, this is only true for an infinitely long simulation run. The error in the total energy of a system given by QMC is an entirely statistical error, which scales as the inverse square root of the number of time steps. It is the extreme computational cost of this method, which limits its usefulness in practice. For very small systems it can, however, be a very valuable tool if very accurate total energies are needed. In praxis, the forces on the ions are hard to obtain and computationally much more expensive than the total energy, which make structural relaxations difficult. A good review on QMC methods has been given by Foulkes *et al.* [21].

18.4
From Total Energies to Materials' Properties

In principle, DFT calculations allow to predict material properties that can directly be derived from total energies and atomic forces of any atomic arrangement of interest. Moreover, electronic properties as described by the KS orbitals are directly accessible. Before we specifically discuss the use of DFT calculations for studying point defects in semiconducting materials, we will first give a brief survey of properties obtainable by considering the electronic system in the ground state.

Properties which are conceptually easy to calculate include those that can be directly derived from total energy differences. To this group belong, for example, cohesive energies, defect formation enthalpies, or energies of various configurations. The latter can be used in the cluster expansion method that, together with Monte Carlo simulations, allows to calculate configurational entropy and free energy differences as well as phase diagrams [23].

Being conceptually relatively simple, however, does not mean that accuracy is easily achieved. The calculation of most material properties involves the use of a supercell, that is, a cell which contains several identical primitive cells in a periodic arrangement. The use of a supercell, however, can introduce finite-size errors, especially if defective structures are considered. In practice, finite-size scaling procedures allow for minimizing these effects. For example, the cohesive energy of a crystal is calculated as the energy difference between the perfect crystal and the free atom. While one does not have to worry about finite-size errors for the perfect crystal, calculations of the free atom involve finite-size effects because the free atom interacts with itself through the periodic boundary conditions. Therefore, one has to make sure that the energy of the free atom is sufficiently converged with respect to the size of the cell, that is, the amount of vacuum around the atom. The same reasoning applies to surface energy calculations, where slab geometries are used and convergence both with respect to the thickness of the slab and to the vacuum layer has to be ensured. Also, in calculation of point defect formation enthalpies, cell-size effects are an important issue [24].

The next class of properties is obtained from derivatives of the total energy with respect to continuous variables. First-order derivatives with respect to atomic positions, strain, and electric fields directly yield atomic forces, stress states and polarization. Lattice constants and elastic moduli are usually calculated by fitting total energies to a thermodynamic equation of state relating energy and volume. From high-order (including mixed) derivates force constants, elastic moduli, polarizabilities, dielectric susceptibilities, or piezoelectric tensors, to name only a few, can be obtained [25]. In principle, derivatives can be obtained by calculating energy variations due to a direct perturbation, like atomic displacements, by applying perturbation theory [26, 27] or by evaluation of time correlations (Green–Kubo formalism) in molecular dynamics (MD) simulations.

These alternative routes can be exemplified by the phonon density of states, which allows to access thermodynamic properties like the heat capacity and differences in free energy. The most common method is the frozen-phonon method, where the energy of all different phonon modes is calculated by displacing atoms in an appropriately sized supercell. By diagonalization of the force constant matrix the eigenfrequencies and thus the phonon dispersion relation and the phonon density of states are subsequently obtained. Alternatively, the density functional perturbation theory (DFPT) [26, 27] can be used. The third method to access the phonon density of states is by means of *ab-initio* MD. *Ab-initio* MD yields the velocities of the atoms and their trajectories as a function of time. By means of spectral analysis, one can then obtain the phonon density of states from the velocity autocorrelation function.

Electronic properties such as the fundamental band gap, valence band offsets, and defect transition levels can be obtained consistently from DFT within the limits defined by the functionals used for exchange and correlation. This is because the only physically meaningful properties directly generated by DFT calculations are the total energy, the ground state electronic density, and the energy of the highest occupied state. A word of caution is in order here because the quasi-particle eigenstate energies of DFT and especially the ones of the unoccupied excited states are sometimes confused with the true energies of excited state electrons. In particular, the optical band gap may not be determined from total energy differences, while the fundamental band gap can. In practice, these two quantities are often close, but this cannot be taken for granted. In case true excited state properties (in particular the optical band gap and absorption spectra) are of interest, one has to resort to methods beyond DFT. The story is a bit different in case of hybrid functionals, for which the unoccupied states may in fact be interpreted as an approximation to excited state energies within a generalized Kohn–Sham framework [16].

18.5
Ab-initio Characterization of Point Defects

One of the most powerful applications of first-principles electronic structure theory is the modeling of point defect in solids, whose presence controls the functional properties of many semiconductors. Therefore, in this final part we will guide the reader through the procedures relevant for calculating defect properties in semiconductors. In order to see the necessity of first-principles calculations in the context of point defects, it is worth mentioning some quantities of interest which can be calculated. The formation energies of intrinsic defects like vacancies, interstitials, anti-sites, or more complex defect arrangements can be compared in order to identify the predominant defect species. In the case of extrinsic impurities, the calculated formation energies provide insights into the solubility of the impurities. The stable charge states of intrinsic or extrinsic defects can be obtained, which in turn classify the defects as acceptors or donors in a specific host material. Furthermore, the activation energies for changing the charge state, that is, the ionization energies of the defects can be obtained and therefore allow to identify appropriate donors and acceptors for device development. In this context, *ab-initio* methods have made significant contributions to the understanding of AX and DX centers[2] in II–VI semiconductors and particularly in GaAs [29–31].

Defect-induced states within the band gap affect the optical properties, for example, in the case of color centers, or can act as recombination centers. The assignment of these levels to specific defects is experimentally difficult but possible with the aid of DFT calculations. Apart from the thermodynamic stability, the mobility of point defects can also be assessed. The free energy of defect migration determines the kinetic stability of intrinsic point defects and impurities under nonequilibrium conditions. In the case of more than just one defect, their interactions can be studied by calculating binding energies. The characteristics of extrinsic point defects are sometimes heavily influenced by the presence of intrinsic point defects, for example, due to defect association.

Also, the association of solely intrinsic point defects is a key for understanding the materials behavior as it is the case for chalcopyrites [32]. Another important application is the determination of doping limits, which can be estimated by first-principles calculations. Alternatively it is possible to choose one specific defect and even focus on a single certain charge state in order to calculate properties of interest which in turn can be used for the experimental identification or characterization of this specific defect.

Various experiments capable of identifying point defects can be simulated from first principles with, however, varying reliability. For example, it is possible to calculate the formation volume of defects to study their impact on the crystal lattice parameter [33]. Calculating ionization energies enables to estimate the conductivities which can then be measured [34]. It is possible to identify the defect related magnetic moment and derive parameters for electron paramagnetic resonance measurements [35]. It is even possible to model the positronic trapping state at a defect location and derive from it the characteristic lifetimes in order to compare them with results obtained by positron annihilation spectroscopy [36].

In the case of localized defect related states, optical spectra are the primary means for defect identification. The calculation of optical properties is, however, often difficult in the case of solids. Using simple LDA/GGA-DFT in most cases it is at least possible to identify whether an absorption feature can be expected or not. The absorption energy on the other side is clearly an excited state property and affected by the difficulties mentioned above. A good example for the calculation of a defect excitation can be found in Ref. [37], where the full methodology of MBPT was applied to the excitation of a negatively charged nitrogen-vacancy color center in diamond.

In the following, we present a list of guidelines that should allow the interested reader to judge the reliability of defect calculations and their particular strengths and drawbacks:

---

2) A DX centre is a substitutional donor which may behave like an acceptor depending on the position of the Fermi level. Similarly, an AX centre is an acceptor, which may behave like a donor. This behaviour is usually associated with large lattice relaxations and bond breaking. In the AX and DX denomination, A stands for acceptor and D for donor. These substitutional donors and acceptors were previously thought to associate with unknown point defects, represented by the symbol X, which was not confirmed later. The denominations AX and DX centre, however, remained [28].

- Energy differences are reliable even within the LDA.
- The determination of the defect geometry (relaxation and volume changes) is reliable.
- Energy differences between structures of similar bonding type and geometry are more reliable than those between different ones (e.g., ionic solid vs. molecule). This is due to error cancellations important for the success of first-principles calculations. For example, the calculation of the migration energy is often more reliable than the calculation of the formation energy (reference energies can involve metals or molecules).
- The characterization of neutral defects is usually more reliable than of charged defects. This is especially true for defects, which involve highly localized defect states. In the case of delocalized defect states, that is, for defects which are naturally charged a neutral calculation often leads to an overestimation of their formation energy.
- In the case of charged defects, a higher charge state is not necessarily less accurately described. The reliability as a function of charge state depends on the specific defect type.
- The characterization of localized defect states is more difficult for small-gap materials than for materials with larger band gaps.
- Excited state properties are especially difficult to obtain in the case of defects in solids. The reliability of such calculations can be even difficult to estimate.

Although in many cases these guidelines are valid, exceptions can always be found. In the following we will focus on how to calculate the formation energies of intrinsic point defects for a two-component system. This covers many aspects of the defect modeling. By using a two-component system, the example provides sufficient generality.

18.5.1
Thermodynamics of Point Defects

The key quantity for accessing the thermodynamics of point defects in solids is the Gibbs free energy of defect formation (ΔG_f). Ultimately, one is interested in defect concentrations which can be calculated from the formation energy for the dilute limit ($c < 10^{-3}$) through

$$c = c_0 \exp\left(-\frac{\Delta G_f}{k_B T}\right) = c_0 \exp\left(-\frac{\Delta E_f + p\Delta V_f - T\Delta S_f}{k_B T}\right) \tag{18.1}$$

Here, c_0 is the concentration of available positions for the defect in the lattice, k_B is the Boltzmann constant, and T the temperature. The quantities ΔE_f, ΔV_f, and ΔS_f are the formation energy, formation volume, and formation entropy of the corresponding defect, respectively. The discussion now focuses on the formation energy, which is usually the dominating contribution and sufficient for low-temperature and low-pressure conditions. The other two contributions can, however, also be calculated within electronic DFT.

At this stage, two systematic paths can be followed. It is possible to construct defect pairs or clusters which are charge neutral and conserve the particle numbers. This method leads to the well-known Kröger–Vink notation of defect reactions. A typical example is the so-called Schottky defect equilibrium which in the case of a simple metal oxide (MO) could be

$$0 \rightleftharpoons V_O^{2+} + V_M^{2-} \tag{18.2}$$

The defect equilibrium and therefore the concentrations can be expressed in terms of Gibbs free energy of defect formation $\Delta G_{Schottky} \approx \Delta E_{Schottky}$. The strength of this method is that by conserving charge as well as particle numbers, the formation energy does not depend on external reservoirs other than the material itself. In the case of a first-principles approach, $\Delta E_{Schottky}$ is calculated by constructing a supercell of the MO, remove a cation and an anion from arbitrary positions, and calculate the formation energy $\Delta E_{Schottky}$ from total energy differences via

$$\Delta E_f = E_{defect}^Z - E_{ideal}^Z. \tag{18.3}$$

Here, E_{ideal}^Z and E_{defect}^Z are the total energies as obtained by a total energy calculation for the supercells containing the ideal and the defect structure, respectively. This approach uses only the DFT total energies and returns relatively accurate numbers. There are, however, some pitfalls using this approach. Apparently, the charge states in Eq. (18.2) were simply guessed and could well differ from $q = \pm 2$. It is of course possible to check the charge state via the electron density output of the DFT code, which is always available. In case the charge state is different for either the oxygen or cation vacancy (the oxygen vacancy could be a color center or single donor), this supercell calculation cannot represent the ground state of the defect. For the case that V_O is only singly charged, one could insert two V_O in order to re-establish the ground state. Then, the conservation of particle numbers has to be omitted. In complex oxides (e.g., in sesquioxides), defect reactions are even more complicated as exemplified by the following reaction:

$$0 \rightleftharpoons 3V_O^{2+} + 2V_M^{3-}. \tag{18.4}$$

It can be seen that the number of defects (five in this case) increases dramatically with the complexity of the materials stoichiometry. Considering typical cell sizes accessible by present DFT calculations ($50 < N_{atoms} < 1000$) the assumption of dilution is not fulfilled, and the resulting numbers will surely depend on the actual arrangement of the defects within the cell. Therefore, this approach is neither unique nor flexible. Following the ideas of Zhang and Northrup [38], it is more efficient to calculate the formation energies of individual defects sequentially and in different charge states so that charge as well as particle conservation do not need to be fulfilled. Defect reactions are constructed at a later stage from the individual defects with the lowest formation energies. Since the constraints of constant particle numbers and charge are neglected, the free energy of defect formation depends on the chemical potentials and electrochemical potential of the reservoirs:

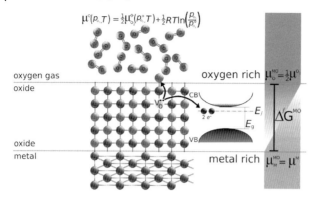

Figure 18.1 Defective metal oxide and possible reservoirs for anions, cations, and electrons. Here, as an example, we show the case of an oxygen vacancy in charge state $q = +2$, which is formed by releasing a neutral oxygen atom to the gas reservoir with chemical potential $\mu_O^{MO} = \frac{1}{2}\mu^{O_2} = \mu^O(p_{O_2}, T)$ and by moving two electrons into the electron reservoir with the Fermi energy E_F.

$$\Delta G_f = \Delta E_f + p\Delta V_f - T\Delta S_f - \sum_i n_i \mu_i + qE_F \qquad (18.5)$$

Here, n_i are the number of exchanged particles of type i in order to construct the defect and μ_i are the corresponding reference chemical potentials. The reference electrochemical potential for the electrons is the Fermi energy E_F, and q is the charge state of the defect.

Figure 18.1 illustrates the process of defect formation as a subsequent exchange of (neutral) atoms and electrons with their respective reservoirs. The chemical potentials of the relevant atomic species μ_i and the Fermi energy E_F are experimental control variables. The Fermi energy is also sometimes referred to as the chemical potential of the electrons. Conceptually, the Fermi energy can be controlled or influenced by dopants and impurities. The control over the chemical potentials is often difficult to establish, experimentally. The chemical potentials μ_i are related to the species activities a_i via $\mu_i = \mu_i^{\ominus} + RT \ln a_i$, where $\mu_i^{\ominus}$ is the reference chemical potential in the chemical standard state. In case of a reference reservoir in the gas phase the chemical potential may be related to the partial pressure p_i via the ideal gas law. For the oxygen gas phase, for example, this allows to relate the chemical potential of oxygen μ^O to the oxygen partial pressure p_{O_2} via

$$\mu^O(p_{O_2}, T) = \frac{1}{2}\mu_{O_2}^{\ominus}(p_{O_2}^{\ominus}, T) + \frac{1}{2}RT \ln\left(\frac{p_{O_2}}{p_{O_2}^{\ominus}}\right)$$

where $p_{O_2}^{\ominus}$ refers to the partial pressure in the standard state. In the case that the reference chemical potentials are the ones of solid phases it is difficult to establish precise experimental control. However, the activities and therefore the chemical potentials are always closely related to the availability of the respective species during production of the material. In applied studies the chemical potential of metallic species is therefore often treated as a free parameter, expressed as deviations from the cohesive energies $\mu_i = \mu_i^{el} + \Delta\mu_i$ of the most stable elemental reference phases μ_i^{el}. The chemical potential of the solid phase is equal to its cohesive energy and serves as a reference for the metal-rich limit ($\mu_M^{MO} = \mu^M$ in Figure 18.1). The maximal deviations from the elemental reference values are restricted by the heat of formation of the compound ΔH_f, which can also be calculated from first principles:

$$\Delta H_f^{MO} = \Delta\mu_M + \Delta\mu_O \tag{18.6}$$

It is common to plot the defect formation energies as a function of the Fermi energy at specific chemical potentials of interest.

At this stage, it may appear irritating that the defect stability, which depends on the Fermi energy, influences the Fermi energy itself, as defects are charged. Once the formation energies of all (predominant) defects are known the determination of the concentrations is achieved as follows: first, a specific environment is chosen by fixing the atomic chemical potentials. Next, the formation energies and concentrations are expressed as a function of the Fermi energy. The actual value of the Fermi energy is obtained by additional physical constraints. At the resulting Fermi energy, the total charge (including also free charge carriers and dopants) should vanish (charge neutrality (CN)), and none of the defects should have a negative formation energy. A detailed description of this process is given in Ref. [34].

Although the Fermi energy always assumes a fixed value in thermodynamic equilibrium, it is instructive to plot the formation energies of all defects as a function of this parameter. In this representation, it is most convenient to discuss the changes of the defect equilibria upon extrinsic doping and changes of the environment.

In Figure 18.2, several well-known defect equilibria are translated into this representation. The first panel (Figure 18.2a) represents the Schottky reaction of

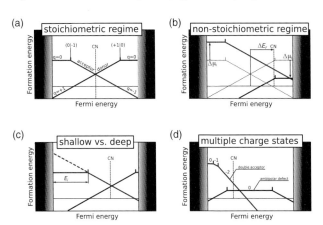

Figure 18.2 Examples for prototype defects: (a) a Schottky/Frenkel defect-pair and stoichiometric defect reaction. (b) Shift toward a nonstoichiometric regime due to changes of the μ_i. (c) A deep acceptor and a shallow donor defect. (d) Examples for multiple charge states, for example, an ambipolar defect and a double acceptor defect. The (approximate) point of CN is indicated for all examples.

Eq. (18.2). The slope of each line represents the charge state of the defect (Eq. (18.5)). In this case, the acceptor and donor formation energies are totally symmetric with respect to each other and the CN is easily found at the point where the formation energy of both defects has the same value. Charge states can change as they are denoted by the changing slopes of the defects. The apparent symmetry between donor and acceptor formation energies is not necessarily present in real materials. The situation changes especially when a different environmental condition ($\Delta\mu_i$) is chosen.

In the case of our metal oxide, the second panel (Figure 18.2b) shows the effect of a lower oxygen pressure and/or higher temperature and a consequently lower chemical potential of oxygen and a correspondingly higher chemical potential of the metal. Under these conditions, the formation energy of the oxygen vacancy is lower and that of the cation vacancy higher. The Fermi energy necessarily shifts to the right (n-type) since negative formation energies are unphysical. The CN is now more difficult to determine since three species significantly contribute to it. Positive charge carriers are the oxygen vacancies whereas cation vacancies and electrons in the conduction band (CB) are negative compensating charges. The exact numbers have to be determined numerically and now also depend on the density of states in the CB. We also see that in this case, the fact that the oxygen vacancy can transform into a neutral color center may further affect the CN.

The third panel (Figure 18.2c) displays the situation when the ionization energy of the acceptor and donor differ from each other. The ionization energy of the defect is given by the distance of the charge-state transition energy to the VBM for acceptors and to the CBM for donors. The donor is shallow (low ionization energy), while the acceptor is deep (high ionization energy). Only very shallow donors are able to keep their charge even at a Fermi energy close to the band edges. These defects are generally good candidates to produce free charge carriers. This is necessary but not a sufficient condition. In order to obtain the free charge carriers the defects of opposite charge should additionally have high formation energy. In Figure 18.2c, for example, this is not the case. The point of CN is well below the CBM. In this situation, even with the help of additional extrinsic donors, the Fermi level can hardly move higher. This is because the formation energy of the acceptor becomes negative for a Fermi energy well below the CBM. This is usually referred to as Fermi-level pinning, a situation which leads to difficulties of n-type doping, for example, in chalcopyrites due to the presence of negatively charged copper vacancies. The probability for the existence of such pinning levels generally increases with larger band gaps. Therefore, the larger the band gap, the fewer materials can be found which are still dopable to a significant extent. This is because the intervals of the Fermi level for which negative defect formation energies occur, are potentially larger for wide-gap materials. While small band-gap materials like silicon or GaAs can usually be doped n- as well as p-type, one type of doping usually predominates for wide-gap materials, while the other is more difficult or even impossible to achieve. Finally, for insulating materials often neither n- nor p-type doping can be achieved.

In the last example (Figure 18.2d) it is shown that defects can potentially exist in multiple charge states. It is possible that donors and acceptors have two or more ionization levels. Consequently, they exhibit two or more different charge states. It is further also possible that defects exist in negative as well as positive charge states, that is, ambipolar defects. Important defects of this class are, for example, vacancies and interstitials in silicon as well as hydrogen in many semiconductors. In the case of hydrogen, the switching from donor to acceptor does not occur via a neutral charge state. The omission of a charge state is called a negative U behavior and is usually related to large structural relaxations and conformational changes of the defect.

18.5.2
Formation Energies from *Ab-Initio* Calculations

After giving some examples, we now show the process of obtaining the formation energies of point defects from the output of first-principles calculations. First, an appropriate variant of first-principles methodology is chosen, which reproduces most of the relevant properties of the bulk material and still allows to calculate at least 100 atoms. Bulk calculations are performed, and the accuracy of the (DFT) method is tested with the semiconductor and all relevant reference phases (metal and the oxygen molecule in the case of an oxide). From a band structure calculation of the host semiconductor, the positions of the VBM and CBM and their corresponding energies are determined. In the next step, neutral defects are placed into a supercell of a sufficient size and the structures are optimized usually at constant volume conditions. The size of the supercell should be as large as possible. However, this sensitively depends on the material, the method, and available computational resources. The calculations are usually conducted within periodic boundary conditions and the defects in the supercells should be separated by at least two neighboring shells. The defects are subsequently charged, and the total energies are calculated for each charge state. When periodic boundary conditions are used, a homogeneous counter-charge is added to avoid energy divergence. Unfortunately, this measure introduces spurious cell-size effects. In the case of potentially magnetic defects, additional spin-polarized calculations have to be conducted in order to avoid spin-contamination. The formation energies for each charge state can now be obtained for a Fermi energy at the VBM by evaluating Eq. (18.5) and setting $E_f = 0$. Normally, only the line segments for the charge states with the lowest formation energy are plotted for each defect (see Figure 18.2). As discussed above, defect-formation energies are usually plotted under specific chemical potential of its constituents, which are of interest. It is often instructive to plot the formation energies under the most extreme conditions, which give upper and lower limits to the formation energies of specific defects. In the case of the metal oxide (see Figure 18.1) this would be $\Delta\mu_O = 0$; $\Delta\mu_M = H_f^{MO}$ in the oxygen-rich (high oxygen pressure) limit whereas $\Delta\mu_O = H_f^{MO}$; $\Delta\mu_M = 0$ in the reducing limit (low oxygen pressure).[3]

At this stage, the formation energies are still affected by so-called supercell size effects. The magnitude of these effects can well be on the order of electron volts.

---

[3] The Gibbs free energy of formation G_f is equal to the formation enthalpy H_f when the entropy of formation is neglected. This is usually a safe approximation at room temperature.

494 | *18 Ab-Initio Modeling of Defects in Semiconductors*

For example, charged defects interact electrostatically with their periodic replicas in calculations with periodic boundary conditions. In the case of very extended defect-localized wavefunctions, it is additionally possible that the defect-related state shows a significant band dispersion, which in turn leads to usually overestimations of the total energy. Several other supercell size effects have been pointed out previously for charged but also for neutral defects. They are reviewed extensively in the literature [39–41].

18.5.3
Case study: Point Defects in ZnO

In order to illustrate the formalism presented before, we show results for ZnO, an important semiconducting material. The data presented in Figure 18.3a and b

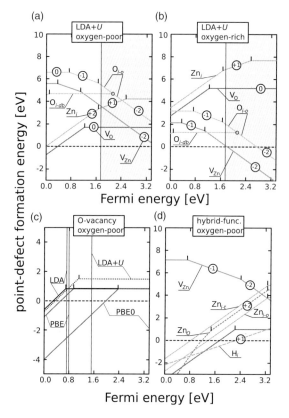

Figure 18.3 Formation energies of intrinsic point defects of ZnO in the LDA with U corrections in the oxygen-poor (a) and oxygen-rich (b) limiting cases according to Ref. [33]. (c) A comparison of different exchange-correlation functionals applied to the oxygen vacancy [42]. Vertical lines denote the band gaps corresponding to the respective functional. (d) Comparison of several donor defects including hydrogen interstitials and using hybrid-functional methodology according to Ref. [43].

corresponds to the calculated formation energies according to Ref. [33]. This study was undertaken using the so-called LDA + U approach in order to calculate the formation energies of all intrinsic defects. In a second step, the formation energies were corrected for supercell size effects by an extrapolation technique, which involves the calculation of the formation energy in supercells with different sizes.

The formation energies are shown for oxygen-poor (Figure 18.3a) and oxygen-rich (Figure 18.3b) conditions. These are the stability limits of ZnO toward decomposition into metal + oxygen gas (oxygen-poor) on the one side and zinc-peroxide formation on the other side (oxygen-rich). The corresponding environmental conditions can be estimated with the help of the ideal gas law and electrochemical tables. Exemplarily, the oxygen pressure range is 10^{-26}–10^{16} Pa for a temperature of 600 °C. In the following, the defect energetics is discussed with respect to the calculated band gap (shaded area), which is generally too low using local exchange-correlation functionals. This deficiency is further discussed below.

In the oxygen-poor regime only oxygen vacancies have low formation energies. This defect is the primary donor of ZnO and exists in the charge states $q = +2/0$. The fact that the charge state $q = +1$ is not present indicates the existence of large lattice relaxations. In the charge state $q = +2$, the energy gain due to lattice relaxations is so high that the removal of an electron from the neutral charge state results in the immediate ejection of the second electron (negative U effect). The Zn interstitials (Zn_i) on the other side have higher transition levels, that is, they are shallow donors with higher formation energies. It is interesting to note that the oxygen vacancy has a relatively high ionization energy (color center) but low formation energy on the one side, whereas the Zn_i has a low ionization energy but high formation energy on the other side. This means that both defects cannot lead to high intrinsic electron concentrations, but due to different reasons. This finding is at odds with the experimental observation of high intrinsic electron concentrations, especially in reduced ZnO samples. This issue has been extensively discussed in the literature and will not be reviewed here. The most commonly cited theories resolving this controversy can be found in Refs. [32, 44].

Among the acceptor defects, the Zn vacancy (V_{Zn}) has the lowest formation energy, which is comparatively large in the oxygen-poor limit but rather low in the oxygen-rich limit. Therefore, by increasing the oxygen partial pressure the formation of V_{Zn} can be favored over V_O. Additionally, at lower Fermi energy in the oxygen-rich limit, oxygen interstitials in a dumbbell configuration have a low formation energy. Similar to the oxygen vacancies, this defect shows the interesting effect of omitting a charge state ($q = -1$). The most stable charge state is $q = 0$. In this dumbbell configuration the oxygen atoms have a net charge of -1 (obtained via direct analysis of the charge density) and can, therefore, be classified as peroxo-ions which is a closed (electronic-) shell configuration of the oxygen dimer. Consequently, additional electrons cannot be accommodated within this geometry. Upon electron addition the dimer breaks and one oxygen reoccupies a regular oxygen site, whereas the other oxygen moves to the more symmetrical octahedral interstitial position within the wurzite lattice. In the final configuration, both anions have a net charge of -2, so that in total two electrons are needed to accomplish this rearrangement. It is important to note that ZnO is often strongly n-type doped. The formation energy of acceptors (especially V_{Zn}) intersect the zero formation energy line. This point depends on the oxygen chemical potential and denotes a pinning level for n-type doping in this material. Under high n-type doping, the V_{Zn} will cause a decay of the doping efficiency in this material. In summary, the defect equilibria of ZnO are mainly determined by oxygen and zinc vacancies similar to situation (b) of Figure 18.2.

Up to this point we have discussed the defect energetics with respect to the calculated band gap, which is strongly underestimated. The use of different exchange-correlation functionals results in different band-gap energies. In Figure 18.3, the calculated band gaps of several functionals are indicated together with the resulting formation energies for the oxygen vacancies in ZnO [42].

While there are significant differences in the band-gap energies for different functionals, it is interesting that the formation energy of the neutral charge state is comparably constant. LDA and GGA-PBE give very similar results and their band-gap energies are very close. In contrast, the hybrid functional (PBE0) results in a decrease in the formation energy of more than 3 eV for the positive charge state. The electronic transition energy is consequently shifted upward while the formation energy of the zero charge state is essentially the same as in LDA calculations. In contrast, the LDA + U approach leads to both a higher formation energy and higher transition energy for the oxygen vacancies. In the LDA + U approach, the band gap is still significantly underestimated, while the PBE0 is able to reproduce the correct band gap. The relatively constant formation energies for the neutral charge state express the fact that total energies are reliable even when obtained with the LDA. Whenever the ionization of a defect is considered, however, complications due to band-gap errors arise. Since the hybrid functional is able to reproduce the band gap of ZnO, it is interesting to reexamine also the energetics of other defects. Figure 18.3d shows the defect energetics of several intrinsic defects and hydrogen interstitials in ZnO. The results are taken from Ref. [43], which were obtained by the hybrid functional method in a plane-wave implementation [45]. Also in this case, the formation energies were corrected for finite-size effects by using increasingly larger supercells up to a size of 784 atoms and an extrapolation technique.

For the common defects, the diagram looks similar to the LDA + U result, apart from the pronounced shift to higher Fermi energy. The order of the defect energetics is essentially the same as in the case of the LDA + U calculations. For example, the energetic difference between V_O and Zn_i is comparable. What changes is mainly the positions of the band edges with respect to the transition energies and to a smaller extent the relative positions of the transition energies with respect to each other. In contrast, the formation energies of the neutral charge states change to a much lesser extent.

As an example for extrinsic defects, hydrogen, a potentially important donor for ZnO, is also displayed in Figure 18.3d. This defect has a positive slope throughout the whole band gap and is therefore a very shallow donor. It can be found as interstitial or substitutional defect. In this case, the formation energy of the defect is related to the solubility of hydrogen.

18.6
Conclusions

Over the last decades, powerful theoretical and numerical methods were developed that have triggered the deployment of computational materials science as an entirely new discipline providing information about materials' properties and processes. Quantum-mechanical methods have gained significant importance, especially in the context of understanding point defects in semiconductors, since they allow to systematically study a variety of material properties including those that are not directly accessible by experiments. Therefore, by systematically combining first-principles methods and experiments further improvement of materials for photovoltaic materials can be expected. This chapter has outlined the principle concepts of electronic DFT and presented its application to intrinsic point defects in oxide materials.

References

1 Mitas, L. (1997) Quantum Monte Carlo. *Curr. Opin. Solid State Mater. Sci.*, **2**, 696–700.

2 Szabo, A. and Ostlund, N.S. (1996) *Modern Quantum Chemistry: Introduction to Advanced Electronic Structure Theory*, Dover, New York.

3 Hohenberg, P. and Kohn, W. (1964) Inhomogeneous electron gas. *Phys. Rev.*, **136**, B864.

4 Kohn, W. and Sham, L.J. (1965) Self-consistent equations including exchange and correlation effects. *Phys. Rev.*, **140**, A1133.

5 Singh, D.J. and Nordstrom, L. (eds) (2006) *Planewaves, Pseudopotentials, and the LAPW Method*, Springer, Berlin.

6 Ceperley, D.M. and Alder, B.J. (1980) Ground state of the electron gas by a stochastic method. *Phys. Rev. Lett.*, **45**, 566–569.

7 Perdew, J.P., Burke, K., and Ernzerhof, M. (1996) Generalized gradient approximation made simple. *Phys. Rev. Lett.*, **77**, 3865.

8 Hubbard, J. (1963) Correlations in narrow energy bands. *Proc. Roy. Soc. A*, **276**, 238.

9 Anisimov, V.I., Zaanen, J., and Andersen, O.K. (1991) Band theory and Mott insulators: Hubbard U instead of Stoner I. *Phys. Rev. B*, **44**, 943.

10 Anisimov, V.I., Aryasetiawan, F., and Lichtenstein, A.I. (1997) First-principles calculations of the electron structure and spectra of strongly correlated systems: The LDA + U method. *J. Phys.: Condens. Matter*, **9**, 767.

11 Becke, A.D. (1993) Density-functional thermochemistry. III: The role of exact exchange. *J. Chem. Phys.*, **98**, 5648.

12 Adamo, C. and Barone, V. (1999) Toward reliable density functional methods without adjustable parameters: The PBE0 model. *J. Chem. Phys.*, **110**, 6158.

13 Heyd, J., Scuseria, G., and Ernzerhof, M. (2003) Hybrid functionals based on a screened Coulomb potential. *J. Chem. Phys.*, **118**, 8207.

14 Heyd, J., Scuseria, G., and Ernzerhof, M. (2006) Hybrid functionals based on a screened Coulomb potential (vol. 118, pg. 8207, 2003). *J. Chem. Phys.*, **124**, 219906.

15 Marsman, M., Paier, J., Stroppa, A., and Kresse, G. (2008) Hybrid functionals applied to extended systems. *J. Phys.: Condens. Matter*, **20**, 064201.

16 Kümmel, S. and Kronik, L. (2008) Orbital-dependent density functionals: Theory and applications. *Rev. Mod. Phys.*, **80**, 3.

17 Payne, M.C., Teter, M.P., Allen, D.C., Arias, T.A., and Joannopoulos, J.D. (1992) Iterative minimization techniques for

ab initio total-energy calculations: Molecular dynamics and conjugate gradients. *Rev. Mod. Phys.*, **64**, 1045.

18 Martin, R.M. (2004) *Electronic Structure: Basic Theory and Practical Methods*, Cambridge University Press, Cambridge.

19 Friedrich, C. and Schindlmayer, A. (2006) Many-body perturbation theory, in *Computational Nanoscience: Do It Yourself!*, vol. 31 (eds J. Grotendorst, S. Blügel, and D. Marx) NIC Series, John von Neumann Institute for Computing, Jülich, Germany, pp. 335–355.

20 Onida, G., Reining, L., and Rubio, A. (2002) Electron excitations:density-functional versus many-body Green's-function approaches. *Rev. Mod. Phys.*, **74**, 601.

21 Foulkes, W.M.C., Mitas, L., Needs, R.J., and Rajagopal, G. (2001) Quantum Monte Carlo simulation of solids. *Rev. Mod. Phys.*, **73**, 33.

22 Hedin., L. (1965) New method for calculating the one-particle Green's function with application to the electron-gas problem. *Phys. Rev.*, **139**, 796.

23 Lerch, D., Wieckhorst, O., Hart, G.L.W., Forcade, R.W., and Müller, S. (2009) Uncle: A code for constructing cluster expansions for arbitrary lattices with minimal user-input. *Mod. Sim. Mater. Sci. Eng.*, **17**, 055003.

24 Erhart, P., Juslin, N., Goy, O., Nordlund, K., Muller, R., and Albe, K. (2006) Analytic bond-order potential for atomistic simulations of zinc oxide. *J. Phys: Condens. Matter*, **18** (29), 6585–6605.

25 Wu, X., Vanderbilt, D., and Hamann, D.R. (2005) Systematic treatment of displacements, strains, and electric fields in density-functional perturbation theory. *Phys. Rev. B*, **72** (3), 035105.

26 Baroni, S., de Gironcoli, S., Dal Corso, A., and Giannozzi, P. (2001) Phonons and related crystal properties from density-functional perturbation theory. *Rev. Mod. Phys.*, **73**, 515.

27 Gonze, X., Allan, D.C., and Teter, M.P. (1992) Dielectric tensor, effective charges, and phonons in alpha-quartz by variational density-functional perturbation-theory. *Phys. Rev. Lett.*, **68**, 3603.

28 Morgan, T.N. (1986) Theory of the DX center in Al_xGa_{1-x} as and gas crystals. *Phys. Rev. B*, **34**, 2664.

29 Chadi, D.J. (1999) Predictor of p-type doping in II–VI semiconductors. *Phys. Rev. Lett.*, **59** (23), 15181.

30 Chadi, D.J. and Chang, D.J. (1989) Energetics of DX-center formation in gas and Al_xGa_{1-x} as alloys. *Phys. Rev. B*, **39**, 10063.

31 Wei, S.-H. and Zhang, S.B. (2002) Chemical trends of defect formation and doping limit in II–VI semiconductors: The case of CdTe. *Phys. Rev. B*, **66**, 155211.

32 Lany, S. and Zunger, A. (2005) Anion vacancies as a source of persistent photoconductivity in II–VI and chalcopyrite semiconductors. *Phys. Rev. B*, **72**, 035215.

33 Erhart, P., Albe, K., and Klein, A. (2006) First-principles study of intrinsic point defects in ZnO: Role of band structure, volume relaxation and finite size effects. *Phys. Rev. B*, **73**, 205203.

34 Erhart, P. and Albe, K. (2008) Modeling the electrical conductivity in $BaTiO_3$ on the basis of first-principles calculations. *J. Appl. Phys.*, **104** (4) 044315.

35 Van de Walle, C.G. and Blöchl, P.E. (1993) First-principles calculations of hyperfine parameters. *Phys. Rev. B*, **47**, 4244.

36 Puska, M.J. and Nieminen, R.M. (1994) Theory of positrons in solids and on solid-surfaces. *Rev. Mod. Phys.*, **66**, 841.

37 Ma, Y.C., Rohlfing, M., and Gali, A. (2010) Excited states of the negatively charged nitrogen-vacancy color center in diamond. *Phys. Rev. B*, **81**, 041204.

38 Zhang, S.B. and Northrup, J.E. (1991) Chemical potential dependence of defect formation energies in GaAs: Application to Ga self-diffusion. *Phys. Rev. Lett.*, **67**, 2339.

39 Persson, C., Zhao, Y.-J., Lany, S., and Zunger, A. (2005) n-type doping of $CuInSe_2$ and $CuGaSe_2$. *Phys. Rev. B*, **72**, 035211.

40 Hine, N.D.M., Frensch, K., Foulkes, W.M.C., and Finnis, M.W. (2009) Supercell size scaling of density functional theory formation energies of charged defects. *Phys. Rev. B*, **79**, 024112.

41 Castleton, C.W.M., Höglund, A., and Mirbt, S. (2006) Managing the supercell approximation for charged defects in semiconductors: Finite-size scaling, charge correction factors, the band-gap problem, and the *ab initio* dielectric constant. *Phys. Rev. B*, **73**, 035215.

42 Ágoston, P., Albe, K., Nieminen, R.M., and Puska, M.J. (2009) Intrinsic n-type behavior in transparent conducting oxides: A comparative hybrid-functional study of In_2O_3, SnO_2, and ZnO. *Phys. Rev. Lett.*, **103**, 245501.

43 Oba, F., Togo, A., Tanaka, I., Paier, J., and Kresse, G. (2008) Defect energetics in ZnO: A hybrid Hartree–Fock density functional study. *Phys. Rev. B*, **77**, 245202.

44 Van de Walle, C.G. (2000) Hydrogen as a cause of doping in zinc oxide. *Phys. Rev. Lett.*, **85**, 1012–1015.

45 Paier, J., Marsman, M., Hummer, K., Kresse, G., Gerber, I.C., and Ángyán, J.G. (2006) Screened hybrid density functionals applied to solids. *J. Chem. Phys.*, **124**, 154709.

19
One-Dimensional Electro-Optical Simulations of Thin-Film Solar Cells

Bart E. Pieters, Koen Decock, Marc Burgelman, Rolf Stangl, and Thomas Kirchartz

19.1
Introduction

In this chapter, we discuss the simulation of thin-film solar cells, that is, the application of detailed mathematical models to describe the physics relevant to the operation of thin-film solar cells. These models should describe both the optical and the electronic processes in the device. We will first discuss the fundamentals of device simulation. In Section 19.3, we discuss the simulation of various thin-film materials used in solar cells, including a-Si:H, μc-Si:H, Cu(In,Ga)Se$_2$, and CdTe. Section 19.4 deals with commonly used optical models. Various popular tools used for the simulation of thin-film devices are discussed in Section 19.5.

19.2
Fundamentals

The semiconductor equations consist of the Poisson equation

$$\nabla \cdot (\varepsilon \nabla \Psi_{vac}) = -\rho \qquad (19.1)$$

and the continuity equations for electrons and holes

$$\frac{\partial n}{\partial t} = \frac{1}{q} \nabla \cdot \vec{J}_n + G - R$$
$$\frac{\partial p}{\partial t} = -\frac{1}{q} \nabla \cdot \vec{J}_p + G - R \qquad (19.2)$$

where Ψ_{vac} is the potential related to the local vacuum level, ρ is the space charge density, n and p refer to electron and hole concentrations in the extended conduction and valence band, respectively, J_n and J_p are, respectively, the electron and hole current density, t is time, G is the generation rate, and R is the recombination rate.

The electron and hole current densities are, respectively, described by

$$\vec{J}_n = \mu_n n \nabla E_{Fn}$$
$$\vec{J}_p = \mu_p p \nabla E_{Fp} \qquad (19.3)$$

where E_{Fn} and E_{Fp} are the quasi-Fermi levels for electrons and holes, respectively.

Using the Maxwell–Boltzmann approximation for the carrier concentrations as a function of the quasi-Fermi levels and the effective density of states in the valence (N_v) and conduction (N_c) band, we can write

$$n = N_c \exp\left(\frac{E_{Fn}-E_c}{kT}\right) = N_c \exp\left(\frac{E_{Fn}-(E_{vac}-q\chi)}{kT}\right)$$
$$p = N_v \exp\left(\frac{E_v-E_{Fp}}{kT}\right) = N_v \exp\left(\frac{(E_{vac}-q\chi-E_\mu)-E_{Fp}}{kT}\right) \qquad (19.4)$$

where E_{vac} is the vacuum potential, χ is the electron affinity, and E_μ is the mobility gap. In a crystalline semiconductor, the mobility gap is identical to the band gap. In disordered semiconductors like a-Si:H or μc-Si:H, the band gap is no longer a well-defined quantity and the mobility gap – defined as the energy range where states are localized and carriers have a negligible mobility – takes over its role.

In equilibrium the product of n and p is equal to

$$n_i^2 = N_c N_v \exp\left(\frac{E_v-E_c}{kT}\right) = N_c N_v \exp\left(-\frac{E_\mu}{kT}\right) \qquad (19.5)$$

where n_i is the intrinsic carrier concentration.

To solve the semiconductor equations, the boundary conditions to the simulated domain need to be specified. Typically device simulators offer the possibility to use an ohmic (flatband) model or a Schottky contact (i.e., including a barrier). In addition to the possible presence of a barrier at the contact, the surface recombination rates can typically be specified. Surface recombination is described by

$$\vec{J}_n = -qR_n^{surf} = -qS_n(n-n_{eq})$$
$$\vec{J}_p = -qR_p^{surf} = -qS_p(p-p_{eq}) \qquad (19.6)$$

where S_n and S_p are the surface recombination velocities of electrons and holes, respectively, and n_{eq} and p_{eq} are the equilibrium concentrations at the contacts for electrons and holes, respectively. Commonly infinite surface recombination velocities are assumed for thin-film silicon devices. However, the theoretical maximum value of the surface recombination velocity is the thermal velocity.

19.3
Modeling Hydrogenated Amorphous and Microcrystalline Silicon

19.3.1
Density of States and Transport Hydrogenated Amorphous Silicon

The schematic density of states in a-Si:H is shown in Figure 19.1. Indicated are the extended states in the valence and conduction band, the valence and conduction band tails, and the states originating from dangling bonds. Due to the localized nature of band-tail states and states from structural defects, the mobility of electrons and holes at these states is much lower as compared to the nonlocalized states in the valence and conduction band, where the carriers are considered free. The energy separation of localized and nonlocalized states is rather sharp [1] and, consequently, this dividing energy has been termed "mobility edge." The energy difference between the valence-band mobility edge and conduction-band mobility edge is the so-called mobility gap E_μ. As the energy separation of localized and nonlocalized states is sharp, electronic transport in a-Si:H is dominated by the carriers in the states just below the valence-band mobility edge and just above the conduction-band mobility edge; hence for electronic transport, the mobility gap is the amorphous equivalent of the band gap. In Figure 19.1, the mobility edges of the valence and conduction band are indicated by E_v and E_c, respectively. A typical value of the mobility gap in a-Si:H is 1.75 eV [2].

In a-Si:H, there are distributions of localized band-tail states. The localized tail states are single-electron states, that is, they can be occupied by either 0 or 1 electron. Tail states belonging to the conduction band exhibit acceptor-like behavior, meaning that the states are neutral when unoccupied and negatively charged when occupied by an electron. Tail states belonging to the valence band, on the other hand, exhibit donor-like behavior and are positively charged when unoccupied and neutral when

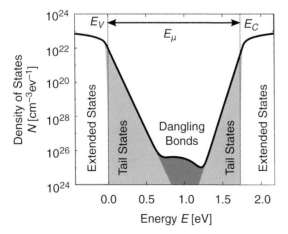

Figure 19.1 Schematic representation of the density of states in a-Si:H.

occupied by an electron. The tail-state densities of the conduction and valence band decay exponentially into the gap [3]. The density of tail states can be described by

$$N_{cbt} = N_c \exp\left(\frac{E-E_c}{E_{c0}}\right)$$
$$N_{vbt} = N_v \exp\left(\frac{E_v-E}{E_{v0}}\right)$$
(19.7)

where N_{v0} and N_{c0} are the density of tail states at the mobility edges of the valence and conduction band, respectively, and E_{v0} and E_{c0} are the characteristic energies (or Urbach energies) of the valence and conduction band tails, respectively. In a-Si:H, the valence band tail is much broader than the conduction band tail. Typical values of the characteristic energies of the tail states in device quality a-Si:H are 45 meV for the valence band tail and 30 meV for the conduction band tail. The characteristic energies of the band tails are, however, temperature dependent [4, 5]. For the valence band tail, the following relation is often used [4]:

$$E_{v0}(T) = \sqrt{E_{v0}(0)^2 + (kT)^2}$$
(19.8)

Aljishi et al. [5] suggested that the temperature dependency of the band tails exhibits a freeze-in temperature, above which the tail slope becomes temperature dependent.

Defect states arising from dangling bonds are *amphoteric* in nature, that is, the dangling bond is multivalent and can have three charged states, namely, positively charged when the state is unoccupied by electrons, neutral when the dangling bond is occupied by one electron, and negatively charged when the dangling bond is occupied by two electrons. A dangling bond, therefore, has two energy levels: the $E^{+/0}$ level related to the transition between the positively and neutrally charged states of the dangling bond, and the $E^{0/-}$ level related to the transition between the neutrally and negatively charged states of the dangling bond. The energy difference between the $E^{+/0}$ and $E^{0/-}$ levels of the dangling bond is the correlation energy, U. It is generally accepted that in a-Si:H, the correlation energy is positive, that is, the $E^{0/-}$ level is higher than the $E^{+/0}$.

For the distribution of dangling-bond states in the mobility gap of a-Si:H and related materials, the most frequently used models are the standard model and the defect-pool model. The standard model is a simple approach where the distribution of dangling-bond states is assumed to be Gaussian. The distribution of dangling-bond states as characterized by their $E^{+/0}$ transition level is

$$N_{db}^{+/0} = N_{db}\frac{1}{\sigma_{db}\sqrt{2\pi}}\exp\left(-\frac{(E-E_{db0}^{+/0})^2}{2\sigma_{db}^2}\right)$$
(19.9)

where $N_{db}^{+/0}$ is the distribution of energy levels arising from the $E^{+/0}$ transition level, N_{db} is the total dangling-bond concentration of which the $E^{+/0}$ are distributed around the energy level $E_{db0}^{+/0}$ with a standard deviation of σ_{db}. Note that Eq. (19.9) can easily be transformed to describe the distribution of $E^{0/-}$ transition levels by replacing $E_{db0}^{+/0}$ with $E_{db0}^{0/-} = E_{db0}^{+/0} + U$.

The defect-pool model, of which several versions exist [6–9], is an elaborate thermodynamical model that describes chemical equilibrium reactions where weak Si–Si bonds break to form two dangling bonds and the reverse reaction (the weak-bond to dangling-bond conversion model [10]). Hydrogen plays a key role in the equilibration processes in a-Si:H. At normal deposition temperatures of a-Si:H, the hydrogen in the material is mobile. The establishment of a chemical equilibrium between weak bonds and dangling bonds requires structural changes in the material, for which the mobile hydrogen provides the required atomic motion [11, 12]. The central idea behind the defect-pool models is that the concentration of dangling-bond states depends on the formation energy of the dangling bonds. Furthermore, the formation energy of a dangling bond depends on charged state of the dangling bond and thereby on the position of the Fermi level and the energy levels of the amphoteric dangling-bond state [13]. In defect-pool models, the energy distribution of the dangling-bond states is computed such that the free energy of the system is minimized [8], resulting in equilibrium defect-state distributions that depend strongly on the position of the Fermi level in the material. An important result is that the defect-pool model can account for the observed differences in energy distribution of defect states in undoped and doped a-Si:H [9].

In the following section, we will briefly introduce the defect-pool model as it was formulated in 1996 by Powell and Deane [8] (in 1993 a similar defect-pool model was published by Powell and Deane [7]). The defect-pool function, $P(E)$, describes the probability distribution that if a dangling bond is created, it is created at energy level E. Usually the defect-pool function is assumed to be a Gaussian distribution:

$$P(E) = \frac{1}{\sigma_{dp}\sqrt{2\pi}} \exp\left(-\frac{(E-E_{dp})^2}{2\sigma_{dp}^2}\right) \tag{19.10}$$

where σ_{dp} is the standard deviation of the Gaussian defect pool and E_{dp} is the mean energy of the Gaussian defect pool. Note that the final distribution of defects is *not* Gaussian as the formation energy, and thus the equilibrium concentration of dangling bonds depends on the energy level. Using this Gaussian defect-pool function and taking into account the influence of the formation energy of dangling bonds on the equilibrium concentration thereof, Powell and Deane derived the following expression for the energy distribution of dangling-bond defect states:

$$N_{db}^{+/0}(E) = \gamma \left(\frac{2}{F_{eq}^0(E)}\right)^{kT/2E_{v0}} P\left(E + \frac{\sigma_{dp}^2}{2E_{v0}}\right)$$

$$\gamma = N_{v0} \left(\frac{H}{N_{SiSi}}\right)^{kT/4E_{v0}} \left(\frac{2E_{v0}^2}{2E_{v0}-kT}\right) \exp\left(-\frac{1}{2E_{v0}}\left[E_p - E_v - \frac{\sigma_{dp}^2}{4E_{v0}}\right]\right) \tag{19.11}$$

where F_{eq}^0 is the equilibrium occupation function for neutral dangling-bond states, N_{SiSi} is the concentration of electrons in Si–Si bonding states; taking four electrons per Si atom N_{SiSi} is approximately 2×10^{29} m^{-3} [8], H is the concentration of hydrogen in the a-Si:H and is approximately 5×10^{27} m^{-3} [8]. The thermal equilibrium occupation functions for amphoteric dangling-bond states are given by [14, 15]

$$F_{eq}^{+} = \frac{1}{1 + 2\exp\left(\frac{E_f - E^{+/0}}{kT}\right) + \exp\left(\frac{2E_f - E^{+/0} - E^{0/-}}{kT}\right)}$$

$$F_{eq}^{0} = \frac{2\exp\left(\frac{E_f - E^{+/0}}{kT}\right)}{1 + 2\exp\left(\frac{E_f - E^{+/0}}{kT}\right) + \exp\left(\frac{2E_f - E^{+/0} - E^{0/-}}{kT}\right)} \quad (19.12)$$

$$F_{eq}^{-} = \frac{\exp\left(\frac{2E_f - E^{+/0} - E^{0/-}}{kT}\right)}{1 + 2\exp\left(\frac{E_f - E^{+/0}}{kT}\right) + \exp\left(\frac{2E_f - E^{+/0} - E^{0/-}}{kT}\right)}$$

where F_{eq}^{+} and F_{eq}^{-} are the equilibrium occupation functions for positively charged and negatively charged dangling-bond states, respectively.

Figure 19.2 shows the defect-state distributions computed with the 1996 defect-pool model for three positions of the Fermi level (as indicated by the arrows).

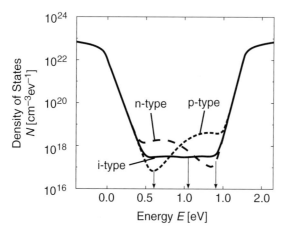

Figure 19.2 Defect-state distributions in a-Si:H according to the 1996 defect-pool model for three positions of the Fermi level, close to the conduction-band mobility edge ("n-type"), the intrinsic position ("i-type"), and close to the valence-band mobility edge ("p-type"). The three positions of the Fermi level are indicated with arrows.

In the case the Fermi level is around mid-gap ("i-type"), the total dangling-bond concentration is lower than that when the Fermi level is close to the conduction band ("n-type") or close to the valence band ("p-type"). The defect-pool model thus predicts a inhomogeneous distribution of defects in the absorber layer as the defect densities are highest near (and in) the doped layers, thus leading to relatively more recombination closer to the p-type and n-type as compared with the assumption of a constant density of dangling bonds through the absorber layer.

19.3.2
Density of States and Transport Hydrogenated Microcrystalline Silicon

In literature μc-Si:H is used as a general term for a mixed-phase material consisting of varying amounts of c-Si nanocrystallites, a-Si:H, and grain boundaries. The presence of crystalline grains, grain boundaries, the columnar structure, and the presence of a-Si:H tissue all influence the transport properties in μc-Si:H. As this chapter deals with one-dimensional device modeling, we cannot take all the effects of such a mixed phase into account. We will, therefore, use effective models, that is, we pretend the material is homogeneous.

It was suggested by Overhoff et al. [16] that conduction through μc-Si:H takes place primarily through the crystalline phase, that is, along a percolation path. From this percolation theory, it is to be expected that the effective medium properties of μc-Si:H are dominated by the properties of the crystalline phase.

The density of states in μc-Si:H looks similar to the density of states in a-Si:H. Beside the extended states, there are band tails and dangling-bond defects. Both time-of-flight and photoluminescence indicate the presence of exponential band tails [17–20]. Both the valence and conduction band tail have about the same characteristic energy, $E_{v0} \approx E_{c0} \approx 31$ meV. It is difficult to infer from the measurements carried out whether the tails are temperature dependent like in a-Si:H [21].

From electron spin resonance (ESR) spectroscopy, it has been found that there are two dangling-bond distributions with different paramagnetic properties in μc-Si.H [22], presumably these two dangling-bond distributions are separated spatially where one distribution is located within the crystallites and the other in the a-Si:H tissue [23]. The total defect density was found to be in the range 5×10^{15}–1×10^{16} cm^{-3}. As we are interested in parameter values for the effective media approximation and we assume transport predominantly to take place through the crystalline part of the material, the effective density of defect states could be lower than the total defect-state density. The total dangling-bond distribution can be considered broad [22, 23]. A Gaussian distribution is assumed with a standard deviation of 150 meV [22].

The mobility gap of μc-Si:H is usually found to be higher than the band gap of crystalline silicon. The mobility gap of μc-Si:H is typically found to be around 1.2 eV [21]. As according to the percolation theory conduction only takes place

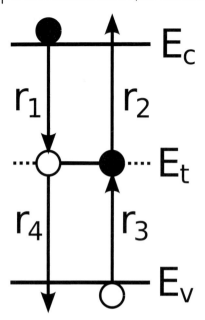

Figure 19.3 Schematic illustration of capture and emission processes on a single-electron trap state.

through the crystalline phase, it follows that the effective density of states in μc-Si:H should scale with the crystalline volume fraction [21].

19.3.3
Modeling Recombination in a-Si:H and μc-Si: H

19.3.3.1 Recombination Statistics for Single-Electron States: Shockley–Read–Hall Recombination

Shockley–Read–Hall (SRH) recombination considers two capture and two emission processes as illustrated in Figure 19.3. Recombination occurs when a trap state occupied by an electron captures a hole or vice versa. The capture and emission rates can be described by the equations in Table 19.1 for both single-electron trap

Table 19.1 Capture and emission rates of single-electron trap states[a].

Process		Rate
Electron capture	r_1	$nv_{th}\sigma_n N_t(1-f)$
Electron emission	r_2	$e_n N_t f$
Hole capture	r_3	$pv_{th}\sigma_p N_t f$
Hole emission	r_4	$e_p N_t(1-f)$

a) The rates $r_1 - r_4$ are defined in Figure 19.3.

states. In Table 19.1, N_t is the concentration of traps with electron occupation probability f, σ_n and σ_p are the electron and hole capture cross sections, respectively, and e_n and e_p are the electron and hole emission coefficients, respectively. In thermal equilibrium, there is no net recombination and the principle of *detailed balance* applies, meaning $r_1 = r_2$ and $r_3 = r_4$. Furthermore, in thermal equilibrium, the electron occupation probability of a trap at energy E_t is described by the Fermi–Dirac distribution, $f = \left(1 + \exp\left(\frac{E_t - E_f}{kT}\right)\right)^{-1}$. Applying the principle of detailed balance and using the Fermi–Dirac distribution for the occupation probability function yields for the emission coefficients

$$e_n = v_{th} \sigma_n N_c \exp\left(\frac{E_t - E_c}{kT}\right)$$

$$e_p = v_{th} \sigma_p N_v \exp\left(\frac{E_v - E_t}{kT}\right) \tag{19.13}$$

where v_{th} is the thermal carrier velocity.

The recombination efficiency, η_R, is defined as the net recombination rate per trap state. In nonequilibrium steady-state conditions, the recombination efficiency is equal to the net rate at which electrons are captured by a trap state. Under steady-state conditions, the net electron capture rate must be equal to the net capture rate of holes (i.e., the average charged state of trap states is not changed); therefore it follows that

$$N_t \eta_R = r_1 - r_2 = r_3 - r_4 \tag{19.14}$$

Using the equations from Table 19.1 and Eq. (19.14), the electron occupation function can be determined as

$$f = \frac{n v_{th} \sigma_n + e_p}{n v_{th} \sigma_n + p v_{th} \sigma_p + e_n + e_p} \tag{19.15}$$

The recombination efficiency of single-electron trap states is then determined as

$$\eta_R = v_{th}^2 \sigma_n \sigma_p \frac{np - n_i^2}{n v_{th} \sigma_n + p v_{th} \sigma_p + e_n + e_p} \tag{19.16}$$

The total recombination on single-electron trap states can be obtained by integrating the recombination over all the single-electron trap states in the band gap

$$R = \int_{E_v}^{E_c} N(E) \eta_R(E) dE \tag{19.17}$$

For the modeling of a-Si:H, the donor-like trap states of the valence band and the acceptor-like states of the conduction band not necessarily have the same capture cross sections. In that case, the integral of Eq. (19.17) should be computed for both tail-state distributions separately. When we refer to capture cross sections of trap states, we usually distinguish between the charged states of the trap. Acceptor-like

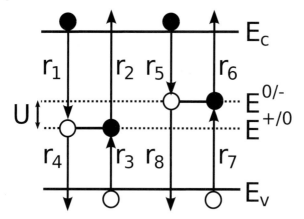

Figure 19.4 Schematic illustration of capture and emission processes on an amphoteric trap state.

traps that capture an electron must be empty, and thus positively charged and acceptor-like traps that capture a hole must be filled and thus neutral. Therefore we refer to the capture cross sections of donor-like states as σ_n^+ and σ_p^0, for the electron and hole capture cross sections, respectively. Likewise, for acceptor-like states, the capture cross sections are referred to as σ_n^0 and σ_p^-, for the electron and hole capture cross sections, respectively.

19.3.3.2 Recombination Statistics for Amphoteric States

Sah and Shockley [24] developed a model that can describe recombination and trapping statistics on amphoteric states. Figure 19.4 illustrates an amphoteric trap level and the capture and emission processes that can occur for an amphoteric trap level. Table 19.2 lists the rates for each capture or emission process of an amphoteric trap level. The symbols $F^{+,0,-}$ denote the occupation functions and represent the probability that the trap is empty, F^+, the trap is occupied by a single electron, F^0, or the trap is occupied by two electrons, F^-.

Table 19.2 Capture and emission rates of amphoteric states[a].

Process		Rate
Electron capture	r_1	$nv_{th}\sigma_n^+ N_{DB} F^+$
Electron emission	r_2	$e_n^0 N_{DB} F^0$
Hole capture	r_3	$nv_{th}\sigma_n^0 N_{DB} F^0$
Hole emission	r_4	$e_n^- N_{DB} F^-$
Electron capture	r_5	$pv_{th}\sigma_p^0 N_{DB} F^0$
Electron emission	r_6	$e_p N_{DB} F^+$
Hole capture	r_7	$pv_{th}\sigma_p^- N_{DB} F^-$
Hole emission	r_8	$e_p N_{DB} F^0$

a) The rates $r_1 - r_8$ are defined in Figure 19.4.

19.3 Modeling Hydrogenated Amorphous and Microcrystalline Silicon

In a similar fashion as for single-electron states, the emission coefficients can be derived from the equilibrium electron, n_{eq}, and hole, p_{eq}, concentrations and the equilibrium occupation probabilities, F_{eq}^+, F_{eq}^0, and F_{eq}^- (see Eq. (19.12)). The emission coefficients are found as

$$e_n^0 = v_{th}\sigma_n^+ n_{eq} \frac{F_{eq}^+}{F_{eq}^0} = \frac{1}{2} v_{th}\sigma_n^+ N_c \exp\left(\frac{E^{+/0} - E_c}{kt}\right)$$

$$e_n^- = v_{th}\sigma_n^- n_{eq} \frac{F_{eq}^0}{F_{eq}^-} = 2v_{th}\sigma_n^- N_c \exp\left(\frac{E^{0/-} - E_c}{kt}\right)$$

(19.18)

$$e_p^+ = v_{th}\sigma_p^0 p_{eq} \frac{F_{eq}^0}{F_{eq}^+} = 2v_{th}\sigma_p^0 N_v \exp\left(\frac{E_v - E^{+/0}}{kt}\right)$$

$$e_p^0 = v_{th}\sigma_p^- p_{eq} \frac{F_{eq}^-}{F_{eq}^0} = \frac{1}{2} v_{th}\sigma_p^- N_v \exp\left(\frac{E_v - E^{0/-}}{kt}\right)$$

In steady-state situations, the net recombination rate is zero and therefore the rate equation of the occupation functions should be zero. The rate equations of the occupation functions are given by

$$\frac{\partial F^+}{\partial t} = -nv_{th}\sigma_n^+ F^+ + e_n^0 F^0 + pv_{th}\sigma_p^0 F^0 - e_p^+ F^+$$

(19.19)

$$\frac{\partial F^-}{\partial t} = nv_{th}\sigma_n^0 F^0 - e_n^- F^- - pv_{th}\sigma_p^- F^- + e_p^0 F^0$$

By further taking into account that the sum of the occupations functions should be unity ($F^+ + F^0 + F^- = 1$), the occupation functions are obtained as

$$F^+ = \frac{P^0 P^-}{N^+ P^- + P^0 P^- + N^+ N^0}$$

$$F^0 = \frac{N^+ P^-}{N^+ P^- + P^0 P^- + N^+ N^0}$$

(19.20)

$$F^0 = \frac{N^0 N^+}{N^+ P^- + P^0 P^- + N^+ N^0}$$

where, for readability the terms P^0, P^-, N^0, and N^+ are introduced, which are defined as

$$P^0 = pv_{th}\sigma_p^0 + e_n^0$$

$$P^- = pv_{th}\sigma_p^- + e_n^-$$

(19.21)

$$N^0 = nv_{th}\sigma_n^0 + e_p^0$$

$$N^+ = nv_{th}\sigma_n^+ + e_p^+$$

The recombination efficiency is given by $\eta_R = (r_1 - r_2 + r_5 - r_6)/N_t$. Using the derived occupation functions, we obtain

$$\eta_R = v_{th}^2 (pn - n_i^2) \frac{\sigma_n^+ \sigma_p^0 P^- + \sigma_n^0 \sigma_p^- N^+}{N^+ P^- - P^0 P^- + N^+ N^0} \tag{19.22}$$

The total recombination rate can be obtained by integration of the recombination efficiency over all dangling-bond states, that is, as done in Eq. (19.17) for single-electron states.

Often amphoteric states are modeled with two, uncorrelated, single-electron states [25–27]. The pair consists of one donor-like state and one acceptor-like state. The energy levels of the uncorrelated states should be slightly shifted compared to the transition levels of the correlated state. By equating the emission coefficient of the acceptor-like state (Eq. (19.13)) to the emission coefficient of the negatively charged amphoteric state (Eq. (19.18)), one can find that the acceptor-like state should be at $E^{0/-} + kT\ln(2)$. Likewise one can find that the donor-like state should be located at $E^{+/0} - kT\ln(2)$. The effective correlation energy of the uncorrelated pair is therefore $U + kT\ln(4)$. Note that modeling a correlated defect with two uncorrelated defects leads to errors as the two states are independently capable of trapping carriers. This leads to higher recombination rates for the pair of uncorrelated states, and may lead to the physically impossible situation that the $E^{0/-}$ (acceptor-like) state is occupied while the $E^{+/0}$ (donor-like) state is unoccupied. Willemen [2] showed that modeling an amphoteric state with a pair of uncorrelated states is only accurate under the condition that (1) the capture cross section of the neutral state is much smaller than the charged states and (2) the correlation energy is positive and considerably larger than kT.

19.3.4
Modeling Cu(In,Ga)Se₂ Solar Cells

A chalcopyrite-based solar cell is a rather complex system, consisting of several layers of various materials. This creates several heterointerfaces, which require quite some effort to model. Not only both adjacent materials are important but also the interface itself and the band line-up at the interface. The defect physics of Cu(In,Ga)(S,Se)₂ exhibits several particularities. They are usually native defects, they can be multivalent [28], and metastable [29]. Additionally Cu(In,Ga)(S,Se)₂ is to be considered as a material system rather than as a material. The exact composition of the material can vary widely, and together with the composition, other properties can be varied. The composition is easily changed, often unintentionally due do diffusion from an adjacent layer. Sometimes, however, the composition is deliberately changed throughout the absorber layer, leading to a graded structure. This section will focus on the modeling aspects of such graded solar cells.

19.3.4.1 Graded Band-Gap Devices
One of the main properties depending on the composition of the Cu(In,Ga)(S,Se)₂ is the band gap. Increasing the Ga/In ratio results in a conduction-band raise,

without much changing the valence band. Increasing the S/Se ratio results in both a conduction-band raise and a valence-band lowering, dominated by the valence-band effect.

Varying the band gap throughout the absorber layer can have two main beneficial effects: (1) the recombination probability can be reduced in regions where this probability is higher; and (2) additional electric fields can be built in by changing the conduction-band level.

1) Increasing the band gap near the junction reduces the junction-recombination rate, which usually dominates the cell recombination and governs the open-circuit voltage. Only increasing the band gap locally ensures that the absorption is not significantly reduced. In order to preserve the band alignment at the heterointerface, the band gap raise is preferred to be a valence-band lowering and thus originating from an increased sulfur content.
2) Varying the conduction-band level leads to the introduction of an additional electric field. This field can be used to push back charge carriers from the back contact, where the recombination probability can be high. This is possible by increasing the Ga-content toward the back of the absorber.

19.3.4.2 Issues when Modeling Graded Band-Gap Devices

When varying the band gap by means of a composition change, it is paramount to know the link between the band gap and the composition. Unfortunately, there is a rather large spread on the reported band-gap values for $Cu(In,Ga)(S,Se)_2$ in literature. Moreover, most of the time a band-gap value is reported at only one composition, or only as a function of the Ga/In or the S/Se ratio. In [30], a general formula for the band gap of $Cu(In_{1-y},Ga_y)(Se_{1-x},S_x)_2$ is proposed combining the reports of several authors and reproduced in Eq. (19.23)

$$E_g(\text{eV}) = 1.04(1-x)(1-y) + 1.68(1-x)y + 1.53x(1-y) + \\ 2.42xy - 0.14x(1-x)(1-y) - 0.15y(1-y) \quad (19.23)$$

A second problem arising when modeling graded structures is to find the absorption ($\alpha(\lambda)$) characteristic of a material which is a mixture of two other materials. There are two properties of importance for a $\alpha(\lambda)$-characteristic: the cut-off wavelength which is determined by the band gap, and the absorption for high-energy photons. The former will be badly reproduced when performing an interpolation between the characteristics of the pure materials along the α-axis, the latter when interpolating along the λ-axis. An adequate interpolation scheme is reported in [31], and is implemented in the solar-cell simulator SCAPS (see Section 19.5.5).

A third issue appears when interpreting the results of simulations for graded band-gap devices. As a high band gap favors the open-circuit voltage and harms the short-circuit current, there exists a compromise between both, leading to an optimum band gap. If the goal of the modeling is to find the influence of the grading on the cell performance, the result will thus be veiled by the band-gap dependence of the IV-characteristics on nongraded solar cells. Hence it is important to compare the results for the graded structure with the results for a uniform reference cell. This reference cell should be designed with great care, however, and is dependent on the investigated structure as well as on the property studied. For example, according whether the defect concentration is thought to originate from the introduction of the grading, the uniform reference should take this defect concentration into account as well. If one wants to investigate the effect of grading on the short-circuit current, one could require the reference to have the same open-circuit voltage as the graded structure. An advanced way of constructing an adequate reference cell, which leads to objective results for most graded structures, involves a combination of two structures carefully selected based on their maximum power point behavior [32].

19.3.4.3 Example

Simulations reveal the principal design problem when grading the absorber of a $Cu(In,Ga)(S,Se)_2$ device, that is, the influence of nongraded and graded parameters is equally important.

In the example below, the influence of a grading at the front of an absorber is investigated. The absorber band gap is locally increased from a certain depth in the absorber, called the *grading depth*, up to the buffer–absorber interface. This increase is assumed to originate from an increasing sulfur concentration. The maximum sulfur concentration (which appears at the buffer–absorber interface) and the grading depth are varied, and the efficiency of a uniform reference structure is subtracted from the resulting efficiency. All details about this model can be found in [33].

Whether the grading is beneficial and what are the optimum values for the grading depth and the maximum sulfur concentration are questions which do not seem to have an unanimous answer. It is largely dependent on parameters not belonging to the grading structure. In Figure 19.5, the simulations have been performed thrice, each time with a different electron-diffusion length in the part of the absorber where no grading is present, thus the back part of the absorber. The optimum parameters governing the grading as well as the absolute value of the efficiency gain/loss are largely different in all three cases. As a conclusion, one could observe that implementing a graded band-gap structure in order to improve the cell performance is not straightforward, but should be done while keeping in mind the entire cell structure.

19.3.5
Modeling of CdTe Solar Cells

Thin-film cadmium telluride solar cells seem to be rather tolerant to imperfections and defects: a decent cell performance is obtained by a wide variety of fabrication technologies, if these include some form of "activation treatment," usually with $CdCl_2$. However, this apparent simplicity should not distract from the complexity in the cell structure and operation. This is illustrated already by the appearance of room temperature I–V measurements under standard illumination: there is a kink in both the dark and illuminated I–V curves at forward bias ("roll-off"), and quite often the light and dark I–V curves intersect ("cross-over"). These phenomena are related with

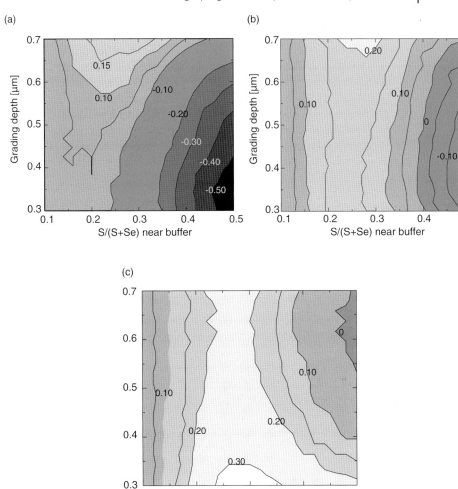

Figure 19.5 Example of the influence of the electron diffusion length in the nongraded part of the absorber on the effect of grading on the cell performance. The electron-diffusion length is the smallest in figure (a) and the largest in figure (c). Figures taken from [33].

the key technology aspects of CdTe cells: the $CdCl_2$ activation treatment, and especially the structure and properties of the back contact to CdTe. A phenomenological two-diode model was set up to explain the gross features of the I–V [34] and C–V measurements [35], but full numerical modeling was necessary to explain all observed phenomena. This has been reviewed in [36]. In this section, we will use numerical modeling to illustrate the interrelation between uncommon aspects observed in I–V and in C–V measurements; these aspects are typical for CdTe cells, but are occasionally also observed in other cells [37].

19.3.5.1 Baseline

An elaborated parameter set was developed [38] for Antec CdTe cells that were treated with $CdCl_2$ in both air and vacuum; see, for example, [39]. In setting up this parameter set, only defects levels were introduced that were confirmed with DLTS, and no unphysical assumptions were made, such as neutral defects or an assumed light or voltage dependence of some parameter. This parameter set could successfully explain a variety of measurements: dark I–$V(T)$, light I–$V(\lambda)$, I_{sc}–$V_{oc}(T)$, C–$V(T)$, C-$f(T)$, and $QE(V)$. It has been used as a baseline set for a numerical parameter exploration [40]. We will use this parameter set here to illustrate the influence of two key parameters: the energy barrier Φ_b at the CdTe back contact, and the acceptor density N_{Ac} in the CdTe layer close to the back contact.

19.3.5.2 The Φ_b – N_{Ac} (Barrier–Doping) Trade-Off

Simulated I–V curves under AM1.5G illumination have been presented in [40]. In Figure 19.6, all parameters had their baseline values but for Φ_b (left) and N_{Ac} (right). It is obvious that the effect of a low doping density at the contact can be erroneously interpreted as an effect of a high contact barrier.

This is further illustrated in Figure 19.7: a too large barrier Φ_b and a too low doping density N_{Ac} can cause a very comparable efficiency loss. From the shape of the illuminated I–V curve in the active quadrant alone, it is not possible to distinguish between the effects of a high Φ_b or of a low N_{Ac}: the I–V curve calculated with $\Phi = 0.4$ eV, $N_{Ac} = 3 \times 10^{14}$ cm^{-3} ("good Φ_b, bad N_{Ac}") almost coincides with the curve calculated for $\Phi_b = 0.55$ eV, $N_{Ac} = 1 \times 10^{17}$ cm^{-3} ("bad Φ_b, good N_{Ac}").

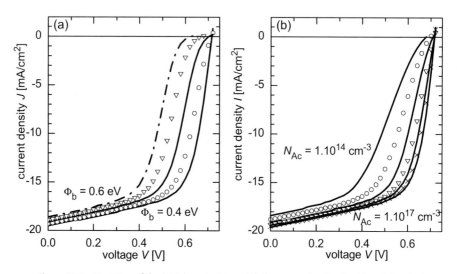

Figure 19.6 Variation of the I–V characteristics with the energy barrier Φ_b (a) and the doping density N_{Ac} at the CdTe back contact (b). The other parameters have their baseline values. (a): $\Phi_b = 0.40, 0.45, \ldots 0.60$ eV. (b): $N_{Ac} = 1 \times 10^{14}, 3 \times 10^{14}, 1 \times 10^{15}, \ldots 1 \times 10^{17}$ cm^{-3}.

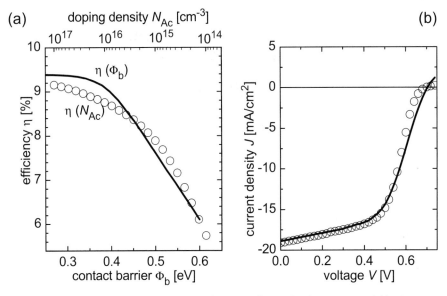

Figure 19.7 (a) dependency of efficiency η on the contact barrier Φ_b (bottom axis; solid line) or on the CdTe doping density N_{Ac} (top axis; open symbols). (b): an I–V curve simulated with $\Phi_b = 0.4$ eV, $N_{Ac} = 3 \times 10^{14}$ cm^{-3} (solid line) and $\Phi_b = 0.55$ eV, $N_{Ac} = 1 \times 10^{17}$ cm^{-3} (open symbols).

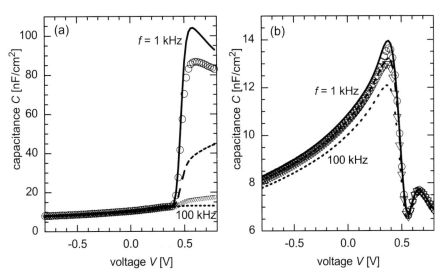

Figure 19.8 (a) C–V curves calculated with "bad Φ_b, good N_{Ac}" (see text). (b) C–V curves calculated with "good Φ_b, bad N_{Ac}" (see text). The parameter is the frequency, ranging from 1 kHz (upper curve) to 100 kHz (lower curve).

19.3.5.3 C–V Analysis as an Interpretation Aid of I–V Curves

Although the two situations, "bad Φ_b, good N_{Ac}" and "good Φ_b, bad N_{Ac}," have very similar I–V curves, they have very different C–V and C–f curves (Figure 19.8). The gross features of C–V curves dominated by contact effects were explained in [35, 36]: the C–V curve in general shows two maxima separated by a minimum. Left of the first maximum, the low-frequency C–V behavior is in circumstances determined by the CdS–CdTe junction capacitance, while right of the second maximum, the CdTe-contact capacitance is dominating the C–V behavior. The shape of the C–V curves at forward voltage clearly is a much better indicator than the I–V curves to distinguish the situation of "bad Φ_b, good N_{Ac}" from "good Φ_b, bad N_{Ac}."

The same holds for the C–f curves measured in an extended temperature range (Figure 19.9). The C–f(T) curves calculated with "good Φ_b, bad N_{Ac}" show an additional step at high-frequency/low-temperature step, which can be ascribed to an RC time constant with the CdS–CdTe junction capacitance and the resistance of the CdTe bulk.

The normal procedure to analyze a possible effect of a too high contact barrier is to measure the temperature-dependent dark I–V(T) curves. The curves for our two situations only differ in the high-voltage range $V > 0.5$ V (no illustration), and even then the shape of the I–V(T) does not allow to distinguish between the two situations. We have shown here that impedance measurements, especially C–V(f) and C–f(T), together with numerical simulation, provide better criteria for assessing the contact effects.

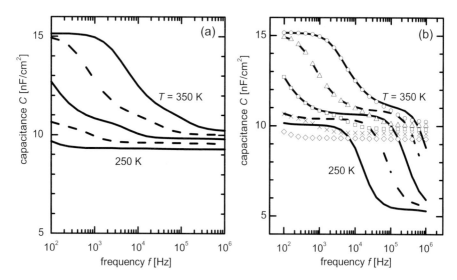

Figure 19.9 (a) C–f curves calculated with "bad Φ_b, good N_{Ac}" (see text). (b) C–f curves calculated with "good Φ_b, bad N_{Ac}"; the data of (a) are also shown in open symbols, for comparison. The parameter is the temperature, ranging from 350 K (upper curve) to 250 K (lower curve), step 25 K. The dc bias voltage is 0 V.

19.4
Optical Modeling of Thin Solar Cells

The optical generation rate in a solar cell is also an important input parameter for electrical modeling. It is determined from the absorption profile of the photons in the solar cell. When assuming that every photon generates one and only one electron–hole pair, the generation rate profile is equal to the absorption profile. Calculation of the absorption profile in a-Si:H-based solar cells is complicated by the commonly used light-trapping techniques making a-Si:H-based solar cells complex optical systems.

19.4.1
Coherent Modeling of Flat Interfaces

For flat solar cells, the multilayer thin-film optics model can be used. The multilayer thin-film optics model uses the complex refractive indices of the media and the effective Fresnel's coefficients to calculate the optical generation rate profile. This model includes multiple internal reflections and interference. See for details, for example, Refs. [41, 42]. Although all practical solar cells use textured substrates and therefore their interfaces are rough, solar-cell structures with flat interfaces are a very useful experimental tool for examining the optical models, extracting unknown optical parameters, and showing the trends in optical behavior.

19.4.2
Modeling of Rough Interfaces

The textured structures used in thin-film solar cells are generally difficult to describe. It is possible to compute the light intensities in a solar cell using Maxwell solvers, see for example reference [43]. However, solving Maxwell's equations for the disordered rough interfaces of the commonly used substrates requires a significant computation effort. Because of this limitation of the Maxwell's equations, the scalar-scattering theory is quite popular [44–46]. In this chapter, we introduce the multi-rough-interface model which is based on the scalar-scattering theory.

In the multi-rough-interface model, it is often assumed that a rough interface reflects (transmits) the same *amount* of light as a flat interface. When light is reflected (transmitted) at a rough interface, a certain amount of light is scattered where the remainder to the reflected (transmitted) light will continue in the specular (non-scattered) direction. In the multi-rough-interface model, the light is assumed to be incoherent. The haze for the reflectance and transmittance can be described by scalar-scattering theory [47, 48]

$$H_R = 1 - \exp\left[-\left(\frac{4\pi\delta_{rms} n_0 \cos(\theta^{in})}{\lambda}\right)^2\right]$$

$$H_T = 1 - \exp\left[-\left(\frac{4\pi\delta_{rms}|n_0 \cos(\theta^{in}) - n_1 \cos(\theta^{out})|}{\lambda}\right)^2\right]$$

(19.24)

where n_0 is the refractive index of the medium of incidence, δ_{rms} is the root-mean-square (rms) roughness of the interface, and θ^{in} and θ^{out} are the incident angle, and the angle of the outbound reflected or transmitted light, respectively.

Unfortunately, with commonly used textured substrates scalar-scattering theory does not work very well to describe the haze. The reason for this is that for scalar-scattering theory to hold it is required that

- the interface should be completely random and have a normal (Gaussian) height distribution; this is generally not the case for commonly used rough substrates used in solar cells, and
- the correlation length of the interface morphology, which is a measure of the lateral feature size, has to be much larger than the wavelength of light; also this condition is often violated for the commonly used substrates.

In order to describe the haze more accurately, it has been proposed to use [49]

$$H_R = 1 - \exp\left[-\left(\frac{4\pi C_R \delta_{rms} n_0 \cos(\theta^{in})}{\lambda}\right)^2\right]$$

$$H_T = 1 - \exp\left[-\left(\frac{4\pi C_T \delta_{rms} |n_0 \cos(\theta^{in}) - n_1 \cos(\theta^{out})|}{\lambda}\right)^3\right]$$

(19.25)

where C_R and C_T are fitting parameters which depend on the two media, and both approach 1 when $|n_0 - n_1|$ is large. Note that this correction is empirical in nature and the appropriate correction depends on your substrate.

The amount of scattered light at a rough interface is incident angle dependent and has an angle distribution. This means that the reflected and transmitted scattered light, R_{scatt} and T_{scatt}, respectively, are proportional to

$$R_{scatt} = H_R f_R^{in}(\theta^{in}) f_R^{out}(\theta^{out}) R_{tot}$$

$$T_{scatt} = H_T f_T^{in}(\theta^{in}) f_T^{out}(\theta^{out}) T_{tot}$$

(19.26)

where $f_{R,T}^{in}$ describes the dependence of scattering on incident angle for reflection and transmittance, $f_{R,T}^{out}$ is the angular distribution of the scattered reflected or transmitted light, θ_{in} and θ_{out} are the incident angle, and the angle of the outbound reflected or transmitted light, respectively, and R_{tot} and T_{tot} are the total reflectance and transmittance of the interface (scattered and specular), respectively. For both angle distribution functions, $f_{R,T}^{in}$ and $f_{R,T}^{out}$, we will assume a $\cos^2(\theta)$ distribution unless otherwise specified.

In practical thin-film solar cells, interference effects are often observed despite the application of surface texture. In order to account also for these interference effects, coherent nature of specular light has to be included in the simulation. In this case, a semicoherent model is required, where specular light is assumed to be coherent and scattered light is incoherent. A semicoherent model can be obtained by combining the multilayer thin-film optics model to analyze the specular light propagation, with the multi-rough-interface model to analyze the scattered light propagation.

19.5
Tools

In this section, we introduce several tools for the simulation of thin-film solar cells. Many 1D simulation tools for solar cells have been developed over the years and as a result we cannot provide a list that is anywhere near to complete. We attempted to provide a list of the better known and readily available programs.

19.5.1
AFORS-HET

AFORS-HET (automat for simulation of heterostructures) is a one-dimensional numerical computer program for modeling multilayer homo- or heterojunction solar cells and solar-cell substructures, as well as a variety of solar-cell characterization methods. Thus different measurements on solar-cell components as well as on the whole solar cell can be compared to the corresponding simulated measurements in order to calibrate the parameters used in the simulations and thus to obtain reliable predictions. Originally, it was developed in order to treat amorphous/crystalline silicon solar cells; however all types of thin-film solar cells (i.e., a-Si, CIS, and CdTe) can as well be appropriately treated. AFORS-HET (current version 2.4, launched December 2009) is an open source on demand program. It is distributed free of charge and it can be downloaded via Internet: http://www.helmholtz-berlin.de/forschung/enma/si-pv/projekte/asicsi/afors-het/index_en.html.

AFORS-HET solves the one-dimensional semiconductor equations (Poisson's equation and the transport and continuity equations for electrons and holes) with the help of finite differences under different conditions: (i) equilibrium mode, (ii) steady-state mode, (iii) steady-state mode with small additional sinusoidal perturbations, (iv) simple transient mode, that is, switching external quantities instantaneously on/off, and (v) general transient mode, that is, allowing for an arbitrary change of external quantities. A multitude of different physical models has been implemented. (1) Optical models: the generation of electron–hole pairs can be described either by a ray traced Lambert–Beer absorption including rough surfaces and using measured reflection and transmission files (neglecting coherence effects), or by calculating the plain-surface incoherent/coherent multiple internal reflections, using the complex indices of reflection for the individual layers (neglecting rough surfaces). (2) Semiconductor bulk models: local profiles within each semiconductor layer can be specified. Radiative recombination, Auger recombination, Shockley–Read–Hall, and/or dangling-bond recombination with arbitrarily distributed defect states within the band gap can be considered. Super-band-gap as well as sub-band-gap generation/recombination can be treated. (3) Interface models for treating heterojunctions: interface currents can be modeled to be either driven by drift diffusion or by thermionic emission. A band to trap tunneling contribution across a heterointerface can be considered. (4) Boundary models: the metallic contacts can be modeled as flatband or Schottky like metal/semiconductor contacts, or as metal/insulator/semiconductor contacts. Furthermore, insulating boundary contacts can also be chosen.

Thus, all internal cell quantities, such as band diagrams, quasi-Fermi energies, local generation/recombination rates, carrier densities, cell currents, and phase shifts can be calculated. Furthermore, a variety of solar-cell characterization methods can be simulated, that is, current voltage, quantum efficiency, transient or quasi-steady-state photo conductance, transient or quasi-steady-state surface photovoltage, spectral resolved steady-state or transient photo- and electroluminescence, impedance/admittance, capacitance-voltage, capacitance-temperature and capacitance-frequency spectroscopy, and electrical detected magnetic resonance. The program allows for arbitrary parameter variations and multidimensional parameter fitting in order to match simulated measurements to real measurements.

AFORS-HET is capable of treating full time-dependent problems and can thus implement a huge variety of different solar-cell characterization methods. Furthermore it offers different physical models in order to adequately describe recombination and transport in thin-film semiconductor layers. It can treat graded layers. However, in order to describe amorphous silicon within pin cells, the defect-pool model should be implemented. Furthermore, a tunnel junction model in order to describe pin/pin tandem cells is still missing. The optical model "coherent/incoherent internal reflections" is very slow if coherent and noncoherent layers have to be considered simultaneously (i.e., treating thin-film layers on glass), and could be significantly improved.

If one expects a numerical computer simulation to give reliable results, a good model calibration, that is, a comparison of simulation results to a variety of different characterization methods, is necessary. The solar cell under different operation conditions should be compared to the simulations. Also different characterization methods for the solar-cell components, that is, for the individual semiconductor layers and for any substacks should be tested against simulation. Only then the adequate physical models as well as the corresponding model input parameters can be satisfactorily chosen. Thus AFORS-HET is especially capable to simulate most of the common characterization methods for solar cells and its components and compare them to real measurements.

19.5.2
AMPS-1D

AMPS-1D is a well-known device simulation program for thin-film (silicon) solar cells developed at Penn State University. The program provides an easy-to-use graphical user interface. The program has a particularly flexible system to define the gap-state distribution of donor and acceptor levels, where apart from continuous distributions like exponential tails and Gaussian distributions of dangling-bond defects, also (banded) discrete levels can be introduced. However, the program does not provide defect-pool models or amphoteric states.

The software is available free of charge. For further information, see http://www.ampsmodeling.org/.

19.5.3
ASA

The *ASA* program was specifically designed for modeling of thin-film silicon solar cells, that is, the program is designed to model both electrical and optical properties of multilayered heterojunction device structures. The *ASA* program is a full-featured and versatile program for the simulation of inorganic semiconductor solar cells in general and thin-film silicon-based solar cells in particular. ASA is written in C and is relatively fast. Furthermore it implements several defect-pool models, Shah and Shockley statistics for amphoteric defects. Several optical models are implemented, including a thin-film optics model and a semicoherent optical model. Although ASA is supplied with a graphical user interface, it is also available as a command-line driven program which can take an ASCII text file as input. The command-line version is particularly flexible as it allows for an easy integration of the ASA program in (mathematical) script languages like GNU Octave [50] and MATLAB® [51]. This allows, for example, using the built-in routines of MATLAB for nonlinear optimization of simulation parameters. A nice example of this is a simulation program for bulk-heterojunction solar cells written in GNU Octave, which makes use of iteratively calls to the command-line version of ASA [52]. This way we effectively added models to the ASA program without modifying the ASA program itself. The ASA program is commercially available from Delft University of Technology.

19.5.4
PC1D

PC1D is de-facto the industry standard for the simulation of solar cells. Its primary focus, however, is on crystalline silicon solar cells, making the program not very suitable for most thin-film technologies. The program was developed starting from 1982 at the University of New South Whales. In 2007, the source code was released under the General Public Licence (GPL), meaning that the program can be freely distributed and modified, provided that the modified source code is made available under the terms of the GPL. The program can be obtained from http://sourceforge.net/projects/pc1d/.

19.5.5
SCAPS

SCAPS is a numerical simulation tool that is developed at the University of Gent [53] that is widespread in the thin-film photovoltaics research community. It was developed especially for thin-film $Cu(In,Ga)Se_2$ and CdTe cells. However, it has also been used for Si and III–V cells, and recently its applicability was further enhanced to amorphous and micromorphous Si cells. Its advantages are that it is fast, interactive, with an intuitive user interface, and that it can simulate various measurements from everyday's PV practice: I–V, C–V, C–f, QE, all under a wide variation of excitations (voltage, illumination intensity, spectrum, wavelength, and frequency). Also, many users experience a low threshold to use it. Since it became available [53, 54], its capabilities have been continuously extended, and now also include intraband tunneling [55], the impurity photovoltaic effect [56], and graded properties ("graded band gaps") [57]. The last extension is the implementation of multivalent defects [58], which is crucial in simulating a-Si cells, and appears to be important in Cu(In,Ga)Se$_2$ cells as well [59].

SCAPS is freely available to the PV research community on request from the authors.

19.5.6
SC-SIMUL

SC-Simul is a program developed by the University of Oldenburg, which is suitable for steady-state and transient simulations of solar cells. The software was developed for simulations of amorphous silicon-crystalline silicon heterojunction solar cells and is therefore capable to simulate amorphous semiconductors. The models for recombination of charge carriers include band-to-band and Auger recombination as well as recombination via tail states and Gaussian dangling-bond distributions. However, a defect-pool model for the calculation of the defect-state distribution in a-Si:H is not implemented. The optical model used to determine the generation rate is a simple Lambert–Beer's law, that is, no interference effects are taken into account. The software is delivered with an intuitive graphical user interface. The user is, however, able to run the simulations from the command line, which allows a similar flexibility as ASA in combining the drift-diffusion solver with external programs as MATLAB.

The software is available free of charge. For further information, see http://www.physik.uni-oldenburg.de/greco/.

References

1 Mott, N. (1987) The mobility edge since 1967. *J. Phys C: Solid State Phys.*, **20**, 3075.
2 Willemen, J.A. (1998) Modelling of Amorphous Silicon Single- and Multi-Junction Solar Cells. Ph.D. thesis, Delft University of Technology.
3 Tiedje, T. (1981) Evidence for exponential band tails in amorphous silicon hydride. *Phys. Rev. Lett.*, **46**, 1425.
4 Stutzmann, M. (1992) A comment on thermal defect creation in hydrogenated amorphous silicon. *Phil. Mag. Lett.*, **66**, 147.
5 Aljishi, S., Cohen, J., Jin, S., and Ley, L. (1990) Band tails in hydrogenated amorphous silicon and silicon–germanium alloys. *Phys. Rev. Lett.*, **64**, 2811.
6 Winer, K. (1990) Defect formation in a-Si:H. *Phys. Rev. B*, **41**, 12150.
7 Powell, M.J. and Deane, S.C. (1993) Improved defect-pool model for charged defects in amorphous silicon. *Phys. Rev. B*, **48**, 10815.
8 Powell, M.J. and Deane, S.C. (1996) Defect-pool model and the hydrogen density of states in hydrogenated amorphous silicon. *Phys. Rev. B*, **53**, 10121.
9 Schumm, G. (1994) Chemical equilibrium description of stable and metastable defect structures in a-Si:H. *Phys. Rev. B*, **49**, 2427.
10 Stutzmann, M. (1987) Weak bond-dangling bond conversion in amorphous silicon. *Phil. Mag. B*, **56**, 63.

11 Kakalios, J., Street, R.A., and Jackson, W.B. (1987) Stretched-exponential relaxation arising from dispersive diffusion of hydrogen in amorphous silicon. *Phys. Rev. Lett.*, **59**, 1037.

12 Street, R.A. (1991) Hydrogen chemical potential and structure of a-Si:H. *Phys. Rev. B*, **43**, 2454.

13 Bar-Yam, Y. and Joannopoulos, J. (1987) Theories of defects in amorphous semiconductors. *J. Non-Cryst. Solids*, **97**, 467.

14 Shockley, W. and Last, J.T. (1957) Statistics of the charge distribution for a localized flaw in a semiconductor. *Phys. Rev.*, **107**, 392.

15 Okamoto, H. and Hamakawa, Y. (1977) Electronic behaviors of the gap states in amorphous semiconductors. *Solid State Commun.*, **24**, 23.

16 Overhof, H., Otte, M., Schmidtke, M., Backhausen, U., and Carius, R. (1998) The transport mechanism in microcrystalline silicon. *J. Non-Cryst. Solids*, **227–230**, 992.

17 Dylla, T., Finger, F., and Schiff, E.A. (2004) Hole drift-mobility measurements and multipletrapping in microcrystalline silicon. *Mater. Res. Soc. Symp. Proc.*, **808**, 109.

18 Dylla, T., Finger, F., and Schiff, E.A. (2005) Hole drift-mobility measurements in microcrystalline silicon. *Appl. Phys. Lett.*, **87**, 032103.

19 Reynolds, S., Smirnov, V., Main, C., Finger, F., and Carius, R. (2004) Interpretation of transient photocurrents in coplanar and sandwich pin microcrystalline silicon structures. *Mater. Res. Soc. Symp. Proc.*, **808**, 127.

20 Merdzhanova, T., Carius, R., Klein, S., Finger, F., and Dimova-Malinovska, D. (2005) A comparison of model calculations and experimental results on photoluminescence energy and open circuit voltage in microcrystalline silicon solar cells. *J. Optoelectron. Adv. Mater.*, **7**, 485.

21 Pieters, B., Stiebig, H., and Zeman, M., and van Swaaij, R. (2009) Determination of the mobility gap of intrinsic μc-Si:H in p–i–n solar cells. *J. Appl. Phys.*, **105**, 044502.

22 Kanschat, P., Mell, H., Lips, K., and Fuhs, W. (2000) Defect and tail states in microcrystalline silicon investigated by pulsed ESR. *Mater. Res. Soc. Symp. Proc.*, **609**, A27.3.

23 Dylla, T. (2005) Electron Spin Resonance and Transient Photocurrent Measurements on Microcrystalline Silicon. Ph.D. thesis, Forschungszentrum Jülich.

24 Sah, C.-T. and Shockley, W. (1958) Electron-hole recombination statistics in semiconductors through flaws with many charge conditions. *Phys. Rev.*, **109**, 1103.

25 Fantoni, A., Vieira, M., and Martins, R. (1999) Simulation of hydrogenated amorphous and microcrystalline silicon optoelectronic devices. *Math. Comput. Simulat.*, **49**, 381.

26 Chatterjee, P. (1994) Photovoltaic performance of a-Si:H homojunction p–i–n solar cells: A computer simulation study. *J. Appl. Phys.*, **76** (2), 1301.

27 Hernandez-Como, N. and Morales-Acevedo, A. (2010) Simulation of heterojunction silicon solar cells with AMPS-1D. *Sol. Energy Mater. Sol. C.*, **94**, 62.

28 Zhang, S.B., Wei, S.-H., and Zunger, A. (1998) Defect physics of the $CuInSe_2$ chalcopyrite semiconductor. *Phys. Rev. B*, **57** (16), 9642.

29 Zabierowski, P., Rau, U., and Igalson, M. (2001) Classification of metastabilities in the electrical characteristics of $ZnO/CdS/Cu(In,Ga)Se_2$ solar cells. *Thin Solid Films*, **387**, 147.

30 Decock, K., Lauwaert, J., and Burgelman, M. (2009) Modelling of thin film solar cells with graded band gap. Proc. of the 45th MIDEM, Postojna, Slovenia, p. 245.

31 Burgelman, M. and Marlein, J. (2008) Analysis of graded band gap solar cells with SCAPS. Proc. of the 23rd EUPVSEC, Valencia, Spain, p. 2151.

32 Decock, K., Khelifi, S., and Burgelman, M. (2010) Uniform reference structures to assess the benefit of grading in thin film solar cell structures. Proc. of the 25th EUPVSEC/WCPEC-5, Valencia, Spain, p. 3323.

33 Decock, K., Lauwaert, J., and Burgelman, M. (2010) Characterization of graded CIGS solar cells, *Energy Procedia*, **2**, 49.
34 Stollwerck, G. and Sites, J. (1995) Analysis of the CdTe back contact barriers. Proc. of the 13th EUPVSEC, Nice, France, p. 2020.
35 Niemegeers, A. and Burgelman, M. (1997) Effects of the Au/CdTe back contact on IV- and CV-characteristics of Au/CdTe/CdS/TCO solar cells. *J. Appl. Phys.*, **81**, 2881.
36 Burgelman, M. (2006) Cadmium telluride thin film solar cells: characterization, fabrication and modeling, in: *Thin Film Solar Cells: Fabrication, Characterization and Applications* (eds J. Poortmans and V. Arkhipov), John Wiley & Sons, Ltd, Chicester, UK, p. 277, Chapter 7.
37 Eisenbarth F T., Unold, T., Cabballero, R., Kaufmann, C.A., and Schock, H.W. (2010) Interpretation of admittance, capacitance-voltage, and current–voltage signatures in Cu(In,Ga)Se$_2$ thin film solar cells. *J. Appl. Phys.*, **107**, 034509.
38 Nollet, P. and Burgelman, M. (2004) Results of consistent numerical simulation of CdTe thin film solar cells. Proc. of the 19th EUPVSEC, Paris, France, p. 1725.
39 Bonnet, D. and Harr, M. (1998) Manufacturing of CdTe solar cells. Proc. of the 2nd WCPEC, Wien, Austria, p. 397.
40 Burgelman, M., Verschraegen, J., Degrave, S., and Nollet, P. (2005) Analysis of CdTe and CIGS solar cells in relation to materials issues. *Thin Solid Films*, **480–481**, 392.
41 Tao, G. (1994) Optical Modeling and Characterization of Hydrogenated Amorphous Silicon Solar Cells. Ph.D. thesis, Delft University of Technology.
42 van den Heuvel, J. (1989) Optical Properties and Transport Properties of Hydrogenated Amorphous Silicon. Ph.D. thesis, Delft University of Technology.
43 Rockstuhl, C., Lederer, F., Bittkau, K., and Carius, R. (2007) Light localization at randomly textured surfaces for solar-cell applications. *Appl. Phys. Lett.*, **91**, 171104.
44 Tao, G., Zeman, M., and Metselaar, J.W. (1994) Accurate generation rate profiles in a-Si:H solar cells with textured TCO substrates. *Sol. Energy Mater. Sol. C.*, **34**, 359.
45 Schropp, R.E.I. and Zeman, M. (1998) *Amorphous and Microcrystalline Solar Cells: Modeling, Materials, and Device Technology*, Kluwer, Dordrecht.
46 Zeman, M., van Swaaij, R.A.C.M.M., Metselaar, J.W., and Schropp, R.E.I. (2000) Optical modeling of a-Si:H solar cells with rough interfaces: Effect of back contact and interface roughness. *J. Appl. Phys.*, **88**, 6436.
47 Bennett, H.E. and Porteus, J.O. (1961) Relation between surface roughness and specular reflectance at normal incidence. *J. Opt. Soc. Am.*, **51**, 123.
48 Carniglia, C.K. (1979) Scalar scattering theory for multilayer optical coatings. *Opt. Eng.*, **18**, 104.
49 Zeman, M., van Swaaij, R.A.C.M.M., Metselaar, J.W., and Schropp, R.E.I. (2000) Optical modelling of a-Si:H solar cells with rough interfaces: Effect of back contact and interface roughness. *J. Appl.Phys.*, **88**, 6436.
50 Eaton, J.W. (2007) GNU Octave Manual. Bristol.
51 http://www.mathworks.com/products/matlab[0]/.
52 Kirchartz, T., Pieters, B.E., Taretto, K., and Rau, U. (2008) Electro-optical modeling of bulk heterojunction solar cells. *J. Appl. Phys.*, **104**, 094513.
53 Burgelman, M., Nollet, P., and Degrave, S. (2000) Modelling polycrystalline semiconductor solar cells. *Thin Solid Films*, **361–362**, 527–532.
54 Burgelman, M., Verschraegen, J., Degrave, S., and Nollet, P. (2004) Modeling thin film PV devices. *Prog. Photovoltaics*, **12**, 143–153.
55 Verschraegen, J. and Burgelman, M. (2007) Numerical modeling of intra-band tunneling for heterojunction solar cells in SCAPS. *Thin Solid Films*, **515** (15), 6276–6279.
56 Verschraegen, J. and Burgelman, M. (2007) Numerical modeling of intra-band tunneling for heterojunction solar cells in scaps. *Thin Solid Films*, **515** (15), 6276–6279.

57 Burgelman, M. and Marlein, J. (2008) Analysis of graded band gap solar cells with scaps. Proceedings of the 23rd European Photovoltaic Solar Energy Conference, Valencia, Spain, pp. 2151–2155.

58 Decock, K., Khelifi, S., and Burgelman, M. (2010) Modelling multivalend defects in thin film solar cells. presented at the E-MRS spring meeting 2010, session M-O12-4, Strasbourg. (To be published in Thin Solid Films).

59 Wei, S.-H., Zhang, S.B., and Zunger, A. (1998) Effects of Ga addition to $CuInSe_2$ on its electronic, structural, and defect properties. *Appl. Phys. Lett.*, **72**, 3199–3201.

20
Two- and Three-Dimensional Electronic Modeling of Thin-Film Solar Cells

Ana Kanevce and Wyatt K. Metzger

20.1
Introduction

Research and development to expand solar cell use is based on finding ways to create more efficient devices, eliminate as many losses as possible, lower material usage, and use cheaper production processes. Empirical optimization is important but can be misguided and require significant time, effort, and cost. Modeling can give insight to the physical processes, has a power to predict the impact of different parameters on the output results, and thus can help guide experimental investigations. The need for modeling solar cells in two and three dimensions is constantly increasing as solar technology matures. A description of multidimensional modeling and its applications is given here.

20.2
Applications

A standard 1D model simulates the electron and hole motion in the direction perpendicular to a planar junction. An enormous amount has been learned from this type of model. However, 1D models must also eliminate the inherent 3D nature of solar cells. Figure 20.1 shows a schematic of a typical thin-film solar cell. The carriers generated in the absorber will not only move perpendicular to the junction plane (the z direction) but also in directions parallel to the junction plane (x and y directions). In a standard thin-film module, individual cells are just several microns thick, but the cell width is usually several millimeters to several centimeters across and can be several feet wide. The lateral distance traveled by the carriers may impose additional requirements on material quality and alter the optimal distance between contacts. Consequently, it is important to consider motion in all directions.

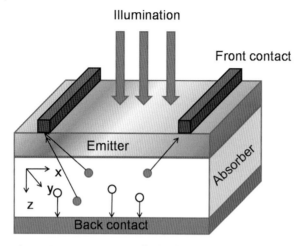

Figure 20.1 A cross section of a thin-film solar cell. The z direction is perpendicular to the planar junction, the x direction runs from left to right, and the y direction comes out of the page. A 1D model takes into account motion only along the z axis.

Modeling in more than one dimension is also necessary for understanding the impact of grain boundaries (GBs) and nonuniformities on device performance. The nonuniformities have been observed for both CdTe and $Cu(In,Ga)Se_2$ (CIGS) cells with many experimental methods such as optical beam induced current (OBIC), electron beam induced current (EBIC), light beam induced current (LBIC), and scanning tunneling microscopy (STM). A review is given in Ref. [1]. For examples of experimentally observed inhomogeneities and GBs, the reader is referred to Chapters 7, 11 and 12. GBs run both vertically and horizontally relative to the solar cell in the absorber layer. Voids and pinholes often appear in different layers of CIGS and CdTe solar cells under certain growth conditions.

Nonuniformities may be manifested in the thickness of all the layers as well as in the electronic parameters such as band gap, carrier density, and carrier lifetime. The impact of inhomogeneity on device performance can be analyzed by a 2D or 3D model.

Although planar cell geometry (as in Figure 20.1) is generally the most amenable to manufacturing, there can be benefits to more complex designs. For example, by positioning the front and back contacts on the same side of a cell, losses connected to the series resistance of the emitter can be avoided and the photocurrent can increase because the shading of the grid is avoided.

Numerical modeling is also useful to simulate different electro-optical experiments, in the presence of GBs and nonplanar junctions and geometries, and to help interpret data obtained in such experiments. In addition, multidimensional models can be used to examine novel grid designs, the effects of nonuniform illumination, and advanced technologies such as nanowires.

Some examples of the above problems and the modeling efforts directed toward their solutions will be presented in Section 20.4.

20.3
Methods

The first step in creating a computational model is defining and providing discretization of a 2D area or a 3D volume. Two main approaches to the discretization have been used for studying solar cells: equivalent-circuit modeling (Section 20.3.1) and solving semiconductor equations (Section 20.3.2).

Equivalent-circuit modeling represents a solar cell device as a collection of circuit elements. Diodes are inserted to the circuit to represent the photovoltaic effect from the semiconductor layers, and resistors are inserted to simulate series and shunt resistance between and within individual cells. The most commonly used software package for circuit simulations is SPICE. This and other packages can handle numerous circuit elements in very little time. This makes circuit modeling amenable to simulating modules, systems, and problems with a large number of discrete units. For example, circuit modeling has often been used to examine how local nonuniformities and shunts impact module performance.

By solving the semiconductor equations, one can get more information about the electron–hole dynamics occurring within a solar cell, but solving very complex problems with many discrete parts is much more cumbersome and can lead to numerical convergence issues. Obtaining these solutions also requires detailed input on the properties of each material in the device. The semiconductor equations can be solved by commercial software packages, generic partial differential equations (PDEs) solver packages, or custom code. In the literature [2–9], one can find studies executed with Sentaurus Device by Synopsys [10], COMSOL Multiphysics [11], Crosslight APSYS [12], and ATLAS by Sylvaco [13]. Sentaurus Device uses finite difference method, while COMSOL Multiphysics [11] and Crosslight APSYS [12] use finite element method to solve the differential equations. Some commercial software packages allow an opportunity of combining both semiconductor and equivalent-circuit modeling.

20.3.1
Equivalent-Circuit Modeling

The equivalent-circuit approach is based on representing regions in a solar cell simply as diodes. A current source parallel with the diode represents light generation, and resistors can be connected to simulate parasitic resistances. A nonuniform device can then be represented as a network of nonuniform diodes. Figure 20.2 shows part of such a network, where six diodes are connected in parallel. The current density J flowing through the different diodes is defined with the diode equation

$$J = J_0 \left(\exp\left(\frac{qV}{n_{id}kT} \right) - 1 \right) - J_L \qquad (20.1)$$

The different diodes are defined with their saturation current densities J_0 and diode quality factors n_{id}. In a nonuniform device, the characteristics of each diode

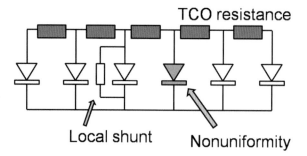

Figure 20.2 A schematic of an equivalent circuit of a nonuniform solar cell represented as a network of diodes. The cells are connected with resistors that represent the TCO.

may be varied. A part of the device with characteristics different than the rest is highlighted in Figure 20.2. A local leakage current, which could be caused by something like a pinhole in CdS, can be simulated with a shunt resistor added to one of the diodes. The parallel diode model is most valid when the fluctuations have a scale greater than several diffusion lengths. The network program SPICE has been widely used for this type of simulation. Although some device physics is lost in representing unit cells with the exponential diode law, this approach has given some important results. Some of these will be reviewed in Section 20.4.1.

20.3.2
Solving Semiconductor Equations

The output of the equivalent-circuit method is limited to calculating the current-density–voltage (J–V) characteristics of the device. To model cells more precisely, and to calculate physical properties such as band diagram, carrier density, photogeneration, recombination, and to simulate different characterization experiments, such as quantum efficiency, photoluminescence, admittance spectroscopy and others, one needs to solve equations that describe the electrostatics and carrier transport in the semiconductor layers. The number and precise form of these equations depend on the specific physics that needs to be captured in the simulations. The most common set of equations applied in thin-film solar cell simulations are three interdependent nonlinear PDEs [14]:

The Poisson equation Eq (20.2)

$$\nabla \cdot (-\varepsilon(\vec{r})\nabla\varphi(\vec{r})) = q(p(\vec{r}) - n(\vec{r}) + N_d(\vec{r}) - N_a(\vec{r})) \qquad (20.2)$$

and the continuity equations for holes Eq (20.3) and electrons Eq (20.4)

$$\nabla \cdot J_p(\vec{r}) = \nabla \cdot (q\mu_p(\vec{r})p(\vec{r})E(\vec{r}) - qD_p(\vec{r})\nabla p(\vec{r})) = q\left(G(\vec{r}) - R(\vec{r}) - \frac{\partial p(\vec{r})}{\partial t}\right)$$

$$(20.3)$$

$$\nabla \cdot J_n(\vec{r}) = \nabla \cdot (q\mu_n(\vec{r})n(\vec{r})E(\vec{r}) + qD_n(\vec{r})\nabla n(\vec{r})) = -q\left(G(\vec{r}) - R(\vec{r}) - \frac{\partial n(\vec{r})}{\partial t}\right)$$
(20.4)

Here φ is electrostatic potential, ε is dielectric constant, p and n are electron and hole densities, N_a and N_d are densities of ionized acceptors and donors, J_n and J_p are electron and hole current densities, E is electric field, and G and R are the generation and recombination rates, respectively. All of the physical properties can vary as a function of the position $\vec{r}$.

20.3.2.1 Creating a Semiconductor Model

To create a structure, one first defines the materials and contacts in the model. Then, the geometry and position for each layer and contact needs to be specified. After the structure is defined, the next step is defining the electronic properties such as doping profiles and band structure. The optical properties, such as the absorption coefficient of different materials involved, as well as reflection coefficient of the layers and contacts also need to be included in the model. The user inputs the incident illumination spectrum, usually the standard spectrum; AM1.5 for terrestrial applications, or AM0 for space applications [15].

For most thin-film materials, the dominant recombination mechanisms are nonradiative. In order to define the recombination mechanisms, the modeler needs to input the density of different types of defects, their spatial and energetic distribution, as well as capture cross sections, or lifetime values that represent the overall recombination rates. Then, the interfaces need to be defined: whether there are interface states, and what is the transport mechanism across the interfaces.

For single crystal materials like Si or GaAs, many material properties are well established in the literature. For thin-film polycrystalline materials, experimental input is sometimes absent or not clear, and the material properties may vary largely based on deposition methods and conditions. The lack of good input can present a difficulty for modeling; conversely, modeling can help determine what values are important to measure and the range of values that are realistic. Once the parameters are set, one needs to approach the solution techniques.

To attain solutions, the device is discretized in two or three dimensions by creating a mesh. Numerical algorithms calculate solutions to the equations at the mesh vertices. Hence, while increasing the mesh density can improve accuracy, it also increases the computational time and memory requirements. Fewer dimensions and less complicated models require far less computational time and effort.

An example of a meshed model is shown in Figure 20.3. The meshing needs to be denser where one expects rapid changes in variables, as well as at the heterointerfaces. In the bulk of material to improve the calculation time, the mesh should be less dense. In this example, the mesh is significantly denser at the junction and at the GBs within the absorber. Equations (20.2)–(20.4) describe transport in the bulk of the device and are solved at every mesh point. In addition to these, boundary conditions must also be specified. The contacts must be specified as either ohmic or Schottky

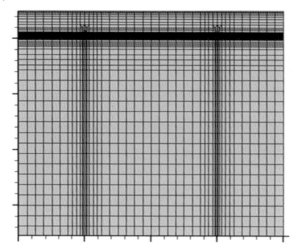

Figure 20.3 An example of a meshed structure used in 2D simulations. A model of a CIGS cell. The cell has two columnar GBs prpendicular to the junction.

with appropriate parameters. Boundary conditions and transport at each material interface (such as CdTe and CdS) must also be specified. Band offsets are typically determined using the Anderson model based on values of the band gap and electron affinities for each material. One must also specify current and energy flux equations and thermionic or tunneling transport if necessary. Boundary conditions must also be specified for the sides of the device. A detailed description of boundary conditions is given in Ref. [16].

20.4
Examples

This section will give some examples of problems that were addressed with 2D or 3D modeling and the knowledge that was gained through them. It does not by any means list all of them.

20.4.1
Equivalent-Circuit Modeling Examples

In an equivalent-circuit model, like the one shown in Figure 20.2, a nonuniform device is represented with diodes with different properties. If the highlighted diode has a lower turn-up voltage (defined as a weak diode) than the rest of the diodes in the device, it will be forward biased by the surrounding diodes [1, 17–20]. The efficiency loss due to nonuniformities depends on the area taken by the weak diodes, the difference in voltages of the weak diode compared with the strong ones, as well as on the resistance in the transparent-conductive oxide (TCO). An example is shown in

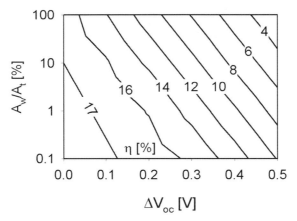

Figure 20.4 Device efficiency as a function of the ratio of weak diode area to total diode area and open-circuit voltage difference between the weak and the strong diodes [20].

Figure 20.4. In this example, PSPICE simulations are used to calculate how the ratio of the weak area A_w versus the total device area A_t affects the device efficiency. The $\Delta V_{oc} = V_{oc}(s) - V_{oc}(w)$ is the difference in open-circuit voltage between the strong and the weak diodes. In case of resistive TCO. as the resistivity increases. the voltage drop across TCO will be higher, which could localize the voltage difference into smaller area.

References [21–23] have reported that it is difficult to avoid potential fluctuations in CIGS cells and that they have a detrimental effect on them. These studies have also reported that moderate series resistance in a window layer (like i-ZnO layer, for example) can reduce the losses due to fluctuations. At the same time, however, large TCO resistance reduces the fill-factor (FF) of the entire device. In certain cases, there is an optimal value of TCO resistance that minimizes the damage due to fluctuations, while maintaining a low FF loss.

Hybrid methods where circuit simulations are combined with device simulations have been used to simulate modules [24–27].

20.4.2
Semiconductor Modeling Examples

One of the problems that calls for modeling in more than one dimension lies in the polycrystalline nature of CIGS and CdTe thin-film cells. The GBs in these materials have been observed by several scanning techniques.

The boundaries between different grains disturb the periodicity of the crystal lattice and can create energy states within the band gap. This can result in electronic charge on the boundaries, and creation of electric fields and potentials within the absorber material. There has been a great deal of research on the importance of GBs in CIGS and CdTe solar cells. Researchers have debated that the GBs are detrimental, innocuous, or beneficial to device performance. Clearly, one motivation for extending modeling beyond one dimension is to understand the impact of GBs.

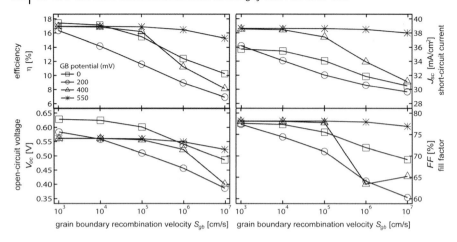

Figure 20.5 Simulated impact of recombination velocity at the GBs on V_{oc}, J_{sc}, FF, and conversion efficiency of a CIGS cell for four different values of GB potential: 0 V (square), 200 mV (circle), 400 mV (triangle), 550 mV (asterisk) [28]. (Reprinted with permission from Ref. [28], Copyright 2005, American Institute of Physics.)

The impact of GBs on the device performance will depend on the type of charges at the boundaries and their energy distribution. Intuitively it is natural to assume that if GBs are positively charged, the charge can create potential that repels holes, attracts electrons, and reduces recombination. However, modeling indicates that recombination and forward current will generally increase due to minority-carrier attractive potentials at GBs [4, 28]. An example of simulated impact of GBs on device performance is shown in Figure 20.5. The results demonstrate the complex relationships between GB potential, GB recombination, and device performance that can be manifested by modeling and difficult to predict by intuition. If positively charged GBs are oriented perpendicular to the plane of the junction, the GB and/or the region adjacent to it may act as a collection channel for minority carriers, thereby increasing photocurrent. But in general, the negative effects on V_{oc} and FF outweigh the positive effects on photocurrent, and efficiency decreases. Further GB studies of CIGS cells can be found in Refs. [4, 7, 9, 28, 29, 30] and for polycrystalline Si in Refs. [3, 8].

In addition to simulating current–voltage curves from solar cells, modeling can also be used to simulate experiments such as quantum efficiency, EBIC, cathodoluminescence, time-resolved photoluminescence, and near-field scanning microscopy performed on polycrystalline films or devices [28]. This is an important component to characterizing the properties of nonuniform material and generating input for solar cell models. These kinds of simulations have been used to evaluate whether large GB potentials are consistent with electro-optical experiments on CIGS material [28], the degree of GB recombination that is present in record-efficiency CIGS cells [31], and how the presence of a junction can affect carrier lifetime measurements [32, 33].

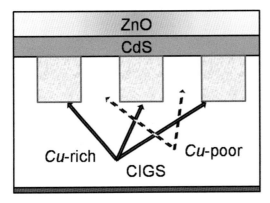

Figure 20.6 A schematic of an interdigitated CIGS solar cell.

During the deposition of Cu(In,Ga)Se$_2$, different phases with different Cu/In/Ga/Se ratios might form. These phases have different band gaps and electronic structures. The impact of band gap fluctuations on the device performance is analyzed in Ref. [34]. The Cu content can influence the conductivity. The higher Cu domains will have p-type conductivity, and the lower Cu domains n-type [35]. Therefore, as opposed to a single p–n junction at the heterointerface, the device could have multiple p–n junctions [35]. A schematic is shown in Figure 20.6. What this means for the device performance is difficult to tell without a model. How do multiple junctions affect carrier transport and device performance?

Analysis of the implications caused by the possible existence of nanodomains has been addressed by 2D modeling [2]. The model can simulate details such as currents and carrier densities for a range of material properties and domain sizes. An example in Figure 20.7 shows electron flow for p-type and n-type domains that extend perpendicular to the junction plane into the absorber material. If the domains are on the nanometer scale and have small relative band offsets, transport will be similar to a uniform material with the average properties of the constituent domains, and the device can be treated as a planar junction. If the domains have strong relative band offsets or are larger than tens of nanometer across, they can form a multidimensional network of p–n junctions. This network may either increase or decrease efficiency depending on the material properties of the constituent domains.

As mentioned in Section 20.2, there are several benefits that can be gained by placing the cathode and anode contacts only on the bottom of the cell. An example of such a design is shown in Figure 20.8. In this case, sunlight is able to enter the top of the cell without shading from top contacts. However, absorption occurs preferentially at the top of the cell. As the current-collecting junction is now at the bottom of the cell, rather than the top, the carriers must travel further in the material before being collected. The doping profile, contact thickness, and distance between contacts will all influence the device performance.

Nichiporuk et al. [36] used a 2D model to calculate the optimal geometry and doping, and to analyze the impact of several parameters on the interdigitated cell with

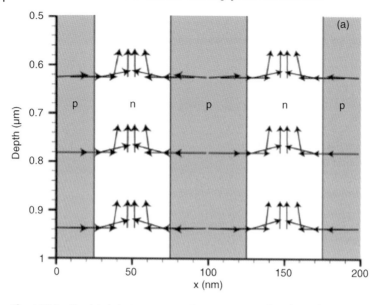

Figure 20.7 Simulated electron transport in a CIGS solar cell. The depth represents the distance into the CIGS-interdigitated region relative to the CIGS/CdS interface, p and n refer to regions doped p- and n-type, respectively [2]. (Reprinted with permission from Ref. [2], Copyright 2008, American Institute of Physics.)

p-type c-Si used as a base. They have found that with their parameters, the optimum width of the emitter should be couple of times larger than the one of the base contact, regardless of the distance between them. Kim *et al.* [37] have found that for minority-carrier lifetimes higher than 50 µs in Si devices, the interdigitated geometry outperforms their standard planar device.

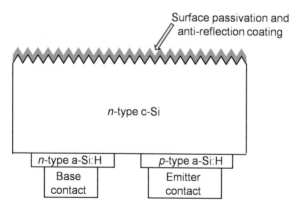

Figure 20.8 A schematic of a silicon heterojunction cell with an interdigitated back contact. Figure courtesy of Hao-Chih Yuan.

20.5
Summary

Modeling can provide physical insight to device operation, help distinguish important material properties from unimportant properties, predict trends, and help interpret experimental data. Its use has expanded rapidly in the past decade and is likely to become more and more common to address complex issues, as PV technology develops. There are numerous applications for multidimensional modeling. The examples in this chapter have shown how modeling has improved our understanding of experimental results and helped predict the optimal designs of devices. As seen by examples in this chapter, the second dimension can shed light to the real structures and help directing the device research and applications more successfully. Understanding gained by numerical modeling can highlight the requirements needed for production of higher efficiency devices.

References

1. Karpov, V.G., Compaan, A.D., and Shvydka, D. (2004) Random diode arrays and mesoscale physics of large-area semiconductor devices. *Phys. Rev. B*, **69** (4), 045325.
2. Metzger, W.K. (2008) The potential and device physics of interdigitated thin-film solar cells. *J. Appl. Phys.*, **103** (9), 094515.
3. Christoffel, E., Rusu, M., Zerga, A., Bourdais, S., Noel, S., and Slaoui, A. (2002) A two-dimensional modeling of the fine-grained polycrystalline silicon thin-film solar cells. *Thin Solid Films*, **403**, 258.
4. Gloeckler, M., Sites, J.R., and Metzger, W.K. (2005) Grain-boundary recombination in Cu(In,Ga)Se$_2$ solar cells. *J. Appl. Phys.*, **98** (11), 113704.
5. Lu, M., Bowden, S., and Birkmire, R. (2007) Two dimensional modeling of interdigitated back contact silicon heterojunction solar cells. Proceedings of the 7th International Conference on Numerical Simulation of Optoelectronic Devices, p. 55.
6. Malm, U. and Edoff, M. (2009) 2D device modeling and finite element simulations for thin-film solar cells. *Sol. Energy Mater. Sol. C.*, **93** (6–7), 1066.
7. Taretto, K. and Rau, U. (2008) Numerical simulation of carrier collection and recombination at grain boundaries in Cu(In,Ga)Se$_2$ solar cells. *J. Appl. Phys.*, **103** (9), 094523.
8. Taretto, K., Rau, U., and Werner, J.H. (2001) Two-dimensional simulations of microcrystalline silicon solar cells. *Polycryst. Semicond. IV Mater., Technol. Large Area Electron.*, **80–81**, 311.
9. Taretto, K., Rau, U., and Werner, J.H. (2005) Numerical simulation of grain boundary effects in Cu(In,Ga)Se$_2$ thin-film solar cells. *Thin Solid Films*, **480**, 8.
10. www.synopsys.com (22. June 2010).
11. www.comsol.com (22. June 2010).
12. www.crosslight.com (22. June 2010).
13. www.silvaco.com (22. June 2010).
14. Streetman, B.G. (1990) *Solid State Electronic Devices*, Prentice-Hall, Englewood Cliffs, NJ.
15. ASTM (2008) Standard G173-03. *Standard Tables for Reference Solar Spectral Irradiances: Direct Normal and Hemispherical on 37° Tilted Surface*. ASTM International, West Conshohocken, PA.
16. Selberherr, S. (1984) *Analysis and Simulation of Semiconductor Devices*, Springer, Wien New York.
17. Shvydka, D., Compaan, A.D., and Karpov, V.G. (2002) Nonlocal response in CdTe photovoltaics. *J. Appl. Phys.*, **91** (11), 9059.
18. Karpov, V.G., Rich, G., Subashiev, A.V., and Dorer, G. (2001) Shunt screening, size

effects and I/V analysis in thin-film photovoltaics. *J. Appl. Phys.*, **89** (9), 4975.
19 Karpov, V.G. (2003) Critical disorder and phase transitions in random diode arrays. *Phys. Rev. Lett.*, **91** (22), 226806.
20 Kanevce, A. and Sites, J.R. (2007) *Impact of Nonuniformities on Thin Cu(In,Ga)Se2 Solar Cell Performance*, vol 1012, (eds T. Gessert, K. Durose, C. Heske, S. Marsillac, and T. Wada). Mater. Res. Soc. Symp. Proc., Warrendale, PA, 1012-Y08-02.
21 Rau, U., Grabitz, P.O., and Werner, J.H. (2004) Resistive limitations to spatially inhomogeneous electronic losses in solar cells. *Appl. Phys. Lett.*, **85** (24), 6010.
22 Werner, J.H., Mattheis, J., and Rau, U. (2005) Efficiency limitations of polycrystalline thin film solar cells: case of Cu(In,Ga)Se$_2$. *Thin Solid Films*, **480**, 399.
23 Grabitz, P.O., Rau, U., and Werner, J.H. (2005) A multi-diode model for spatially inhomogeneous solar cells. *Thin Solid Films*, **487** (1–2), 14.
24 Koishiyev, G.T. and Sites, J.R. (2009) Impact of sheet resistance on 2-D modeling of thin-film solar cells. *Sol. Energy. Mater. Sol. C.*, **93** (3), 350.
25 Galiana, B., Algora, C., Rey-Stolle, I., and Vara, I.G. (2005) A 3-D model for concentrator solar cells based on distributed circuit units. *IEEE Trans. Electron. Dev.*, **52**, 2552.
26 Brecl, K. and Topic, M. (2008) Simulation of losses in thin-film silicon modules for different configurations and front contacts. *Prog. Photovoltaics*, **16** (6), 479.
27 Koishiyev, G.T. and Sites, J.R. (2009) Effect of shunts on thin-film cdte module performance. Mater. Res. Soc. Symp. Proc. Warrendale, PA, 1165, 1165-M05-22.
28 Metzger, W.K. and Gloeckler, M. (2005) The impact of charged grain boundaries on thin-film solar cells and characterization. *J. Appl. Phys.*, **98** (6), 063701.
29 Nerat, M., Cernivec, G., Smole, F., and Topic, M. (2008) Simulation study of the effects of grain shape and size on the performance of Cu(In,Ga)Se$_2$ solar cells. *J. Appl. Phys.*, **104** (8), 083706.
30 Rau, U., Taretto, K., and Siebentritt, S. (2009) Grain boundaries in Cu(In,Ga)(Se, S)$_{(2)}$ thin-film solar cells. *Appl. Phys. A-Mater.*, **96** (1), 221.
31 Metzger, W.K., Repins, I.L., Romero, M., Dippo, P., Contreras, M., Noufi, R., and Levi, D. (2009) Recombination kinetics and stability in polycrystalline Cu(In,Ga)Se$_2$ solar cells. *Thin Solid Films*, **517** (7), 2360.
32 Metzger, W.K., Ahrenkiel, R.K., Dashdorj, J., and Friedman, D.J. (2005) Analysis of charge separation dynamics in a semiconductor junction. *Phys. Rev. B*, **71** (3), 035301.
33 Metzger, W.K. (2008) How lifetime fluctuations, grain-boundary recombination, and junctions affect lifetime measurements and their correlation to silicon solar cell performance. *Sol. Energy Mater. Sol. C.*, **92** (9), 1123.
34 Rau, U. and Werner, J.H. (2004) Radiative efficiency limits of solar cells with lateral band-gap fluctuations. *Appl. Phys. Lett.*, **84** (19), 3735.
35 Yan, Y.F., Roufi, R., Jones, K.M., Ramanathan, K., Al-Jassim, M.M., and Stanbery, B.J. (2005) Chemical fluctuation-induced nanodomains in Cu(In,Ga)Se$_2$ films. *Appl. Phys. Lett.*, **87** (12), 121904.
36 Nichiporuk, O., Kaminski, A., Lemiti, M., Fave, A., and Skryshevsky, V. (2005) Optimisation of interdigitated back contacts solar cells by two-dimensional numerical simulation. *Sol. Energy Mater. Sol. C.*, **86** (4), 517.
37 Kim, D.S., Meemongkolkiat, V., Ebong, A., Rounsaville, B., Upadhyaya, V., Das, A., and Rohatgi, A. (2006) 2D-Modeling and development of interdigitated back contact solar cells on low-cost substrates. Proceedings of the 4th World Conference on Photovoltaic Energy Conversion, Waikoloa, Hawaii, 2006, p. 1417.

Index

a

ab-initio total energy
– formation of   493–494
– overview of   483
absorptance   6, 9, 10, 11, 28, 48, 53, 55, 56, 58
absorption coefficient   10, 11, 12, 20, 65, 67, 180
absorptivity   153, 170
activation energy   44–46, 89, 94
Adiabatic local density approximation   484
admittance   81
admittance spectroscopy   81, 95–97
AFORS-HET   521, 522
AM1.5 spectrum   6, 42, 516, 533
ambipolar transport   181
amorphous fraction   381
amphoteric   251
amphoteric state   504
amplitude modulation   279
AMPS-1D   522
angle distribution functions   520
angle-dispersive X-ray diffraction   357
angularly resolved scattering   114, 116, 120, 121
annular dark-field detector   327
anomalous dispersion   219
anti-Stokes band   368
anti-Stokes process   369
APSYS   531
Arrhenius dependence   460
Arrhenius plot   94
ASA   13, 523
ATLAS   531
atomic force microscope   110, 275
atomic force microscopy   318
– conductive   276, 279
– non-contact   278
attempt-to-escape frequency   221
Auger electron spectroscopy   411, 427
Auger electrons   301
Auger process   427, 428
Auger recombination   16
autocompensation   254

b

back surface field   292
backscattered electrons   300
band alignment   399, 400, 403, 404
band bending   13, 14, 88, 285, 288, 289
band diagram   291
band-gap energies   496
band-gap fluctuations   9
band-tail width   220
band-band transitions   163
bandtail. *see* tail
Bethe–Salpeter equation   484
bilayer process   21
bimolecular recombination   258
biplate   130
black body spectrum   41
black body   6, 63
Boeing process   21
Bohr magnetron   232, 234
Bragg diffraction   306, 327, 332, 336
Bragg's law   351
bremsstrahlung   333, 334
Bruggeman effective medium approximation   136
built-in electric field   14, 15, 19, 24, 26, 28, 212, 286
built-in voltage   82, 213, 227

Advanced Characterization Techniques for Thin Film Solar Cells,
Edited by Daniel Abou-Ras, Thomas Kirchartz and Uwe Rau.
© 2011 Wiley-VCH Verlag GmbH & Co. KGaA. Published 2011 by Wiley-VCH Verlag GmbH & Co. KGaA

c

capacitance   81, 210
– dc   82
– differential   82
capacitance-voltage measurement   281
capacitance-voltage profiling   81
capture cross section   17, 82, 94, 509, 510, 512, 533
capture rate   86
carrier collection
– voltage dependent   15
carrier lifetime   181
cathodoluminescence image   312
cathodoluminescence   305, 312, 536
$CdCl_2$ activation   24
Chandezon method   114
charge extraction   10, 12–14, 18, 19
chemical bath deposition   21
close space sublimation   315
co-evaporation   21
coherence length   130
coincidence-site lattice   305
collection function   308–311
collection probability   5, 49
COMSOL   531
concentric hemispherical analyser   429
conduction band   492
conductive AFM see atomic force microscopy, -conductive
confocal microscope   155
continuity equations   501, 521, 532
conventional TEM see transmission electron microscopy, -conventional
convergent-beam electron diffraction   329
Curie law   249, 252
current/voltage curve   35, 83
cylindrical mirror analyser   429

d

dangling bond   4, 26, 193, 231, 503
dark conductivity   184
dark current density   36
Debye length   86, 290
deep level optical spectroscopy   98, 99
deep level transient spectroscopy   81, 95, 462
deep state   81
defect   16, 27, 37, 39, 52, 180, 194, 207, 231, 285, 288, 365, 377, 504–507, 512, 514, 521–524, 533
– metastable   27
defect-pool model   504–507
demarcation energy   89, 206, 222

density of states   81, 89, 178, 198, 221, 222, 225, 257, 502, 503, 507, 508
– hydrogen   462, 463, 471
– phonon   381
density-functional perturbation theory   486
density functional theory   244, 251, 269, 480
– band-gap problem of   483
– basis sets   480, 481
– Hohenberg–Kohn theorem   480
– Jacobs ladder of   482
– Kohn–Sham method   480
– local-density approximation/generalized gradient approximation functionals   481–483
– material properties   485–486
– time-dependent   484
– – adiabatic local density approximation   484
depletion approximation   85
desorption   450, 454, 455, 459, 462, 463, 468
detailed balance   6, 9, 10, 87, 509
deuterium   454
diamagnetic   257
dielectric constant   180, 533
dielectric function   128, 133
dielectric relaxation time   187, 209, 217
diffraction   112
diffraction pattern   325
diffusion coefficient   181, 185
– hydrogen   460
diffusion constant   309
diffusion length   14, 26, 65, 69–71, 109, 177, 181–186, 191, 195, 197, 291, 308–312, 321
– hydrogen   456
diode equation   7, 531
diode-quality factor   531. see also ideality factor
direct semiconductor   10
disorder activated modes   377
disorder   81
dispersion
– optical   126
dispersion parameter   220
dispersion   205
– anomalous   206
dispersive transport   206
displacability   217, 225
displacement current   205
donor-acceptor pair recombination   160
drive-level capacitance profiling   81, 97

e

effusion   449–472
Einstein relationship   435
electrically detected magnetic resonance   232, 264
electrodeposition   374, 375
electroluminescence   61, 264, 522
electron backscatter diffraction   288, 302, 322
– pattern   302–304, 322, 323
electron beam induced current   307, 530
electron density   480
electron energy-loss spectroscopy   323, 329
electron gas   480
electron holography   335
electron microscopy   299–345
electron paramagnetic resonance   231
electron spin resonance   231, 507
electron-beam induced current   307, 530
electronegativity   421
electronic structure methods   483
electron-nuclear double resonance   238
electron-spin echo envelope modulations   238
ellipsometry   125
energy-dispersive X-ray diffraction   291, 334, 358, 418
energy-dispersive X-ray elemental distribution maps   323
 energy-filtered transmission electron microscopy *see* transmission electron microscopy, -energy-filtered
energy-loss near-edge structure   332
enthalpy   459
entropy   459
equivalent circuit   41, 531
equivalent circuit modeling   531
evanescent light   109, 113
exact numerical inversion   133
exchange-correlation energy   481
exciton   5, 313
excitonic   144, 158
extended energy-loss fine-structure   333

f

fast Fourier transformation   109, 118
Fermi energy   81, 82, 95, 97, 491
Fermi level   39, 45, 178, 194, 207, 231, 252, 257, 284, 285, 309, 469
Fermi's golden rule   151, 392, 393
Fermi–Dirac distribution   509
finite-difference time-domain   116
flatband   13–15
fluorescence   372
focused ion beam   305, 341

four point probe technique   41
Fourier analysis   130
Fourier coefficients   130
Fourier transform   85
Fourier-ratio deconvolution   333
free energy   459
free bound transition   159
frequency modulation   279
Fresnel fringes   337
Fresnel's coefficients   519
full potential linearized augmented plane wave method   481

g

Gaussian/numerical atomic orbital basis function   481
generalized gradient approximation   481
generalized Kirchhoff's law   153
generalized Planck's law   153
generation function   48
gettering   291
Gibbs free enthalpy   100
glow discharge mass spectroscopy   411, 413, 417, 418
glow-discharge optical emission   411, 413
glow-discharge optical emission spectroscopy   416, 418
goniometer   188
grain boundary   288, 305, 311, 530
grating   112, 179, 187
grazing incidence X-ray diffraction   354
Green–Kubo formalism   485
gyromagnetic ratio   232

h

Hall effect   288, 289
Hall measurement   203
Hamaker constant   277
Hamiltonian   240, 241
Hartree–Fock calculation   251
Hartree–Fock exchange   482
Hartree–Fock method   482
haze   114
Hecht plot   213
high angle annular dark field   327
high injection condition   152
Hooke's law   277
hopping   205, 264, 265, 268
Hough space   304
Hubbard model   482
hydrogen dilution   137
hydrogen passivation   4
hydrogen   449, 450, 453, 454, 456–472
hyperfine interaction   233, 234, 240

i

ideality factor   37, 38, 39, 44
inelastic scattering   388
interface recombination   44
interfacial barrier   47
interference   11, 48, 50, 54, 55, 115, 118, 178, 348, 349, 351, 519, 520
intrinsic carrier concentration   502
inverse photoelectron spectroscopy   390, 391
ion bombardment   433, 436
ion-beam sputtering   427, 433

k

Kelvin probe force microscope   276, 282
Kelvin probe technique   275
Kikuchi pattern   302
Kirchhoff's law   6
– Würfel's generalization   6
Kohn–Sham framework   486
Kramers–Heisenberg formalism   392, 393
Kramers–Kronig rule   133
Kramers–Kronig transformation   96

l

Lambert–Beer law   355, 521
Lambertian distribution   10
Landé factor   232
Laplace transform   226
Larmor frequency   245
laser   70, 209, 370
– dye   209
– gas   370
– solid state   371
least squares regression   134
lifetime   14, 37, 40, 57, 152, 186, 533, 538
light beam induced current   530
light scattering   109
light trapping   4, 10, 11, 12, 18, 19, 109, 116, 519
light-induced degradation   194, 248, 257
light-induced electron spin resonance   237
light-soaking   195
local density approximation
– ZnO, point defects, formation energies   494
localised vibrational modes   377
long-range order   207
Lorentzian broadening   143
LO-TO splitting   370, 374
low pass filter   83

m

magnetic moment   232, 240, 249
many-body perturbation theory   484
Maxwell–Boltzmann approximation   502
mean-inner potential   337
Meyer–Neldel rule   97, 99
Miller indices   350
mobility edge   219
mobility gap   25, 502–504, 507
mobility   9, 14, 15, 17, 20, 25, 26, 27, 36, 81, 180, 185, 186, 205, 212, 231, 252, 254, 255, 258, 265, 502–504, 506, 507
mobility-lifetime product   178
molecular dynamics simulations   485
monoplate   130
Monte Carlo simulation   309
Mott insulators   482
multidimensional modeling   529
multijunction solar cell   51, 68
multiple-trapping model   215, 219

n

near-field scanning optical microscopy   318
neutron scattering   347–349, 351–354
Newton–Raphson algorithm   133
non-contact AFM   see atomic force microscopy, -non-contact
nuclear magnetic resonance   235
numerical modeling   530

o

Ohm's law   180
open-circuit voltage   7, 8, 15, 16, 17, 20, 43, 46, 513, 514
optical beam induced current   530
optical near-field   109
optical transitions   151, 158
ordered vacancy compound   375
oxygen vacancy   489

p

paramagnetic   231, 232, 237, 239, 241–246, 248, 251, 252, 254, 255, 257, 260, 261, 263, 264, 268
parasitic absorption   48, 57
passage effect   236, 238
Pauli principle   252, 265
PC1D   523
percolation theory   507
phase locked loop   279
phasor diagram   82
photoluminescence calibration   156
phonon   367
photocapacitance measurement   81, 98
photoconductivity   178
photocurrent   7, 12, 36, 37, 40, 41, 52, 55, 58

photocurrent decay   205
photoelectron spectroscopy   291, 388, 390
photoluminescence   151, 264, 313
– micro-PL   171
photoluminescence setup   154
photomultiplier tube   129, 372, 415
photon flux   153
photothermal deflection spectroscopy   55
physical vapour deposition   374
plasma enhanced chemical vapor deposition   135
plasmon   428
point defects
– *ab-initio* characterization of   486–488
– case study
– – in ZnO   494–496
– formation energies of   493, 494
– formation process   490
– Gibbs free energy   488
– Kröger–Vink notation of   489
– thermodynamics of   488–493
– *vs.* Fermi energy   494
Poisson equation   92, 177, 336, 501, 532
polarization   180, 188, 366, 367, 374
– optical   125
Poole–Frenkel effect   94
post-transit   206, 223
potential fluctuations   162
pre-transit   206
profilometer   425

q

quadrupole   421, 422, 450
quantum efficiency of solar cell   47–56, 58, 63–66, 292, 536
– of a light emitting diode   66, 68
quantum Monte Carlo method   484
quasi-Fermi level   38, 94, 111, 502
quasi-Fermi level splitting   152, 153

r

radiative lifetime   152
Raman cross section   371
Raman scattering   367
Raman spectroscopy   365
Rayleigh scattering   367
RC time constant   518
reciprocity   62, 63, 65, 66
recombination   36–39, 41, 44–48, 52, 55, 57, 58, 61–66, 68–71, 194, 291, 313, 501, 502, 507–513, 521, 522, 533
recombination center   81

recombination current density   36
reflection   10, 12, 18, 19, 48, 54
refractive index   10, 112, 126, 179, 355, 520
relative sensitivity factor   417
remote electron beam induced current   307
resonant inelastic X-ray scattering   390
reverse bias deep level transient spectroscopy   95
Rietveld analysis   351, 352, 353
Ritter–Zeldov–Weiser analysis   181
Rutherford scattering   327

s

SAED. *see* selected-area electron diffraction
saturation current density   36, 39, 46
Savitzky–Golay filter   430
scanning capacitance microscopy   280
scanning electron microscopy   299–323
scanning near-field optical microscopy   109
scanning spreading resistance microscopy   280
scanning tunneling microscopy   275, 284, 318, 530
SCAPS   90, 513, 523, 524
scattering cross section   347, 349
Scherrer equation   360
Schrödinger equation   479
– *ab-initio* methods   479
– Hamiltonian of   479
SC-Simul   524
secondary electrons   300, 388
secondary ion mass spectrometry   411, 420, 454
secondary neutral mass spectrometry   422
selected-area electron diffraction   327, 328
selenisation   21
Sentaurus   531
series resistance   37, 39, 40, 42–44, 70, 71, 75, 77, 83
Sah and Shockley statistics   523
shallow state   81
sheet resistance   73–77
Shockley–Queisser theory   5, 7, 36, 47
Shockley–Read–Hall recombination   14, 37, 52, 508
short-circuit current density   6, 7, 12, 14, 15, 17, 36, 40, 42–44, 46, 47
shunt   40, 44, 72, 75–77, 531
Snell's law   128, 179
solar simulator   42

space charge region   13, 37–39, 41, 45
spatial inhomogeneities   170
spectral absorptivity   153
spectral response   47, 52
spectroscopic ellipsometry   48, 125
spectrum imaging   314
SPICE   40, 76, 531
spin   231–234, 236–241, 243–246, 248–252, 254, 255, 264–266, 268, 269
spin-lattice relaxation time   234
spin-orbit interaction   241
spontaneous emission rate   152
sputter deposition   21
sputtered neutral mass spectroscopy   411
S-shape   15
Staebler–Wronski effect   27
steady-state photocarrier grating   177
stoichiometric defect reaction   491
Stokes process   369
strain   365
stray light   372
substrate   18, 19, 20, 21, 22, 26
superposition principle   52
superposition   7
superstrate   4, 18, 19, 26
surface photovoltage   522
surface recombination   14–18, 184, 193, 502
surface recombination velocity   65, 69, 309, 310, 502
susceptibility tensor   374
susceptibility   366
synchrotron radiation   387, 390

t

tail   9, 16, 71, 81, 184, 195, 252, 503, 504, 507
tandem cell   51, 52
temperature-programmed desorption   450
texture
– geometric   109
thermal desorption spectroscopy   450
thermal velocity   17, 87, 502
thermalization   5, 27
thermodynamic stability   487
three-stage process   21
time-dependent density functional theory *see* density-functional theory, -time-dependent
time-of-flight   203
time-of-flight mass spectrometer   416

time-resolved photoluminescence   536
topography   111, 112, 120, 278, 283, 311
total internal reflection   10, 109, 113
transient photocapacitance spectroscopy   99
transit time   205, 206, 214, 216
transmission electron microscopy   323–338
– bright-field   324
– conventional   323, 340
– dark-field   324
– energy-filtered   329
– high resolution   326
– scanning   323, 324
transparent conductive oxides   23
trap   38, 81, 87, 93, 205, 219, 220, 223, 231, 280, 285, 288, 313, 508–510, 512, 521
trapped charge   180, 184
trap state   81
tripod polishing   340
tritium   454
tunneling luminescence microscopy   319
tunnelling enhanced recombination   38
two-diode model   39
type inversion   25

u

Urbach energy   504
UV photoelectron spectroscopy   390

v

van der Waals force   276
virtual interface analysis   133, 134, 136
void   461, 466, 469–471

w

wavelength-dispersive X-ray spectrometry *see* X-ray spectrometry, -wavelength-dispersive
work function   282, 286, 288, 291, 435, 436

x

X-ray absorption spectroscopy   390, 391
X-ray diffraction   286, 328, 350, 354, 357, 358, 373, 376, 378
X-ray emission spectroscopy   389, 390
X-ray excited Auger electron spectroscopy   390

X-ray fluorescence   334, 358, 391, 436, 441
X-ray photoelectron spectroscopy   390, 411, 435
X-ray scattering   347, 348
X-ray spectrometry
– energy-dispersive   306, 411, 440
– wavelength-dispersive   306, 353
X-rays   347–359, 387

z

Z contrast imaging   302, 327
Zeeman levels   232, 244